Green Consensus and High Quality Development

China Council for International Cooperation on Environment and Development (CCICED) Secretariat

Green Consensus and High Quality Development

CCICED Annual Policy Report 2020

China Council for International Cooperation on
Environment and Development (CCICED)
Secretariat
Beijing, China

ISBN 978-981-16-4798-7   ISBN 978-981-16-4799-4 (eBook)
https://doi.org/10.1007/978-981-16-4799-4

Jointly published with China Environment Publishing Group Co., Ltd.
The print edition is not for sale in China (Mainland). Customers from China (Mainland) please order the print book from: China Environment Publishing Group Co., Ltd.

This Springer imprint is published by the registered company Springer Nature Singapore Pte Ltd.
The registered company address is: 152 Beach Road, #21-01/04 Gateway East, Singapore 189721, Singapore

Preface

2020 was a very unusual year: the sudden outbreak of the COVID-19 pandemic plunged the global economy into severe recession. It was meant to be a "super-year", with significant expectations for important actions on environmental protection, but many ambitious plans to tackle climate change and nature conservation had to be postponed. In 2020, China made important announcements to address climate change, aiming to peak carbon dioxide (CO_2) emissions by 2030 and achieve carbon neutrality by 2060, giving new impetus to global climate governance.

In 2020, the China Council for International Cooperation on Environment and Development (CCICED) adjusted its work methods to overcome the negative impacts of the COVID-19 pandemic. It has carried out the plan as approved by the Executive Members under the annual theme of "Green Consensus and High-Quality Development". It has also provided advice and recommendations on issues such as green recovery, ecological civilization, and the 14th Five-year Plan (14th FYP). The aim of this period is to focus on building an innovative platform for international cooperation on the environment and development that embraces global inclusiveness, open cooperation, and mutual benefits and development.

The CCICED Secretariat publishes Annual Policy Report every year, in order to better reach a consensus on green development and promote green transition and transformation. The Annual Policy Report is a flagship publication that presents research outputs, paper, report by Chinese and international CCICED teams, as well as outlining policy recommendations, the CCICED has proposed to the Chinese government. It shares the observations and thinking of the Chinese and international Members, Special Advisors and experts on hot issues in the environment and development, for the reference of decision makers at all levels, scholars, and the public.

The 2020 CCICED Annual Policy Report incorporates the policy recommendation proposed to the Chinese government "From Recovery to Green Prosperity: The Pathway to China's High-quality Development during the 14th Five-Year Plan period"; the 2020 issues paper "Recovering Forward", which is written by CCICED's Chief Advisors; and research reports from all Special Policy Studies, including "Global Climate Governance and China's Role", "Post-2020 Global Biodiversity Conservation", "Global Ocean Governance and Ecological Civilization", "Green

Urbanization Strategy and Pathways towards Regional Integrated Development",
"Ecological Compensation and Green Development Institutional Reform in the
Yangtze River Economic Belt", "Green Transition and Sustainable Social Gover-
nance", "Major Green Technology Innovation and Implementation Mechanisms",
and "Green BRI and 2030 Agenda for Sustainable Development", "Global Green
Value Chains", "Green Finance".

Beijing, China   China Council for International Cooperation on
                                            Environment and Development (CCICED)
                                                                        Secretariat

Acknowledgements

China Council for International Cooperation on Environment and Development (CCICED) conducted a series of research projects in 2020 with the support of Chinese and international experts (including Council Members and Special Advisors) and partners, including Global Climate Governance and China's Role, Post-2020 Global Biodiversity Conservation, Global Ocean Governance and Ecological Civilization, Green Urbanization Strategy and Pathways towards Regional Integrated Development, Ecological Compensation and Green Development Institutional Reform in the Yangtze River Economic Belt (YREB), Green Transition and Sustainable Social Governance, Major Green Innovation Technologies and Implementation Mechanisms, Green BRI and 2030 Agenda for Sustainable Development, Global Green Value Chain, and Green Finance. The book is based on the outputs of these research projects. Acknowledgement is given to the following experts engaged in these projects:

Overview: Scott Vaughan, Liu Shijin, Li Yonghong, Zhang Jianyu, Knut Alfsen, Dimitri de Boer, Robyn Kruk, Zhang Huiyong, Liu Kan.

Chapter 1 Xie Zhenhua, Liu Shijin, Scott Vaughan, Arthur Hanson, Knut Halvor Alfsen, Guo Jing, Li Gao, Li Yonghong, Kate Hampton, Zou Ji, Wang Yi, Lei Hongpeng, Liu Qiang, Zhong Lijin, Zhao Xiao, Gu Baihe, Dong Yue, Xin Jianan, Zhang Xiaohan, Zhao Wenbo, An Yan, Zhai Hanbing, Zhang Huiyong, Liu Kan, Li Ying, Yao Ying, Hugh Outhred, Maria Retnanestri, Yudiandra Yuwono, Septia Buntara, Monika Merdekawati, Philip Andrews-Speed.

Chapter 2 Arthur Hanson, Li Lin, Ma Keping, Gao Jixi, Shen Xiaoli, Wei Wei, Zou Changxin, Alice Hughes, Marcel Theodorus Johannes Kok, Liu Dong, Xu Mengjia, Zhang Kun, Zhu Yingying, Wu Qiong, Liu Yinan, Luo Maofang, Harvey Locke, Eliane Ubalijoro, Beate Jessel, Dominic Waughray, Hideki Minamikawa, Guido Schmidt-Traub, Lennart Kuemper-Schlake, Wang Ran.

Chapter 3 Jan-Gunnar Winther, Su Jilan, Dai Minhan, Wang Juying, Sun Song, Liu Hui, Han Baoxin, Birgit Njåstad, John Mimikakis, Lisa Svensson, Nishan Degnarian, Fei Chengbo.

Chapter 4 Zhang Yongsheng, Zheng Siqi, Bob Moseley, Sander van der Leeuw, Lin Jiang, Yue (Nina) Chen, Li Xiaojiang, Zhang Jianyu, Xu Wei, Wei Ping, Yang

Jidong, Li Dong, Liu Lu, Li Ting, Qiu Xi, Yu Xiang, Zhang Ying, Cong Xiaonan, Zhao Yong, Zhang Min, Zhao Haishan.

Chapter 5 Wang Jinnan, Ahmed M. Saeed, Li Huayou, Zhang Qingfeng, Arthur Hanson, Robert Costanza, Brendan Gillespie, Annette T. Huber-Lee, Ouyang Zhiyun, Ma Jun, Li Junsheng, Ge Chazhong, Yu Fang, Ma Guoxia, Au Shion Yee, Guo Dongmei, Yang Weishan, Cheng Cuiyun, Li Yuanyuan, Qiu Qiong, Shi Yinghua, Lan Hong, Jiang Lahai, Yang Wenjie, Song Xiaoyu, Fan Mingyuan, Liang Kangheng, Bob Tansey, Isao Endo, Eva Abal, Yao Ying.

Chapter 6 Ren Yong, Åsa Romson, Zhang Yong, Fan Bi, Zhang Jianyu, Zhou Hongchun, Yu Hai, Wang Zhongying, Guo Jiaofeng, Zhang Xiaodan, Li Jifeng, Huang Yonghe, Zhao Fang, Eva Ahlner, Ulf Dietmar Jaeckel, Lewis Akenji, Hideki Minamikawa, Chen Weidong, Charles Arden-Clarke, Miranda Schreurs, Mushtaq Ahmed Memon, Vanessa Timmer, Chen Gang, Han Guoyi, Zhao Yongqiang, Liu Bin, Li Nan, Lv Jing, Fang Ying, Liu Hanwu, Dong Yao, Huo Lulu, Wang Jia, Liu Haidong, Gan Hui, Qian Lihua, Wang Ying, Cai Zipei, Liu Qingzhi, Zhou Caihua, Wang Yong, Meng Lingbo, Yan Fei, Cao Dandan, Li Gongtao.

Chapter 7 Li Xiaojiang, Dominic Waughray, Zhai Qi, Ye Qing, Zhu Rongyuan, Zhan Kun, Zhang Chun, Zhang Yongbo, Lv Xiaobei, Ren Xiyan, Wei Baojun, Wu Sufeng, Zhou Jun, Guo Yongcong, Claudia Sadoff, Lin Jiang, Susan Bazilli, Christian Hochfeld, Arjan Harbers, Charles Godfray, Fan Shenggen, KheePoh Lam, Lin Boqiang, Charlo Ratti, Li Yi, Guo Jifu, Song Yehao, Fu Lin, Fang Li, Zheng Degao, Zhang Jing, Sun Liming, Ding Shineng, Hu Jingjing, James Pennington, Fei Chengbo.

Chapter 8 Zhou Guomei, Shi Yulong, Kevin P. Gallagher, Ge Chazhong, Lan Yan, Rebecca Ray, Dong Liang, Wang Lixia, Ni Biye, Peng Ning, Li Panwen, Zhang Min, Fei Chengbo.

Chapter 9 Manish Bapna, Marjorie Yang, Chen Ming, Craig Hanson, Rod Taylor, Charles Victor Barber, Li Bo, Fu Xiaotian, Liu Ting, Yuan Yu, Dong Xin, Erik Solheim, John Hancock, Niu Hongwei, James Leape, Joaquim Levy, Guillermo Castilleja, Ren Yong, Ye Yanfei, Zhang Jianyu, Zhou Guomei, Zhang Jianping, Tang Dingding, Ai Luming, Liu Shijin, Scott Vaughan, Arthur Hanson, Dimitri De Boer, Knut Halvor Alfsen, Chris Elliott, Cristianne Close, David Cleary, Elizabeth Economy, Fang Li, Guido Schmidt-Traub, Li Xiaoliang, Jocelyn Blériot, John Ehrmann, Justin Adams, Leonardo Fleck, Margot Wood, Melissa Pinfield, Michael Obersteiner, Chen Jie, Chen Wenming, Chen Ying, Lin Meng, Mao Tao, Qu Fengjie, Tan Lin, Wang Ying, Yu Jie, Zhu Chunquan, Zhang Huiyong, Joe Zhang, Brice Li, Peng Ning, Li Gongtao, Jin Zhonghao, Yu Xin, Xu Jin, Dong Ke, Zhang Yu, Qi Yue, Li Fei, Fang Lifeng, Wan Jian, Ma Lichao, Natalie Elwell, Jun Geng, Brian Lipinski, Ayushi Trivedi, Sarah Stettner, Courtney McComber, Corey Park, Zeng Hui, Chen Haiying, Yang Li.

Chapter 10 Zhang Chenghui, Stephen P. Groff, Zhang Junjie, Sagarika Chatterjee, Mark Halle, Wang Yang, Yu Xiaowen, Chen Jianpeng, Margaret Kuhlow, Stephan Contius, Robin Smale, Anne-Mareike Vanselow, Nathalie Lhayani, Liu Donghui, Liu Shijie, Tang Lan, Qiu Jingyi, Ding Han, Hou Dingrui, Ma Junran, Luo Shiyi, Chen Yaqin, Wang Ran.

Meanwhile, we would also like to thank donors and partners for their firm support, including the government of Canada, Norway, Sweden, Germany, Netherlands, and Italy; the European Union, the United Nations Environment Programme, the United Nations Development Programme, the United Nations Industrial Development Organization, World Bank, Asian Development Bank, World Economic Forum, World Wide Fund for Nature, Environmental Defense Fund, Energy Foundation, Rockefeller Brothers Fund, The Nature Conservancy, World Resources Institute, International Institute for Sustainable Development, ClientEarth, and other countries, international organizations and NGOs. Their financial and intellectual support laid a solid foundation for the progress of our research projects.

Special thanks goes to the "Sino-German Environmental Partnership II" project, which is implemented by Deutsche Gesellschaft fuer Internationale Zusammenarbeit (GIZ) GmbH on behalf of the Federal Ministry for the Environment, Nature Conservation and Nuclear Safety of the Federal Republic of Germany (BMU).

Besides, we would also like to thank the following colleagues and all those who have contributed to the publication of this book, including Gao Lingyun, Zhu Jianlei, Hao Xiaoran, Liu Qi, Chen Xinying and Huang Ying.

Contents

Overview: Recovering Forward

In 2018, President Xi cautioned that if "mankind conquers nature with science and creativity, nature will take revenge on mankind". In 2020, COVID-19 has reflected the extent of nature's revenge on human and economic health. Plans for the "2020 super-year" of climate ambition and nature protection have, like all else, been postponed.

It is likely that COVID-19 impacts and recovery will last far longer than the event and aftermath of the 2008 Great Recession. The legacy of that economic shock included productivity and innovation lags that lingered for nearly a decade, while global supply chains as a percentage of GDP never recovered from their 2008 levels. The structural aftershocks of COVID-19 are very likely to endure throughout the entire period of the 14th Five-Year Plan.

This Note argues that the direction of the economic recovery should not point backward to previously rigid economic models characterized by unsustainability, inequality, and inequity, but rather forward, towards high-quality green development, the Sustainable Development Goals (SDGs), de-carbonization pathways, and an Ecological Civilization construction. The 14th Five-Year Plan represents a critical roadmap to enhance sustainable development, green innovation, and green technology within China, and advance renewed or new forms of international cooperation.

This Note examines four issues related to the recovery: (i) public health protection, (ii) a green economic recovery, (iii) trade policy and debt, and (iv) integrated policy.

Public Health

The coronavirus underscores the importance of strengthening risk assessment, preparedness, prevention, surveillance, and monitoring to provide accurate and early warnings of threats to human health, as well as maintaining active disease epidemiology programmes and technical and human resource capacity to respond to epidemiological investigations. Strong surveillance systems for communicable diseases, environmental hazards, and key health status data are central to assessing and minimizing health risks and strengthening emergency response activities, including emergency regulations and enforcement to support public health decisions.

Lessons from COVID-19 include the importance of international cooperation and integrated, holistic approaches that include public health, animal health, land-use change, animal husbandry, zoonotic risk management, ecosystem change including climate change, and other factors. Public health education, reliance on good science and transparency have never been more important, not only in controlling acute communicable diseases, but also in lowering chronic public health as well as environmental challenges related to air, water, and food safety. New approaches to enhancing public health measures and environmental monitoring should be explored, including a greater role for public monitoring and reporting of pollution, freshwater quality, and nature protection. Examples of freshwater monitoring in Finland illustrate innovative new approaches to engaging the public, schools, and others in environmental stewardship.

Trade protectionism affecting medicine, personal protective equipment, ventilators, and food have surged during the first three months of the pandemic, leaving developing countries most exposed now and creating insecurity about access to current supplies and a future vaccine. China's commitment to a rules based, multilateral system of cooperation is ever more urgently needed.

Green Economic Recovery

Previous economic slowdowns coincide with or have caused the weakening of environmental protection. Policy attention during economic turbulence tends to focus on GDP, unemployment, balance-of-payments, and export competitiveness. Environmental regulations have been viewed as incurring sunken costs and dampening economic recovery. Public support for environmental action has declined during past periods of economic downturn, as households focus on wages, job security, and savings.

Some of these previous patterns are being repeated today. Some governments have, for example, suspended environmental regulatory inspections, delayed new regulations, and stopped the surveillance of rainforests and other ecosystems. While air pollution and greenhouse gas emissions have declined in the short term and nature has been left relatively undisturbed during COVID-19, other pressures have been rising, from reports of increased wildlife poaching to a surge in medical waste and single-use plastic pollution to major projects being accelerated with potentially diminished oversight.

However, past conflicts pitting an economic recovery against environmental stewardship are diminishing, as new approaches emphasize stronger win-win results by strengthening the nexus between public health, pollution abatement, climate action, nature protection, social equity, and economic prosperity. Four related reasons why a green economic recovery will avoid past patterns are noted briefly below.

First, Science: A growing body of robust scientific research confirms accelerating rates of ecological degradation and destruction. In addition to more accurate modelling, the *empirical* evidence of ecological change and its consequences is expanding. In 2019, sea levels continued to rise due to warmer average surface ocean temperatures, resulting in melting Greenland ice and retreating glaciers. Ocean acidity is increasing more quickly than anticipated. Heatwaves and prolonged drought increased in 2019, including unprecedented wildfires in Australia as well as in Siberia and other Arctic regions. Global average temperatures in 2019 have risen by 1.1 °C above the preindustrial level, just 0.4 °C short of the Paris Climate Agreement lower-bound objective. All of this and more have economic costs and must be managed to secure a sustained economic recovery.

Second, People: Before COVID-19, public support for ambitious climate action was increasing. During COVID-19, while climate demonstrations have been suspended, public support has not. Polls conducted *during* COVID-19 suggest that support for climate action has increased, with young people as the strongest advocates for ambitious, transformative action. Moreover, behavioural changes during the pandemic will lead to behavioural transformations ahead. While COVID-19 shows that governments can impose strict measures, it has been the actions of people that have made the difference. There is hope that this new sense of community and solidarity will create a new, more equitable world ahead. A century ago, the English author D. H. Lawrence wrote *Look! We Have Come Through!* to describe new personal wisdom following alienation, personal loss, and uncertainty. The Great Lockdown may lead to new practices, including working more from home, fewer face-to-face meetings, less air travel, additional savings, and less debt-driven over-consumption splurges. The end of the crisis may open a new, ethical understanding of the importance of public health, as well as the value of fair wage compensation for labour in general and a fair wage for women's labour in particular.

Third, Economics: Three years after the Great Recession, Achim Steiner noted that a green economy "can catalyze economic activity of at least a comparable size to business as usual, but with a reduced risk of the crisis and shocks inherent in the existing model". Green development over the past decade has not only reinforced that view but also demonstrated that it can outperform business-as-usual economic practices. A recent empirical study that underscores the win-win benefits of green development is the May 2020 Oxford University report, *Building Back Better: A Net-Zero Emissions Recovery*, authored by Nobel Prize economist Joseph Stiglitz, climate economist Nicholas Stern and others. After reviewing 700 economic recovery plans and interviewing scores of financial, central bank, treasury and other officials and experts, the authors conclude that green, low-carbon, and climate-friendly economic projects produce "better results" for the economy and the environment compared to business-as-usual investments. Of critical importance, the research concludes that green, low-carbon projects create more jobs compared to brown or neutral projects. For every USD 1 million spent on clean energy infrastructures, such as renewable energy or green building construction and retrofitting, an average of 7.49 jobs were created in the early stages. This compares to 2.65 jobs for every USD 1 million spent on fossil fuel-based energy systems like coal. The report reinforces other work highlighting the immediate employment benefits, energy savings, avoided carbon emissions, and improved freshwater uses through green building retrofitting and refurbishment.

Such findings reinforce the benefits of ambitious and comprehensive green development approaches. The European Union Green Deal includes a strategic focus on green employment and green job retraining to compensate for the job impacts associated with closing some 230 coal plants, expanding investments in renewable energy and net-zero, circular-economy heavy industry approaches.[1] The European Commission's Circular Economy Implementation Plan similarly highlights the net job benefits stemming from ambitious, comprehensive circular economy approaches.

Perhaps one of the biggest short-term green job creation opportunities is associated with green recovery investments in natural capital by government and industry. These include hiring non-skilled and semi-skilled workers for ambitious afforestation and reforestation projects, wetland restoration and remediation, cleaning waterways and beaches, and creating community collection sites for plastic and other recycling. Other examples of immediate green job creation include the remediation of contaminated soil and contaminated waste sites, which also reduce public liabilities.[2] In April 2020, Canada announced a job creation program to restore abandoned oil and gas wells, leading to thousands of immediate jobs.

A green recovery has emerged as a central feature of economic policy prescriptions. In April 2020, the International Monetary Fund (IMF) recommended that countries implement a green economic recovery, with a focus on five strategic priorities:

- Climate-smart technologies such as renewable energy, green technologies (i.e. battery/hydrogen/carbon capture), and green infrastructure
- Climate adaptation, such as flood protection, resilient roads, and buildings
- Avoiding carbon-intensive investments such as fossil fuel power and high-emission vehicles
- Supporting public work programmes that provide income support
- Extending debt guarantees and other support to green industries/activities in preference to brown industries/activities.[3]

---

[1] The European Union Green Deal adopted in early 2020 includes specific funding to ensure a green transition. These funds have been further strengthened as part of the European Commission post-crisis economic recovery measures proposal. The Just Transition Fund aiming to facilitate the transition towards climate neutrality has been beefed-up from € 7.5 billion to €40 billion. The investment window of Invest EU dedicated to sustainable infrastructures has been almost doubled to reach € 20 billion of guarantee.

[2] The U.S. Department of Energy has estimated that its contaminated soil and hazardous waste liabilities are currently USD 494 billion, so the economic benefits of supporting workers would entail win-win effects.

[3] The IMF also notes that, in those cases where governments do provide recovery support to carbon-intensive activities like coal or airlines, industries and firms should be required to commit to binding emissions reduction targets.

The economic rationale for these investments is compelling. Since public spending on infrastructure, including buildings, will be a pillar of most countries' recovery plans, implementing green, low-carbon, and resilient infrastructure promises to create jobs and large-scale capital investments while lowering the combined carbon footprint of energy, transport, building and water infrastructure. Together, these account for 60% of global greenhouse gas emissions. Examples of successful sustainable infrastructure projects, supported by innovative green financing instruments, are growing. South Korea has deployed intelligent traffic management systems to decrease traffic congestion. The South East Water Authority (UK) provides free water-saving technologies, leading to improved freshwater management. The US Corps of Engineers has supported successful natural infrastructure freshwater and coastal marine buffers.

The investment choices in energy systems made during the economic recovery are of critical importance. Prior to COVID-19, the absolute costs of renewable energy at scale continued to decline (International Renewable Energy Agency [IRENA], 2019), while comparative costs of renewables to coal continued to favour clean energy sources. In 2019, some 75% of all US coal production was more expensive than renewable energy; that number is projected to reach 100% by 2025. During the COVID-19 crisis, renewable energy has experienced a remarkable 5% demand growth, while overall fossil use has plummeted, and total global energy demand has dropped by 6%. The post-COVID-19 energy landscape will undergo even swifter and deeper structural changes: the head of the International Energy Agency (2020) recently predicted a "significantly different" energy landscape in which renewable energy outperforms fossil fuels. In addition to large-scale energy systems, all countries should support new generations of efficient air conditioners and cooling systems that exclude climate-potent HFCs and other short-lived climate pollutants. The climate dividends from tackling short-lived climate pollutants alone entail avoiding up to 0.6 °C warming.

Fourth, Green Finance and Investment Trends: Before the crisis, big investors and blue-chip companies had stepped up commitments to low-carbon and net-carbon investments. In January 2020, the European Investment Bank announced that it was halting all financing of fossil fuels by 2021 and investing EUR 1 trillion in clean energy projects in the coming decade. Goldman Sachs promised USD 750 billion in lending to low-carbon and sustainability ventures. The biggest renewable energy purchases in US markets in 2019 included Amazon, Walmart, Apple, and Facebook. In early 2020, Microsoft went beyond net-zero carbon pledges by announcing not only negative carbon operations but actions to compensate for its "unpaid carbon debt" of the past. In January 2020, the head of Blackrock characterized climate as driving a "fundamental reshaping of finance". Also, in early 2020, the Bank for International Settlements (the Basel bank) cautioned that climate change could affect "every single agent in the economy and every single asset price" that could trigger the stranding of fossil fuel assets, especially thermal coal. In late 2019, Australia's Reserve Bank warned that climate change poses risks to Australia's financial stability, and banking and corporate regulators have become proactive in managing carbon risks.

Trade and Debt

Prior to COVID-19, international trade was turbulent and uncertain. Structural changes included a decline in global supply chains and an increase in regional hubs concentrated in China, the USA, and Germany. Institutional conflicts at the World Trade Organization (WTO) regarding the mandate of its Appellate body have weakened the multilateral system, while protectionism has been increasing, with tariffs, trade remedies, anti-dumping measures, non-tariff barriers, and import bans rising.

During COVID-19, a pronounced spike has occurred in protectionism, as countries scrambled to procure medical equipment, personal protection equipment, and related supplies. Protectionism has widened to include trade in food, as concerns about food security have heightened. Both trends will hurt developing countries the most, and both trends pose serious questions about what will happen if and when a vaccine is discovered. The heads of the WTO and the IMF recently pleaded with governments to stop trade protectionism and support greater international cooperation.

Following the crisis, an important area of international cooperation should be trade. Reverting back to Great Depression era practices of protectionism and mercantilism will stall overall growth and leave developing countries among the worst affected. Yet, like other systems, from energy to finance to development, the crisis is an opportunity to reshape a new rules based trading order fit for the purpose of supporting a green recovery and the SDGs. Short-term options include:

- Build on China's earlier waste import ban to ban trade in single-use plastics.
- Support greater linkages between trade and climate mitigation. The September 2019 Agreement on Climate Change, Trade and Sustainability advanced by New Zealand and others call for an accelerated tariff liberalization of climate-related goods and services, as well as the elimination of fossil fuel subsidies in accordance with G20 commitments and compatible with WTO provisions.
- Consider joining border carbon adjustment measures to accelerate a shift towards de-carbonization, while adhering to WTO principles on non-discrimination.
- Complement new rules to stop illegal trade in wildlife with complementary policies, including enforcement, public education, monitoring, public education and development aid.

Of all the issues facing the COVID-19 recovery, debt management will likely be the most challenging, complex, urgent, and in need of bold new approaches. Since the 2008 global recession, higher risk leveraged private debt markets have expanded rapidly, reaching USD 9 trillion globally. While cheap private credit increased, borrowers' credit quality, insurance and underwriting rules, and other safeguards weakened. This dramatic shift has rightly been dubbed the "privatization of Keynesian economics", in which the engine of global growth has been increasingly unstable private lending, in which record rates of leveraging coincide with weakening regulatory oversight and new rights protecting creditors.

In 2018, the Paris Club warned that the "landscape of public debt is undergoing profound change, characterized by growing vulnerabilities, increasingly more diverse creditors and more complex financial investment". In 2019, the global debt-to-GDP ratio reached the highest level ever, peaking at over 322% of GDP in the third quarter of 2019, representing a total debt of almost USD 253 trillion (or over USD 32,000 for every person on the planet). For the past decade, developing country debt has more than doubled, reaching USD 72 trillion, of which non-financial corporate debt now exceeds USD 31 trillion. While a limited number of countries were categorized as debt distressed in 2019, other countries face high-risk debt management challenges, prompting Gang Yi, Governor of the People's Bank of China, in 2019 to note that China would need to "consider a country's complete debt-servicing capabilities".

The economic crisis of 2020 suggests many or most developing countries are facing or will face debt distress. One indicator of this is the request by over 80 developing countries in March 2020 for emergency IMF financing. A month later, the *IMF Global Financial Stability Report* (2020) warned that emerging markets had experienced the sharpest portfolio flow reversal ever recorded, with cascading risks of bankruptcies, the freezing of credit markets, and a looming threat of banking failures.

One of the agreed carve-outs by the G20 regarding debt servicing is to allow sufficient fiscal space to advance the SDGs. This provision underscores the importance of integrating green provisions within meaningful, sustainable debt management strategies.

China should consider at least three options. First, it should examine how current and new green finance instruments, including scaled-up green, climate, and conservation bonds, can alleviate debt. Second, China should engage with a new generation of innovative, cooperative financing deals in which leading conservation groups, governments, and private sector actors advance representative, sustainably financed, and durable protected area systems around the world, a central pillar of the Convention on Biological Diversity 15th Conference of the Parties (CBD COP15). Third, given the inevitable discounting of sovereign debt in the coming months, China should work with other countries, leading investors, and conservation groups in structuring debt-for-climate adaptation and debt-for-conservation arrangements/swaps. China could co-convene a meeting with France and other countries before the CBD COP15 to examine new, bold debt conservation arrangements that can be supported by international financial institutions to help alleviate debt-distressed and at-risk developing country debt.

Integrated Policies

An important emphasis of CCICED's 2020 work is supporting policy coherence and integrated policy planning and implementation that moves beyond single-issue policy approaches, thereby reinforcing the foundation for Ecological Civilization objectives. There are renewed suggestions to enhance the link between the three Rio Conventions to realize climate and nature objectives and advance comprehensive approaches of three major zones—wilderness areas, cities and farms, and shared lands. Several CCICED Special Policy Studies have highlighted the importance of tools and platforms to support concrete, integrated approaches. Notable examples include large-scale spatial planning that traverses multiple jurisdictions, nature-based solutions linked to Ecological Redline approaches, jurisdictional sustainable commodity sourcing, and third-party certification systems to mainstream ecosystem stewardship within and beyond protected areas to include agriculture, oceans and fisheries, forestry, resource extraction and other sectors. Large-scale spatial planning and Ecological Redline approaches are being used effectively in the Yangtze River Economic Belt and can be extended to green the BRI.

Another important tool is using integrated or comprehensive wealth indices that go beyond GDP to help balance investments and genuine returns from natural capital, human capital, and produced capital investments. Given the importance of enhancing human health and well-being and proactively advancing gender equality, composite indicators are important to measure progress and hold those responsible accountable.

COVID-19 magnifies the importance of integrated and interdisciplinary approaches. The One Health platform is a notable example by its emphasis on integrated risk prevention and emergency response capacities that include climate change and other risks. The platform stresses the importance of adaptive, holistic, and forward-looking approaches to detect, prevent, monitor, control, and mitigate communicable and non-communicable diseases—and improve health outcomes more broadly. Among the priorities in holistic approaches is assessing complex interconnections among species, the environment, and human society, including climate impacts. China may wish to lead a new forum to support integrated risk management to coordinate the full multilateral system moving forward.

Future risks that are unknown, nonlinear, and cascading will also be present. COVID-19 and climate risk have these traits in common. In January 2020, the Bank for International Settlements warned that its current menu of risk assessment quantitative economic models are ill-suited to anticipating climate change impacts, which they warn can have sudden whiplash characteristics. They therefore recommend complementing standard quantitative economic modelling tools with scenarios and foresight analysis. Such tools, together with system dynamics tools, can help policymakers best navigate uncertainty and assist in minimizing potential risks and impacts and in maximizing investment returns.

Chapter 1
Global Climate Governance and China's Role

The Special Policy Study on *Global Climate Governance and China's Contribution*, under the CCICED umbrella research of the Global Governance and Ecological Civilization Task Force, was launched in July 2018 and will last for three and a half years. The research project has four primary focuses: (1) the impact of Chinese cabinet restructuring on climate change policy and recommendations; (2) China's contribution and leadership in the global climate governance system and its overall mid- and long-term strategy and roadmap for combating climate change; (3) green investment in infrastructure, and climate investment and financing in the context of the Belt and Road Initiative; and (4) key takeaways for improving the effectiveness of carbon pricing policies.

In 2019, while continuing to study China's mid- and long-term strategy for climate change, the team also embarked on research on low-carbon power infrastructure development in Southeast Asia, an issue that draws considerable international attention.

Southeast Asia represents a hotbed of investment and development for the Belt and Road Initiative. As the region's economy took off to become one of the most vibrant globally, its demand for energy and electricity and coal consumption has also soared. Amid mounting pressures in the global fight against climate change and local environmental pollution, the fast-expanding coal-fired power generation capacity in the region has captured global attention.

While China has supported Southeast Asian countries in increasing electricity supply and achieving their energy access goals, China has received criticism for its participation in financing and building coal-fired power plants. The international community has voiced its concern that the projects increase the risk of climate change and run counter to the principle of a green Belt and Road.

This report consists of three sections. The first section focuses on Chinese efforts to build low-carbon power infrastructure in Southeast Asia; the second section deals

© The Author(s) 2022
China Council for International Cooperation on Environment
and Development (CCICED) Secretariat,
Green Consensus and High Quality Development,
https://doi.org/10.1007/978-981-16-4799-4_1

with gender issues in climate change; and the third section contains a comprehensive set of annual policy recommendations.

1.1 China Promotes Power Infrastructure Development in the Belt and Road Region and Tackles Climate Change: Southeast Asia as a Case Study

1.1.1 Belt and Road Initiative and China's Policy Support to Overseas Infrastructure Development

1.1.1.1 Green Development and Achieving SDGs Are Part and Parcel of the Belt and Road Initiative

When visiting Kazakhstan and Indonesia in September and October of 2013, Chinese President Xi Jinping outlined the notions of building the Silk Road Economic Belt ("Belt") and the twenty-first century Maritime Silk Road ("Road"), respectively, and set in motion the Belt and Road Initiative (BRI). BRI seeks to harness existing bilateral and multilateral mechanisms between China and other countries to actively foster economic partnerships with the countries along the route and build "a community of shared interests, future and responsibility featuring mutual political trust, economic integration and cultural inclusiveness" [1]. As of the end of January 2020, the Chinese government had signed 200 intergovernmental cooperation agreements with 138 countries and 30 international organizations.[1] The geographical scope of cooperation has expanded beyond the Eurasian continent to include countries from Africa, Latin America, the South Pacific, and Western Europe.

Green development represents the consensus of countries jointly building the Belt and Road. President Xi has called for the building of a green Belt and Road on multiple occasions. The *Vision and Actions on Jointly Building the Silk Road Economic Belt and the Twenty-First Century Maritime Silk Road* issued in March 2015 proposed that efforts should be made to incorporate and integrate the green concept, environmental protection, and the principles of sustainable development into the five areas of connectivity advocated by the BRI.[2] Speaking at the opening ceremony of the Belt and Road Forum for International Cooperation in May 2017, President Xi called on all parties to build the Belt and Road into a road of innovation, pursue green development, and strengthen cooperation in ecological and environmental protection to attain the 2030 Sustainable Development Agenda. At the second Belt and Road Forum for International Cooperation in April 2019, President Xi once again stressed the need for open, green, and clean cooperation in a people-centered approach.

---

[1] Belt and Road Portal: https://www.yidaiyilu.gov.cn/xwzx/roll/77298.htm.

[2] Namely policy coordination, facilities connectivity, unimpeded trade, financial integration and people-to-people bonds.

Countries along the BRI now account for 67% of the global population and roughly 34% of the world's GDP. The vast majority of them are developing countries and nations with economies in transition. The per capita GDP of approximately two thirds of countries along the BRI are lower than the global average [2, 3]. A deficit in infrastructure and fragile ecological environment expose these countries to the full brunt of climate change. Losses from climate disasters in the BRI countries are more than double the global average. Seven of the top 10 countries most affected by global climate disasters between 1995 and 2005 belonged to the Belt and Road region [2]. In the meantime, due to the potential for economic development and increased energy intensity, countries along the BRI will likely become the largest growing energy consumers and greenhouse gas emitters globally. In the broader context of global low-carbon development and climate-resilient transformation, the green development needs of the BRI region are growing ever more urgent.

In a word, promoting green and sustainable development along the Belt and Road is not only an inherent requirement of the BRI but also an inevitable choice for the countries along the route. Creating a green and low-carbon community of shared future along the Belt and Road is of utmost significance to building a community of shared future for mankind, facilitating the achievement of the 2030 Sustainable Development Goals and the vision of a clean and beautiful world.

1.1.1.2 China's Overseas Investment Policy Has Shifted with Surging Investment and Improving Management

Since China's reform and opening-up, the country's policy on outbound direct investment (ODI) has undergone a paradigm shift from restriction to encouragement. Prior to 2000, the Chinese policy was to keep a tight rein on outbound direct investment; after 2000, the country began to adopt a "going global" strategy by gradually removing restrictions on the approval of ODI and actively encouraging investment abroad [4]. The BRI provides a strong strategic platform for the implementation of the "going global" strategy and created wider channels for ODI.

China's ODI rank among the top three in the world in terms of its flows and existing stock, and the Association of Southeast Asian Nations (ASEAN) has witnessed the fastest growth in ODI flows from China. The country's ODI stood at USD 143.04 billion in 2018, a year-on-year (YoY) decrease of 9.6% (see Fig. 1.1). With a 29% slash in global ODI, which had fallen for three consecutive years, China rose to become the second-largest ODI originator slightly behind Japan (USD 143.16 billion). At the end of 2018, China's existing stock of ODI reached USD 1.98 trillion, 66.3 times the amount at the end of 2002 and trailing only the United States and the Netherlands. An examination of the regional distribution of China's ODI flows (see Table 1.1) reveals that Asia remains the most popular destination for ODI from China and that ASEAN, in particular, is where Chinese FDI has grown most rapidly.

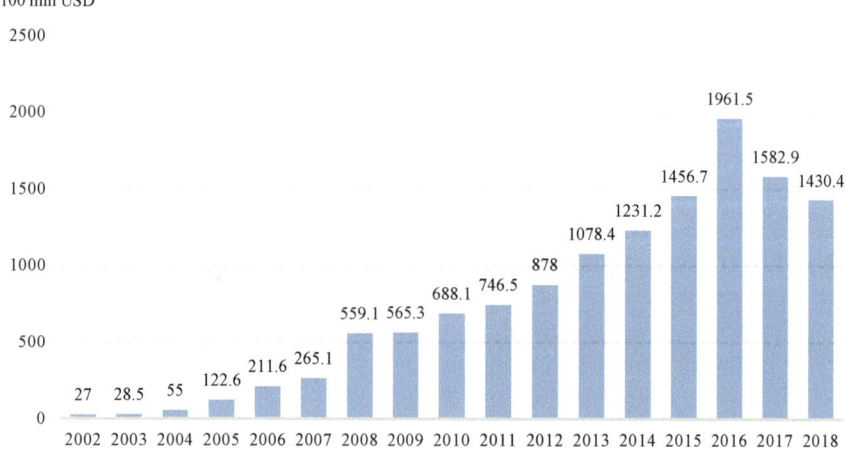

Fig. 1.1 China ODI between 2002 and 2018. *Data source* Ref. [5]

Table 1.1 China ODI flows to major economies in 2017

Economy	Investment (USD 100 million)	YoY growth (%)	Weight in total investment (%)
Hong Kong SAR	911.5	−20.2	57.6
ASEAN	141.2	37.4	8.9
European Union	102.7	2.7	6.5
United States	64.3	−62.2	4.0
Australia	42.4	1.3	2.7
Russia	15.5	19.7	1.0
Total	1277.5	−18.6	80.7

Data source Ref. [6]

1.1.1.3   China Has yet to Place More Stringent Requirements on Environmental Protection in Outbound Investment Policies

Multiple Chinese government agencies implement and manage the policies on overseas investment. Large foreign investments exceeding USD 2 billion must seek approval from the State Council. Other important government agencies include the People's Bank of China (PBOC), the National Development and Reform Commission (NDRC), the Ministry of Commerce (MOC), the Ministry of Finance (MOF), and the China Banking and Insurance Regulatory Commission. The MOC is tasked with the review and approval of direct investment by Chinese enterprises overseas, and the approval authority for most items is delegated to provincial-level commerce authorities with the exception of special projects. The NDRC is mainly responsible for evaluating overseas resource development projects, projects requiring large-value foreign exchange transactions, overseas acquisitions and bidding projects, and some special projects. The State Administration of Foreign Exchange (forex) supervises and manages foreign exchange receipts and payments and forex registration associated with domestic enterprises' direct overseas investment. The PBOC and the State Administration of Foreign Exchange are in charge of managing the pilot CNY settlement of overseas direct investment. The National Agency for International Development Cooperation (NADICA) is responsible for formulating strategic guidelines, plans, and policies for foreign assistance; coordinating and making recommendations on major issues of foreign assistance; promoting the reform of foreign assistance modalities; preparing foreign assistance programs and plans; determining foreign assistance projects; and supervising and evaluating their implementation.

There are few provisions in China's foreign investment policies that lay particular emphasis on environmental protection. The Guidelines for Environmental Protection in Foreign Investment and Cooperation promulgated by the MOC and Ministry of Environmental Protection in 2013 remains the only policy document that seeks to reduce the environmental impact of Chinese companies' overseas operations. It aims to guide the companies to promptly identify and prevent environmental risks. However, it is not legally enforceable and relies on voluntary compliance. The Opinions on Further Guiding and Regulating the Direction of Overseas Investment, issued jointly by the MOC, the PBOC, and the Ministry of Foreign Affairs in August 2017, classifies foreign investment projects into encouraged, restricted, and prohibited categories. The Administrative Measures for Overseas Investment by Enterprises, unveiled by the NDRC in November 2017, stipulates that overseas investments by Chinese enterprises in projects deemed sensitive would be subject to prior recordation and approval, as well as close monitoring and scrutiny afterwards. Subsequently, in January 2018, the NDRC released the Catalogue of Sensitive Industries for Overseas Investment (2018 Edition), listing cross-border water resources development and utilization as a sensitive industry. The Interim Measures for the Reporting of Outbound Investments Subject to Record-filing or Approval, published by seven ministries and commissions including the MOC, the PBOC, and SASAC in January 2018, provides that outbound investments by Chinese investors will be guided and regulated based on the principle of "encouraging development and a negative list" and that an outbound "blacklist" will be researched and established.

Compared with domestic investment policies (see Annex 1), the policies on foreign investment do not explicitly forbid projects with high pollution and carbon emissions, high resource consumption, and outdated and polluting technology but, instead, place a greater priority on safeguarding national economic and security interests. The environmental considerations only occur for projects that fail to meet the technological, environmental, and energy consumption standards of the investment destinations. However, target countries have varying standards. Developed countries, such as members of the European Union (EU) and a number of developing countries have stricter emission control protocols and higher environmental performance than China. Meanwhile, environmental governance in many countries along the Belt and Road is much weaker than that in China, and Chinese companies are subject to fewer environmental regulations and constraints in these countries.

1.1.1.4   Chinese Financial Institutions Are Still Rendering Financing Services for Overseas Coal Power and Related Industries

To achieve the Paris Agreement emission reduction targets, multilateral development agencies and many foreign financial institutes have halted financial support to coal power projects, while Chinese financial institutions are still playing a pivotal role in endorsing coal power projects both at home and abroad. China's four largest commercial banks (Agricultural Bank of China, Bank of China, China Construction Bank, and Industrial and Commercial Bank of China) have invested far more in coal-related assets than their international competitors. According to the *Banking on Climate Change 2019* report, in 2018, 71% of the global coal mining financing and 55% of the financing for coal-fired power generation came from China's four largest commercial banks. The four banks also topped the global list of banks that funded coal projects between 2016 and 2018, as shown in Table 1.2. By examining their financing policies, one could discern the important factor for their strong backing of coal power: the four banks are yet to draw up restrictive or prohibitive provisions against coal power financing, which falls short of their mainstream international peers. Apart from the increased environmental risks of funding these projects, the banks are subject to potentially exorbitant financial risks as a result of the future tightening of climate and environmental protection policies.

Data from Boston University's Global Development Policy Center reveals that China's two major policy banks (China Development Bank and China Export–Import Bank) provided approximately USD 3.2 billion to foreign countries in financing energy projects in 2019, bringing the total since 2000 to USD 250 billion. Between 2007 and 2014, China Development Bank and the Export–Import Bank of China invested more in the energy sector overseas than the World Bank, Asian Development Bank, African Development Bank and Inter-American Development Bank combined. Between 2013 and 2017, 41% of the loans to the power industry were invested in coal power, 57% in large hydropower, and 2% in non-hydro renewable energy projects.

Table 1.2 Ranking of coal power financing by Chinese commercial banks

Ranking	Bank	2016	2017	2018
1	Bank of China	47.44	49.88	63.69
2	Industrial and Commercial Bank of China	51.96	55.79	53.21
3	China Construction Bank	56.36	31.88	28.72
4	Agricultural Bank of China	43.40	26.15	26.33

Unit USD 100 million
Data source Ref. [7]

1.1.1.5 China is Promoting Green Foreign Investment Policies

China has crafted a path toward green overseas investment through an enhanced policy framework. In September 2015, the Central Committee of the Communist Party of China and the State Council distributed the *Master Plan for Reforming the System of Ecological Civilization*, which called on the country to "strive for faster progress in eco-civilization, ensure that resources are used more efficiently, and step up efforts to promote the new pattern of modernization in which man develops in harmony with nature" [8]. In 2017, the *Guidance on Promoting Green Belt and Road*, jointly published by the Ministry of Environmental Protection, the Ministry of Foreign Affairs, the National Development and Reform Commission, and the Ministry of Commerce, proposed to "mainstream ecological civilization in the Belt and Road Initiative, bolster green development, strengthen eco-environment protection, and jointly build a green silk road" [9]. In 2018, suggestions on promoting sustainable finance and green finance, drafted by the G20 Sustainable Finance Study Group with the People's Bank of China as the lead, were incorporated into the Buenos Aires G20 Leaders' Declaration, driving a global consensus on green finance. The Network of Central Banks and Supervisors for Greening the Financial System (NGFS), of which China is one of eight founding members, has seen its membership and influence grow steadily.

Through building a green Belt and Road, China will put into practice the green development philosophy, adopt green as a way of life, build an ecological civilization, and achieve the 2030 SDGs. At the two Belt and Road Forums, China and its partners have launched the Belt and Road Sustainable Cities Alliance and the BRI International Green Development Coalition, formulated the Green Investment Principles for the Belt and Road Development, kicked off the joint establishment of the BRI Environmental Big Data Platform, and worked with other countries to implement the Belt and Road South-South Cooperation Initiative on Climate Change. On 25 April 2019, the Green Investment Principles for Belt and Road Development was put into the List of Deliverables of the Second Belt and Road Forum for International Cooperation. Twenty-seven large international financial institutions participated in the ceremony, which marks an emerging consensus on green investment under the BRI framework.

Chinese financial institutions and companies are beginning to realize the risks associated with coal investments and have started to act accordingly and divert from coal. In March 2019, Wang Huisheng, chairman of the State Development and Investment Corporation (SDIC), announced that the company had completely withdrawn from coal-related operations and will focus future investments in new energies. This makes the SDIC the first Chinese state-owned enterprise to completely divest from coal.

Though China is yet to be equipped with all-round expertise for building a green Belt and Road, the country has gained a wealth of practical experience in green low-carbon transformation, which can be used by other developing nations for their own green transformation. China's influence in green global governance will be enhanced, thereby raising its global stature in other arenas.

1.1.2 Socioeconomic Development and Power Infrastructure in ASEAN

1.1.2.1 Positive Economic Growth but Uneven Development in ASEAN

Southeast Asia is one of the most economically vibrant regions in the world. The average annual growth of the regional GDP has amounted to 5.4% over the past decade, much higher than the global average of 3.3%. In 2018, the total GDP of the 11 Southeast Asian countries accounted for 3.65% of the world, making the region a key driver of world economic growth (see Fig. 1.2). According to the e-Conomy SEA report jointly produced by Google and Temasek, "Southeast Asia's Internet economy continues to grow at an unprecedented pace, and it has soared to USD 100 billion for the first time in 2019." This momentum will continue in 2020. The report predicts that, by 2025, the region's Internet economy will triple in size to USD 300 billion [10].

In 2018, the per capita GDP of Southeast Asian countries stood at roughly USD 4783.8 [11]. According to the World Bank's classification of countries by income levels, the vast majority of countries in Southeast Asia, with the exception of Singapore, Brunei, and Malaysia, are lower-middle-income countries. Singapore's per capita GDP of over USD 60,000 places the country comfortably at the top of the standings in the region. Brunei, which boasts the second-highest per capita GDP in the region, has a comparable per capita GDP to that of the developed world. However, the per capita GDP of most countries in Southeast Asia are still below USD 4000, with the lowest being Myanmar's USD 1330. There are still notable disparities in

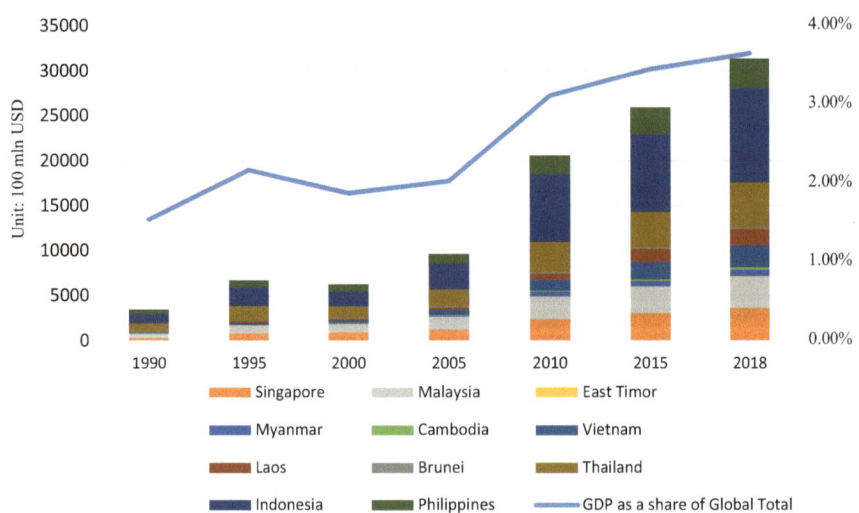

Fig. 1.2 GDP of Southeast Asian countries and their global share, 1990–2018. *Data source* Ref. [11]

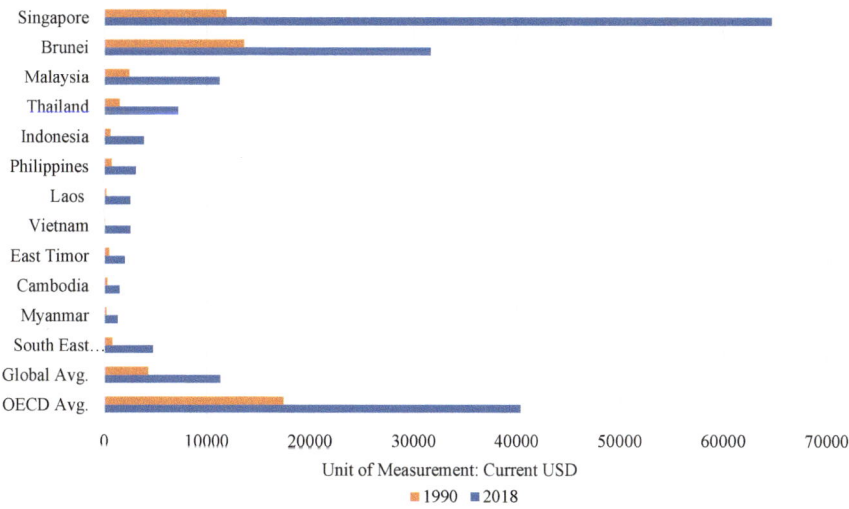

Fig. 1.3 Per capita GDP of Southeast Asian countries. *Note* 1990 data of East Timor, Myanmar and Cambodia were missing and replaced by those from 2000. *Data source* Ref. [11]

development among the various countries in the region. Figure 1.3 shows the per capita GDP by country [11].

1.1.2.2 Rich in Energy Resources but Uneven in Energy Distribution Across Countries

Fossil-based energy remains the most important energy resource in Southeast Asia, but its geographic distribution is uneven. The technically recoverable natural gas, crude oil, hard coal, and lignite resources in Southeast Asia stand at 6.46 trillion cubic metres, 1.82 billion tonnes, 37.53 billion tonnes, and 10.23 billion tonnes, respectively, accounting for 39, 29, 38, and 3% of their total reserves. **The distribution of fossil-based energy resources in ASEAN countries is extremely uneven** and primarily concentrated in four countries: Indonesia, Malaysia, Vietnam, and Thailand. Among them, Indonesia is the fifth-largest producer and second-largest exporter of coal in the world [12].

The region boasts a wide variety of renewable energy resources with considerable potential for development. There are abundant and diverse resources of renewable energy, including geothermal, hydro, biomass, solar, wind, and ocean energy. Resource endowments and conditions for development vary greatly from country to country. Hydropower is the primary source of renewable energy in Southeast Asia. Except for Singapore, where there are no large rivers, ASEAN countries have rich hydropower resources. Wind is mainly concentrated in Vietnam, Laos, Thailand, the

Table 1.3 Distribution of renewable energy in Southeast Asia

Country	Biomass (GW)	Geothermal (GW)	Hydro (GW)	Wind (GW)	Tidal (GW)	Solar (kWh/m^2/day)
Brunei			0.07			9.6–12
Cambodia			10			5
Indonesia	32.6	28.9	75		49	4.8
Laos	1.2	0.05	26			3.6–5.3
Malaysia	0.6		29			4.5
Myanmar			40.4	4		5
Philippines	0.24	4	10.5	76	170	5
Singapore					0.03–0.07	3.15
Thailand	2.5		15			5–5.6
Vietnam	0.56	0.34	35	7	0.1–0.2	4.5

northern Philippines, and parts of the coasts of other countries. In the region, the Philippines, Thailand, and Malaysia in particular receive a considerable amount of solar radiation, with an annual average of 5 kWh per m^2 per day. Indonesia and the Philippines are also rich in geothermal resources. Indonesia is estimated to contain the world's largest geothermal energy reserves, accounting for approximately 40% of the global total. The Philippines and Indonesia are endowed with abundant wave and tidal energy resources due to the large number of islands. Indonesia is the richest among its peers in Southeast Asia in biomass energy. The distribution of renewable energy in the region is shown in Table 1.3.

1.1.2.3 Consumption of Primary Energy and Power Increasing Rapidly, Predominated by Fossil Fuels

The region has witnessed a rapid growth in energy use and has become a net importer of fossil fuels. The development of infrastructure and industrial bases in the region, coupled with higher incomes and the emergence and rise of the consumer class, have pushed up the demand for coal and natural gas, especially for the purpose of power generation. Consequently, primary energy consumption in Southeast Asia has grown substantially. Primary energy demand in the region has surged by more than 80% since 2000, with average annual growth of 3.4%, far exceeding the global average of 2% (see Fig. 1.4). Rising fuel demand, especially for oil, has far outstripped the region's own production. For the first time, Southeast Asia as a whole will become a net importer of fossil fuels in the next few years.

The energy consumption in Southeast Asia has long been dominated by fossil fuels, with renewable energy accounting for a small fraction. Fossil fuel consumption took up three quarters of the region's primary energy consumption in 2018. Oil

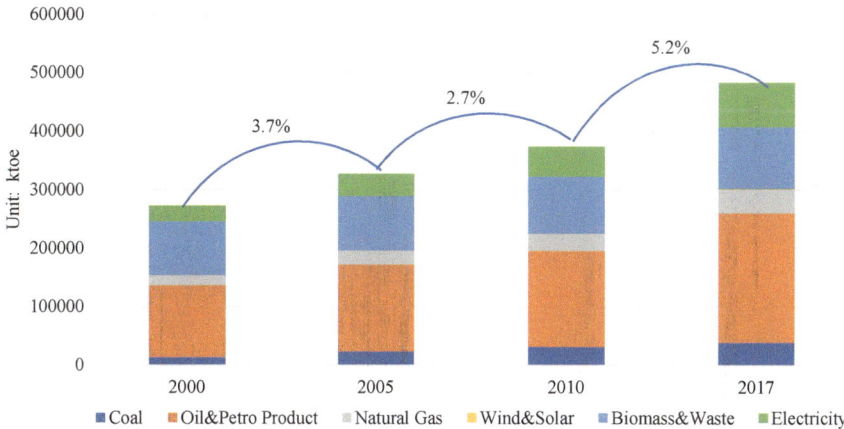

Fig. 1.4 Primary energy consumption in Southeast Asian countries. *Data source* Ref. [14] (data missing for East Timor and Laos)

is the most important component in the region's energy mix, while coal has been the fastest growing and most available source since 2000. Renewable energy (excluding solid biomass utilized for cooking) currently meets only 15% of the region's energy needs [13]. The amount of hydropower has quadrupled since 2000, and the use of modern biomass energy for heating and transportation has also seen tremendous growth. Despite declining costs, the contribution of solar photovoltaic (PV) and wind to the total energy consumption remains low.

Electricity use has registered high growth, but consumption per capita remains low at approximately half of the global average. Electricity demand in the region has been climbing at an average of 6% per year, which far exceeds the global average, making it one of the world's fastest-growing electricity consumers. In addition, population growth and economic development will continue to spur power consumption in the region. In 2017, total power consumption in the region stood at roughly 937 billion kWh/year (data is missing for East Timor and Laos), and the per capita amount was 1445 kWh/year. **Power consumption per capita varies greatly among countries. The average power consumption per capita in the region made up only half of the global average** (roughly 3200 kWh/year) (see Fig. 1.5)

The construction sector will overtake the industrial sector as the largest electricity consumer in the future. In the *Southeast Asia Energy Outlook 2019*, the International Energy Agency (IEA) forecasts that electricity consumption in the region will double by 2040, growing nearly 4% on an annual basis, twice as fast as the rest of the world. Electricity currently represents just 18% of total final energy use in Southeast Asia, lower than in most other regions. However, that proportion is expected to hit 26% in 2040, comparable to the global average. The construction sector (residential facilities and services) will experience the fastest increase in power

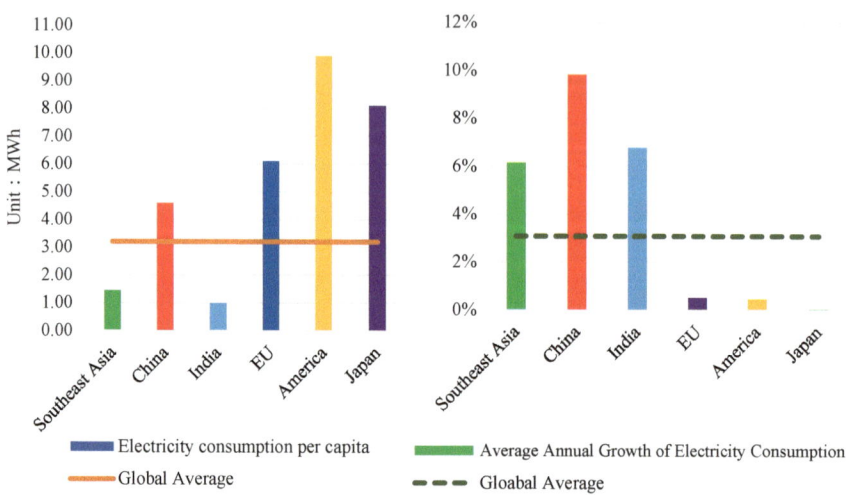

Fig. 1.5 Average annual growth of electricity consumption and per capita electricity consumption, 2000–2017. *Data source* Ref. [14] (data missing for East Timor and Laos)

use and jump by 250% by 2040 to reach over 1200 TWh, making it the largest final consumer of electricity ahead of the industrial sector.

1.1.2.4 There is a Significant Gap in Power Infrastructure, and Thermal and Hydropower Plants Provide the Majority of the Installed Power Generation Capacity

The installed electricity generation capacity of Southeast Asian countries has risen steadily in recent years. The region has experienced a major shift in the source of electric power from oil-fired to coal-fired power generation. In the meantime, the installed renewable power generation capacity is also growing continuously. In 2018, the installed capacity of coal, gas, and oil-fired power plants stood at 95 GW, 75 GW, and 25 GW, respectively [13] Fig. 1.6. **The installed capacity of renewable energy reached 64.31 million kW**, of which hydropower (including small hydropower) accounted for 72.49%, followed by biomass at 11.78%, geothermal at 6.05%, solar at 6.95%, and wind at 2.73% (see Fig. 1.7) [15].

Total electricity generation in Southeast Asia registered at 1001.213 billion kWh in 2017 (Note: East Timor not taken into account for lack of data and 2015 data is used for Laos). **Of the total power production, 76.6% was derived from fossil fuel**, of which natural gas and coal contributed 37.76% and 36.19%, respectively, and oil merely 2.65%. The amount of **power produced from renewable sources** stood at 234.247 billion kWh **or 23.4% in the region. Within the category of renewable energy, hydropower (including small hydropower) made up 18.25%**, geothermal 2.3%, biomass 1.78%, wind 0.63%, and solar 0.25% [13]. The gross production of

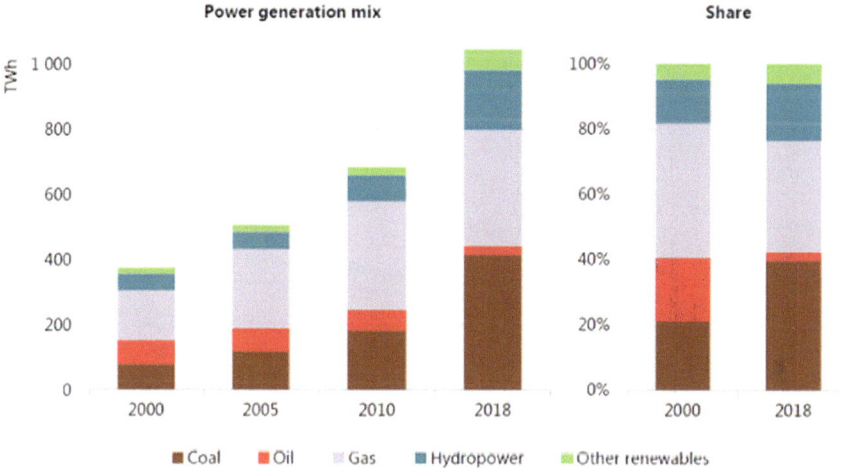

Fig. 1.6 2000–2018 installed power generation mix of Southeast Asia. *Data source* IEA [13]

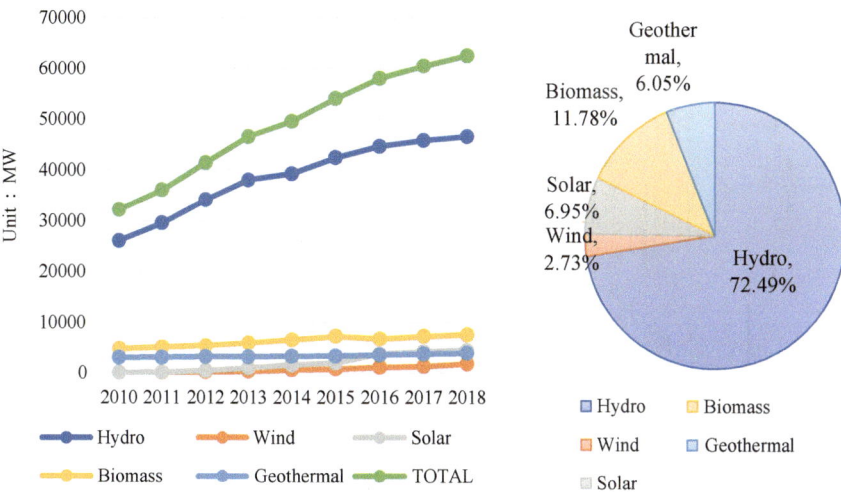

Fig. 1.7 Total power generation in Southeast Asian countries from 1990 to 2017. *Data source* IEA [14]

electricity by Southeast Asian countries between 1990 and 2017 is shown in Fig. 1.8; the power generation mix between 2000 and 2018 is presented in Fig. 1.6; and the amount and sources of power generated in 2017 can be found in Fig. 1.9.

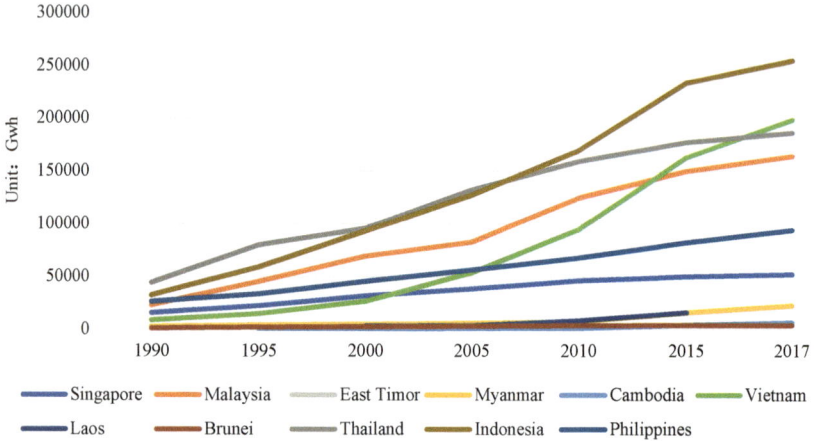

Fig. 1.8 Development of installed renewable power capacity in Southeast Asia. *Data source* IRENA [15]

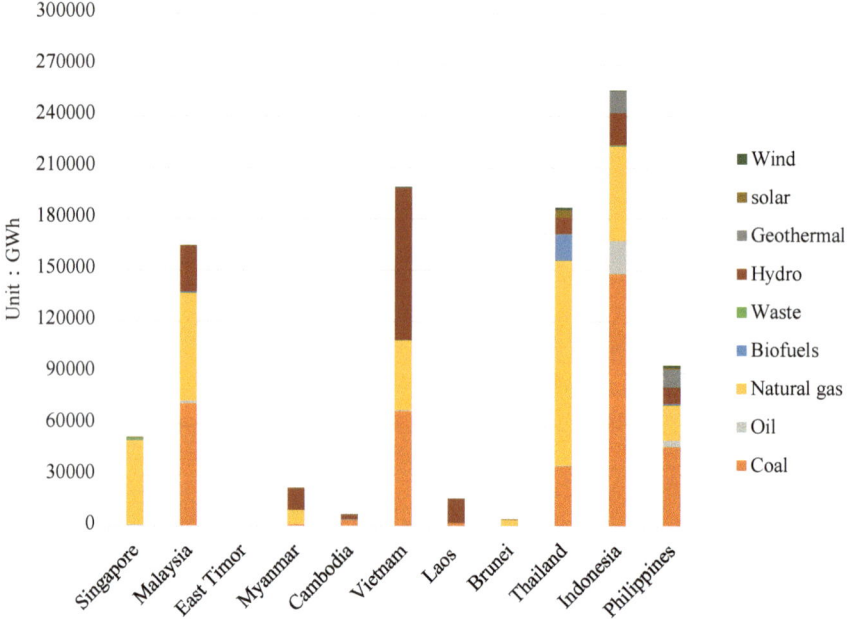

Fig. 1.9 The amount and sources of electric power generated in 2017. *Data source* IEA [14]

1.1.3 Power Management Policies and Clean and Low-Carbon Development Goals in Southeast Asia

1.1.3.1 The Power Sectors in Most Southeast Asian Countries Are not Market-Based and Lack Incentives for Competition

The vertical integration/single buyer model is the most common form of power market of Southeast Asia, where power plants are not completely separated from the grid and a vertically integrated management model is commonplace (see Table 1.4). The only exceptions are the Philippines and Singapore, where market forces are more mature. The vertical integration is prone to market monopoly and the superpowers of certain market players.

Examples can be found in EGAT—the only power system operator and the biggest power enterprise in Thailand. It manages and controls the power supply through a national control center and five regional control centers. It also possesses a power transmission network covering the entire country, including transmission lines and high-voltage substations of varying voltage classes. The power markets in Singapore and the Philippines are characterized by liberal competition between retailers, and independent power producers play a crucial role in the market with total installed capacity making up over 50% of the national total.

The power price in Southeast Asia is heavily subsidized. The average sales price of electricity in major Southeast Asian countries is higher than in China, with Vietnam as the only exception (see Fig. 1.10). The manufacturing sector in Southeast Asia is hampered by excessive power costs. For instance, based on the progressive tariff system in Myanmar's industrial and commercial sectors, the unit price is USD 0.1608 for power consumption between 10,000 and 50,000 kWh, USD 0.1281 between 50,000 and 200,000 kWh, USD 0.1067 between 200,000 and 300,000 kWh, and USD 0.0855 for consumption over 300,000 kWh. In Laos, three different power prices are defined for the industrial and commercial sectors, with a maximum of USD 0.1484 for the entertainment industry, a minimum of USD 0.07675 for high-voltage industries, and USD 0.1005 for other services.

The Philippines ranks the highest in power prices among all Southeast Asian countries due to its lack of power supply and has among the world's highest electricity bills for both residential and industrial purposes. The Philippines has adopted a categorized power pricing system with approximate rates of USD 0.2067 per kWh for residential consumption and USD 0.1144 per kWh for industrial use [16]. Thailand mainly relies on natural gas for its power plants, which represents roughly 60–70% of its generated power. This has led to its high electricity pricing, at USD 0.1111 on average in 2016.

Vietnam enjoys the lowest power price in Southeast Asia at USD 0.0685 per kWh in 2016. Yet it is noteworthy that the affordability is accompanied by the instability of its power supply, with blackouts occurring now and then.

Table 1.4 Power management mechanisms in Southeast Asian countries

Country	Market structure	Power generation	Power transmission and distribution	Power consumption
Cambodia	Vertical integration/single buyer	Independent power plant (IPP) National power company Rural power enterprises	National power company Rural power enterprises	Phnom Penh, provincial capitals, rural power enterprises
Brunei	Vertical integration/single buyer	Department of Electrical Services Berakas Power Company	Department of Electrical Services (operation) Berakas Power Company (maintenance and development)	End users
Indonesia	Vertical integration/single buyer	PT Pembangkitan Jawa-Bali (PTPJB) Indonesia Power IPP Leasing power plants	Indonesia Power	Residential Industry Commerce Others
Laos	Vertical integration/single buyer	National power company IPP	National power company	Key accounts End users
Malaysia	Vertical integration/single buyer	National power company IPP	National power company (in various regions) Sabah Energy Sarawak Energy	End users
Myanmar	Vertical integration/single buyer	Myanmar Electric Power Enterprise Hydropower enterprises IPP (hydropower)	Myanmar Electric Power Enterprise	End users

(continued)

Table 1.4   (continued)

Country	Market structure	Power generation	Power transmission and distribution	Power consumption
Thailand	Vertical integration/single buyer	Thai Power Authority IPP Small power plants Micro power plants	Thai Power Authority	Direct users End users Industrial zones
Vietnam	Cost pool	Vietnam Electricity IPP	Vietnam Electricity	End users
Singapore	Price pool	Domestic IPPs	Singapore Power Ltd	End users
Philippines	Price pool	National power corporations—small power companies Independent power plants	National Transmission Corporation (TransCo)	Power supply market—monopolized Power supply market—non-monopolized

Note Cost pool On-grid order, size and price determined by the variable cost of power generation
Price pool Liberal and competitive retail market for power sector
Single buyer Buyer monopoly, minimum power bills for a given load
Data source Ref. [12]

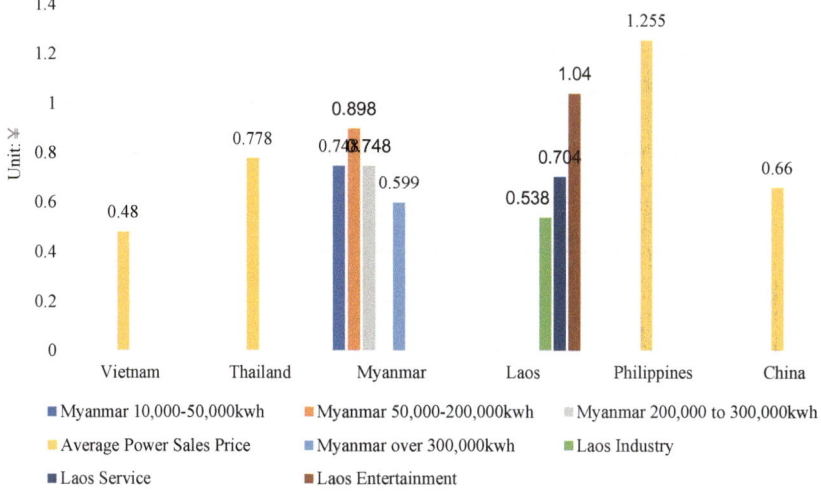

Fig. 1.10   Power sales prices in Southeast Asian countries and China. *Data source* Ref. [16]

1.1.3.2 Increasing Energy Supply and Electrification Remains the Strategic Priority of Power Development in Most Southeast Asian Countries

Southeast Asia is one of the most dynamic players in the global energy system. Its members are in different economic development phases with varying energy resource endowments and consumption models. **Yet they face the shared challenge of satisfying the growing need in a safe, affordable and sustainable manner.** In recent years, great efforts have been made in this region to upgrade the policy framework, reform the consumption subsidy for fossil fuels, intensify regional cooperation, and encourage investment in renewable energy.

Southeast Asian countries have spared no effort in expanding access to electricity and securing rural electrification. At the 33rd ASEAN Ministers on Energy Meeting and a series of related events in October 2015 in Kuala Lumpur, the *ASEAN Plan of Action for Energy Cooperation (APAEC)* was adopted, with enhanced accessibility to electricity and modern energy being one of the key goals. It is noteworthy that ASEAN countries have long been dedicated to improving electricity access through renewable energy technologies and enabling rural electrification through distributive solar PV and micro-grids. **Yet many hurdles remain for renewable energy development due to technical and financial constraints, lack of fiscal arrangements, and weak political mutual trust, resulting in much-delayed action compared to expected targets. In the end, results remain uneven between countries. The most challenging bottlenecks seem to be the lack of awareness, financial tools, and the lobbying power of incumbent companies.**

Facilitate regional grid interconnection and energy integration. The interconnection of the power grid and infrastructure is a key factor in clean energy development. Therefore, a clean energy power transmission grid shall be a priority for energy and power development. Currently, most trade in power on the grid of Southeast Asia remains on a bilateral basis. Though cross-border grids exist in most countries, trade in power is largely non-directional (such as purchase agreements). To fully harness these grids, the region is planning to boost multilateral power trade. The cross-border interconnection and multilateral power trade will promote asset utilization and resource sharing, improve the flexibility of the power sector in ASEAN, and ensure greater uptake of renewable energy, especially solar and wind.

China has launched trials in power grid interconnection with Southeast Asian countries and is especially active in power project cooperation and development in the Greater Mekong Subregion (GMS). Power interconnection in GMS will eventually enable large-scale, long-distance stable power transmission and reception, large-scale power exchange between multiple countries and minor power exchange in border regions between neighbouring countries, with total exchange expected to reach approximately 50 million kW. The EU's experience with electricity market integration could be applied in ASEAN. Directive 90/377/EEC of the European Union regulating electricity market reform imposes transparency in the prices of electricity and natural gas for all industrial users, and Directive 90/547/EEC on electricity transmission aims to remove barriers between member states and establish a single EU electricity market.

1.1.3.3 Renewable Energy Gets Attention with a Multi-pronged Approach to Boost Renewable Sources of Electricity

In recent years, Southeast Asian countries have turned their eyes to renewable energy and set corresponding targets in national plans and strategies. According to ASEAN plans, renewable energy shall represent 23% of the primary energy mix by 2025. Specific 2030 targets of different countries are listed in Table 1.5.

As for Nationally Determined Contributions (NDCs), despite these countries' varying emission reduction targets, renewable energy constitutes an important component (see Table 1.6). Singapore and Malaysia have opted for emission intensity as their emission reduction indicator; Myanmar and Laos have proposed policy actions without quantifiable mitigation targets. Brunei, the Philippines, and Cambodia, on the other hand, have pledged 60–70% absolute emission reduction based on the baseline scenario, as the three countries are extremely prone to climate change impact, thus highly motivated to reduce emissions.

To facilitate renewable energy development, ASEAN members have rolled out supportive policies and incentives. As is seen in Table 1.7, from the industrial perspective, support policies offered by ASEAN members include: setting renewable energy targets, introducing a feed-in tariff (FIT) policy, self-consumption plan and competitive tendering (auctioning). Policy incentives include tax incentives, preferential loans, capital subsidy, tradable renewable energy certificates, etc. **Among many incentives,**

Table 1.5 Renewable energy development targets of Some Southeast Asian countries

Country	Target
Brunei	Power from renewable energy accounting for 10% by 2035
Cambodia	Installed capacity of hydropower increasing to 2241 MW by 2020
Indonesia	Share of new energy and renewable energy increasing to 23% of primary energy supply by 2025 and 31% by 2050
Laos	Renewable energy making up 30% of primary energy supply by 2025
Malaysia	Installed capacity of renewable energy reaching 2080 MW by 2020 and 4000 MW by 2030
Philippines	Energy consumption decreasing by 16% annually based on the baseline forecast by 2030
Singapore	Installed capacity of solar PV reaching 350 MW by 2020 and 1 GW afterwards
Thailand	Increased shared of renewable energy in end consumption, reaching 30% by 2036; installed capacity of renewable energy reaching 36% and power generation from renewables reaching 20% by 2037
Vietnam	Installed capacity of non-hydro renewables reaching 12.5% by 2025 and 21% by 2030

Data source Ref. [13]

Table 1.6 Climate change mitigation targets pledged in country NDC

Country	Climate change mitigation targets	Target year	Description ("conditional" refers to availability of national aid and technical support)
Brunei	Absolute emissions reduction under the relative baseline scenario	2035	(1) Energy sector: energy consumption reduction by 63% (from business as usual [BAU]), share of renewable energy reaching 10% in total energy; (2) Land transport sector: carbon dioxide emissions cut by 40% (BAU); (3) Forest coverage reaching 55% with 34% increase (based on current figure, 2015)
Philippines	Absolute emissions reduction under the relative baseline scenario	2030	By 2030, greenhouse gas emissions will be reduced by about 70% compared with the baseline scenario
Malaysia	Carbon intensity	2030	Unconditional carbon emissions intensity reduction by 35% + conditional emission cut by 10% (2005 as baseline year)
Cambodia	Absolute emissions reduction under the relative baseline scenario	2030	Unconditional: except for land use, land-use change, and forestry (LULUCF), the total emissions decreased by 27% compared with the baseline scenario. Conditional: forest coverage reaching 60%, LULUCF emission cut by 57%
Singapore	Carbon intensity	2030	Carbon emissions intensity reduction by 36% in 2030 from 2005
Thailand	Absolute emissions reduction under the relative baseline scenario	2030	Unconditional: 20% reduction by 2030 compared to baseline scenario. Conditional: increase to 25%
Vietnam	Absolute emissions reduction under the relative baseline scenario	2030	Unconditional: 8% reduction by 2030 compared to baseline scenario. Conditional: increase to 25%
Indonesia	Absolute emissions reduction under the relative baseline scenario	2030	Unconditional: 29% reduction by 2030 compared to baseline scenario. Conditional: increase to 41%
Myanmar	Policy action, no quantifiable mitigation target	–	–
Laos	Policy action, no quantifiable mitigation target	–	–

Data source Ref. [17]

Table 1.7 Incentive measures of renewable energy development in ASEAN countries

ASEAN countries	Renewable energy target	FIT	self-consump. plan	Competitive tendering (auctioning)	Tax incentives	Preferential loans	Capital subsidies	Tradable renewable energy credit
Brunei	✓							
Cambodia	✓				✓			
Indonesia	✓	✓	✓	✓	✓			
Laos	✓				✓			
Malaysia	✓	✓	✓	✓	✓			
Myanmar	✓				✓			
Philippines	✓	✓	✓		✓			✓
Singapore	✓			✓	✓			
Thailand	✓	✓	✓	✓	✓	✓	✓	
Vietnam	✓	✓	✓		✓			

Data source Ref. [18]

Table 1.8 FIT mechanisms by country

Country	FIT mechanism
Indonesia	The FIT is based on energy production cost rather than technical cost and requires a comparison between the local production cost of energy (LPCE) and the national production cost of energy (NPCE). For solar, wind, biomass, biogas, and tidal, maximum FIT shall be 85% of LPCE if LPCE is higher than NPCE. For hydro, solid waste, and geothermal power, FIT shall be equal to LPCE if LPCE is higher than NPCE. For all energy types, if LPCE is not higher than NPCE, FIT shall be determined based on agreement by all stakeholders (state-owned PLN and IPPs)
Malaysia	The FIT of renewable energy companies is determined by an annual quota granted by the government
Philippines	The FIT is fixed rather than determined by the type of renewable energy, specific regions, or scope of capacity
Thailand	Renewable energy technologies are classified into two types: natural energy (hydropower, wind power, solar PV) and bioenergy (urban solid waste, biomass, biogas) There are two types of natural energy FITs: fixed FITs and extra subsidy (subsidy premium of three southern provinces) The FIT of bioenergy consists of two parts: fixed FIT and variable FIT (the variable depends on the inflation rate)
Vietnam	A nationally uniform and fixed FIT is adopted for all types of renewable energy rather than setting an FIT based on specific regions or installed capacity

Data source Ref. [19]

the FIT has gradually emerged as the centerpiece for boosting renewable power. The FIT has been generally adopted in countries with high growth in the installed capacity of renewable energy, such as Indonesia, Malaysia, the Philippines, Thailand, and Vietnam. Among these countries, Indonesia has identified the price cap of regional and national power costs as the standard while the rest of the countries have opted for the levelled power cost plus extra subsidy for investment returns of different technologies as the standard (see Table 1.8). In general, FIT policies in Southeast Asian countries are subject to frequent changes with immature design in the specific mechanism, hence the need for improvement according to feedback on existing policies.

1.1.3.4   Environmental Standards and Institutional Regulations Are Taking Shape with Growing Public Awareness for Environmental Protection

In recent years, a stringent environmental standard framework has emerged in Southeast Asian countries. Taking the environmental quality standard as an example, Southeast Asian countries have all promulgated their national standards for air quality, which are regularly reviewed and updated, with Myanmar as the only exception. Despite the gap in air quality standards, some countries have narrowed the gap in conventional pollutants such as sulphur dioxide (SO_2) and nitrogen dioxide (NO_2), compared to the United States and the EU, by updating their own standards. The average daily limit is 105 ug/m^3 for SO_2 and 75 ug/m^3 for NO_2 in Malaysia, both of which are higher than Grade II standards of China; however, Malaysia's SO_2 standard remains lower than Western standards, such as that of the EU [19].

The environmental charge, as per existing laws in Southeast Asian, countries is mainly a pollution discharge fee, which is directly levied from polluters according to the type and quantity of pollutants legally ratified by environmental protection authorities. In addition, an environmental protection fee is also charged from enterprises in some countries. Vietnam, for instance, applies such fees on manufacturers of petroleum, diesel, lubricant, coal, HCFC solution, nylon bag (taxable), herbicide, formicide, forestry product preservative, solid disinfectant, etc. Environmental protection fees levied on oil extraction, gas exploitation, and nylon bag production are 4.306 USD/tonne, 8.611 USD/tonne, and 1.292–2.153 USD/kg, respectively [19].

However, the implementation and enforcement of environmental laws is less commendable in Southeast Asian countries due to the following reasons: (1) structural defects exist in the legal system with inconsistencies in legal authorization, multi-layered structure and forms of existence, as well as overlapping or conflicting authorities that compromise the execution of the law; (2) the vertical segmentation of the institutional structure makes horizontal coordination difficult among ministries and departments; (3) the lack of public trust for law and the judicial system [20]. In recent years, strengthened penalties for environmental violations, integration of environmental and resource conservation authorities, enhanced efficiency in environmental administration, and rigorous law execution have improved the situation.

The voices of residents and environmental protection organizations are seldomly heard. Investments such as coal-fired and hydropower plants have triggered a backlash from the local community due to severe social and environmental impacts, potential damage to people's quality of life, and the possibility of violating the national pledge for carbon emission reduction. In Indonesia, residents of Bengkulu Province in Sumatra reacted against the Bengkulu 2 * 100 MW PLTU steam coal-fired power plant for lacking local community involvement during the environmental impact evaluation and analysis prior to project kick-off and the negative impacts on marine life and mangroves. Civil societies were also critical of the coal-fired power plant constructed by China on the grounds of potential air and water pollution in Java-Bali and impairment to tourism in Bali. However, the outcry of residents and environmental NGOs were not taken seriously or examined. In most cases, only a small portion of the local communities are solicited for their opinions. The information transparency concerning environmental and social impacts is undermined by too few opinion solicitations from local communities or meaningful engagement of residents in project decision-making.

1.1.4   There Exists Enormous Demand for Power Infrastructure Investment in Southeast Asian Countries

1.1.4.1   Power Infrastructure Investments in Southeast Asia Promises May Hit the Trillion Threshold While Renewable Power May Be the Investment Focus

Southeast Asia is an important destination for the relocation of Chinese industries. With its per capita power consumption around half of the world average, the region is expected to maintain steady growth in installed capacity in the next 10 years, supported by the growing population and industrial development. According to a forecast under the existing policy framework, the installed power capacity in the region will see a net increase of 90 GW for coal-fired energy and over 180 GW for renewable energy [13]. Taking transport, labour, and construction experience into consideration, the investment costs will be higher than those of China. At a unit construction cost of 5000 CNY/kW, the market size of energy (power plant project) investment in the region will reach CNY 1 trillion in the next two decades.

Air pollution is Southeast Asia has become a major environmental issue during the expansion of power infrastructures. At present, most Southeast Asian countries have developed their goals for renewable energy and come up with supporting policies. From the perspectives of technological cost, energy security, environmental constraints, and international trends, further development of renewable energy is indispensable for the region.

1.1.4.2   China's Involvement in Power Infrastructure Development in Southeast Asia

Apart from Coal-Fired Power Plants, China is Actively Engaged in Renewable Energy Projects in Southeast Asia

South Asia and Southeast Asia are the main destinations of China's investment in coal-fired power plants. A statistical analysis by Greenpeace based on disclosed information shows that, by the end of 2018, Chinese enterprises had built 10.8 GW of coal-fired power plants overseas through equity investment, nearly 94% of which are in South Asia and Southeast Asia; another 23.1 GW is in the planning or construction phase. Based on the estimated construction cost of coal-fired plants in China recently (unit construction cost of thermal power stations launched in 2016–2017 was 3593 CNY/kW), the total investment in these uncompleted projects is close to CNY 83 billion [21].

While investing in coal-fired projects, China is also actively involved in renewable energy projects in South Asia and Southeast Asia. According to statistics and analysis by Greenpeace, from 2014 to 2018, wind power and solar PV projects involving China via equity investment are mainly in South Asia and Southeast Asia. During this period, in countries including Pakistan, India, Malaysia, and Thailand, the total completed installed solar PV capacity with an equity investment from Chinese enterprises reached 1185 MW, accounting for 93% of total investment in countries along the Belt and Road in the same period. In addition, another 996 MW of installed solar PV capacity is in the planning or construction phase, which will bring the total Chinese contribution to 2181 MW in these countries. The installed capacity of solar PV power stations invested in or planned for investment by China in Bangladesh, Afghanistan, Vietnam, and Pakistan had exceeded 30% of total solar PV installed capacity in these countries by the end of 2018. Apart from equity investment, from 2014 to 2018, China participated in constructing 8440 MV of solar PV power stations in countries along the Belt and Road by exporting equipment. During this period, three of the top five destinations of China's solar PV equipment export were in South Asia and Southeast Asia, namely India (5800 MW), Thailand (1060 MW), and the Philippines (250 MW). In the same period, approximately 80% of wind power projects in Belt and Road countries involving Chinese enterprises via equity investment were located in South Asia and Southeast Asia, with an installed capacity of 397.5 MW already completed and 1362 MW under construction or planning, adding up to 1759.5 MW [22].

Furthermore, Southeast Asia is also a hub for overseas solar PV bases launched by Chinese enterprises. A total of 12 Chinese solar PV enterprises have participated in building solar PV component factories in the manufacturing base hub in Southeast Asia, especially in Vietnam and Thailand, with an announced capacity of 7 GW.

Engineering, Procurement, and Construction (EPC) Are the Most Common Chinese Involvement in Overseas Power Infrastructure Projects, yet China is Now Migrating from EPC to Equity Investment

China's participation in foreign power infrastructure projects is mainly through equity investment, financial support, EPC, and equipment export, etc. Each coal-fired power project may involve one or more of the above approaches, and the dominating approach will determine if Chinese enterprises and financial institutions are in a decision-making status and if long-term economic gains would be possible. China has gone through periods from project assistance to EPC to the current integrated project development for overseas coal-fired power plant investment, which has allowed Chinese equipment, technologies, and capital to gradually make their way to foreign markets.

From 2009 to 2018, China participated in building 74.3 GW of overseas coal-fired projects via EPC and 10.8 GW via equity investment, meaning that EPC is the most common for China in foreign coal-fired power projects. This makes Chinese enterprises merely "constructors" or "equipment providers," who are only entitled to short- and mid-term economic gains, rather than decision-makers. In this case, Chinese businesses are primarily driven by market forces, the demand of the host countries for energy development, and enterprises' pursuit of profit. The relationship between China and the host country and the policies of the Chinese government are only secondary factors.

But the situation is now changing. The first batch of overseas coal-fired power plants with Chinese enterprises as equity investors was put into operation in 2012. According to Greenpeace, overseas coal-fired power projects with China as an equity investor added to an installed capacity of only 0.4 GW before 2013, but the figure soared to 10.4 GW from 2014 to 2018, 26 times that of the previous 5 years. The year **2018 marked the first time that the installed capacity built by Chinese enterprises through equity investment exceeded that of EPC, reaching 3.5 GW, and the role of Chinese investors in foreign coal-fired power projects is gradually shifting from EPC to equity holder**. In the 5 years from 2019 to 2023, the capacity of coal-fired power projects with Chinese enterprises as an equity investor, whether built, under construction or in planning, is on track to reach 39.8 GW. Another 24.1 GW will be added through EPC. **Equity investment will overtake EPC in terms of installed capacity of coal-fired power projects in the future as China's mainstream foreign coal-fired power investment** [21].

Chinese State-Owned Banks and Major State-Owned Enterprises Are Strong Backings for Overseas Coal-Fired Power Investments, While Most Private Enterprises Favour Renewable Energy Projects

Most Chinese investors and developers in the coal-fired power sector are state-owned enterprises. Policy banks such as China Development Bank and China Import–Export Bank are the leading players among financial institutions, followed by commercial banks such as Bank of China and Industrial and Commercial Bank of China. Major state-owned enterprises—including State Grid, which monopolizes China's public utilities market, Sinomec in infrastructure, State Power Investment Corporation and China Huadian Co., Ltd.—are the giants in the power sector and take the biggest share among all competitors in the field. **In contrast, about two thirds (64%) of outbound energy investment by private enterprises goes to renewables** [23].

1.1.4.3   Evaluation of China's Participation in Power Infrastructure Construction by Southeast Asian Countries

Chinese bids for power plants are cheaper and respond to local demand for cheap solutions. According to a local expert interviewed for this report, local actors are divided on the Chinese construction of local power infrastructure, and some local actors are satisfied with the Chinese companies bringing additional power generation capacity to support the local economy's growth.

However, Chinese plants are sometimes subject to quality issues and might affect the effectiveness of the plant and the cost-effectiveness. In general, according to local experts, the global reputation of Chinese manufacturing in infrastructure is not very high.

Local populations are increasingly sensitive to air quality, water usage, and environmental pollution issues, and the impacts on health. According to a local expert, air pollution, both from traffic and new coal power plants, is increasingly becoming a public issue. Air pollution is already reaching levels previously seen in China, given the same pathway of growing industry and unabated coal power, although the use of ultra-supercritical technologies might lead to lower levels of air pollution. Air pollution is becoming an issue of public concern in the already high-density industrial and road traffic areas where the new plants are installed. These new investments are unlikely to be modified in the short term, given the limited available cash, and therefore the health impacts are highly likely to last for the next decades.

Chinese Companies Often Bring Their Own Workforce, Which Brings Limited Development Opportunities, Capacity Building, and Independence for the Local Community

The involvement of the local population—notably, through consultations with the local community, NGOs and civil society—is lacking when Chinese actors implement power projects. In comparison, according to local experts, projects led by BP, for instance, used to inform and consult local NGOs on a weekly basis when conducting infrastructure projects. The foreign-led projects would push forward best international practices in terms of environmental, social, and corporate governance (ESG), safety, and environment to ensure a sustainable local economy.

A local bribery case[3] in 2019 involving a power purchase agreement between PLN, Indonesia's local energy producer and distributor; Blackgold, an Indonesia-focused coal mining company; and China Huadian Engineering Co. Ltd. led to the arrest of nine officials, including the chief executive of PLN and a former Social Affairs Minister. Experts in local power investments estimate that the risk of corruption is higher in relation to high-growth infrastructure projects. These kinds of cases risk eroding public trust in the coal sector. In 2016, the Corruption Eradication Commission (KPK) found that 40% of 10,992 coal sector licences issued in four Indonesian provinces had failed to meet all legal requirements, including the payment of taxes, land rents, and other royalties. As a consequence, over 2000 of these permits have been revoked or allowed to expire.

---

[3] https://www.chinadialogue.net/article/show/single/en/11375-Corruption-and-coal-dug-up-in-Indonesia.

1.1.5   Case Study of Low-Carbon Transformation of Power Infrastructure in Southeast Asia: Indonesia[4]

1.1.5.1   Despite the Sustained Rapid Increase in Power Supply in Indonesia, the Per Capita Power Usage is Fairly Low

Indonesia is a vast archipelago comprising more than 17,000 islands. The five major islands are Papua, Kalimantan, Sumatra, Sulawesi, and Java. Indonesian territory covers 1.9 million km^2, comparable to one fifth of China or the United States. It is administratively divided into 34 provinces, from Aceh on the western tip of Sumatra to Papua in the east. At 264 million in 2018, the Indonesian population ranks fourth after China, India, and the United States, and it is the most populous nation in Southeast Asia. The population is spread unevenly among the 6000 inhabited islands, with 57% residing in Java and Bali and 43% spread across the remaining islands.

Recent years have witnessed a dramatic improvement in electricity access in Indonesia. The total installed electricity capacity has grown from 46,613 MW in 2013 to 56,510 MW in 2018, with an average growth of 4.1% per year. Electricity generation climbed from 216,189 GWh in 2013 to 267,085 GWh in 2018, up 4.93% YOY. **With expanding power supply, electrification in Indonesia has continuously improved,** up from 78% in 2013 to 97% in 2018, connecting 12.8 million people per year on average and reducing the number without electricity supply from 54 million in 2013 to 8 million in 2018. But the country's power access is unevenly distributed, with nearly 100% in the western region (DKI Jakarta, Ba, Banten, West Jave, DI Yogyakarta) and only 59.85% in southeastern region (Nusa Tenggara Timur [NTT]) [24].

Between 2013 and 2018, total electricity consumption grew from 187.5 TWh to 234.6 TWh, up 5.1% YOY. The household sector consumes the largest share of electricity, followed by industry, business, and public service, with shares averaging 42%, 33%, 18%, and 6%, respectively (Annex 2). But the per capita power use is relatively low at only 888 kWh per person in 2018, which was way below the world average and even a far cry from the average in Southeast Asia (1507 kWh per person in 2015).

1.1.5.2   Renewable Energy Has Potential, but Fossil Fuels Predominate the Generation Mix

Between 2010 and 2017, power produced from fossil fuels accounted for 85–90%. In that period, oil declined from 22 to 5.81%, gas remained stable, and **coal climbed from 38 to 57.22%** [24] (Fig. 1.11). This data suggests that fossil fuels, especially coal, are of vital importance in the power industry in Indonesia, which boasts rich coal resources with a reserve-to-production ratio of 61 years. In terms of coal-fired power generation technologies (Annex 3), **though subcritical technology is still in**

---

[4] This section is mainly contributed by Dr. Maria Retnanestri, Director of IPEN Pty. Ltd and an Indonesian energy specialist.

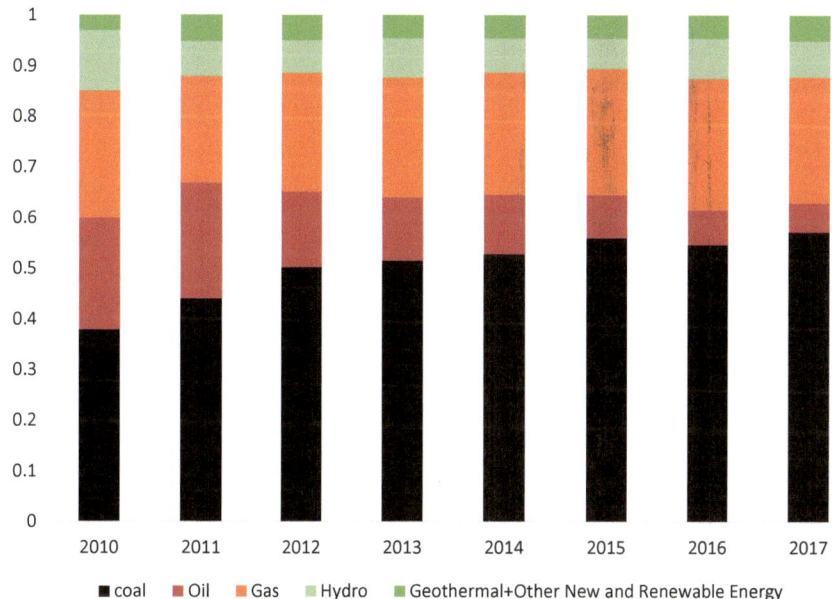

Fig. 1.11 Share of electricity generation capacity by energy type. *Data source* Ref. [24]

use, large and new power plants (mainly on Java Island) primarily use super critical or ultra-super critical technologies.

From 2010 to 2017, the share of hydropower in the generation mix dropped from 12 to 7.06%, while other renewable energy except hydropower grew from 3 to 5.09%. As a whole, the share of renewables fell from 15 to 12.5%. This data shows the relatively low percentage of renewable energy in the generation mix [24].

The plan for additional electricity generation capacity from 2020 to 2028 (Annex 4) indicates that fossil fuel will decline in its percentage of additional power generation capacity to 70%, but coal will remain high at 48%. This points to the irreversible importance of coal in the short term. The share of renewables will increase to 30%.

Indonesia's electricity supply systems are not integrated into one interconnected system due to its archipelagic nature. To attain the 100% power supply goal, a distributed renewable energy power system provides one of the viable solutions.

1.1.5.3   Indonesia Has Set Targets for Renewable Energy Development, but Coal Is Still Deemed as an Indispensable Tool for Achieving the 100% Electrification Goal

The Goal of Power Development

Government Regulation 79/2014 regarding National Energy Policy stipulates the following electricity targets: close to 100% electrification ratio by 2020; per capita electricity consumption of 2500 kWh by 2025 and 7000 kWh by 2050; installed capacity of 115 GW by 2025 and 430 GW by 2050.

In 2019, the Indonesian government unveiled the target of renewable energy development: by 2025, the power generated by renewables should reach 23% of total electricity generation in the country. According to the planned investment of renewable energy in 2025, it is estimated that the investment in solar PV generation will be USD 17.45 billion; hydropower and micro hydropower will be USD 14.58 billion; wind power generation will be USD 1.69 billion; waste treatment power generation will be USD 1.67 billion; biomass power generation will be USD 1.37 billion; and hybrid power generation will be USD 260 million [25].

Policies for Promoting Renewable Energy Development

The Indonesian government has rolled out a range of policies to spur the development of renewable energy.

- Government Regulation 79/2014, replacing Presidential Regulation 5/2006, on National Energy Policy, setting renewable energy target of 23% by 2025 and 31% by 2050.
- Ministerial Regulation 50/2017, replacing Ministerial Regulation 12/2017, on mechanism and pricing of renewable electricity purchase by PLN:

    - Mechanism: Build, Own, Operate and Transfer (BOOT).
    - Electicity purchase price: Solar PV power, wind power, biomass power, biogas power, and ocean power should be no more than 85% of local average generation cost (BPP) and hydropower, waste power, and geothermal power can be up to 100%.

- Presidential Regulation 35/2018, replacing Presidential Regulation 18/2016, on the acceleration of waste power development for electricity in 12 major cities in Indonesia.
- Ministerial Regulation 41/2018, replacing Ministerial Regulation 26/16, on biodiesel financing for palm oil businesses.
- Ministerial Regulation 49/2018 on rooftop PV.

In 2018, the Indonesian Parliament called for speeding up the legislative process of renewable energy power generation and actively invited stakeholders and academic representatives to participate in legislative discussions.

1.1.5.4   Challenges for Low-Carbon Transformation of the Power Sector in Indonesia

Lack of a Full-Fledged and Robust Renewable Energy System and Policy

First, the policy lacks consistency and adequacy. Frequent policy changes undermine investor confidence and increase project development risks, and the inadequate policy makes it difficult to grow the percentage of renewables in power generation, a case being the Ministerial Regulation 13/2019 on Rooftop PV. While the government argued that the policy would allow solar PV owners to save 30% on their energy bills, other voices said the 65% scaling of energy outflow to the grid is deemed to discourage public willingness to invest in solar PV. **Second, it takes complicated procedures to acquire a land-use permit.** For example, geothermal resources are often located in a protected forest or conservation forest, making it complicated to obtain a development permit. **Third, certain systems and policies are absent.** There are currently no incentives and investment attraction regulations to boost the uptake of renewable power generation.

Renewable Energy is not Attractive for Investment

First, renewable energy pricing is not competitive. The power purchase price by PLN at 85% of BPP is considered unattractive, as developers may be unable to recover their investments and make a reasonable profit. Such pricing is seen as placing renewables at the unfavourable position of being unsubsidized while competing with subsidized coal electricity.

Second, renewable energy subsidies are unclear. Appropriate subsidies make investment more appealing in renewable energy power generation, but Indonesia currently lacks clarity on renewable subsidies for the buyer.

Fossil Fuels Are Easily Available

The abundant reserves and lower prices of fossil fuels make it hard to move Indonesia toward renewables for power generation in the short term. Coal resources are available in Indonesia with a reserve-to-production ratio of 61 years. Annual coal power station installation is planned to peak in 2020–2023 and slow down to 2028. The 2019–2028 total additional installation of 27,064 MW remains the largest in proportion at 48%. **Gas resources are available in Indonesia with a reserve-to-production ratio of 49 years.** Additional gas generating capacity installation is scheduled to peak by 2022, and, in 2019–2028, the additional capacity is projected to reach 12,416 MW, accounting for 22%. Since turning into a net oil importer in 2003, Indonesia has reduced the use of oil in power generation. Diesel

Table 1.9 National energy master plan of Indonesia	Renewable power	Potential (GW)
	1. Geothermal	29.5
	2. Hydro	75.1
	3. Mini and micro hydro	19.4
	4. Bioenergy	32.7
	5. Solar	207.9
	6. Wind	60.6
	7. Ocean	18.0
	Total	443.2

generation is reserved for areas where other options are not available or only for stand-by operation to brace for emergencies.

Renewable Energy Has Great Potential but High Cost for Power Generation

Despite its tremendous potential (Table 1.9), the utilization of renewables for power generation is still low at less than 1% of its potential. Renewable resources with relatively high capacity, such as geothermal and hydro, are very site specific, thus only possible to be developed in certain provinces. Otherwise, the higher cost of renewables compared to fossil fuel is the main reason (Annexes 5 and 6). **The construction/investment of renewable energy power generation is generally more expensive than fossil fuels**. The construction/investment cost of hydro, thermal, and solar PV power stations is 1500 USD/kW, 1750 USD/kW, and 1200 USD/kW, respectively; for coal, diesel, cogeneration and gas it is only 1250 USD/kW, 900 USD/kW, 680 USD/kW, and 400 USD/kW. **From the operation cost point of view, renewable energy doesn't stand out either**. The unit operation cost of hydro, thermal, and solar PV power stations is 18 USD/MWh, 106 USD/MWh, and 411 USD/MWh, respectively, while coal, diesel, cogeneration, and gas is 51 USD/MWh, 179 USD/MWh, 86 USD/MWh, and 344 USD/MWh.

Renewable Energy Development is Set to Hurt Vested Interest

In the power market, the producers are limited in number and face high entry barriers. Diverse producers in the power generation industry offer the same product. Coal and other conventional fossil fuel generators have long been established in the market, and their interests would be impaired if renewables got a big boost. In Indonesia, the state firm PLN has a market monopoly, and, given the high uptake of coal power in its electricity portfolio, the company would spare no effort to avoid stranded coal assets. Furthermore, it would prioritize grid stability and curb the size of renewables connected to the grid.

1.1.5.5 International Cooperation of Coal Power Development in Indonesia

Overview of Power Cooperation

To fill the funding gap in power investment, Indonesia strongly encourages international cooperation on coal power. Between 2015 and 2019, the government could only secure 41% of the infrastructure funding needed; the rest was expected to be financed by the private sector, along with knowledge and experience sharing in the development, operation, and management of infrastructure services. Collaboration models include IPP or Kerjasama Pemerintah Swasta (KPS) (known as public–private partnerships [PPP] in English).

China, Japan, Indonesia, and Malaysia are the biggest investors in the country's coal power sector. Incomplete statistics suggest that China participated, in various forms, in 32 coal power projects in Indonesia, involving 20,169 MW of installation, among which 12,197 MW are in-service units and 7972 MW are planned and signed. Annex 7 introduces some of the coal-fired power plants under international cooperation.

The decision-making process for procuring a coal-fired power project with foreign investment includes four scenarios: direct assignment, direct selection between competing proposals, open tender, and PPP.

The limit on foreign investment share: For projects with power generation of less than 1 MW, only domestic investment is allowed; for 1–10 MW, the foreign share is up to 49%; for above 10 MW, the maximum foreign investment is 95%, although it might be greater under certain conditions.

Differences Between International and Local Investment in Terms of Project Selection

The future 10-year power station projects (2019–2028) indicate that larger power installed capacity, greater capital intensity, and more state-of-the-art technology will lead to greater private (local and/or foreign) investments, whether in coal or renewables. Annex 8 illustrates the percentage of PLN and private investors in power generation projects between 2019 and 2028. **First, over half are joint projects**. Of the allocated projects (49.9 GW), 16.2 GW (32.5%) is designated to be fully owned by PLN, while 33.67 GW (67.5%) is joint projects of various kinds. **Second, joint projects and those assigned to PLN have their own distinctive features**. The greater allocation for joint projects involves coal, mine-mouth coal, geothermal, mini hydro, hydro, and other renewable power projects. PLN features more combined-cycle/cogeneration and gas power, while diesel generation and pumped storage power generation are reserved for PLN only.

The Positive Impact of International Cooperation on Coal Power
Development in Indonesia

First, international cooperation will introduce advanced technologies. Both the 2
× 1000 MW PLTU Jawa 7 in Serang Regency, West Java, and the 2 × 1000 MW PLTU
Jawa Tengah in Batang Regency, Central Java, use ultra-supercritical (USC) technology, facilitating technology transfer and creating thousands of local jobs during
project construction. **Second, it will bring in top-notch management expertise.** In
the PLTU 2 × 15 MW project in Deli Serdang (North Sumatera) and the PLTU 2
× 20 MW in Gorontalo (Sulawesi Island) undertaken by Shanghai Electric Power
Construction (SEPC), Shanghai University of Electric Power Engineering (SUEP)
has provided training for the local employees. **Third, it will make up the shortfall
in funding.** Domestic funding sources are inadequate to finance coal power station
development in the next 10 years, and international cooperation fills the gap. **Fourth,
it strengthens research and development (R&D) cooperation on coal power technology.** Shenhua Guohua Electric Power Company has collaborated with universities
in China and is exploring R&D collaboration with universities in Indonesia.

1.1.6 Opportunities and Challenges for Low-Carbon and Clean Power in Southeast Asian Countries

1.1.6.1 Opportunity 1: Pressing Demand and Vast Market Potentials for Power in Southeast Asia

Southeast Asia reported average annual growth of over 5% in power demand between
2010 and 2018, twice that of the world's average. Under the scenario of IEA's
predetermined policy, power demand in the region would double by 2040, reaching
2000 TWh, with an annual increase of nearly 4% or twice of that in the rest of the
world [13]. Currently, power accounts for 18% in end energy consumption, which
is lower than most other regions; yet this percentage is expected to hike and hit the
world average in 2040, reaching 26%. The pressing need for power is real in Southeast
Asia, where coal-fired power is considerably losing its appeal under the pressure of
global climate change, carbon emissions, and air pollution, while renewable energy
is gaining popularity with the governments and the public.

1.1.6.2 Opportunity 2: Diverse Renewable Energy Resources with Tremendous Potential for Development

Indonesia and Thailand are early runners on the fast track of development in renewable energy. As the world's biggest archipelagic state with a distinctive tropical rainforest climate, Indonesia is naturally blessed with rich geothermal, wind, solar, and
hydropower and ranks second in installed capacity of geothermal [26]. The country

also boasts considerable land availability and favourable resource conditions for building power stations.

Malaysia, the Philippines, and Vietnam started early in renewable energy development, with a focus on hydropower that is highly market-based and demonstrating active momentum. A narrow piece of land higher in the west and lower in the east, Vietnam features a tropical monsoon climate with 3260 km of coastline (excluding islands) and is rich in wind resources throughout the year, with an average wind velocity of 7.3 and 9–10 m/s in coastal regions in the south. The Philippines, on the other hand, owns the world's third-largest installed capacity of geothermal power and significant untapped reserves [26].

Singapore, Brunei, Cambodia, Laos, and Myanmar are trailing behind in renewable energy development due to constraints in historical background, geographical conditions, economic development, and natural resources. Myanmar is endowed with abundant resources in hydro, wind, and geothermal power with significant potential for development. Progress is on track to feed renewable energy into the grid, establish a corresponding power market, and advance power system reform. Singapore and Brunei are highly developed economies with small populations and territories and poor hydro, wind, and geothermal power potential. However, with ample solar power resources, these countries may learn from Japan's experience in solar power development.

1.1.6.3 Opportunity 3: Active Renewable Power Development Goals Are Set in Southeast Asia Countries with Supports to Clean Power

Strategy and target setting will be the key drivers of renewable energy in Southeast Asia. Major Southeast Asian countries have mapped out their goals in power generation from renewable energy. ASEAN has set its target in the *ASEAN Plan of Action for Energy Cooperation 2016–2025*, working toward a 23% share of renewables in total energy supply. Correspondingly, ASEAN countries have respectively defined their national targets: Laos (59%), Philippines (41%), Indonesia (26%), Cambodia (35%), Myanmar (29%), and Thailand (24%) have developed more ambitious goals than ASEAN as a whole. As per information released by the International Renewable Energy Agency (IRENA), in order to increase the share of renewables to 23%, ASEAN needs to invest USD 27 billion (i.e., 1% of GDP) annually in the next 8 years.

Governments of Southeast Asian countries have changed their tune by announcing policies in support of clean power development. With mounting pressure from environmental pollution and backlash from the public, the capacity of newly launched coal-fired power projects in the region has slumped from a whopping 12.92 GW at its peak in 2016 to a mere 1.5 GW in the first half of 2019. Governments have also adopted multiple synergized policies, such as FIT and tax incentives, to endorse renewable energy. Malaysia has implemented a Green Investment Tax Allowance (GITA), with a maximum of 100% tax exemption for investment in green assets.

1.1.6.4   Opportunity 4: Significant and Hopefully Continuous Reduction in the Cost of Renewable Energy

In light of economic viability, Southeast Asian countries are most likely to opt for fossil fuels with higher economic efficiency in the short run; however, with further depletion of coal resources and the inclusion of carbon emissions into cost, the cost of coal-fired power may be on the rise. Based on the latest IRENA report (2017), except for solar power, the global levelized cost of energy (LCOE) of most renewables has fallen within the range of fossil fuel cost. With the exception of hydropower and geothermal power, the LCOE of renewable energy projects (biomass power, geothermal power, hydropower, onshore wind power, offshore wind power, solar power, and ground solar PV) launched since 2017 has been reduced since 2010. The LCOE of solar PV alone fell by over 70% [27] (see Fig. 1.12).

The past 5 years have seen the LCOE decline in renewable energy in Southeast Asia, though the degree of reduction may vary. In particular, the LCOE of solar PV slid enormously, down by 42–52% in Indonesia. In Thailand and Vietnam, the LCOE of onshore wind power also decreased by 16–43% during the same period [28] (Figs. 1.13 and 1.14). The lower cost of renewable energy has made cheap power access possible and will prove especially effective for regions highly dependent on diesel power or power grid extension at a high cost, thereby boosting the uptake of renewable energy.

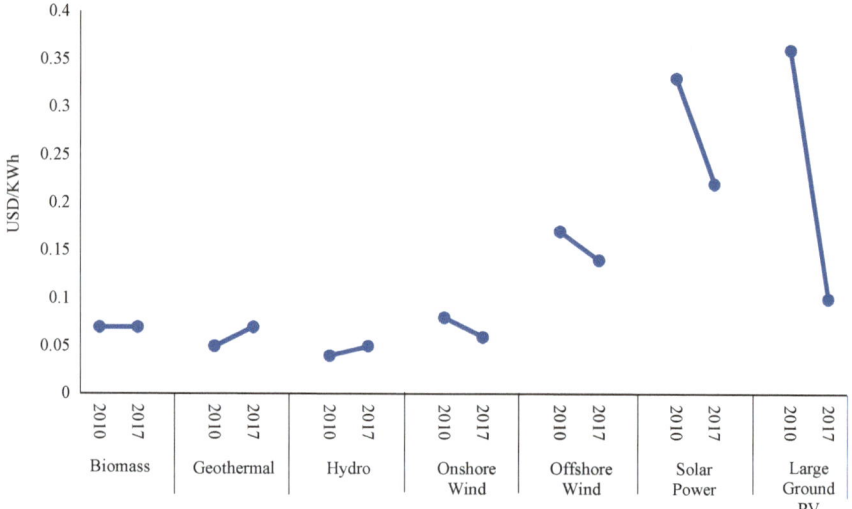

Fig. 1.12   LCOE changes in the world's renewable energy. *Data source* Ref. [27]

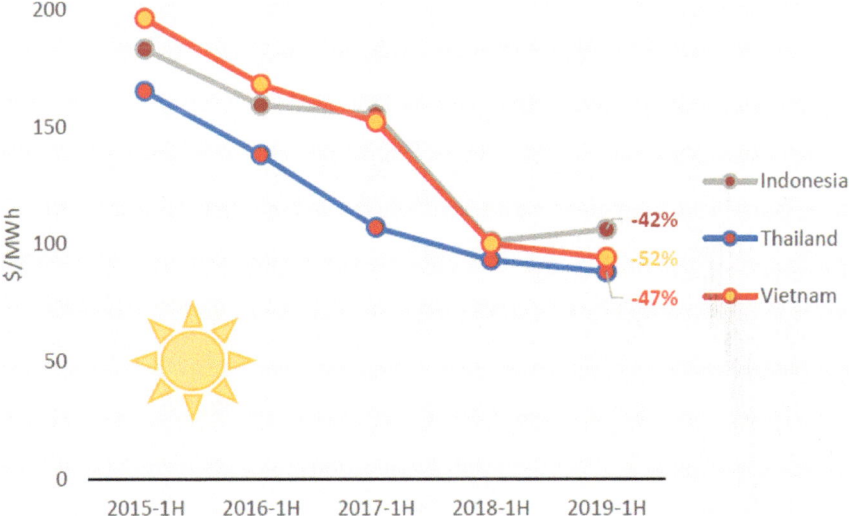

Fig. 1.13 Tendency of LCOE changes in PV in Indonesia, Thailand and Vietnam (First Half of 2015—First Half of 2019). *Data source* Ref. [28]

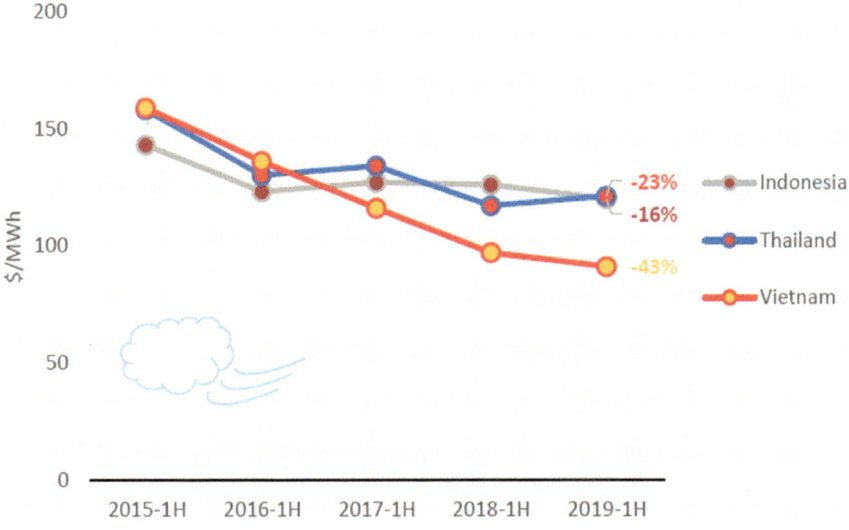

Fig. 1.14 Tendency of LCOE changes in on-shore wind power in Indonesia, Thailand and Vietnam (First Half of 2015—First Half of 2019). *Data source* Ref. [28]

1.1.6.5 Challenge 1: A Powerful Coal-Fired Power Lobby and Lack of Market Competition in Power Sectors in Most Countries

Other than Singapore and the Philippines, where market forces are dominant, the rest of Southeast Asia is yet to separate grids from power plants and operates the power sector in a vertically integrated fashion. Furthermore, most power plants are coal-fired, and, with a monopoly in the market, there is a lack of incentives to move toward renewable energy. The vertically integrated model is prone to monopoly, leading to poor economic efficiency. In some countries, the government still plays the principal role in power pricing due to the absence of a market pricing mechanism. Besides, as the prospect for investment attraction is dampened by low marketization, the region seems unaffected by the record low price of renewable energy in the rest of the world.

1.1.6.6 Challenge 2: Human Resources and Indigenous Innovation Fall Short

For all its rich labour resources, Southeast Asia has long been troubled by a lack of R&D investment and top-notch professionals in the field of renewable energy. R&D places high demands on financial investment and the competence of skilled talents. The situation is worsened by the tightly controlled key technology transfer from developed countries that make the region unlikely to acquire core technologies even at high prices. Countries are unable to translate their resource advantages into energy advantages. At present, the region still needs to step up international cooperation in renewable energy in terms of technical standards and protocols, pilot projects of advanced technologies, science and technology cooperation bases, and joint professional training, etc. This is not only a viable approach for obtaining renewable energy technologies but also a major avenue for filling the funding gap.

1.1.6.7 Challenge 3: Governments Are Under Huge Financial Strain and Lack Effective Market Financing Channels

To attain the target of renewable energy development and boost economic and environmental sustainability, ASEAN members are expected to invest USD 2.36 trillion into the energy sector from 2016 to 2040. Yet, most ASEAN members are developing countries with per capita GDP below USD 5000, the exceptions being Singapore, Brunei, Malaysia, and Thailand. Countries such as Cambodia, Vietnam, Laos, and Myanmar are unable to raise sufficient funds to fuel the energy sector due to their underdeveloped national economy and strapped public purse.

Management and operation of financial institutions in Southeast Asian countries are far from mature. Many ASEAN countries are not familiar with the business model of renewable energy and tend to overestimate the risks, which makes financing a major challenge for renewable energy projects. More often than not, these financial institutions show more preference for well-understood new energy projects. For instance, banks in Malaysia are generally more interested in solar PV projects with little enthusiasm for biogas or biomass. This has further aggravated the financing pressure for non-PV projects [29].

Moreover, **some Southeast Asian countries are haunted by the problem of debt sustainability**. At present, renewable energy development in Southeast Asia is mainly financed through lending from multilateral banks such as the World Bank and Asia Development Bank. Many debt-ridden countries tend to be more conservative and discreet in borrowing to fund clean energy.

1.1.6.8   Challenge 4: Underdeveloped Grid Infrastructure Hampers Connection of Renewable Energy

Aside from developed economies such as Singapore and Brunei and the relatively developed Malaysia, Southeast Asian countries are hindered by underdeveloped grid infrastructure, especially for the four newcomers of ASEAN—Cambodia, Laos, Myanmar, and Vietnam, whose grid infrastructures are severely unsufficient. Electricity accessibility is only 61% in Cambodia and 56% in Myanmar.

With the massive connection of wind and solar PV power plants to the grid, the intermittence, randomness, and instability of wind and solar will result in fluctuations of voltage, current, and frequency of the grid, thus affecting the quality of power. In order to mitigate the negative impacts, grid operators need to leave some spinning reserve capacity, which would increase the operational cost and place indirect constraints on new energy development. Currently, most Southeast Asian countries are hobbled by poor grid structure, few high-voltage lines, and deficient cross-border grid interconnectivity. In addition, the scale of pumped-storage hydropower stations with high adjustability in the region is rather limited, with inadequate peak-shaving capacity, which, in some measure, has impeded the development of renewable energy [30].

1.1.6.9   Challenge 5: Power Generation from Renewables is Less Competitive Than Fossil Fuels in Terms of Cost

Despite the lowered cost of power generation from renewable energy, it is still more expensive than coal power. Reference [28] argues that economic viability is the key impediment for renewables in Southeast Asia. **At the national level**, LCOE analysis was conducted on coal-fired power and renewable energy in Indonesia, Malaysia, Philippines, Thailand, and Vietnam. Apparently, the LCOE of solar PV, onshore wind, geothermal, biomass, small hydro, and combined cycle is generally higher than that of coal power (Fig. 1.15). **At the project level**, most competitive projects of five renewable energy sources (solar PV, onshore wind, geothermal, biomass, and small hydro) were selected from several Southeast Asian countries (Indonesia, Malaysia,

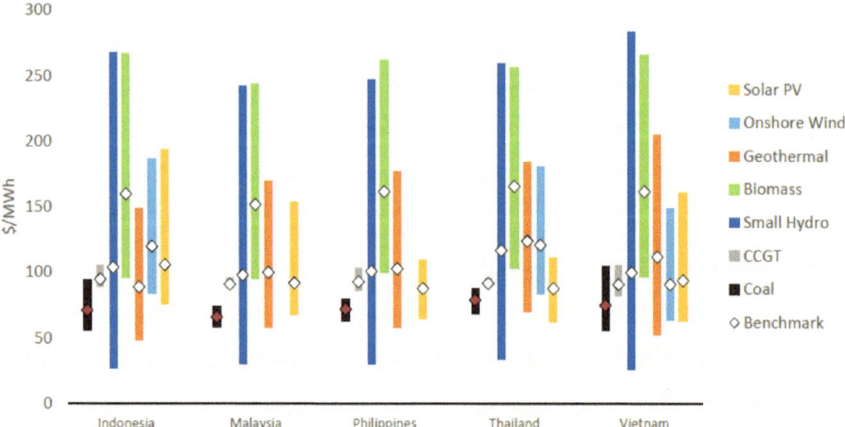

Fig. 1.15 LCOE of power technologies in Some Southeast Asian countries (First Half of 2019). *Data source* Ref. [28]

Philippines, Thailand, and Vietnam) to compare with the most competitive coal-fired power projects. Results show that the cost advantage of power from renewable energy is only reflected in small hydro projects (in Thailand and Vietnam) with LCOE *significantly* lower than that of coal power (25–35 USD/MWh) and some geothermal projects (in Thailand and Indonesia) with LCOE *slightly* lower than that of coal power. LCOE of biomass, onshore wind, and solar PV remain generally higher than that of coal power (Fig. 1.16).

1.1.6.10   Challenge 6: Grid Connection of Power from Renewables Will Boost Electricity Bills and Compromise Affordability

The grid connection of more expensive renewable power would drive up electricity prices. Considering the large population living under the poverty line, the affordability for end users cannot be neglected—and renewable energy should not be developed at the expense of users. Annex 9 suggests that the electricity bill for 100 kWh accounts for over 5% of minimum monthly wages in many Southeast Asian countries, running as high as 8.5 and 7.8% in the Philippines and Cambodia, respectively. Connecting renewable power means a higher economic burden for consumers, which, to some degree, puts a brake on the development of renewables [29]. In view of the geographical characteristics of Southeast Asia, smart microgrid layout can expand the function of renewable energy and provide power solutions for remote and island areas.

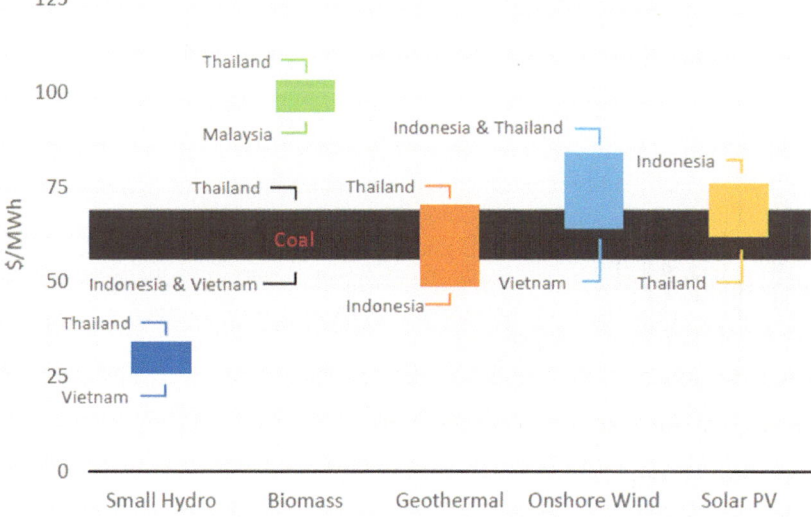

Fig. 1.16 LCOE of most competitive renewable energy projects and coal-fired power projects in Southeast Asia (First Half of 2019). *Data source* Ref. [28]

1.1.7   Recommendations for China's Engagement in Low-Carbon Transformation of Power Infrastructure in Southeast Asia

1.1.7.1   Revise Overseas Investment Policies, Taking into Account the Environmental and Climate Impacts as Crucial Factors

In the environmental management of foreign investment projects, China often adopts the standards of target countries, most of which lack a full-fledged environmental management framework and environmental protection standards. This has prompted a series of high-carbon projects in developing countries, giving rise to massive greenhouse gas emissions and environmental pollution. Therefore, the Chinese government should make environmental protection and climate change mitigation factors mandatory requirements for foreign aid policies and the overseas investment policies of financial institutions. They should also draw up a negative list, restrict high-carbon lock-in projects (e.g., coal power), encourage low-carbon investment, implement the green BRI, and advance the win–win strategy of opening up.

1.1.7.2 Enhance Strategic Cooperation with Southeast Asian Countries to Provide Technical and Funding Assistance to the Planning and Roadmap of Local Renewable Energy Development

Most Southeast Asian countries are in the infant stage of industrialization with a fairly high demand for electricity supply. They are inclined to deploy low-cost coal power projects with readily accessible resources. Some countries have recognized the superiority of clean power, but their limitations in planning, funding, and technical expertise render it unlikely to drive the transformation of energy systems in a systematic and efficient manner. As the world's largest producer of renewable energy, China boasts extensive expertise in the low-carbon transformation of the energy system and should carry out in-depth cooperation with target countries in strategic planning. It should also leverage multi-level governmental dialogues in energy and electricity macro planning and bolster policy exchanges with Southeast Asia in terms of clean energy and power. Research on cooperation should be strengthened to jointly promote technological advancement and reduce the cost of clean energy development. The varied platforms should be fully harnessed for the aforementioned purpose.

1.1.7.3 Chinese Businesses Should Prioritize Impact Assessment of Overseas Investment Projects and Secure the Sustainability of the Local Economy, Society, and Environment

Chinese enterprises primarily adopt the EPC model for power infrastructure projects in Southeast Asia and are unlikely to empower the local labour force and industrial chains, etc. Apart from investment income, Chinese investors should be more conscious of the impact on the sustainability of the local economy, society, and the environment; pursue both economic and social benefits; operate with more local headcounts; and actively participate in public welfare activities. It is important to conduct a systematic assessment prior to, during, and after projects; examine the pros and cons for the macroeconomy, job creation, and eco-environment; make an effort to minimize negativity; and make the most of the benefits.

1.1.7.4 Boost Conversations Between Chinese Government and Businesses and Investment Recipients

The Chinese government should reinforce communication with Southeast Asian countries at the central and local levels by harnessing bilateral and multilateral platforms, including the Clean Energy Forum of East Asia Summit and the Greater Mekong Subregional Energy Cooperation. China could maximize its strength in clean power technology to offer more technical assistance and project demonstration, etc. Chinese enterprises should carry out more cross-cultural conversations with local governments, labour unions, social organizations, and media outlets, and better communicate their contributions to local society and the economy to obtain more understanding. They should stay alert to public opinion risks and make prompt media announcements and clarifications about misinformation regarding enterprises.

1.2 Gender Issue

Achieving SDG 5, gender equality and empowering all women and girls, will have positive cascading effects on the achievement of the other SDGs, notably energy access and climate action. Some key synergies between SDG5, SDG7 and SDG13 and corresponding policy recommendations are presented below.

1. Implement Systematic Information and Consultation of Local Communities and Improve Involvement of Women to Unlock Life Quality-Improving Choices for Clean Energy Access

Women are often more involved in household decision making and have the knowledge of what is needed to adapt to changing environmental circumstances in order to determine practical solutions for all, notably the children and elderly. But they remain a largely untapped knowledge resource and have limited access to decision making at the local or political level.

When preparing investments in host countries, investors should implement systematic consultation of local communities, including civil society and populations with a conscientious effort towards gender equality. In regions where women are household leads, women are key decision-makers and drivers of behavior regarding pollution and health conditions of the household. Their concern for a more integrated consideration of negative externalities of pollution should be relayed to the regional and national level.

In recent years, leading women figures have often appeared as bringers of change in Asia. One such pioneer is Ms. Wandee Khunchornyakon,[5] founder and chief executive of Solar Power Company Group, which is the first company to develop a solar farm for commercial purposes in Thailand. Despite heavy difficulties for accessing traditional credits, her company now operates 19 solar farms in Thailand with a total generation capacity of 96.98 MW, reducing an equivalent of 200,000 tons of CO_2 and generating 20,000 new permanent and local jobs.

2. Create Innovative Financing Schemes that Involve Local Communities and Women in Particular from Assessment, Planning, Revenues Benefits and Management of Distributed Renewable Energy and Energy Storage, Creating Income Complements.

Distributed renewable energies are better for the planet, but difficult to deploy in part due to the lack of training of local population on operating and maintaining

---

[5] Roots for the future, The Landscape and Way Forward on Gender and Climate Change, UICN, GGCA.

equipment, whereas higher polluting concentrated fossil fuel-based energies appear easier to deploy, notably on isolated islands. However, distributed renewables could bring more energy and financial independence for local vulnerable communities.

Engaging women as active actors of clean energy and improved energy access can improve the efficiency by rooting implementation in local communities as well as contribute to their economic independence. A gender lens should be applied to the inclusion of local communities in the planning, economic assessment, financial revenues sharing, maintenance and management of distributed renewable energies such as small-medium scale PV and energy storage units.

With financing mechanisms that could be supported by foreign aid and micro-credits, such locally planned and managed projects will bring revenue complements to the community, creating self-reinforcing practical and financial incentives, ensuring the sustainable deployment of renewables. At the same time, by involving women and the local communities, these projects would provide better livelihood prospects through access of electricity, improving their education opportunities, shielding them from harmful health effects of indoor pollution (through burning of biomass) while making them active vocal ambassadors of climate change mitigation.

Successful examples of such projects include the Grameen[6] technology centers' actions in Bangladesh, the Barefoot College in Rajasthan in India, and the Solar Sister's in Africa conducting trainings for women as technicians and engineers, teaching them to build, install and maintain solar energy sources. In these cases, it had multiple impacts on women's health, role in the community, education and overall wellbeing.

3. Raise Awareness on Impacts of Climate Change on Women

We are already seeing some of the devastating effects of climate change, with increasing floods, hurricanes and other natural disasters. Women are the most vulnerable in these situations, facing the maximum risk due to their socio-economic status. With 70% living in poverty, women are disproportionately affected by extreme weather events, loss of agricultural productivity, destruction of life and property and so on, all of which stem from the climate crisis. In its efforts to support the development of countries along South-East Asia, China should raise awareness of the impacts of climate change, notably on vulnerable populations and women.

1.3 Annual Policy Recommendations

Given the multi-dimensional pressure from the novel coronavirus outbreak, the economic downturn, the global climate security challenges and a fragile domestic ecological environment, China is in dire need of new economic growth engines and innovative drivers for its economic transition to secure the green and low-carbon transformation of its economy by 2030.

---

[6] Applying a gender lens to science-based development, GenderInSite, 2017.

The 14th Five-Year Plan period marks a pivotal time for China's high-quality development, which should embody and support the vision to achieve a Beautiful China by 2035 and the global long-term strategy for tackling climate change as outlined in the Paris Agreement. The country needs to make the fight against climate change the new driver of its economic transformation, stay committed to clean and low-carbon development, enhance its confidence to accelerate the economic and energy transformation. This Special Policy Study makes the following recommendations:

1. **China should innovate its development pathway through green and low-carbon transformation. This transformation is the inevitable requirement to achieve the grand goal of its modernization, the upgrading of its economic pathway of growth and the new driving force of long-term economic prosperity and will catalyze supply side structural reform, and mitigate the current downward pressure of the economy.**

   1.1 China should further integrate the green and low-carbon transformation into the top-level strategic planning of China's social and economic development as well as the specific strategies of various departments and specialized fields, as an important measure to improve total factor productivity, hence contributing to maintain China's medium to high economic growth rate in the long-term;

   1.2 China's entrance into an aging and high-income society must be fully recognized and taken into account. This societal change will bring about huge changes in the structure, nature and volume of socio-economic demands regarding energy consumption, air quality demands, and overall higher quality of life. The stability of the strategic determination should be kept for a long time so that the supply side structural reform can continuously adapt to the demand side changes, promoting the upgrading of industrial structure and product structure and technological advancement. The total energy and resource use should be continuous curbed, by improving energy and resource efficiency per unit output value, gradually decoupling the growth of income per capita (GDP) from the emission of pollutants and greenhouse gases;

   1.3 Continuously reducing coal consumption is a critical part of China's energy revolution. It will support the long-term decarbonization of the energy structure and reduce the emission of greenhouse gases and conventional pollutants per unit of energy. Therefore, China's strategic and institutional objectives of deep economic reform should recognize and integrate more clearly the relevance and pertinence of addressing energy, environment and climate change. To this end China should set clear goals and paths for controlling and reducing coal consumption, developing non fossil energies and improving energy efficiency;

   1.4 In the new form and stage of socio-economic development, comprehensive policy and management dimensions such as price signal, tax revenues, public finance, financial mechanisms, industry, market economy, investment, employment, social security, environment, and energy should be integrated in the efforts for the aforementioned strategies by deploying policy tools such as regulatory measures, emissions trading or taxation, and public communication and education. To be more specific, first, benefitting from the tailwinds of the country's fast-tracked implementation of its innovation and development strategy, China should intensify the investments in emission reduction technologies (such as energy storage, low-carbon cooling, carbon capture and storage, blockchain, etc.) that will drive the development of new industries, achieving the double dividend of pulling economic growth and reducing GHG emissions; second, China should actively promote and improve the green investment and green finance ecosystem, in coordination with the credit reporting system of financial institutions to establish a low carbon technology or low-carbon project library.

2.   **The critical time window of 2020 must not be missed. Despite the downward pressure on the economy, China must avoid relaxing targets and efforts for environmental protection and climate change during the 14th Five-Year Plan period. On the contrary, China should formulate more aggressive targets for carbon emission reduction, enhance China's Nationally Determined Contribution, formulate mid-century low-carbon development strategies, and implement China's green economic transformation and high-quality development**.

2.1   Formulate and implement a total carbon dioxide emission cap, using a combination of three approaches: total carbon emissions, emission intensity and energy structure adjustment. China should strive to not reduce the level of climate ambition for the 14th Five-Year-Plan and improve upon and exceed the current 2015 NDC, making a greater contribution to address climate change;

2.2   Propose the strategic deployment of "Leading the emission peaking" by publishing guidelines of the State Council to promote emission peaking in key industries and key regions (such as the relatively more developed areas of Eastern China, and the energy and material intensive industry sectors) to promote industrial transformative upgrading and high-quality economic development;

2.3   Swiftly improve the carbon emission trading system and the associated market mechanism. Not only should carbon pricing be made the core instrument promote the development of green, low-carbon, circular economy and trigger the energy revolution, but also make it possible for renewable energies to be included in the national carbon market (applied to power generation). Through the carbon market, low-carbon power supply infrastructures would be incentivized to avoid the future carbon-intensive infrastructure and mitigate stranded asset costs. In case of limited legislative resources, the necessary legislative basis for the national carbon market should be included in the amendment agenda for the Environmental Protection Law.

3. **Accelerate the pace of energy transformation and upgrading, especially in the coal-reliant economy, speed up the research and establishment of a new generation of policy environment to support a high share of renewable energy, promote technological innovation and industrial modernization, peak the total coal consumption as soon as possible, and ensure a zero growth rule for coal power plants.**

3.1 Promote alternative actions to reduce coal consumption, expand the scope of pilot projects focused on switching to natural gas and clean and renewable energy sources, such as geothermal and solar energy, according to local conditions. By encouraging innovation in local system and mechanism, guide social capital to increase investment in "coal to electricity" or "coal to gas". Comprehensively adopt and improve on peak-valley price mechanism and resident tiered pricing policy, expand dispatch-based transactions, reduce the cost of cleaner solutions such as heating gas and electricity, attract market interest and support clean heating;

3.2. Set up the next generation of renewable energy policies and management system, including policies to further reduce the financing cost of renewable energy enterprises, encourage the development of new distributed renewable energy technologies by creating favorable market conditions, especially in terms of land allocation, IPO fast-tracking, easing the access to preferential loans, etc. Accelerate the reform of electric power system, implement the renewable energy quota system, overcome the existing policy barriers, and improve the flexibility of renewable energy grid connection;

3.3. Formulate a national strategy of socio-economic transition for the coal economy in support of economic growth. Transform China's engine of growth by renewing its industry. In the coal-dependent regions and cities, economic diversification should be faster developed and coal-free industries taking benefits in regional advantages should be promoted, giving priority to the development of strategic emerging industries such as by new materials, Internet plus (e-commerce and data centers) and tourism. Optimize the market environment, provide favorable fiscal and tax policies for new industries and technologies, and boost the decisive role played by the market. Public finance and private funds should jointly set up a just transformation fund. The fund should invest in green finance and emerging green industries such as cultural tourism, part of the proceeds should be redistributed to support the economic transformation of coal dependent regions, mainly focusing on social issues such as resettlement compensation and employment transition, and developing skill formation related to emerging industries;

3.4   Strengthen experience exchange with economies which have successfully
      implement the transition of their economy out of coal, such as the United
      Kingdom, Germany, Spain, the United States, etc.

4. **Comprehensively improve the coordinated management of economic transformation, energy revolution, environmental governance, climate change
and public health**.

4.1   During the 14th Five-Year Plan, it is necessary to break sector bound-
      aries, achieve synergies between economic development goals, climate
      and environmental goals, and strengthen the coordination of energy and
      industrial goals;

4.2   Learning from the experience of the novel Coronavirus outbreak in
      early 2020, and attach greater importance to the tremendous social and
      economic risks associated with the system public environmental risk of
      climate change. Capacity building on environmental quality, climate risk
      assessment and risk management research should be carried out relent-
      lessly to accumulate data and case studies. First, we need continuous and
      long-term institutional arrangements in order to address climate change.
      Second, we need to accurately disclose climate change related information
      and achieve high quality MRV. Third, we need to further strengthen the
      public dissemination of climate change-related scientific research, with
      its certain results and uncertainty levels;

4.3   Starting from the 14th Five-Year Plan, China should attach greater impor-
      tance to the relationship between environment, climate, and long-term
      economic development and public health. Both symptoms and root causes
      should be addressed, starting from health and economic risk assessment,
      environmental quality standards and governance targets, and near-term,
      mid-term and long-term planning for environmental and climate gover-
      nance measures. Systematical review of strategies and plans for compre-
      hensive prevention and control of environmental, climate and health risks
      should be implemented;

4.4   By integrating greenhouse gases into the existing environmental moni-
      toring and control system, explore the links between the emission trading
      scheme and the administrative emission permits system, and accelerate
      the establishment and improvement of greenhouse gas performance emis-
      sion standards in key industries. Gradually update the ultra-low emission
      strategy of the power and metallurgical industries by integrating the green-
      house gases emissions and conventional pollutants into the comprehen-
      sive emission reduction target. Improve the carbon pricing mechanism,
      strengthen the incentive of carbon price to deliver effective emission
      reduction and low-carbon technology R&D innovation, and drive low-
      carbon investment from carbon market. Promote policy study on China's
      carbon tax policy, and incorporate a carbon tax into the environmental tax
      system;

4.5   Establish a social and environmental impact assessment system of the policy, and provide recommendations according to the long-term social and environmental impacts to policy makers;

4.6   Encourage and support cross border exchanges, joint research and data sharing among think tanks, professional associations and non-governmental organizations, and provide technical support for cross-sector cooperation and collaboration.

5.   **Global climate governance is facing a new geopolitical situation. The expectation of major actors for China's leadership role in the global climate governance is expected to grow in the long-term. Against the background of complex Sino-US relations and the challenge posed by the US withdrawal from the Paris Agreement, the Chinese government should proactively work with European and major developing country governments to build a new global climate political leadership based on multilateralism and promote the implementation of the Paris Agreement. In cooperation with non-Party actors supporting climate change (such as some US state governments, businesses, NGOs, etc.), through track 1.5 or track 2 dialogues explore opportunities to expand leadership areas of global governance (such as the green "Belt and Road" Initiative).**

5.1   The Chinese government should actively respond to the EU's "Green New Deal", strengthen cooperation initiatives on the 15th Conference of the Parties to the Convention on Biological Diversity, and consider reaching an agreement to cooperate with the EU at the 2020 China-EU Summit. "Third-party market cooperation" related to Belt and Road countries will effectively combine the experience and expertise of China and the support of developed countries (such as funding, technology, and capacity building) to respond to the mitigation and adaptation needs of developing countries, jointly promoting a global low-carbon transition and broadening action and support to tackle climate change. The European Union's "Green Deal" is the first to propose ambitious long-term goals of "climate neutrality" and a more aggressive mid-term NDC by 2030. It calls for a green low-carbon pathway for the economy achieving by 2050 net zero greenhouse gases emissions. It clearly decouples economic growth and resource consumption. The European Union is looking forward to an alliance with China and in-depth climate cooperation, and highly regards the alignment between the objectives of its "Green Deal" and China's strategic plan to promote low-carbon high-quality economic development, the construction of a beautiful China. Climate change cooperation is likely to become the shining beacon that will enrich the strategic partnership between China and Europe. Where the withdrawal of the U.S. Republican federal government has adversely affected the global climate governance environment, accelerating the establishment of a China–Europe joint leadership in global climate momentum is conducive to maintaining China's advantageous strategic position in the governance of global issues such as climate change since the Paris Agreement. China's proactive position will also reveal advantageous in hedging potential future pressures to reduce emissions from a Democratic US government.

5.2 In the future, the overall relationship between China and the United States will pursue to be of strategic competition in nature. The Chinese government should recognize that climate change has been and will increasingly be an important topic for China-US relations. There remain large political, economic, and technological forces in the US politics, business, and think tanks that actively advocate for climate change. If America's efforts to combat climate change prevail in different ways and on different occasions, climate change, which was the highlight of Sino-US relations, would turn into a new friction and adversarial point in Sino-US relations. In effect, China and the United States would form a larger multi-dimensional gap in their trajectories in the fields of economic structure, energy, technological research and development. By then China's active, one-step-ahead position in global climate governance would have turned into a passive, reactive position. The Chinese government should proactively promote a low-carbon transition strategy and engage in dialogue and exchange with relevant US entities (provincial and state governments, non-party actors such as enterprises).

5.3 Actively make use of the CCICED platform in combination with the 2020 international agenda (e.g., the China-EU Summit in June, the annual meeting of the CCICED in 2021, the China-EU Summit in Leipzig in late 2020, the Kunming Biodiversity Conference in May 2021, and the COP in Glasgow in November 2021, etc.) to organize outstanding think tanks and experts from China and Europe and other regions to discuss topic-oriented, pragmatic issues such as the European "Green Deal" and its international cooperation, the promotion of climate goals, the rationale behind new economic growth, the pathway towards a low-carbon economic transition, the rapid and low-carbon transition of the energy system, the transformation for the coal economy, transport electrification, the coordination between dealing with climate change and the protection of biodiversity. Multi-level China-EU track 2 dialogues should be launched, to deepen mutual understanding between China and Europe, and promote the comprehensive and effective implementation of the Paris Agreement.

5.4 China should strengthen the top-level design of the "Belt and Road" climate cooperation, actively support the "Belt and Road" partner countries to formulate low-carbon development plans and action roadmaps, and change from a mere commercial project deployment model to that of strategic cooperation to work with "Belt and Road" partner countries on climate change, supporting the "Belt and Road" countries to update their NDC and formulate and implement mid-century low-greenhouse gas emission development strategies, therefore attracting broader support from the international community. First, China should support the development of the "Belt and Road" developing countries, especially the least developed countries, landlocked developing countries, and small island States, in formulating low-carbon development plans, roadmaps, and action plans. Second, "Belt and Road" countries can share China's best practices in tackling climate change, focusing on areas such as renewable energy and energy, air quality and greenhouse gas co-management, climate investment and financing, agriculture, and nature-based solutions. Third, domestic banks and financial institutions should improve their risk assessment system, formulate relevant policies, and align with the relevant standards of international financial institutions as soon as possible, and gradually stop providing funding for carbon-intensive projects such as overseas coal-fired power generation.

Appendix

Annex 1: Comparison Between China's Domestic and Outbound Investment

Catalog of restricted industries in China	The catalog of restricted industries mainly include production capacity, techniques, technologies, equipment and products which are obsolete, incompliant with industrial criteria of entry or relevant regulations, and are prohibited from launching or expansion, or requiring renovation For example: coal mines below 300,000 tons/year, wet cooling generator unit of coal consumption above 300 g standard coal/kwh, ethylene production by naphtha cracking below 800,000 tons/year, soda ash, caustic soda, yellow phosphorus, nitrogen fertilizer produced from oil or gas, coking projects of steel plant without coke dry quenching, coal filling and coke pushing/dedusting equipment, sintering machine of less than 180 m^2, electrolysis of aluminum project, clinker production line of new dry method below (excl.) 2000 s ton/day, general filament lamp, etc[a]
Restricted outbound investment	1. Investment in sensitive countries and regions with which China has not established diplomatic ties, experiencing wars or are restricted for investment by bilateral or multilateral treaties signed by China 2. Real estate, hotels, the entertainment industry, movie studios and sports clubs 3. Overseas equity investment funds or platforms with no real industrial project 4. Investment involving obsolete equipment that does not comply with the technical standards of the target country 5. Investment that violates the target country's environmental, energy or safety standards Whereas the first three categories require approval from authorities of outbound investment[b]

(continued)

(continued)

Catalog of eliminated industries in China	Eliminated category mainly includes obsolete techniques, technologies, equipment and products that do not comply with appropriate laws and regulations or safety production requirements, cause severe waste of resources or environmental pollution, thus should be eliminated For example, small coal mines overlapping with planar projection of major coal mines, coal mines below 300,000 tons/year in Shanxi, Inner Mongolia, Shaanxi and Ningxia, regular coal-fired power units with capacity of 300,000 kW or below that fail to comply with standards, atmospheric and vacuum petrochemical units of 2,000,000 tons/year or below, diaphragm caustic soda production units, HCFCs, coking by indigenous method, coke oven of steel plants without coke dry quenching installation, self-baked aluminum electrolysis cell and pre-baking cell under 160 kA, dry method hollow kiln, shaft kiln for cement, Lepol kiln, wet method kiln, etc[c]
Prohibited outbound investment	Domestic enterprises are prohibited from involving in outbound investments that harm or may harm China's interests or national security, including: 1. Projects involving the export of China's core military technology and products 2. Investments with techniques or products that are prohibited from export 3. Gambling and sex industry 4. Investment banned by the international treaties signed or joined by China 5. Other outbound investment that harm or may harm China's interests or national security
Strategic emerging industries (encouraged in China)	1. New generation of IT industry 2. High-end equipment and new materials 3. Bio industry 4. New energy vehicles, new energy, energy conservation and environmental protection industries 5. Digital creative industry 6. Cutting-edge technology R&D and industrialization, with the focus on core areas such as aerospace, ocean, information network, life science, nuclear technology, etc[d]
Encouraged outbound investment	1. Infrastructure projects that facilitate the Belt and Road Initiative and peripheral infrastructure interconnectivity 2. Projects promoting the export of China's advanced production capacity, equipments and technical standards 3. Strengthen investment cooperation with overseas enterprises of high and new technologies or advanced manufacturing, encourage establishment of overseas R&D centers 4. Prudently participate in overseas energy resources exploration and development such as oil, gas and mining with thorough analysis of economic viability 5. Expand international cooperation of agriculture, encourage mutually beneficial and win–win cooperation in agricultural, forestry, livestock husbandry, side-line production and fishery 6. Promote outbound investment in trade and commerce, culture, logistics and other sectors in the service industry in an orderly manner, support competent financial institutions to establish overseas branches and service networks to run business abiding by law and compliance requirements

[a]Guiding Catalog for Industrial Restructuring (2019)
[b]Notice of the General Office of the State Council on Forwarding the Guiding Opinions of the National Development and Reform Commission, the Ministry of Commerce, the People's Bank of

China and the Ministry of Foreign Affairs on Further Directing and Regulating the Direction of Overseas Investments [2017] No. 74
cGuiding Catalog for Industrial Restructuring
dNotice of Publishing the Development Plan for National Strategic Emerging Industries during the 13th 5-year Period. [2016] No. 67

Annex 2: Electricity Consumption (GWh) and Share of Consumption (%) by Sector and Year 2013–2018

Sector and average share	2013	2014	2015	2016	2017	2018	Avg Gr %
Household, Ave. 42% (GWh)	77,211	84,086	88,682	93,635	94,457	97,823	5.24
Industry, Ave. 33% (GWh)	61,381	65,909	64,079	68,145	72,238	76,947	4.24
Business, Ave. 18% (GWh)	34,498	36,282	36,978	40,074	41,695	44,027	6.07
Public, Ave. 6% (GWh)	11,451	12,3246	13,106	14,150	14,743	15,812	6.74
Total (GWh)	187,541	198,602	202,846	216,004	223,134	234,609	5.13

Data source: Ref. [32]

Annex 3: Generation Technologies of Existing Coal Power Stations

Coal power station	Province	Capacity (MW)	COD
Subcritical technology			
1. PLTU* Paiton 3 Unit 1[34]	East Java	815	2012
2. PLTU Tanjung Kasam Unit 1–2[35]	Riau Islands	2 × 55	2012
3. PLTU Sumsel 5 Unit 1–2[36]	South Sumatera	2 × 150	2015
4. PLTU Kalteng 1 Unit 1–2[37]	Central Kalimantan	2 × 100	2019
5. PLTU Tanjung Power, Tabalong[38]	South Kalimantan	2 × 100	2019
Supercritical			
1. PLTU Cirebon Unit 1[39]	West Java	660	2012
2. PLTU Banten Serang Unit 1[40]	Banten	660	2017
3. PLTU Cilacap Sumber Unit 3[41]	Central Java	660	2019
4. PLTU Bangko Tengah/Sumsel 8 Unit 1–2[42]	South Sumatera	2 × 620	2021
5. PLTU Indramayu Unit 4–5 PLN[43]	West Java	2 × 1000	2021
Ultra-supercritical			
1. PLTU Celukan Bawang Unit 1, 2, 3[44]	Bali	3 × 142	2015
2. PLTU Lontar Unit 4[45]	Banten	315	2019

(continued)

(continued)

Coal power station	Province	Capacity (MW)	COD
3. PLTU Jawa 7 Unit 1–2[46]	Banten	2×1000	2019
4. PLTU Batang Jawa Tengah Unit 1–2[47]	Central Java	2×1000	2020
5. PLTU Tanjung Jati B2 Unit 5–6[48]	Central Java	2×1000	2021

Note *PLTU = Pusat Listrik Tenaga Uap (Steam Coal-Fired Power Plant)

Annex 4: Plan for Additional Electricity Generation Capacity, 2020–2028, MW/year

Gen	2019	2020	2021	2022	2023	2024	2025	2026	2027	2028	Total	%
Coal	1569	6047	3641	2780	4590	3090	1184	1695	1375	1093	27,064	48
Gas	1592	3073	1011	3155	1535	845	40	280	400	485	12,416	22
Diesel	138	8	2	3	47	3	–	–	–	–	201	0.36
RE	559	932	1697	1501	1065	2287	6252	199	648	1574	16,714	30
Total	3858	10,060	6351	7439	7237	6225	7476	2174	2423	3152	56,395	100

Annex 5: Construction Cost of Power Stations by Type of Generation, USD/kW

Generation	Construction/investment cost
Renewables	*USD/kW*
Hydro	1500
Geothermal	1750
PV	1200
Thermal power station	*USD/kW*
Coal	1250
Diesel	900
Cogeneration	680
Gas	400

Annex 6: Operating Cost of Power Stations by Type of Generation, USD/MWh

Generation	Operating cost
Renewables	*USD/MWh*
Hydro	18
Geothermal	106
PV	411
Thermal power station	*USD/MWh*
Coal	51
Diesel	179
Cogeneration	86
Gas	344

Annex 7: Existing investment in Coal Power Station

Coal power station, capacity, location, year of financial close	Country of investment	Investment USD million	Bank
Subcritical technology			
1. PLTU Paiton 3 Unit 1, 815 MW, East Java, 2010	Japan	1215	Japan Bank for International Corporation, Bank of Tokyo-Mitsubishi UFJ, Sumitomo Mitsui Banking Corporation, Mizuho Financial Group, Credit Agricole Group, ING Group, BNP Paribas, Sumitomo Mitsui Trust Holdings
2. PLTU Tanjung Kasam Unit 1–2, 2 × 55 MW, Riau Islands, 2011	China	150	Export-lmpot Bank of China
3. PLTU Sumsel 5 Unit 1–2, 2 × 150 MW, South Sumatera, 2012	China	318	China Development Bank
4. PLTU Kalteng 1 Unit 1–2, 2 × 1 00 MW, Central Kalimantan, 2016	Indonesia	316	Bank Mandiri (Indonesia)
5. PLTU Tanjung Power, Tabalong, 2 × 1 00 MW, South Kalimantan, 2017	Japan	430	Bank of Tokyo-Mitsubishi UFJ, DBS Bank, HSBC, Mizuho Financial Group, Sumitomo Mitsui Corporation, Korea Development Bank
Supercritical technology			
1. PLTU Cirebon Unit 1, 660 MW, West Java, 2010	Japan	595	Japan Bank for International Corporation, Export–Import Bank of Korea, Bank of Tokyo-Mitsubishi UFJ, ING Group, Mizuho Financial Group, Sumitomo Mitsui Banking Corporation, Bank of Tokyo-Mitsubishi UFJ
2. PLTU Banten Serang Unit 1, 660 MW, Banten, 2013	Malaysia	730	Maybank, Export–Import Bank of Malaysia, CIMB Group, RHB Group, Citigroup
3. PLTU Cilacap Sumber Unit 3, 660 MW, Central Java, 2013	China	700	China Development Bank
4. PLTU Bangko Tengah/Sumsel 8 Unit 1–2, 2 × 620 MW, South Sumatera, 2015	China	1200	Export-lmpot Bank of China
5. PLTU Indramayu Unit 4–5 PLN, 2 × 1 000 MW, West Java, 2017	Japan	2000	Japan International Cooperation Agency
Ultra-supercritical technology			

(continued)

(continued)

Coal power station, capacity, location, year of financial close	Country of investment	Investment USD million	Bank
1. PLTU Celukan Bawang Unit 1, 2, 3, 3 × 142 MW (426 MW), Bali, 2013	China	571	China Development Bank
2. PLTU Lontar Unit 4, 315 MW, Banten, 2016	Japan	323	Japan Bank for International Corporation, Sumitomo Mitsui Banking Corporation
3. PLTU Jawa 7 Unit 1–2, 2 × 1000 MW, Banten, 2016	China	1839	China Development Bank, Bank of China, ICBC, China Construction Bank
4. PLTU Batang Jawa Tengah Unit 1–2, 2 × 1 000 MW, Central Java, 2012	Japan	3421	Sumitomo Mitsui Trust Holdings, Bank of Tokyo-Mitsubishi UFJ, DBS Bank, Mizuho Financial Group, OCB Bank, Sumitomo Mitsui Banking Corporation, Mitsubishi Trust, Shinsei, Norinchukin
5. PLTU Tanjung Jati B2 Unit 5–6, 2 × 1 000 MW, Central Java, 2017	Japan	3355	Bank of Tokyo Mitsubishi UFJ, Mizuho Financial Group, Mitsubishi UFJ Financial Group, OCBC Bank, Sumitomo Mitsui Banking Corporation, Sumitomo Mistui Trust Holdings, Norinchukin Bank, Japan Bank for International Cooperation

Annex 8: Planned Additional Power Project Installation 2019–2028

	PLN	IPP	Partnership	Unallocated	Total
1. PLTU (coal)	4704	14,929		1740	21,373
2. PLTU MT (mine mouth coal)		5660	300		5690
3. PLTP (geothermal)	617	3060		930	4607
4. PLTGU (combined cycle)	4603	4220		310	9133
5. PLTG MG (gas power)	3260	20		3	3283
6. PLTD (diesel power)	201				201
7. PLTM (Minihydro)	69	1422		43	1534
8. PLTA (large hydro)	1200	3139		187	4526
9. PS (pumped storage hydro)	1540			1943	3483
10. Other renewable power	49	1186		1330	2565
Total	16,243	33,366	300	6486	56,395

Annex 9: Electricity Bill of 100 kWh as a Percentage of the Monthly Income of a Recipient of Minimum Wages (10 ASEAN Countries)

Country	Percentage of electricity bill of 100 kWh in monthly income of a recipient of minimum wages (%)	Standards
Vietnam	6.2	Based on minimum wages of 2,760,000 Vietnamese dong in Category IV (lowest wages) region in Vietnam
Thailand	3.2	Based on minimum daily wages of 308 baht in Thailand in 2018
Laos	3.7	Based on minimum wages of 1,100,000 Laotian kip in 2018
Indonesia	4.4	Based on minimum monthly wages of 3,350,000 Indonesian rupiah in industrial zone adjacent to Jakarta in 2017
Myanmar	6.2	Based on the national uniform minimum daily wages of 4800 Myanmar kyat
Malaysia	2.1	Based on the minimum wages of 1000 Malaysia riggit in 2017
Cambodia	7.8	Based on the minimum monthly wages of 690,000 Cambodian riel for apparel and shoe-making industries in 2017
Philippines	8.5	Based on the average daily wages of 408 pesos for average worker in the Philippines in 2017
Singapore	2	Based on the minimum wages of 1100 Singapore dollars in 2017
Brunei	0.16	Based on monthly wages of 600 Brunei dollars for primary-level laborers

Data source Yunlu [31]

Annex 10: Electricity Industry Authority and their Roles and Responsibilities

No.	Institution	Roles and responsibilities
1	**Dewan Perwakilan Rakyat (DPR**, House of Representatives)	Commission VII of the DPR is responsible for the approval of energy-related legislation (including electricity) and the supervision of energy-related Government policy http://www.dDr.qo.id/akd/index/id/Tentanq-Komisi-VII
2	**Kementerian Energi and Sumber Daya Mineral (ESDM**. Ministry of Mines and Energy/MoEMR)	Direktorat Jendral Ketenagalistrikan (Ditjen Gatrik, DG of Electricity), as discussed above http://aatrik.esdm.ao.id

(continued)

(continued)

No.	Institution	Roles and responsibilities
3	**Perusahaan Listrik Negara** (PLN, State-owned Electricity Enterprise)	• PLN is responsible for the majority of Indonesia's power generation and has an exclusive role in the transmission, distribution and supply of electricity to the public • PLN produces Rencana Umum Penyediaan Tenaga Listrik Nasional (RUPTL, a ten year planning for electricity generation, transmission and distribution). The current RUPTL is for the 2019–2028 period www.pln.co.id
4	**Kementerian Perencanaan Pembanguan Nasional (Bappenas**, Ministry of National Development and Planning)	Direktorat Kerjasama Pemerintah-Swasta dan Rancang Bangun (Directorate for Public Private Partnership) is tasked with facilitating cooperation on infrastructure projects between the Government and private investors www.bappenas.qo.id
5	**Badan Koordinasi Penanaman Modal (BKPM** Investment Coordinating Board)	BKPM provides a one-stop integrated service for business start-up, licensing procedures, and information for existing and potential investors https://www9.bkpm.qo.id
6	**Kementerian Keuangan Republik Indonesia** (Kemenkeu, Ministry of Finance/MoF)	MoF provides recommendation for electricity subsidy to PLN and approval of tax incentives for a power project https://www.kemenkeu.oo.id

Annex 11: Policies for Attracting International Cooperation in Indonesia

Year	Policies
2007	The Law 25/2007 on Investment states the importance of both domestic and foreign investment to support national development. It regulates the type of businesses that are open to foreign investment, employment, rights and responsibilities, facilities (tax and fiscal incentives, import license, immigration), etc.
2014	The Ministerial Regulation 35/2014 on the delegation of authority to produce electricity business permit from ESDM to the BKPM, to simplify the process of acquiring electricity business license under the BKPM integrated one-stop service
2015	The Presidential Regulation 38/2015 (Collaboration Between the Government and Business Enterprise in Infrastructure Development) includes foreign holding companies in the development of infrastructure projects, replacing previous regulations
2019	The Presidential Regulation 5/2019 or Government Regulation 24/2018 on Investment Guidelines and Facilities

Annex 12: Ditjen Gatrik (DG of Electricity)

Roles[a]	National Electricity Master Plan (Rencana Umum Ketenagalistrikan Nasional) 2018–2037
(1) Formulation of electricity policy (planning, regulation, investment, interconnection, energy supply and security, tariff, etc.)	(1) National electricity policy regarding supply, generation energy mix, investment, permit, tariff, subsidy, cross-border electricity, village electrification, consumer protection, legal aspects, safety and environmental protection
(2) Implementation of electricity policy	(2) Electricity development plan regarding electrification ratio improvement, generation, transmission and distribution, sales and village electrification
(3) Formulate norms, standard, procedures and criteria for electricity undertakings	(3) Current electricity situation by province regarding supply, consumption, installed capacity, generation, transmission, distribution, village electrification
(4) Provision of electrical technical assistance/coaching and evaluation	(4) Projection of electricity demand by province
	(5) Electricity investment

Note The current National Electricity Master Plan (Rencana Umum Ketenagalistrikan Nasional) 2018–2037 was prepared by Ditjen Gatrik
[a]Dirjen Gatrik ESDM, Tugas dan Fungsi, http://gatrik.esdm.go.id/frontend/tugas_fungsi, accessed 19 January 2020

References

1. Hu Jian, Zhang Weiqun, Xing Fang, Geng Hongqiang. Research on the measurement and evaluation of national economic and social development from the perspective of the Belt of Road Initiative, 2017.
2. Chai Qimin, Fu Sha, Qi Yue, Wen Xinyuan. Discussion of climate change factors in the Belt and Road finance, 2019.
3. State Information Center. Report on data of trade cooperation along the Belt and Road, 2018.
4. Li Feng. A study on China's foreign direct investment Policy, 2016.
5. Ministry of Commerce of the People's Republic of China, National Bureau of Statistics, State Administration of Foreign Exchange. Statistical bulletin of China's foreign direct investment in 2018, 2019.
6. Ministry of Commerce of the People's Republic of China, National Bureau of Statistics, State Administration of Foreign Exchange. Statistical bulletin of China's foreign direct investment in 2017, 2018.
7. Bank Track, Rainforest Action Network, Sierra Club, OilChange, Honor Earth, Banking on Climate Change 2019: Fossil fuel finance report card, 2019.
8. CPC Central Committee, State Council. Master plan for reforming the system of ecological civilization, 2015.
9. Ministry of Environmental Protection of the People's Republic of China, Ministry of Foreign Affairs of the People's Republic of China, National Development and Reform Commission, Ministry of Commerce of the People's Republic of China. Guidance on promoting Green Belt and Road, 2017.

10. Google, Temasek, Bain. e-Conomy SEA 2019, 2019.
11. World Bank. database, 2020.
12. ASEAN Center for Energy. The ASEAN report for energy and power cooperation, 2017.
13. IEA. Southeast Asia energy outlook, 2019.
14. IEA. data&statistic, 2020.
15. IRENA. Renewable energy statistics, 2020.
16. Energy Observer. The six truths of Chinese electricity price, 2016.
17. UNFCCC. All NDCs, 2020.
18. ASEAN Center for Energy, China Renewable Energy Engineering Institute. Report on FIT mechanism of renewable energy in ASEAN, 2018.
19. Liu Yi. A study on the environmental legal risks in overseas investment in Southeast Asia, 2018.
20. Fan Chun. Environmental issues in Southeast Asia and legal response to them, 2008.
21. Greenpeace, Shanxi University of Finance & Economics. The tendency and risk analysis of overseas equity investment on coal power, 2019.
22. Greenpeace. Our projects, 2020.
23. Zhou Lihuan, Gilbert Sean, Wang Ye, Muñoz Cabre Miquel, Gallagher Kevin P.. Moving the Green Belt and Road Initiative: from words to actions. 2018.
24. PWC. Power in Indonesia: investment and taxation guide, 2018.
25. Wang Yingbin. Indonesia: renewable energy to reach 23% in 2025, 2019.
26. Courtney Weatherby. It's decision time for Southeast Asia as power demand soars, 2019.
27. IRENA. Report on renewable power generation costs, 2017.
28. Zissler Romain. Renewable energy to replace coal power in Southeast Asia, 2019.
29. Esther Lew Swee Yoong. Renewable energy in Malaysia, 2019.
30. Natural Resources Defense Council (NRDC). Towards green growth in Southeast Asia, 2019.
31. Long Yunlu. Suggestions on power investment in Southeast Asia, 2019.
32. Perusahaan Listrik Negara in Indonesia. Electricity development in Indonesia, 2019.

Chapter 2
Post-2020 Global Biodiversity Conservation

2.1 Leadership and Engagement: China's Roles for CBD COP 15 Success

Since the end of 2019, COVID-19 has become a pandemic, attacking well over 200 countries and territories. Deadly viruses are a biological disaster that human beings have fought throughout history. The unfolding pandemic highlights how vulnerable we still are to nature. Furthermore, major disease outbreaks—such as the 1918 flu pandemic, 2002–2004 SARS, 2009 H1N1 flu pandemic, and 2014–2016 Ebola—were all zoonotic viruses. There is a human hand in pandemic emergences. A single species—humans—has exacerbated pressure to ecosystems, which are the foundation of our survival, through rampant deforestation, uncontrolled expansion of agriculture, intensive farming, mining, and infrastructure development, along with the exploitation of wild species. Our relationship with nature has been distorted. Amazonian fires are largely due to encroaching human activities; the fires in Australia have wiped out homes, vegetation, and countless wildlife; record-breaking temperatures are being recorded in the Antarctic and Arctic; locust swarms ravage Africa and Asia; and on a huge scale, the Great Barrier coral reef is bleaching. These are cries from nature—and warnings to humans.

While these highly destructive events may appear to be one-time or episodic events, careful examination shows they are part of global patterns with serious immediate and long-term impacts. There are also structured and systemic changes in our disturbance of planet Earth, mostly unseen and even more threatening to our existence: climate change, accelerating biodiversity loss, and the ecological impacts of production and consumption (plastic waste being a critical one). The average number of native species in most major land-based habitats has fallen by at least 20%, mostly since 1900. More than 40% of amphibian species, almost 33% of reef-forming corals, and more than a third of all marine mammals are threatened.

© The Author(s) 2022
China Council for International Cooperation on Environment
and Development (CCICED) Secretariat,
Green Consensus and High Quality Development,
https://doi.org/10.1007/978-981-16-4799-4_2

Human actions have altered more than three quarters of the Earth's land surface, destroyed more than 85% of wetlands, and converted more than a third of all land and almost 75% of available freshwater to crops and livestock production. Unregulated trade in wild animals and the explosive growth of global air travel have spread deadly diseases, brought untold human suffering, and halted economies and societies around the world. Nature has given us a signal that it is time to rethink and realign our behaviour and economy with nature. At the right time, and with attention paid to the current emergency, there is an opportunity to highlight that our health, food, climate, and nature are all connected.

We need to pay close attention to all of these systematic transformations even when we are fighting an immediate pandemic crisis. We need to take serious steps now to reduce environmental risks and improve ecological, economic, and social resilience levels locally, nationally, and globally for the decades ahead. The jolt from COVID-19 reminds us that the complex 2030 Sustainable Development Goals (UN 2030 SDGs) must be achieved in a time frame of less than a decade. An ecological civilization in China is to be achieved during the next three FYPs. Full restoration of global biodiversity by 2050 is a long-standing aspiration. Holding global temperature rise to 1.5 or 2.0 °C and achieving zero-carbon society are essential. These and other pressing needs, especially poverty elimination, provide the backdrop to the CBD COP 15 meeting to be hosted by China and rescheduled from October 2020 in Kunming to a date to be determined in 2021. This meeting will set the direction of global biodiversity efforts to 2030, with implications for action to mid-century.

The world that we know today might enter a new era—almost beyond our current recognition as recovery from the COVID-19 pandemic plays out, possibly over a period of five to years or more. Even with its unfortunate start, 2020 remains a critical year when we can continue the process of rethinking and realigning our relationship with nature. Strong political will, determination, and actions will be needed. With China's efforts to control the COVID-19 outbreak, multilateralism facing challenges (but international cooperation needed as never before to address planetary emergencies), and the key decision making moments on multilateralism (UNGA75), nature (CBD) and climate change (UNFCCC) approaching, China can play a stronger leadership role in year 2020 and beyond to collectively secure a safe future—nature-positive and carbon–neutral—for current generations as well as those to come.

2.1.1 Strong Political Will

Recently, the world has seen significant momentum built around and toward elevating a nature agenda among state and non-state actors. Strong visible political will and support are critical to galvanizing the momentum to catalyze actions needed to reverse the loss of nature by 2030.

UN Secretary-General António Guterres has laid out the need for action in the year 2020, including action at the highest level. On February 12, 2020, he observed that.

> biodiversity offers solutions for many global challenges. From climate change to food and water security, from decent jobs to gender equity, healthy ecosystems are critical. The time has come to put nature at the heart of sustainable development and to invest in restoring the earth's nature support ecosystems. This year brings many opportunities: the biodiversity summit in New York, the biodiversity conference in Kunming, as well as the COP26 on climate change in Glasgow.

Guterres calls on "all leaders to show ambition and urgency as we strive to reverse biodiversity loss, conserve and sustainable use natural resources, and share the benefits fairly."

Political leadership can take different shapes and forms.

2.1.1.1 Heads of State (HoS) Biodiversity Summit

When facing multiple global emergencies as we now do, strong political signals from the highest levels on the global stage and within a state/government is urgently needed. Thus, 2020 offers a unique opportunity to act using an integrated approach for health, climate change, nature, and development agendas. The critical HoS Summit at the margin of UNGA75 will be a key moment to send a strong collective political signal to the world in the form of emergency declarations, or calls to commitment, or calls to action, or voluntary leaders' declarations, for nature and people. There are other moments when enlightened HoS can give their political signals. At the global level, these include the November 2020 G20 Riyadh summit, and the HoS moment during the International Union for Conservation of Nature (IUCN) World Conservation Congress "One Planet Summit" (now postponed to January 2021) in France. There are regional moments that can be stepping stones: the ASEAN Conference for Biodiversity, EU–China Summit, etc., which are postponed, with new dates to be decided. The G20 virtual dialogue on March 26, 2020, although focused on the emergency matters of COVID-19, also showed signs that global leaders can rethink and realign our relationship with nature. In this critical historical time, China can play a role in reshaping a robust process that does justice to new realities and new priorities.

Heads of state and government can declare a planetary emergency, or a call to action, that gives strong political commitment, decisions, and urgent actions to build a nature-positive future for all life on Earth. This is needed even more now for **halting and reversing biodiversity loss and putting nature on a path to recovery for the benefit and health of all people and the planet by 2030**. The HoS, based on scientific research, can issue calls to:

- Build a stable climate and diverse nature future as the foundation needed to meet SDGs through whole-of-government and whole-of-society actions;

- Put nature restoration, climate stabilization, and achieving UN2030 SDGs as the foundation for a whole-of-government priority and whole-of-society set of actions;
- Secure an ambitious and transformative post-2020 Global Biodiversity Framework (GBF) and ensure its immediate implementation once adopted;
- Address direct and indirect drivers of nature/biodiversity loss and climate change on land and in the ocean;
- Mainstream nature (biodiversity and ecosystem services) into all relevant economic sectors to significantly reduce the negative footprint of production and consumption;
- Secure and fairly share the benefits derived from conservation and sustainable use of nature;
- Implement economic and financial reform to realign and increase financial resources to address the double challenges that we are facing: nature/biodiversity loss and climate change;
- Ensure that the economic recovery and stimulus measures post-COVID-19 will: (1) promote delivery of SDGs (health benefits to people, job creation, and poverty alleviation, etc.), (2) do no harm to ecological systems and climate change, and (3) promote green transition to a carbon–neutral, nature-positive, and healthy-people future;
- Build actions that reverse nature loss by 2030 and align our national strategies and action plans on climate, nature, and sustainable development commensurate with the challenges we face;
- Work with business, investors, academics, civil societies, women, youth, Indigenous peoples and local communities (IPLCs), cities and other sub-state and non-state actors to join forces to reverse the loss of nature and biodiversity; and
- Enhance policy coherence and synergies across all relevant environmental conventions.

China's leaders should work in coordination with other heads of state, the CBD Secretariat, and other key stakeholders to propose approaches and initiatives to engage biodiversity issues at the highest level. Utilizing and creating opportunities so that UN Secretary-General, heads of state, business front-runners, and key opinion leaders will show their determination and collective efforts to jointly fight this global crisis multilaterally. Specifically, the world leaders should:

(1) **Put nature high on the political agenda** and recognize the fundamental relationship between nature, a stable climate, human well-being, and sustainable development for all:

- Make public statements of commitment for a nature-positive and carbon–neutral sustainable development, including aligning the post-COVID-19 economy recovery plan with biodiversity and climate goals;
- Discuss proposed CBD COP 15 targets in government sessions and provide feedback;
- Advocate in international forums for the targets.

(2) **Seek an emergency declaration to call for ambition, commitment, and action** to reverse the loss of nature by 2030 in order to create a sense of urgency at the highest political levels, and increase the pressure for both short- and longer-term action and impacts.

- Support and participate in a Summit on Biodiversity at or around UNGA75 or before CBD COP 15 with a strong Emergency Declaration outcome;
- Ensure the emergency declaration incorporates strong language on elements of the "New Deal for Nature and People" and sets the ambition at the right level, with targets and time-bound commitments.

(3) Advocate and set ambitious targets for *Nature-Positive by 2030* for the benefit of people and the planet. Specifically:

- Take action and be the first to commit to the targets;
- At COP 15, advocate for the targets to be part of the CBD post-2020 framework;
- Communicate the commitment to the targets on social media, intergovernmental, and other channels.

(4) Adopt and enforce effective implementation and accountability mechanisms for biodiversity. Specifically:

- Increase investments, including within post-COVID-19 recovery plans for a green and just recovery, for nature conservation that protects global and other critical biologically diverse areas, key biodiversity components, and areas likely to host unique aggregations of biodiversity;
- Start formulating the National Voluntary Commitments (NVCs) and update NBSAPs;
- Introduce a ratchet mechanism that includes regular stocktaking to track action progress in order to allow periodic uplift of ambition and implementation;
- Provide an enabling environment so that businesses, investors, academics, civil societies, women, youth, IPLCs, cities, and other sub-state and non-state actors can take action.

2.1.1.2 Emerging Leaders

Actions to elevate the nature agenda at various levels have been emerging—and to some extent surging.

China's President Xi Jinping and French President Macron have made strong commitments to reverse the loss of biodiversity in their Beijing Call for Biodiversity Conservation and Climate Change [1]. China and France are jointly committed to addressing the threats and drivers of loss of nature for global peace and stability of food security, human health, and SDGs.

The two presidents committed to "working together on the link between climate change and biodiversity, and determined to support and work together with other political leaders to prompt a global and effective response to climate change and biodiversity loss in the COP 15 of CBD."

They "call on all countries and, when relevant, subnational authorities, companies, NGOs and citizens to: Encourage concrete and ascertainable commitments and contributions to biodiversity conservation from actors and stakeholders across all sectors to stimulate and support government action in the promotion of a robust post-2020 global biodiversity framework in the frame of the Sharm El-Sheikh to Kunming Action Agenda for Nature and People."

They determined to "promote active engagement of political leaders at the highest level in advocacy for biodiversity at CBD COP 15 with the theme: Ecological Civilization-Building a Shared Future for All Life on Earth."

They committed to work together to "Capitalize on the Nature-Based Solutions Coalition co-led by China and leverage nature-based solutions to coherently address biodiversity loss, mitigation and adaptation to climate change, and land and ecosystems degradation."

They called to "Mobilize additional resources from all sources, both public and private, at the domestic and at the international level, toward both climate adaptation and mitigation; make finance flows consistent with pathways toward low greenhouse emissions and climate-resilient development, as well as for the conservation and sustainable use of biodiversity, the conservation of oceans, land degradation amongst others; ensure that international financing, particularly in the infrastructure field, is compatible with the Sustainable Development Goals (SDGs) and the Paris Climate Agreement."

Other groups of countries are taking actions in various timeframes and levels. A summary describing some of these coalitions can be found in Annex 1.

Clusters of countries such as these are increasingly demonstrating their willingness and leadership to conserve nature. The trend is getting stronger, and changes are happening quickly. These leaders are increasingly calling for actions to address nature loss. About 50 national governments—and the numbers are growing—have called or signed on to call to actions through Montreal Nature Champions' Summit, G7's Metz Biodiversity Charter and corresponding International Leaders initiatives, Trondheim Conference, and the Leader's for Nature and People event during the UNGA74 high-level week:

- Africa (12)—Burkina Faso, Cameroon, CAR, Egypt, Gabon, Kenya, Niger, Rwanda, Senegal, Seychelles, South Africa, Uganda
- Asia Pacific (12)—Australia, Bhutan, China, Fiji, India, Indonesia, Japan, New Zealand, Palau, UAE, Vanuatu, Vietnam
- Europe (13 + EU)—Austria, Belgium, EU, Finland, France, Germany, Italy, Monaco, Netherlands, Norway, Portugal, Serbia, Spain, UK
- North America (2)—Canada, United States (*Metz Charter*)
- Latin America and the Caribbean (11)—Belize, Bolivia, Brazil, Chile, Colombia, Costa Rica, Ecuador, Granada, Guyana, Mexico, Peru.

These leaders, depending on the issues that they care about the most, are calling for urgent action to address the planetary emergency. It is a great opportunity for China, as CBD COP 15 host country, to join forces with them and to play a leadership role. It is timely that China joins in at least some of the movements, garners the momentum and energy, and shows the type of leadership to fulfill China's commitment made together with France.

With the immediate and long-term impacts of COVID-19, the priorities of these leaders may be adjusted. Our team will continuously monitor the progress and provide timely updates and analysis.

2.1.2 Building Momentum

More and more countries, institutions, and non-state and subnational actors are calling for, or taking, actions to elevate the nature agenda, taking a "whole-of-society approach" as noted below:

- **The Sharm El-Sheikh to Kunming Action Agenda for Nature and People**: Launched by China, Egypt, and the secretariat of the CBD. So far, only a few commitments have been registered, and many of them existed before COP 14. Here, political and practical support from China is important. One important step is to make the Action Agenda an integral part of the post-2020 GBF, as this will provide some clarity about the long-term direction for stronger non-state involvement in the CBD. A clear signal from China about this encouraging non-state actors to participate and contribute would be helpful.
- **The Nature Champions Summit, Montreal**: Where ministers from government gathered with CEOs from business and NGOs in Montreal in April 2019 and began a global mobilization committing jointly to take a different, better path that puts nature first, recognizing that it sets the context for all life—including human life—and accordingly requires our full respect and care in return. Collaboratively, these Nature Champions commit to placing nature's needs at the heart of all global agendas.
- **Trondheim Conference**: The ninth Trondheim Conference on Biodiversity was held in Trondheim, Norway, in July 2019, created opportunities for increasing the understanding amongst stakeholders about issues on the biodiversity agenda.
- **High-Ambition Coalition for Nature**: An intergovernmental group championing a global deal for nature and people that can halt the accelerating loss of species and protect vital ecosystems (e.g., 30 × 30 movement to protect 30% of the earth by 2030) that are the source of our economic security. Two co-chairs, France and Costa Rica, planned to formally launch it at the IUCN WCC.[1]
- **"United for #Biodiversity" Coalition**: Made up of zoos, aquariums, botanical gardens, national parks, and natural history and science museums from around the world, launched on World Wildlife Day in 2020 by the European Commission.

---

[1] With the changing date of the IUCN, there is a discussion that the launch will be at the UNGA 75.

2.1.3 Evidence Supporting the Need for Stronger Biological Diversity Decision Making

Efforts to gather evidence for decision making are growing, focusing especially on the role of nature from an economic point of view.[2]

Recent assessments presented in the Global Assessment Report on Biodiversity and Ecosystem Services, released in May 2019 by the Intergovernmental Science-Policy Platform on Biodiversity and Ecosystem Services (IPBES), have estimated that extinction rates are estimated to be 1000 times the background rate, and that 75% of the Earth's land surface is significantly altered, 66% of the ocean area is experiencing increasing cumulative impacts, and over 85% of wetlands have been lost.

In addition, more and more studies focus on the enhancement of natural capital, evidence for action from new perspectives and synergies, and developing improved mechanisms for implementable changes. Some have been released after the CCICED AGM 2019, and some are to be released in the coming months. Here is a non-exhaustive list:

- **Food and Land Use Report. 10 Critical Transitions to Transform Food and Land Use**
  Launched in October, 2019, the Food and Land Use Coalition (FOLU) report is the first to assess the benefits of transforming global food and land-use systems as well as costs of inaction. The report reveals benefits that far outweigh the costs: it also proposes actionable solutions. It is estimated USD 12 trillion a year in hidden costs relate to how we produce and consume food and use land currently. The benefits stand to unlock USD 4.5 trillion in new business opportunities each year by 2030, at the same time saving USD 5.7 trillion a year in damage to people and the planet by 2030, more than 15 times the investment cost of up to USD 350 billion a year. See also the 2019 FAO report on the *State of the World's Biodiversity for Food and Agriculture.*
- **The Global Risks Report 2020**: Launched by the World Economic Forum (WEF) in January 2020, shows that for the first time in 10 years, **the top five global risks in terms of likelihood are all environmental**. The report points to a need for policy-makers to match targets for protecting the Earth with ones for boosting economies—and for companies to avoid the risks of potentially disastrous future losses by adjusting to science-based targets.
- **WEF's** "New Nature Economy Report Series's" first report *Nature Risk Rising: Why the Crisis Engulfing Nature Matters for Business and the Economy* was launched in January 2020. This report highlights "that $44 trillion of economic value generation—more than half of the world's total GDP—is moderately or

---

[2] See for example: OECD [10] (prepared for G7 Presidency and Environment Minister's Meeting). May 2019; United Kingdom. April 2020. *The Dasgupta Review—Independent Review on the Economics of Biodiversity Interim Report.* The Economist World Ocean Initiative. June 2020. *A Sustainable Ocean Economy in 2030: Opportunities and Challenges.* The Netherlands Bank (DNB) June 2020. *Indebted to Nature—Exploring Biodiversity Risks for the Dutch Financial Sector.*

highly dependent on nature and its services and is therefore exposed to nature loss."

- **WWF's The Nature of Risk: A Framework for Understanding Nature-related Risk to Business** was launched in September 2019, intended to give a clear understanding of risks related to nature and climate change.
- *Nature Is Too Big to Fail—Biodiversity: the Next Frontier in Financial Risk Management*, jointly launched by PwC Switzerland and WWF Switzerland in January 2020, calls for at least USD half a trillion per year to cover funding gaps for biodiversity conservation and restoration.
- *Economic and financial systems and tools to develop biodiversity conservation*, jointly published by WWF France AXA in May 2019, identifies best practices and the most promising technical and political perspectives and also proposes a roadmap to develop biodiversity finance commensurate with the current biodiversity crisis.
- *Biodiversity—Opportunities and risks for the financial sector*, a report prepared by the cooperation of a few Dutch banks (Rabobank, ACTIAM, ASN, FMO, Robeco etc.) as part of the Dutch Sustainable Finance Platform. The report outlines what risks and opportunities exist for financial institutions around biodiversity. It estimates that the long-term economic damages of greenhouse gas emissions, based on 2008 figures, would be around USD 1.7 trillion per year. Those from biodiversity loss are estimated to range between USD 2–USD 4.5 trillion per year. This comparison provides a clear message that both phenomena are equally urgent, also for financial institutions, and require immediate action. It calls financial sector to play an important role to realize these opportunities and halt the global loss of biodiversity. It highlights a few case studies from Netherlands banks showing that people, planet, and profit can work in tandem and create positive biodiversity outcomes. These can be used when engaging in CBD negotiation processes.

The number of emerging studies on the links between climate change adaptation and mitigation, biodiversity, and land is increasing rapidly, including those noted below.

IPCC Special Report on Climate Change, Desertification, Land Degradation, Sustainable Land Management, Food Security, and Greenhouse Gas Fluxes in Terrestrial Ecosystems.
The Ocean and Cryosphere in a Changing Climate, September 2019 *GBO5* draft (now at the review stage) (https://www.cbd.int/doc/c/bba0/d84c/e02639e37191 f353553e513d/sbstta-23-02-add3-en.pdf)
Local Biodiversity Outlook—LBO2 (https://beta.localbiodiversityoutlooks.net)
IPCC AR6 Climate Change 2021: Impacts, Adaptation and Vulnerability—Draft under review
IPCC AR6 Climate Change 2021: Mitigation of Climate Change (July 2021— Draft under open peer review until March 2020)
FAO Commission on Genetic Resources for Food and Agriculture. 2019. *The State of the World's Biodiversity for Food and Agriculture*. FAO, Rome.

There are still more studies to be released relatively soon, looking at the relationship between humanity and nature from both nature and economic perspectives. An important example is the UK Government Review of the Economics of Biodiversity under the leadership of Professor Sir Partha Dasgupta.

2.1.4 Significant Events (June 2019 to 2021)

Since the time of CCICED's last AGM in June 2019, there have been several milestone events:

- Trondheim Meeting in July 2019:
- Davos in January 2020.

There are still key events in the coming months that China can utilize for its nature-related diplomacy:

- ASEAN Conference on Biodiversity 2020 and Mega Diverse countries in Kuala Lumpur, Malaysia[3]
- IUCN WCC in Marseille, France[4]
- HoS Summit on Biodiversity in New York City, September 22–23, 2020[5]
- China–EU 22nd Summit in Leipzig, Germany[6]
- CBD COP 15 in Kunming[7]
- UNFCCC COP 26, Glasgow[8]
- G20 summit in Riyadh, Nov. 2020
- UNEA 5, Feb. 2021.

These are the moments when China's green diplomacy and high-level political engagement can demonstrate its impacts. The eyes of the world will be on Kunming, China, hoping that an ambitious and implementable post-2020 GBF can be agreed upon. China's leadership is greatly expected and needed.

These are also the moments when China can engage with key countries in the world to discuss issues related to nature conservation, combating climate change even while fighting COVID-19 and building a nature-friendly economic recovery. Relevant issues include greening the BRI, deforestation-free supply chains, sustainable use and governance of ocean ecosystems (including combating marine plastic litter), as well as biodiversity conservation and health. A focus on nature-based solutions (NBSs) might deliver more than one third of climate solutions. Integrating both GBF

---

[3] Postponed from March 2020 to unknown date, due to COVID-19 pandemic.

[4] Postponed from June 2020 to January 7–15, 2021, due to COVID-19 pandemic.

[5] Format will be virtual. The final dates will be decided at the end of June.

[6] Videoconference Meeting held on June 22, 2020 with some discussion on SDGs and on climate change, and the need for green and inclusive recovery from COVID-19.

[7] Postponed from Oct. 2020 to Q2 2021.

[8] Postponed from Nov. 2020 to Nov. 1–12, 2021.

and the Paris Agreement, as well as the United Nations Convention to Combat Desertification's (UNCCD's) land degradation neutrality plan, into a country's Nationally Determined Contributions (NDCs) and NBSAPs can help China and the world move toward the UN Decade of Restoration and the direction of reversing the loss of nature by 2030. Overall, there is an important opportunity to achieve the UN SDGs and help China reach its 2035 ecological civilization goal. These are all aspects of the potential benefit of building the "Community of Human Destiny."

In light of the postponement of many biodiversity-related events this year, China should consider a high- level follow-up to the UNGA75 Biodiversity Summit. This could take the form of an Opening Session in Beijing for COP 15 for heads of state/government level. An additional approach might be to invite HoS from like-minded, progressive, and high-ambition countries to have an HoS/government session in Beijing, prior to the CBD COP 15 in Kunming, in order to build the strongest political signal to recognize the role of nature to human well-being and to show a willingness to take urgently needed transformative actions for nature, climate, land use, and sustainable development.

2.1.5 Implication of the COVID-19 Pandemic

The outbreak of COVID-19 is having a profound impact on China's diplomacy and political engagement. The OEWG 2 Committee was originally scheduled to be held in Kunming on Feb. 24–29, 2020, for negotiation of a "Zero Draft of the Post-2020 GBF" but was relocated on short notice to Rome due to COVID-19. This restricted attendance, including attendance of Chinese participants and many others. It reduced the opportunity to build momentum and opportunity for important dialogue. With almost all planned events in the first half of 2020 postponed or cancelled, China needs to turn crisis into opportunities by elevating the efforts of green diplomacy. China should significantly increase, mostly virtually, the communications and dialogues with "promoter," "swing," and "blocker" countries to understand their concerns and aspirations, to share China's views, and to gain consensus and convergence toward an ambitious post-2020 GBF and a path to build on the COP 15 Theme: *Ecological Civilization: Building a Shared Future for All Life on Earth.* The postponement due to the Covid-19 outbreak actually provides China some needed time for intensive diplomatic mediation prior to COP 15.

2.2 Stocktaking on Parties' View Regarding Post 2020 Global Biodiversity Framework (GBF) and Its Implementation

2.2.1 Proposed Changes to the CBD Open-Ended Working Group Zero Draft Document of January 2020

Regarding the zero draft of post-2020 GBF [2], we pay special attention to several key issues as follows:

- Global Apex Goal and 2030 Mission should be more ambitious such as bending the curve of biodiversity loss in 2030;
- The theme of COP 15 Ecological Civilization—Building a shared future for all life on earth should be incorporated into the GBF;
- Zero loss of natural habitat should be promoted;
- Sustainable production and consumption, and green supply chains for a transformational change should be promoted;
- Culture diversity and nature diversity should be closely linked.

We also propose the following improvements/revisions:

Background Part of the Zero DraftBackground Part of the Zero Draft

(1) In past discussions, many interested groups, including parties and non-parties and other stakeholders, agreed to set up Specific, Measurable, Achievable, Realistic, and Timely (SMART) goals in the post-2020 GBF [3]. This is not reflected fully in the text. We support the need to emphasize a SMART approach that targets/goals should have measurable and communicable targets which will benefit the conservation action plan and the monitoring and assessment of conservation. This is especially useful for protected area targets.

Introduction of Annex 1 of the Draft Framework Document

A. Background

(2) To reflect the theme of COP 15 in the post-2020 global biodiversity framework (GBF), we propose revising the last sentence of the paragraph by adding *"to realize Ecological Civilization—Building a Shared Future for All Life on Earth"* in the sentence and then connect to the original text. The term "a shared future for all life" is consistent with the 2050 vision of "Living in Harmony with Nature" but with social and political sense for governance to ensure the fulfillment.

The full sentence could be improved to read as below:

The post-2020 global biodiversity framework builds on the Strategic Plan for Biodiversity 2011–2020 and sets out an ambitious plan to implement broad-based action to bring about a transformation in society's relationship with biodiversity, *to realize Ecological Civilization – Building a Shared Future for All Life on Earth*, and to ensure that, by 2050, the shared vision of Living in Harmony with Nature is fulfilled.

B   Theory of Change

(3)   We support emphasis on transformational change, which should be centered within the scope of the post-2020 GBF. "Ecological Civilization" would be a good example that specifies conservation strategies and action plans to implement the GBF. The term "ecological civilization" integrates political, social, and economic elements together and is incorporated into relevant laws/regulations/strategies/action plans for biodiversity conservation in China Although this term has been mainly promoted by China so far, the concept of this term highlights global concerns, emphasizes the role of transformational changes within the scope of GBF, and calls for transboundary cooperation in biodiversity conservation.

In addition, we shall have to make profound changes to reduce/eliminate the direct/indirect drivers of biodiversity loss (such as the threats coming from unsustainable production and consumption) in the transformational changes.

II. Framework of Annex 1

A.   2050 Vision

(4)   In order to achieve ecological civilization—Building a shared future of all life on earth, to assist for the stability of the planet's life support system, and to set up a global compass to halt and reverse climate change and biodiversity loss, we propose the development of a motivational, communicable, science-based, and measurable Global Apex Goal for nature action agenda.

The three elements of the Global Apex Goal will be added as a new paragraph:

A Global Apex Goal will be developed to align with the 2050 Vision which calls for living in harmony with nature: 1) The baseline of 2020 (recommended) will serve as a reference for zero net loss of nature and biodiversity; 2) By 2030, biodiversity and nature are recovering at a global scale; 3) By 2050 nature and biodiversity will be fully recovered and restored. At this point, we will have achieved sufficient functioning ecosystems to support future generations of people and help avoid dangerous climate change.

B.   2030 and 2050 Goals of the Framework part 10(a) of the framework

(5)   Regarding no net loss of ecosystems, we support limiting the use of offset concepts where utilization/loss of ecosystems in one place is compensated by reducing losses elsewhere. We propose to have measures that would ensure high-quality ecosystems are being protected, and to restore degraded

or damaged ecosystems. Low-diversity ecosystems should be used instead to meet targets of no net loss of protection area without sacrificing high-quality ecosystems. This goal can be met by trying new approaches such as setting a baseline using the "Three Global Conditions" approach [4], i.e., to set up different goals adapting to various habitat situations: (1) Farms and Cities, (2) Shared Landscapes and (3) Large Wild Areas.

B   10 (d) and other places

(6)   The global population is currently 7.6 billion, reaching 8.6 billion by 2030, and 9.8 billion by 2050 [5]. In addition, most people are living in developing countries, such as Brazil, China, and India. Therefore, the number of people needing improvements in living standards should be counted in "billions" rather than "millions." This will make the number simple to understand and the goal more communicable to the public.

C.   2030 mission

(7)   We support an ambitious mission for the 10 years post-2020. We propose that it shall be to halt the loss of biodiversity and to reverse the curve of loss.

12.   (a) 2.

(8)   The goal for protected areas should be defined at global and national levels and national goals depending on the nature-related potential of parties. We may have a global conservation area target, but for the target at national level, different conditions shall be taken into consideration. We suggest countries consider common but differentiated responsibilities and associated implementation mechanisms should be developed, such as the Three Conditions approach[9] [4] emphasizing the uneven distribution of land-use drivers and human pressure, and suggest different conservation strategies be taken for the conservation of the three conditions. This will need to be based on scientific research.

D.   12 (b) 7

(9)   Regarding the recent outbreak of the novel coronavirus pneumonia, we need to take human health into consideration for the use, captive breeding, trade and consumption of wildlife. Therefore, we propose to add a sentence at the end of D(12)(b)7:

> to reduce/avoid the risk of disease transmission from animals to humans during the use and consumption of wildlife, to maintain healthy habitats and to develop systems to ensure wildlife trade and consumption are sustainable and well monitored.

---

[9] "Three Global Conditions for Biodiversity Conservation and Sustainable Use (3Cs) is an implementation framework suitable for use in the post 2020 [Strategic Plan for Biodiversity Conservation]". The framework establishes the baseline state of three broad terrestrial conditions: (C1) Cities and farms (18% of global land), (C2) Shared lands (56%), and (C3) Large wild areas (26%). Antarctica is not included.

D.   12 (c) 17

(10)   Sustainable production and consumption related to transformational change
        should be well emphasized in the text.

E.   Implementation-support mechanisms

(11)   The current action targets and measures are not sufficient to ensure complete
        transformational changes to halt and reverse biodiversity loss. Thus, we need
        to establish a mechanism for immediate action to ensure that transformational
        changes are within the scope of the GBF. In addition, taking into consideration
        that any potential biodiversity funding mechanism that will be set up to support
        the implementation of the GBF, we propose to add one point:

> To establish mechanisms, including a financial mechanism, to ensure the parties
> to take immediate policy actions globally, regionally and nationally to transform
> economic, social and financial models to realize conservation goals and to halt
> biodiversity loss.

The proposal for a financial mechanism is provided in Part 5.3.

In response to the discussion during the Open-Ended Working Group second
meeting in Rome, Italy during February 24–29, 2020 [6], two further comments are
proposed:

(12)   **Baseline**. Any biodiversity target requires a baseline from which to measure
        changes and to aim at as a target. Yet deciding on an appropriate baseline
        against which to measure biodiversity change is contentious. Proposed base-
        lines during the Open-Ended Working Group second meeting in Rome ranged
        from Pre-Industrial times to the date of the 2020 CBD meeting. Neither of
        these proposed dates meets the core aims of the Global Biodiversity Frame-
        work as both practical and ambitious. Developing a practicable target requires
        a nuanced understanding which integrates the development needs of devel-
        oping countries to conserve diversity. **Thus, a baseline should include all
        natural habitats, including degraded areas and marginal land, which
        have the potential for restoration**.

(13)   **Payment for ecosystem services (PES)**. In many cases, those who benefit
        from ecosystem services are at significant distances from the sources. In these
        cases, it may be that those at the river source or basin are responsible for the
        continued provision of a service. Thus, this separation between beneficiaries
        and providers often requires a mechanism which rewards those maintaining
        the service to enable them to continue to safeguard that process; such as the
        conservation of river catchments to ensure clean water provision for those
        in settlements downstream. While this provides the most common example
        of payment for an ecosystem service, many other examples exist, from the
        conservation of areas for ecotourism, pollination or cultural services, or on a
        global scale, oxygen provision. On a global scale, climate funds and REDD
        can be looked upon as a form of PES, with the service provision being the

generation of oxygen. This form of payment was mentioned a great deal within the OEWG framework.

(14) **Linking culture diversity and natural diversity.** A holistic approach for biodiversity conservation will embrace the indivisible linkages between nature, people, and culture. It will recognize and enhance the critical role of a wide range of stakeholders such as IPLCs and women in protecting nature, culture, and identity, and integrating traditional knowledge and good practices into decision making.

2.2.2 Potential Analysis for Protected Area Expansion for Parties of the Convention on Biological Diversity for 2021–2030

The Aichi Target 11 of the CBD is composed of several interrelated conditions to protect areas of particular importance for biodiversity and ecosystem services. However, the implementation of Aichi Target 11 has not mitigated the ongoing decline of biodiversity and ecosystem services [7]. Because of varying natural and social conditions among countries, the potential for protected area conservation can be expected to be variable. The responsibility for global biodiversity conservation, the demand and suitable area for protected area (PA) expansion, and the ability of biodiversity conservation under various development or other threats are quite different among nations. If the targets of PA coverage remain equally set, the overall (unified) percentage goal of PAs will not be achieved. Thus, making explicit PA coverage targets for each party is urgently needed.

The Aichi Target 11 asking for 17% is far from adequate to safeguard global biodiversity [8]. To halt global biodiversity loss effectively, previous studies set post-2020 PA targets of about 20–50%. IUCN has recently published a massive literature review, the authors of which concluded the range of protection required is from a low end of 30–70% and higher. A large global scientific survey showed strong support from scientists for protecting up to half the world (e.g., [9]).

Regarding threatened species protection, conservation of 20.2% of the global terrestrial area was proposed. In order to conserve the entire terrestrial species, ecoregions, Important Bird and Biodiversity Areas and Alliance for Zero Extinction sites, a target of 27.9% was put forward. For the conservation of important global areas for biodiversity and ecosystem services such as carbon storage, 31% was set as the bottom line of the post-2020 target. This number becomes even larger in the context of wilderness conservation or the "Half-Earth" plan (see [10]). Therefore, it is urgently necessary to address the issue of feasible and effective conservation in developing the post-2020 global biodiversity framework.

Cost-effective zones for PAs designation were identified and used for setting protected areas coverage targets at global and national levels. The results show the obvious gaps for biodiversity conservation and protected areas designation. Three types of targets were proposed: Ambitious Target, Moderate Target, and Conservative

Target. They called for the protection of 43, 26, and 19% of the total terrestrial areas, respectively, over the next 10 years (2020–2030) [10]. The potential for protected areas expansion varies significantly across countries, indicating the necessity to set different targets for different countries. The total number and proportion of 195 parties (excluding the European Union) of the CBD can be divided into six categories [11] defined by percentage range protected.

The Three Conditions approach provides for target-differentiated settings appropriate to the different conditions of the world. In the Cities and Farms target, a goal of 10–20% protection would be ambitious and would require substantial restoration; in the shared lands of the world, a target of 25–75% is appropriate depending on the situation; in the large wild areas the target should be to keep at least 80% intact [12]. A global target can be set—such as at least 30% by 2030. Such a target should contain a clear call for action simultaneously across each of the three conditions so that all areas receive attention.

2.3 China's Showcase Efforts for Ecological Conservation

Since reform and opening up, China has undergone rapid economic development, with many ecosystems seriously damaged and polluted. In the 1990s, the Chinese government began to pay attention to the coordinated development of economy, society, and the environment. In 1996, the Chinese government proposed changing the mode of economic growth and implementing a strategy of sustainable development in the Ninth FYP. At the beginning of the twenty-first century, with the continuous growth and expansion of the economy, natural resource usage, energy consumption, and waste emissions were growing at the same time. Therefore, the 17th National Congress of the Communist Party of China (CPC) (2007) formally proposed the construction of an ecological civilization. The 18th National Congress of the CPC has further decided to promote ecological civilization progress and never sacrifice the environment for economic growth. Adhering to the priority of ecological and environmental protection has become an important principle for China in formulating major development strategies. Chinese President Jinping Xi also attached great importance to the construction of ecological civilization, and put forward a series of new ideas and new strategies for ecological civilization construction. At the same time, ecological civilization has been set as an overarching governing program objective of the CPC [13].

Under the Ecological Civilization approach, China has formulated the system of main functional areas and ECR. These efforts are driven by the need to optimize land and water space usage and protection of important ecological spaces. As a result, even though major problems remain, China is making remarkable progress in protecting its ecological environment, setting an example for global ecological protection.

2.3.1 Establishing Ecological Civilization System and Formulating Top-Level Design of National Ecological Protection

In 2007, the report of the 17th National Congress of the CPC decided to promote a conservation culture, which seeks to establish an awareness of conservation theoretically, ideologically, and culturally [14]. In 2012, the 18th National Congress of the CPC made the strategic decision of "vigorously promoting the construction of ecological civilization," and incorporated the construction of ecological civilization into the national "five in one" overall layout strategy [15]. In May 2015, the CPC Central Committee and the State Council successively issued the *Opinions on Accelerating the Construction of Ecological Civilization* and the *Master Plan of Ecological Civilization System Reform*, which comprehensively and systematically arranged the construction of ecological civilization in terms of the overall objectives, basic concepts, main principles, key tasks, and a system guarantee, becoming the national top-level design of the ecological civilization system. *The Master Plan of Ecological Civilization System Reform* clearly proposed the ecological civilization system, including eight aspects: natural resources property rights regimes, land spatial development and protection system, spatial planning system, natural resources management and overall saving system, paid use system of resources and ecological compensation system, environmental governance system, market system of environmental governance and ecological protection, performance evaluation, and accountability system of ecological civilization. The core point is to respect nature by saving resources, improving the environment, and protecting ecology, while improving prosperity and the well-being of people.

In terms of ecological protection, the important measures are meant to protect the important ecosystems by optimizing the land spatial pattern. Specifically, there are three main approaches:

- **From the perspective of protection and development, China has put forward the strategy of main functional areas.** In June 2011, the National Main Functional Area Plan was officially released, which was the first master planning for land spatial development in China. The core point is that based on different regions of the carrying capacity of resources and environment, the current development density and development potential, land and space can be divided into three types according to the development mode: optimized development area, key development area, and restricted and prohibited development areas. In November 2017, the CPC Central Committee and the State Council issued their *Opinions on Improving the Strategy and System of the Main Functional Area*. It proposed that on the basis of strict implementation of the planning of main functional areas, the strategic pattern of the main functional areas at the national and provincial levels should be accurately implemented at the municipal and county levels, which can give full play to the basic and key role of the main functional areas in promoting the construction of ecological civilization and the construction of national spatial governance systems to improve a land spatial development and protection system with Chinese characteristics.

- **From the perspective of land spatial usage, China has put forward the strategy of "three zones and three lines."** Since the 18th National Congress of the CPC, a series of central meetings and documents have proposed establishing a land spatial planning system in China, promoting the work of "multiple compliance and integration," and scientifically delimiting "three zones and three lines" (namely, urban space, agricultural space, and ecological space); and ECR, the red line of permanent basic farmland protection, the red line of urban development boundary. In 2015, China's document *Overall Plan for the Reform of Ecological Civilization System* proposed "to build a national unified, interconnected and hierarchical spatial planning system, with spatial governance and spatial structure optimization as the main content." Later, in the 19th National Congress of the CPC, it was affirmed that China definitely has to demarcate three lines. In addition, in November 2019, the General Office of the CPC Central Committee and the General Office of the State Council jointly issued the *Guiding Opinions on the Three Control Lines in Land Spatial Planning*, which made detailed provisions on how to define and implement the three lines in the land spatial planning. It can be seen that with the gradual establishment of the land spatial planning system, the three control lines will be the core elements and mandatory content of the land spatial planning, as an important basis for land space usage and ecological restoration.

- **From the perspective of protecting important ecosystems, China has put forward a national ECR Strategy.** As an important part of "three zones and three lines," China has put forward the delimitation of ECRs. In November 2011, the State Council issued the *Opinions of the State Council on Strengthening the Key Work of Environmental Protection*, which proposed that "the ECR should be defined in the important ecological functional areas, ecological sensitive and vulnerable areas in land and marine, and the corresponding environmental standards and environmental policies should be formulated for various main functional areas respectively." In May 2015, the CPC Central Committee and the State Council issued the *Opinions on Accelerating the Construction of Ecological Civilization*, which clearly stated that "the ECR should be defined in the important ecological functional areas, ecological sensitive and vulnerable areas in land and marine, to ensure that the ecological function, ECR area, and its nature remain unchanged." Subsequently, the delimitation of ECR has been elevated to the legislative level. Accordingly, in the national security law (Article 30), "the State shall improve the system of ecological environment protection, strengthen the ecological construction and environmental protection, delimit ECR, and strengthen the early warning and prevention of ecological risks."

At the same time, China has put forward a series of effective policies regarding the protection and utilization of biodiversity to strengthen land space optimization and ecosystem protection. Among them, a lot of innovative work is underway on

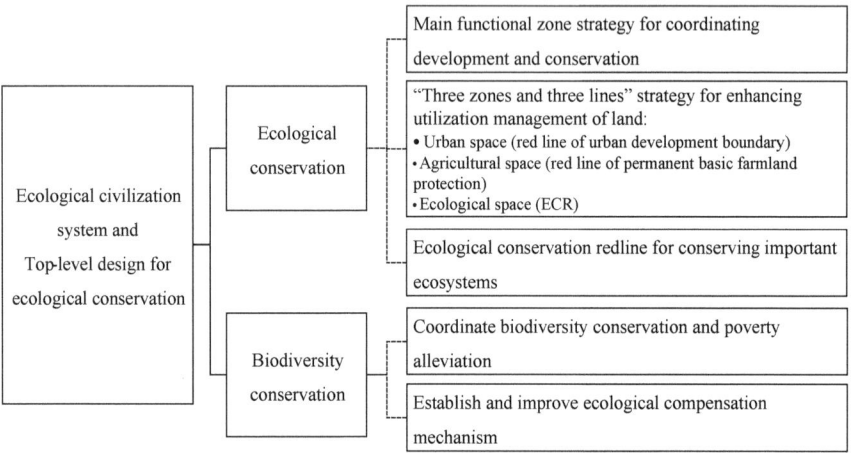

Fig. 2.1 National overall design of ecological protection in China

biodiversity and poverty alleviation, the ecological compensation system, carried out practice and demonstration in many places (Fig. 2.1).

2.3.2 Delimiting ECRs and Control of Important Ecological Space

Delimiting ECR is a major decision of the Chinese government [16]. Compared with existing protected areas at home and abroad, the ECR system is based on ecological service supply, disaster mitigation control, and biodiversity conservation. It integrates existing types of protected areas and supplements the regions where the function of ecological services is extremely important or the ecological environment is extremely sensitive and fragile, so the composition is more comprehensive, the distribution pattern is more scientific, the regional functions are more prominent, and the control constraints are more rigid. It is a major improvement and innovation in the construction of the protected areas system [17]. In June 2018, the Central Committee of the CPC issued *Opinions on strengthening ecological environment protection and resolutely winning in pollution prevention and control*, and further proposed the goal that the area of ECR should account for 25% of China's total land area. In order to better carry out the delimitation of ECRs, MEE has established a coordination mechanism, taken the lead in setting up a leading group for inter-ministerial coordination of ECR. In February 2017, the General Office of the CPC Central Committee and the General Office of the State Council issued *Several Opinions on Delimiting and Strictly Managing the Ecological Conservation Redline*, which clarified the overall requirements and specific tasks of ECR. Subsequently, MEE has developed guidance documents such as the guidelines for the ECR delimitation, opinions and suggestions on the distribution of ECR in provinces (districts and cities), and technical regulations on ECR demarcation (pilot) to guide orderly ECR delimitation in various regions in China.

Currently, China has made the following progress in delimiting ECR: In February 2018, the State Council approved the plan of ECR delimitation in 15 provinces, including the Beijing–Tianjin–Hebei Region, the Yangtze River Economic Belt, and the Ningxia Autonomous Region. ECR delimitation of all the 15 provinces has been issued and implemented by provincial governments. The other 16 provinces have formed preliminary plans for ECR delimitating. To sum up, China's preliminary ECR areas cover about 25% of the total land area of the country. It is estimated that the preliminary ECR can protect more than 95% of the rare and endangered species and their habitats, nearly 40% of the national water conservation, flood regulation, and storage functions, and about 32% of the wind and sand fixation functions. In addition, in order to strengthen the supervision of ECR, China has launched the construction of the national ECR supervision platform, organized for operation, and hence improved the integrated "ground-air-space" monitoring network. In addition, China has started to work out the management measures for ECR, trying to establish the management system for ECR in terms of the rule of law, fiscal and tax policies, standard formulation, and law enforcement, thus strengthening ECR management.

2.3.3 Promoting Biodiversity Protection Through Ecological Poverty Alleviation

In China's thinking regarding ecological Civilization, ecological poverty allevia-tion is essentially a people-centered green development approach. The Chinese government tries to adhere to eco-environmental situations in poor areas, aiming to realize, maintain, and develop the ecological rights and interests of the people in poor areas. The Chinese government has closely linked the issues of ecology and poverty, ecology and civilization, ecology and sustainable development, so as to effectively and in an orderly way, promote the sustainable development of poor areas. The Chinese government has recognized that, on the one hand, biodiversity protection must be combined with utilization to promote the long-term protection of biodiversity, while on the other hand, many areas with rich species resources are also poor. If the livelihood of local residents cannot be effectively improved, the biodiversity protection invested by the government cannot be maintained over the long term [18].

Biodiversity and poverty are key topics of global concern. However, the rela-tionship between biodiversity protection and poverty alleviation is sometimes the unity of opposites [19]. China's biodiversity-rich regions are mainly concentrated in the poor regions of the central and western regions [20]. In the past, residents of poor areas were highly dependent on natural resources, and the excessive use of wild biological resources has a great impact on biodiversity [21]. For this reason, in recent years, China has also explored and promoted the coordinated development of biodiversity protection and poverty reduction [22]. The Biodiversity Conservation Strategy and Action Plan identified 35 priority areas, some of which overlap with poor areas. At present, the ecological poverty alleviation work in some pilot areas has had good results. In the process of ecological poverty alleviation, local communities have reduced their dependence on wild animal and plant resources as much as possible. They have reduced their dependence on local resources and promoted poverty alleviation through livelihood substitution and ecotourism, which have achieved good results.

Guizhou Province is one of China's heaviest poverty-alleviation provinces. It has a large poor population, and 50 of the province's 88 county-level administrative units are national key poverty alleviation and development counties. At the same time, Guizhou Province is also one of the Chinese provinces with extremely rich biodiversity. There are 27 county-level administrative units located in national priority areas for biodiversity protection, and 25 county-level administrative units located in national key ecological function areas. There is a high degree of connection between key areas of poverty and key areas of biodiversity conservation. From the perspective of biodiversity protection and poverty reduction, Guizhou Province has explored ways to break the bottlenecks restricting the development of the poor, keep the two bottom lines of development and ecology, and ensure the goal of poverty alleviation. This includes the top-level design of poverty alleviation; the development of ecological industries; the promotion of ecology resettlement projects; and establishing long-term mechanisms for ecological poverty alleviation, etc. Especially for the problems of soil degradation and rock desertification in poor karst areas in Guizhou, many measures have been carried out, such as researching and developing technologies for ecological restoration and economic optimization of degraded vegetation in karst systems, and building a highly efficient ecosystem industrial technology system; developing characteristic forest industries in rocky desertification areas, grassland eco-animal husbandry, soil–water-fertilizer coupled eco-agriculture, rural clean energy and low-carbon economic development and other technical systems and demonstrations; systematic development of special karst features of rural ecotourism boutique lines in karst areas; and implementation of ecological compensation to enhance ecological service functions and improve people's livelihood.

The Chishui Alsophila National Nature Reserve in Guizhou Province covers seven natural villages. The local residents are mainly engaged in traditional agricultural production with a single economic structure and source. The management committee of the protected area guides local residents based on the characteristics of biological resources, take local product development as a breakthrough, use organic cultivation technology to improve the yield and quality of the original bayberry garden; use the traditional beekeeping culture of the Indigenous people, and build a scientific beekeeping demonstration base. The standardized breeding and large-scale production of characteristic biological resources have played a significant role in creating distinctive brands, carrying out scientific and standardized management, keeping stable prices, and ensuring income alleviation of poverty. They have also reduced the destructive and disorderly exploitation of other resources and played a role in protecting local biodiversity resources.

2.3.4 Establishing and Improving Eco-Compensation Mechanisms

The implementation of eco-compensation is an important means to mobilize the enthusiasm of all parties and protect the ecological environment. Over the past decade, the central government and local governments have actively promoted eco-compensation, pushing forward the construction of a compensation mechanism for orderly ecological protection. However, on the whole, the scope of eco-compensation is still too small, the standards are too low, and the mechanisms connecting protector and beneficiary are not perfect, which affects the impact of ecological environmental protection measures. In order to further improve the eco-compensation mechanism, in 2016, the Chinese government put forward that "by 2020, a complete coverage of important regional eco-compensation such as forest, grassland, wetland, desert, sea, river, farmland and other key areas and prohibited development areas, key ecological function areas, will be implemented. The compensation level will adapt to economic and social development and cross-regional and cross river compensation pilot demonstration will achieve significant progress."

The compensation system related to biodiversity protection includes: public welfare forest compensation; rewards for stopping commercial logging of natural forests; rewards for returning grazing land to grassland; subsidies for grazing prohibition and rewards for balancing grazing and livestock; giving out free seeds or funding the growth of non-native monocultures; important wetland eco-compensation; pilot project of land closure protection and compensation for desertification; subsidies for the breeding, releasing, and ecological environment restoration of aquaculture; compensation of aquatic germplasm resources reserve; compensation for ecological protection in national marine nature reserves and marine special reserves. Various compensatory measures have been promoted in an orderly manner by different authorities and have played an important role in the protection of biodiversity [23].

The establishment of upstream and downstream eco-compensation mechanisms not only ensures the water environment quality of the downstream regions, but also promotes the protection of vegetation and habitat environment in the upstream regions. In 2012, the Ministry of Finance and the Ministry of Environmental Protection coordinated Anhui and Zhejiang provinces in the joint implementation of the cross-provincial eco-compensation mechanism for the Xinan River. On the basis of the success of the first three-year pilot program, a second pilot program was launched in 2015, with a total investment of RMB 700 million for the ecological and environmental protection of the Xinan River. In 2018, the provincial finance department, provincial environmental protection department, provincial development and reform

Table 2.1 Exploration and practice of ecological compensation in China

Type	Contents	Mode
Compensation for ecosystem (vertical ecological compensation)	Services provided by ecosystems such as forest, grassland, wetland, ocean, and farmland	National compensation financial transfer payment; Ecological Compensation Fund; Market transaction
Compensation for river basin (horizontal ecological compensation)	The compensation of multi-provincial river basin; river basin under local administration	Financial transfer payments; Market transactions; Local government coordination
Compensation for different regions (horizontal ecological compensation)	Compensation for the west from east region of China	Financial transfer payments; Market transactions; Local government coordination
Compensation for resource exploitation	The development of mining industry; Land reclamation; Vegetation recover	Beneficiary Pays; Polluter Pays; Developer Pays

commission and the provincial water resources department jointly issued implementation opinions on the establishment of a horizontal ("same level") ecological protection compensation mechanism for upstream and downstream river basins in Zhejiang province, making Zhejiang the first province to implement a horizontal ecological protection compensation mechanism for river basins in China.

Eco-compensation also could be used to strengthen the breeding research of wild resources, innovate the technology of biological resources development and utilization, and reduce the utilization of wild resources according to the principle of "protection first and sustainable utilization." Through the sustainable utilization of biological resources, the development and use of biological diversity resources will become a new growth point of economic development and a new means for residents to escape poverty (Table 2.1).

2.3.5 Promoting Ecological Civilization with the Construction of Demonstration Districts

The Fifth Plenary Session of the 18th CPC Central Committee and the 13th FYP outline clearly proposed the establishment of a unified and standardized national ecological civilization demonstration district. Several districts were selected to carry out innovation experiments for major reform measures, exploring replicable management systems and effective models, thus leading the construction of national ecological civilization reform in China. In 2016, the General Office of the CPC Central Committee and the General Office of the State Council issued the *Opinions on the Establishment of a Unified and Standardized National Ecological Civilization Demonstration District*, proposing the first batch of pilot districts in the provinces of Fujian, Jiangxi, and Guizhou. Since October 2017, the three provinces have all issued specific implementation plans. Fujian has made remarkable achievements in a number of reform measures, such as the target responsibility system of ecological environmental protection, ecological compensation for the whole basin, comprehensive renovation of small basins, ecological judicial protection, auditing outgoing officials' natural resource asset management, environmental rights and interests transaction, and green finance. Jiangxi Province has established ECRs in water resources and land resources, improved the systems of nature resource property rights and land spatial planning. Jiangxi Province comprehensively implemented the river chief system and the ecological compensation of the whole basin, improved the evaluation system of the goal of ecological civilization construction, and assessment system of ecological environment damage. Guizhou Province has carried out a lot of exploration in improving the systems of nature resource property rights, the development and protection system of land space, payment for environmental resources services and the ecological compensation system, etc. China is actively summarizing the good experience and practice of these demonstration districts, so that their models can be copied and extended throughout the whole country.

In addition, in order to give full play to the typical leading role of the ecological civilization demonstration at the municipal and county levels, MEE has carried out construction activities of the national ecological civilization demonstration city (county) and the practice innovation base of "lucid waters and lush mountains are invaluable assets." Since 2017, MEE has carried out the selection of the first batch of national ecological civilization demonstration cities (counties). By the end of 2019, MEE has carried out three batches, naming 175 national ecological civilization demonstration cities (counties) and 52 "lucid waters and lush mountains are invaluable assets" practice innovation bases. The demonstration work of ecological civilization has the following characteristics: firstly, the demonstration work has been conducted in the eastern, central, and western regions of China. The proportion of the eastern, central and western regions is 43%, 28% and 29%, respectively. Secondly, the multi-level demonstration system has been further enriched. Among the three batches of demonstration cities (counties), there are 17 cities, 158 counties. There are nine cities, 35 counties, two towns, two villages and other main bodies such as forest farms in the "lucid waters and lush mountains are invaluable assets" practice innovation bases. Thirdly, the demonstration work has covered different ecosystems and regions (such as mountainous, plains, forests, pastoral areas, coastal areas, islands, ethnic minority areas), providing a diversified, vivid, and valuable reference for the construction of national ecological civilization.

2.4 Post-2020 Biosecurity/Biosafety, Biodiversity and COVID-19 Working Paper

The 2020 crisis created by the COVID-19 coronavirus spreading disease reminds us once again that even the smallest forms of biodiversity can bring about devastating impacts for people, our globalized economies, and society. By the end of May 2020, all nations are facing massive expenditure to control the disease and a global economic turndown of historical significance. The world we know will change—perhaps beyond our imagination. However, action on the global environmental emergency must not be sidelined as a consequence. Indeed, action must be strengthened even as damage from COVID-19 continues to spread. There are various concerns that will need to be addressed in relation to ecological and biodiversity matters. These range from environmental impacts that may be related to disease outbreak control; the overriding attention of governments, business and international organizations to address the costly social and economic recovery issues requiring immediate attention and huge financial resources; mechanisms for medium- and longer-term transformative change are going to be needed to meet the full range of emergency recovery—linking environmental, health, economic, and globalization issues. Where will ecology and biodiversity fit into this complex agenda? The question is especially critical at the start of a decade-long effort to accelerate progress on green economy and development, plus transformative change on the UN 2030 SDGs. It is likely that the COVID-19 emergency and recovery will take place over a prolonged period, and therefore need to be taken into account as major factors for environmental and development action throughout the world.

A working paper regarding the biodiversity, ecological, and environmental implications of COVID-19 (*Post 2020 Biosecurity: Global Emergency to Ecological Civilization*) has been prepared for CCICED as a working paper from the Biodiversity SPS. It takes a global perspective, but with special attention given to China's situation and needs. The paper covers some of the main perspectives and scientific views, drawing from a review of valuable concepts and knowledge generated primarily over the last 10–20 years. Also covered are some of the urgent efforts underway at present on scientific, socioeconomic, and policy matters. The latest draft document was completed in mid-July 2020 and submitted to the CCICED Secretariat and MEE. Some of the findings have been incorporated into the recommendations in this SPS report. The document is available in Chinese and English language versions on the CCICED-IISD website: https://cciced.eco/wp-con tent/uploads/2020/07/cciced-2020-cn-post-2020-biosecurity-global-emergency-to-ecological-civilization.pdf; https://cciced.eco/wp-content/uploads/2020/07/cciced-2020-en-post-2020-biosecurity-global-emergency-to-ecological-civilization.pdf.

As well, a brief on *Biodiversity and Pandemic Risk Reduction* (see Annex 3) provides a number of important considerations related to animal health and disease as they relate to humans, guidelines on what needs to be improved, and several recommendations.

2.5 Recommendations

Six recommendations are provided. These focus on enhancing successful outcomes for the CBD COP 15 to be held in 2021, including both multilateral and national approaches. They draw upon analysis in the current report and previous SPS reports submitted in 2018 and 2019. In this pandemic time it is important to recognize the potential opportunities of future COVID-19 recovery efforts now being proposed. These are still at an early stage and require further attention over the coming months. Recommendation 6 regarding ecological conservation suggestions for China's 14th FYP is included as a contribution to CCICED's input to the State Council.

2.5.1 China's Global Leadership and Engagement

The global effort required in order to address existing trends of massive biodiversity loss needs strong political leadership. Year 2020, and now also 2021 provides a clear opportunity for humanity to better protect our natural capital. China, as the host country for the CBD COP 15, needs to:

- Play a strong leadership role by engaging with the world leaders to send strong political signals that the international community and national governments (taking a whole-of-government and whole-of-society approach) must realign our relationship with nature, taking actions domestically and globally to bend the biodiversity conservation curve even as they work to flatten the COVID-19 curve. This leadership can take the form of China's top leaders bringing these messages and commitment to the UNGA and heads of state (HoS) Biodiversity Summit and G20 Summit;
- Consider having an HoS segment prior to the CBD COP 15 to ensure that the strong political will and signal can be baked into the CBD negotiations for the post-2020 Global Biodiversity Framework;
- Engage with key parties of the CBD COP 15 to conduct green diplomacy to better understand core concerns, seek implementable agreements, bridge diverging views, and propose ways forward for the world to achieve an ambitious post 2020 GBF.

With the impact of COVID-19, the world we know now will change beyond our imagination. China can emerge as an enabling world leader. The process and platform for agreeing on a robust post-2020 GBF is an unmissable opportunity for China to show green leadership in its efforts with concepts and practices based on over 5000 years of its civilization—and now for its own and global green and sustainable development. Well-organized international communication through virtual meetings is becoming the new normal for both information exchange and negotiation. This is a tremendous opportunity for COP 15 preparations.

2.5.2 Recommendation for the Post-2020 Global Biodiversity Framework

The following seven points deserve careful attention as key matters for inclusion in the framework, though some may be controversial.

Global Apex Goal: To develop a motivational, communicable, science-based, and measurable apex global goal for a nature action agenda to stop losing and start restoring nature and biodiversity, such that by 2030, biodiversity and nature are recovering at a global scale and by 2050 they will be fully recovered and restored.

Transformative change: Exert the role of ecological civilization to centre conservation strategies and action plans within the scope of the post-2020 GBF.

Common but differentiated responsibilities: Prioritized conservation goals/targets of individual parties according to their national conditions, in particular to differentiating responsibilities for developing and developed parties.

National voluntary commitments (NVCs) and National Biodiversity Strategy and Action Plans (NBSAPs): All parties and stakeholders could develop and publicly present their own voluntary biodiversity commitments (if appropriate) that are integrated into or in addition to their NBSAPs with the aim to support and increase the level of ambition needed to achieve 2030 Mission and 2050 Vision. Ensure parties and stakeholders, if appropriate, to reflect or integrate the goals and targets of post-2020 GBF in their own NBSAPs.

Protected and conservation area and baseline for progress assessment: Accept the *Three Global Conditions* approach in protected areas design and target application by considering differentiated needs for different categories of habitats (Farms and Cities, Shared Landscapes, and Large Wild Areas). All three categories should be managed sustainably, according to criteria differentiated by category and region. The baseline of 2020 might serve as the reference point for zero net loss of nature and biodiversity.

Innovative multilateral financing mechanism: Develop a multilateral financing mechanism for biodiversity and NBSs. Initiated by China (perhaps in collaboration with the hosts of the UNFCCC COP26 to ensure alignment with the climate agenda). The mechanism would finance the strengthening of national policy frameworks for biodiversity conservation and restoration, including integrated land-use planning, and finance initiatives in support of countries' commitments under the CBD and UNFCCC. It would invite other countries to join and seek to leverage private financing.

International collaboration and technology transfer: Develop better global strategies for international collaboration and technology transfer, and for providing training to assist in capacity building for implementation of the established NBSAPs in developing countries.

2.5.3 Proposal for a Multilateral Nature-Based Solutions Fund Initiated by China to Be Put Forward at CBD COP 15 and UNFCCC COP 26

All parties realize and agree during international consultation and negotiation that financial resources are the key guarantee to fulfill the goals and targets of the post-2020 global biodiversity framework to be negotiated at the COP 15 in Kunming. As emphasized throughout this document, this framework will be intrinsically linked with the Paris Agreement. Financing NBSs is central to achieving the objectives of the Rio Conventions—i.e., the CBD, CCD, and UNFCCC—so financial resources must be mobilized and deployed with a view toward meeting the objectives of both conventions.

It has been said that the annual funding gap for biodiversity conservation and restoration may approach half a trillion USD. Much of the funding will have to come from private investors, but public sector sources are also needed, and removing environmentally harmful subsidies could be a potential source. As past host countries, Japan and South Korea had previously established two funds to support global biodiversity conservation, especially for developing countries, at COP 10 and COP 12, respectively. The Green Climate Fund and the Global Environment Facility are two multilateral financing mechanisms.

In order to facilitate the implementation of the post-2020 global biodiversity framework, we propose that China's government initiate a multilateral fund for nature-based solutions ("NBS Fund") and invite other countries to join. The objective of the NBS Fund will be to enhance the implementation of agreed global conservation and restoration goals for biodiversity and to promote other nature-based solutions in support of the CBD, the CCD, and the UNFCCC. Special focus will be given to support NBSs in developing countries.

Drawing on lessons from China and many other countries, the NBS Fund will support policy and project actions concurrently:

- At the policy level, the NBS Fund will support countries in strengthening integrated national policy frameworks for biodiversity conservation and restoration, along with other nature-based solutions. Among other things, this will include support for land-use zoning and management frameworks drawing on lessons from China's ECR and similar policy frameworks in other countries. International resources, such as the Nature Map, can support such work.
- At the project level, the NBS Fund will co-finance large-scale biodiversity conservation and restoration initiatives and other nature-based solutions. To ensure its long-term success and alignment with other development priorities of recipient countries, project support from the NBS Fund will be closely aligned with policy support.

To be successful, the NBS Fund should be structured to pursue two key objectives. First, it will aim to mobilize maximum resources from public and private sources. To this end, the NBS Fund will be designed with maximum transparency and a shared governance model that ensures effective operations and broad buy-in. A particular focus will be on enabling private donors to support project activities in the context of improved national policy frameworks. This can be achieved through transparent co-investment modalities, as exist today, for example, for the Global Fund to Fight AIDS, Tuberculosis, and Malaria ("Global Fund"). Similarly, the Asian Infrastructure Investment Bank (AIIB) has, under Chinese leadership, been successful in attracting financing from a large number of partners. Such successes must be replicated for the NBS Fund.

Second, the NBS Fund aims to address both financing and implementation gaps in interested countries. In addition to more financing, we need greater clarity on the policy tools and project mechanisms that can best meet the objectives of the CBD and the UNFCCC. This will require unprecedented innovation and learning. Lessons from other sectors suggest that such innovation and learning can be fostered by combining country-leadership (i.e., recipient countries take the lead in developing funding proposals) and rigorous, independent, technical review of proposals (to avoid any political conditionalities and to ensure that the best, technically sound proposals are financed). The Global Fund has pioneered these governance principles, which have generated tremendous successes in fighting the three major infectious diseases.

The initial scale of the fund should be at least USD 10 billion with periodic replenishment mechanisms contributed by parties, public and private sectors as well as global finance enterprises. China can work with various parties such as Canada, France, Germany, Norway, Switzerland, and the UK, and other countries who might have the willingness to join and donate, as well as other enterprises and stakeholders, to advance the design and implementation of the NBS Fund, including its governance principles.

In light of recovery from COVID-19 pandemic and economic crashing, China could propose that a portion from countries' stimulus packages be allocated to help prevent zoonotic diseases and enhance the protection and sustainable use of wildlife, which will subsequently contribute to the conservation of biodiversity and nature.

In addition, regarding other resource mobilization to enhance and facilitate fundraising, we propose including the following key elements as comprehensive resource mobilization components of the post-2020 framework:

- Redirecting and aligning all public and private financial flows to be in line with pathways directed toward halting ecosystem degradation and restoring nature/biodiversity.
- Defining funding needs and targets for mobilizing additional financial resources to achieve the goals and targets of the post-2020 framework.

In addition, we recommend to parties as elements of the long-term approach on mainstreaming of biodiversity into the financial sector:

- Support the creation of a financial sector Task Force on Nature-Related Financial Disclosures (TNFD), to support financial institutions and businesses to measure and disclose their nature-related risks and impacts.

- Support a Global Natural Resource Initiative that would encourage countries to take responsibility for their environmental impacts on other countries through import and consumption policies and investments.

2.5.4 Recommendations on Improving and Popularizing China's Main Ecological Protection Practices and Experiences Sharing

The considerable efforts of China to protect, improve, and restore its natural areas and their biodiversity; to respect the integrity of ecosystems and their services; and to do so in ways that provide economic benefits especially for rural people will need to be strengthened in the coming 14th FYP and later periods. Certainly, such ongoing efforts are of considerable interest for audiences both within and outside of China. At COP 15 they should be highlighted and used to demonstrate not only how challenges can be met but also how new opportunities for economic and social well-being emerge. The innovation of China's ECRs is particularly important.

2.5.4.1 Integrating ECR to NBSs for Climate Change Adaptation and Mitigation

"Nature-based solution" is an effective approach to addressing climate change. Delimiting the ECR is not only conducive to enhance ecosystem stability and resilience, and adaptability to climate change, but also enhances the carbon sequestration function of the ecosystem, thereby mitigating climate change impacts [24]. In September 2019, on the climate change summit of the 74th UN General Assembly, the Chinese government submitted to the General Assembly the action initiative of "delimiting ECR, to mitigate and adapt to climate change—nature-based solutions." According to this initiative, existing practices have proved that the designation of protected areas by ECR can achieve "greater carbon sequestration services provided by a smaller area."

We suggest that the Chinese government should further promote the implementation of the initiative. It can invite improved synergies among the parties of the UN Convention on Climate Change, the Convention on Biological Diversity, the Convention to Combat Desertification and other international organizations, non-governmental organizations and the private sector involved in biodiversity conservation. China can call on all parties to take active actions to integrate ECR into the nature-based solution to adapt to climate change. Thus, it can provide schemes for achieving the goal of the Convention on Climate Change and the Convention on Biological Diversity, make positive contributions to global climate change and the achievement of the objectives of the post Biodiversity Convention.

2.5.4.2 Introducing Important Carbon Sink Ecological Function Areas to Improve the Method and Result of ECR Delimitation

The ECR protects areas with important ecological services, including water conservation, soil conservation, wind protection, and sand fixation, as well as areas with ecological fragility, including soil erosion, land desertification, and rock desertification, which basically covers areas providing important ecological functions. Nevertheless, the delimitation method does not consider the carbon sink function, resulting in a situation where some important areas are not covered by ECR [25]. According to the current ECR delimitation results, only about 45% of the important carbon sink ecological function areas are covered in the protection scope, which is relatively low. In addition, marine and coastal carbon sinks can be enhanced through ecological redlining along mud banks, in mangrove areas, and various offshore marine reserve areas.

The carbon sink has as an essential climate change mitigate function and offers a natural response to climate change, from both international and domestic perspectives [26]. Therefore, we recommend that, in the future delimitation of the ECR, research and establish the delimitation method based on carbon sequestration, in order to incorporate important carbon sink areas into protection status, and further improve the delimitation results accordingly. The improved method and result could contribute to address climate change and advance the United Nations Framework Convention on Climate Change.

2.5.4.3 Integrating ECR into Green Belt and Road Initiatives (GBRI) to Prevent Ecological Damage Caused by Development Activities from Happening in the First Place

Most BRI countries are emerging economies and are at a critical stage in their development. Therefore, to a certain extent, it is also a critical period of balancing development and ecological protection for those countries. If a large number of infrastructure projects are green, not only will they promote economic development, but also by their design, they can protect—instead of damaging—the environment. Therefore, preplanning to protect important ecosystems is an important means of avoiding ecological disruption, and the delimitation of ECR can solve this problem while reducing the ecological footprint of BRI countries.

Delimiting ECR is an important component of NBSs, and it has been highly recognized by the international community. It features China's experience for improving PA systems proposed by the World Conservation Union (IUCN), and also provides China's scheme for fulfilling the two major conventions on biodiversity and climate change. In recent years, China has delimited and implemented its ECR system nationwide, and it has played an important role in protecting biodiversity, maintaining important ecosystem services, ensuring the safety of human settlements, mitigating the impacts of climate change, promoting the sustainable development of both the economy and society [27].

Therefore, BRI countries can avoid the development of important biodiversity areas, especially nature reserves, virgin forests, and other regions with high biodiversity and uniqueness, by delimiting the ECR. Therefore, we recommend promoting the experience and practice of China's ECR to BRI countries, encouraging BRI countries to develop ECR-based policy frameworks and submit such policy frameworks as countries' national strategies under the CBD and UNFCCC, in order to jointly establish an effective ecological protection network.

Specific measures can be taken in three steps: first, China can establish an expert group both at home and abroad to guide BRI countries to delimit their ECR. Second, carry out training for natural protection personnel and relevant management personnel in the BRI countries. Third, launch ECR delimitation work within the BRI countries, and with other countries that have decided not to participate in the BRI but who want to support better biodiversity and climate policies, making this approach an important element of their ecological protection policy.

2.5.4.4 Promoting Chinese Ecological Restoration Concepts and Practices to Increase Integrity and Connectivity of Habitat

Ecological restoration is an important measure for addressing climate change and biodiversity loss, improving ecosystem structure and function, and consolidating national ecological security in China. Over the past few decades, China has planned and implemented a series of ecological restoration projects, including the "Grain to Green" project and natural forest protection projects, which have achieved positive results and played an important role in guaranteeing regional ecological security and sustainable development. However, these previous projects have focused on specific ecological problems at local scale, rather than being systemic and integral in design and implementation, resulting in insufficient optimization and improvement of the entire ecosystem.

Thus, since 2016, China has initiated the ecological restoration practices with greater attention to linkages: mountains-rivers-forests-farmlands-lakes-grasslands. The Ministry of Finance, the former Ministry of Land and Resources, and the former Ministry of Environmental Protection jointly issued the "Notice on Promoting the Ecological Protection and Restoration about these areas. It clearly implemented the pilot work on ecological restoration. This document clearly stipulates that all localities "must uphold the idea of respecting, conforming and protecting of nature, guide ecological restoration practice with the theory of 'mountains-rivers-forests-farmlands-lakes-grasslands' as a life community," fully integrate the funding and policy to launch the overall protection, system restoration, and comprehensive management to up and down the mountain, above and below ground, land and ocean, and upstream and downstream of the basin.

As shown by the practice of restoration, this approach has important effects on improving habitat integrity and connectivity, scientifically responding to climate change, and mitigating habitat fragmentation caused by human development activities. Therefore, the Chinese government could actively promote China's concepts and practices of ecological restoration to the international community, especially to the BRI countries, in order to form a more complete regional or global ecological protection network.

2.5.4.5 Establishing Global Ecological Protection and Risk Early Warning Mechanisms to Protect the Common Interests of All Countries

In recent decades, globalization appeared to be an almost inevitable trend of world development. While it brings development opportunities to countries, globalization can also increase environmental and ecological risks. Therefore, it is necessary to establish global risk early warning mechanisms that include ecological protection. In recent years, many ecological disasters have occurred in the world, seriously threatening regional and even global ecological security, such as the forest fires in Australia, coral reef destruction in Southeast Asia, and various disease outbreaks, including the COVID-19 pandemic.

With the deepening of globalization, the ecological destruction that occurs in a country will inevitably affect the neighbouring countries and even the world. Therefore, we recommend that China and/or the Euro-American countries should take the lead in establishing global ecological protection and risk protection with early warning mechanisms to regularly report the ecological protection status of every country in the world, especially major ecological destruction events; organize global experts on ecological protection to build early warning models and methods for different ecological destruction, and gradually form early warning mechanisms covering all countries.

2.5.5 While Addressing the COVID-19 Pandemic Emergency, and National and Global Economic Recovery, Ensure that Adequate Attention and Financial Support Is Given to Addressing the Eco-Environmental Emergencies Affecting Biodiversity and Climate Change

We have little choice nationally and globally but to ensure that existing investments for environment and development are protected and enhanced in the years ahead. Therefore we must shift our thinking to a strategy consistent with a "Super-Decade of the Environment."

COVID-19 economic recovery plans should not backtrack to an unsustainable state of economic and environmental affairs. It should be a time for sparking innovation on many fronts. Plans should consider how to live within planetary boundaries, and comprehensively reduce environmental risks, including those to both human health and ecosystem health. This is essential if we are to reach a new level of global biosecurity involving all forms of life and to meet the goals of sustainable development. Only then will we achieve the vision of "Harmony between People and Nature" and, as noted in the CBD COP 15, the theme of *Ecological Civilization: Building a Shared Future for All Life on Earth.*

Our post-pandemic recommendations for ecology and environment cover distinctive time frames: China's 14th FYP (2021–2025); medium-term plans (2020 to 2030/2035) covering target periods such as UN SDGs and China's efforts for having a basic national ecological civilization in place; and longer-term to 2050 consistent with various targets related to decarbonization, full realization of biodiversity recovery, and China's ambitions for a prosperous society and a "Beautiful China."

2.5.5.1 During the 14th FYP Period, Significantly Reduce the Level of Environmental and Ecological Risks that Can Lead to Human, Plant, or Animal Disease Outbreaks, Epidemics, or Pandemics

The "One Health" approach linking the health of ecosystems, plant and animal health, and public health should be more strongly supported in China. This will require taking an integrated approach to preventing disease outbreaks. Also, adequate screening of health risks as part of environmental assessments, green development initiatives, and in any stimulus packages with components that seriously increase pollution and greenhouse gases, and disrupt intact ecosystems. Biosecurity for agriculture and animal husbandry requires ongoing review and major improvements.

Scientific research and monitoring need to be greatly improved, especially of disease passed from animals to humans (zoonoses), involving either domesticated animals or wildlife. The recent law with a ban on hunting, the possible prohibition of wet markets, and the revoking of many licences for wildlife husbandry will help to reduce probability of future cross-species disease outbreaks if strictly enforced but is still not complete enough to reduce risk sufficiently.

The Traditional Chinese Medicine (TCM) exemption from the new law intended to reduce wildlife commerce will weaken the effort to reduce the threat of new disease outbreaks in a number of ways. Therefore, establish a "nature-friendly" 21st Century approach. Various TCM products require attention regarding ecological impacts related to their sourcing. Also, whether rising demand might be met in different ways, for example, by advanced biotechnology applications for animal tissue culture, and substitution strategies to avoid threatened and endangered species such as the pangolin.

2.5.5.2 Modifying the Wildlife Conservation Laws to Improve the Capability of Biological Safety Risks Prevention and Control

The novel coronavirus that is currently ravaging the world once again warns the world that protecting wild animals means protecting all human beings. China's Wildlife

Protection Law was revised once in 2016, establishing the principles of protection priority, standardized use, and strict management. It strictly regulates all aspects of hunting, trading, utilization, transportation, and consumption of wild animals: in particular, a series of scientific and rational systems have been established in response to prominent problems such as overeating wild animals. With the implementation of the revised law, the protection of wildlife has improved. However, in various aspects, there are still some problems. It is necessary to further supplement and improve the Wildlife Protection Law in order to increase the intensity of cracking down and punishing the eating of many forms of wildlife.

Especially after the outbreak of COVID-19, it is urgent to integrate biosafety into the national security system, and systematically plan the construction of biosafety risk prevention and control (along with relevant governance systems) in order to improve national biosafety governance capabilities. It is important to introduce biosafety laws as soon as possible, initiate the revision of the Wildlife Protection Law, and accelerate the establishment of a biosafety legal and regulatory system and an institutional support system.

2.5.5.3 Strengthen China's Commitment to Building Ecological Resilience as a Medium- and Long-Term Transformative Approach Toward National Biosecurity

China's significant investment in ecological construction of forests, grasslands, and wetlands, improvement in the management of parks and nature reserves, and integrated management of river basins and coastal areas should be strengthened by setting site-specific ecological resilience goals throughout the country. These can be related to specific needs related to human, plant, and animal health, ecological services, or other needs such as strengthening ecological corridors used as migration routes.

Use ecological redlining as a key mechanism to reduce ecosystem disruptions that are an important factor in disease outbreaks. This would be a means of ensuring full ecological restoration of damaged habitats and maintenance of high biodiversity. Develop criteria related to specific health-related needs for use in determining the location and management of redlined areas.

2.5.5.4 Establish and Lock in New Baseline or Reference Levels of Pollution, Taking into Account Air, Water, Soil and Perhaps Other Forms of Pollution Reduction Experienced During the Current Coronavirus Pandemic

Evidence is mounting that the economic downturn and health measures related to the COVID-19 pandemic have significantly improved air and water quality, effects of noise pollution, etc., with favourable public reaction. This has been observed not only in China but also in Europe and elsewhere. Every effort should be made to protect these gains, in some cases making the reductions the "new normal" and seeking transformative objectives in stimulus packages where necessary. This is a one-time opportunity to accelerate environmental quality progress, starting now but with a cascading effect that can last into the medium- and longer-term periods. In order to get the full positive impact, it may be necessary to include tailored green incentives within the stimulus packages.

2.5.5.5 Ensure Economic Stimulus Packages Support Green Development and Protection of Nature. Also, Do Not Relax Environmental and Ecological Standards either Nationally or in Areas Hard Hit by the Disease Outbreak. If Necessary, Provide Subsidies or Other Incentives on A Temporary Basis. Green Stimulus Packages Specifically Aimed at Biodiversity or Climate Change Should Generally Be Longer Term (5–15 Years) and Dovetail with the Short-Term Efforts on Economic Recovery

There are several points to consider for national-level stimulus packages for economic recovery from COVID-19:

- Screening criteria need to be considered for all recovery projects to avoid environmentally damaging investments.
- Focus greater attention on green infrastructure, decarbonization efforts, further stimulus for the transition to renewable energy, public transportation involving transition to electric buses, etc.
- Green employment in various sectors and improved eco-compensation packages. Attention should be given to vulnerable groups in society along with gender issues.
- Maintain green development incentives that enhance ecological services.

2.6 Recommendations on Ecological Conservation and Restoration for the 14th FYP for China's National Economic and Social Development

The 14th FYP period will be key years for China to build a moderately prosperous society in an all-round way, and also the key period for ecological protection. Therefore we suggest the following recommendations on ecological conservation and restoration for the 14th FYP.

2.6.1   Add Ecological Indicators in the Indicator Session of the Plan

The protection rate of key ecological space should be taken as one of the indicators in the plan. According to the delimitation of the national ecological conservation redline, it is suggested that the protection rate of ecological space should be set at 32% in the 14th FYP.

2.6.2   In View of the Chapter "Strengthening Ecological Protection and Restoration," It Is Suggested to Add or Further Emphasize the Following Contents

2.6.2.1   Strengthen Biodiversity Conservation and Take It as An Important Part of Ecological Protection

- For conservation targets, pay attention to the protection of different types of habitats, including farmlands and cities, shared landscapes, and large areas of wilderness.
- For conservation approaches, pay special attention to the synergy between biodiversity and climate change, as well as the coupling effect of biodiversity and green development.

2.6.2.2   Ecological Conservation Redline Should Be a Critical Part of Ecological Protection and Restoration. Detailed Suggestions Include

- Monitoring, evaluation, and early warning of ecological conservation redline should be incorporated into the plan as key parts.
- Integrate ECR and NBSs to achieve synergy between climate change and biodiversity.
- It is suggested that the Chinese government should share the concept and approaches of ECR with the international community, especially BRI countries, so as to improve the global ecological conservation network.

2.6.2.3   The Construction of Ecological Corridors and Optimization of Homeland Ecological Security Should Be Included in the 14th FYP

In the report of the 19th National Congress, it has been clearly proposed to build ecological corridors. Therefore, we suggest adding relevant content in the 14th FYP, such as constructing ecological corridors based on ECR and protected areas, to build an efficient and stable ecological security network and enhance ecological integrity and connectivity.

2.6.2.4 Strengthen Wildlife Conservation and Risk Control, Including

- Strengthen the prohibition of the illegal wildlife trade, strictly prohibit the habit of eating wild animals, and maintain human health and biological safety.
- Protect wildlife, reduce the risk of human or animal disease outbreaks from the source, and control the probability of environmental and ecological risk of epidemic or pandemic.
- Pay attention to the concept of public health, strengthen the concept of "One Health" which connects ecosystem health, animal and plant health, and public health.

2.6.2.5 Continue to Implement the Major Ecological Restoration Projects of the Mountain-River-Forest-Farmland-Lake-Grass System

Ecological restoration is an important measure for China to address climate change and biodiversity loss, improve the structure and function of ecosystems, and consolidate national ecological security. During the 13th FYP and for a long time in the past, China has implemented a series of ecological restoration projects, such as returning farmland to forest projects, natural forest resource protection projects, etc. These projects have had positive effects and played an important role in ensuring regional ecological security and sustainable development. In particular, the Mountain-River-Forest-Farmland-Lake-Grass ecological restoration practices since 2016 have played an important role in systematically restoring large-scale ecological environment. Hence, we recommend that these projects should be continued in the 14th FYP.

2.6.3 We Suggest Strengthening Ecological Protection Projects Regarding Major Projects, Including

- Investigation, monitoring, evaluation, and early warning projects of ecological conservation redline.
- Ecological corridor and biodiversity conservation network construction projects.
- Mountain-River-Forest-Farmland-Lake-Grass ecological restoration projects.
- Wildlife protection and risk control projects.
- Ecological protection, restoration, and monitoring capacity-building projects.

Appendix

Annex 1: Momentum Built for Actions on Nature and People

1.1   The Sharm El-Sheikh to Kunming Action Agenda for Nature and People

Contributed by Marcel T. J. Kok of PBL Netherlands Environmental Assessment Agency, Bezuidenhoutseweg 30, 2594 AV The Hague, The Netherlands.

The Sharm El-Sheikh to Kunming Action Agenda for Nature and People, was launched by China, Egypt and the secretariat of the CBD with the aim to raise public awareness about the need to halt biodiversity loss and to restore nature; to inspire and help implement NBSs to meet key global challenges; to catalyze cooperative initiatives across sectors and stakeholders in support of the global biodiversity goals. The CBD Secretariat has subsequently established a platform on its website where commitments can be registered.

So far only a few commitments have been registered and many of them already existed before COP 14. At the same time many new coalitions are emerging in business, cities, and local communities, suggesting that indeed a groundswell of action might be emerging. China could tap in on these coalitions, networks and non-state commitments at COP-15 and beyond. Furthermore, the Action Agendas that have been set up in other policy domains like oceans, SDGs and climate, are also highly relevant for showing a growing momentum for biodiversity and showing interlinked efforts in other policy domains.

In the run up to COP 15 the Action Agenda could help building confidence for governments to be ambitious at COP 15, knowing that multiple non-state and subnational actors support stronger action. The Action Agenda as it currently stands does require leadership from China together with other leading countries in cooperation with leaders of societal coalitions to support the further development of the Action Agenda, to show the groundswell of action and further catalyze the momentum.

Here political and practical support from China is important. Some suggestions for this would be: to develop narrative Action Agenda and make clear where commitments need to be about; to link the Action Agenda for Nature and People to other action agendas and highlight biodiversity commitments made in other policy domains; generate and showcase as many new commitments as possible; focus on actors currently less/not involved such as financial sector, agrobusiness, landscape and supply chain initiatives.

The longer-term challenge is to establish credible and ambitious commitments beyond 2020, and this requires establishing a system for measuring and reporting progress as part of the broader accountability framework for CBD Parties in the post-2020 GBF.

One important step is to make the Action Agenda becomes an integral part of the post-2020 GBF, as this will provide some clarity about the long-term direction for stronger non-state involvement in the CBD. A clear signal by China about this and encouraging non-state actors to participate and contribute might be important.

1.2   The Nature Champions Summit, Montreal

We **Nature Champions** gathered in Montreal are beginning a global mobilization with this **Call to Action**, jointly committing to take a different, better path that puts nature first, recognizing it is the context for all life including human life and protecting it accordingly. Collaboratively, we the Nature Champions commit to place nature's needs at the heart of all global agendas including:

- Recognizing the fundamental link between nature, a stable climate, human well-being, and sustainable development for all;
- Uniting nature conservation objectives with addressing climate change and developing NBSs that are effective for both;
- Promoting an ambitious set of new targets for the UN Convention on Biodiversity (CBD) that has clear and measurable objectives for 2030 and effectively enables the world to reach the 2050 Vision of Living in Harmony with Nature;
- Widening the participation in the Convention on Biological Diversity Strategic Plan beyond governments to include commitments and actions by a wide range of actors;
- Addressing nature's needs by increasing the proportion of land and ocean that we protect and conserve around the world and improve the way we manage and restore it;
- Addressing the key drivers of nature loss across the world by enhancing concrete action on:
  - reducing habitat-loss and deforestation;
  - curbing terrestrial and marine pollution; and
  - developing and strengthening sustainable supply and value chain management;
- Embracing nature-based decision making in all key political, economic, cultural, and social decisions;
- Increasing investment in nature conservation and leveraging existing commitments to mobilize new resources; and
- Recognizing and enhancing the role of subnational governments, cities and other local authorities as well as of Indigenous peoples, local communities, women and youth in the protection of nature.

1.3   High-Ambition Coalition for Nature

This is an intergovernmental group championing a global deal for nature and people that can halt the accelerating loss of species, and protect vital ecosystems that are the source of our economic security. It is launched at the UNGA 74. It is co-lead by the government of France and Costa Rica. The High-Ambition Coalition (HAC) will utilize the upcoming UN Framework Convention on Climate Change (UNFCCC) COPs and the 15th meeting of the Conference of the Parties to the Convention on Biological Diversity in 2020, to push for ambitious, science-driven global action to safeguard nature and humanity's future.

As the HAC for Nature and People gathers momentum and grows it will work toward a global deal that includes the following key elements:

- Sustainable management. The entire planet must be managed sustainably with no net loss of natural habitats, supported by a circular economy, and managed for the sustainable and equitable sharing of benefits from nature.
- New spatial targets to protect biodiversity. There must be increased spatial targets to protect or effectively conserve at least 30% of the planet—land and sea—by 2030. Efforts should promote indigenous-led conservation and focus on areas most important for biodiversity. The resulting network of conserved areas should be ecologically representative, well-connected, effectively, equitably managed, and help to maintain species diversity.
- Improved management of existing protected areas. Management must be improved for the entire system of protected and conserved areas around the world. The best available science should be used and sufficient resources made available to deliver the desired conservation outcomes.
- Increased funding. Additional public funding and private financing from corporations and philanthropists must be mobilized to support the long-term management and local governance of protected and conserved areas around the world.
- Implementation: Once these pillars are adopted, the HAC needs to have a clear implementation mechanism that is gradually enforceable and can be incorporated into national development strategies and key economic sectors.

1.4 Trondheim Conference

The ninth Trondheim Conference on Biodiversity was held in Trondheim, Norway, from July 2 to 5 2019. Since 1993 Trondheim Conferences on Biodiversity have created opportunities for increasing understanding amongst stakeholders about issues on the biodiversity agenda. They allow those involved in setting the agenda to learn and to share views and experiences with their peers. The ninth Trondheim Conference brought together decision-makers and experts from around the world to learn about and discuss knowledge and know-how for the global post-2020 biodiversity framework. The Conference seeks to support the process established by the Convention on Biological Diversity for preparing this framework, with opportunities for major players to discuss key issues informally outside of the negotiation process.

This SPS has two team leads (Prof. Ma Keping and Dr. Li Lin) who took part in the conference. The SPS gained insights from interactions with actors at various related events. We have organized two dialogues with representatives from developing countries and with ministers from a few developed countries. The detailed notes can be found in Annexes 1 and 2.

We noticed both common issues of importance as well as divergence on some matters.

The issues of common concerns are:

- "Nature" is a better term than "biodiversity" when communicating to society.

- Manage space for conservation alone will not achieve conservation goals. Need to address quality of protected areas and more importantly drivers behind biodiversity loss.
- Ecosystem based adaptation and restoration is necessary.
- Climate change and biodiversity conservation are inseparable.
- Whole government and whole society approach is needed.
- The importance of CBD needs to be elevated.
- Local communities and indigenous knowledge should be considered in policy discussions.
- Strong and effective implementation is essential in delivering the needed outcomes.

The matters of divergences or non-agreement are:

- Conservation burdens are not shared equally among countries.
- Funding and coordination from developed countries are expected.
- Approaches and responsibilities are different between countries.
- Capacity building and knowledge sharing are needed.

Our general impression is that

(1)   Some developing countries are proactive in taking the actions in their own country's hand, while some others felt hands tied
(2)   Developing countries should take the leadership
(3)   ASEAN Center of Biodiversity is interested in exchange and cooperation with China, linked the ecological conservation redlines. ASEAN Environmental summit in March 2020
(4)   How can we bring ministries of environment and finance together to discuss nature
(5)   Asking CCICED to consider how to support developing countries on post-2020 biodiversity framework development and conservation
(6)   Utilizing the global and national commitment on SDGs to enhance the role of nature and to mobilize in country other more powerful ministries and players.

These are the issues that China needs to pay attention to while engaging with global actors.

Annex 2: Evidence for Decision Making

2.1   Food and Land Use Report. Ten Critical Transitions to Transform Food and Land Use

Launched in Oct. 2019, the Food and Land Use Coalition (FOLU) report is the first to assess the benefits of transforming global food and land-use systems as well as costs of inaction, reveals benefits that far outweigh the costs and proposes actionable solutions, many already in existence. The report calls on global leaders to act now and advance the economic case for change.

It is estimated that the ways in which people produce and consume food and use land currently account for USD 12 trillion a year in hidden costs to the environment, human health and development, costs that are set to rise to USD 16 trillion by 2050 if current trends continue.

The report discloses benefits that far outweigh the costs, proposing a concrete reform agenda centered around ten critical transitions. These stand to unlock USD 4.5 trillion in new business opportunities each year by 2030, at the same time as saving costs of USD 5.7 trillion a year in damage to people and the planet by 2030, more than 15 times the investment cost of up to USD 350 billion a year.

The report's ten transitions include—but are not limited to—measures to protect and restore nature and climate, empower and protect indigenous communities, finance NBSs, promote a diverse and healthy diet, reduce waste, and strengthen rural economies.

The report calls for collective action to unlock the potential for better food and land-use systems, including through policy reform, country-led action and individual engagement in support of the critical transitions. Many solutions are already in existence, in need of support and funding to scale up.

2.2 The Global Risks Report 2020

Launched in Jan. 2020, this is a WEF report that receivedinput from over 750 global experts and decision-makers to rank their biggest concerns in terms of likelihood and impact of global risks. For the first time in the survey's 10-year outlook, **the top five global risks in terms of likelihood are all environmental**. The report forecasts a year of increased domestic and international divisions and economic slowdown. Geopolitical turbulence is propelling us toward an "unsettled" unilateral world of great power rivalries at a time when business and government leaders must focus urgently on working together to tackle shared risks.

This would prove catastrophic, particularly for addressing urgent challenges like the climate crisis, biodiversity loss and record species decline. The report points to a need for policy-makers to match targets for protecting the Earth with ones for boosting economies—and for companies to avoid the risks of potentially disastrous future losses by adjusting to science-based targets.

The report sounds the alarm on:

- Extreme weather events with major damage to property, infrastructure and loss of human life.
- Failure of climate-change mitigation and adaptation by governments and businesses.
- Human-made environmental damage and disasters, including environmental crime, such as oil spills, and radioactive contamination.
- Major biodiversity loss and ecosystem collapse (terrestrial or marine) with irreversible consequences for the environment, resulting in severely depleted resources for humankind as well as industries.
- Major natural disasters such as earthquakes, tsunamis, volcanic eruptions, and geomagnetic storms.

It adds that unless stakeholders adapt to "today's epochal power-shift" and geopolitical turbulence—while still preparing for the future—time will run out to address some of the most pressing economic, environmental and technological challenges. This signals where action by business and policy-makers is most needed.

2.3 New Nature Economy Report

WEF is planning a New Nature Economy (NNE) report series that will make the case for why the nature crisis is crucial to business and the economy; identify a set of priority socioeconomic systems for transformation; and scope the market and investment opportunities for NBSs to environmental challenges.

The first report discusses the nature emergency, the risks of nature loss to businesses, how to manage these risks and how to take action.

2.4 Nature Risk Rising: Why the Crisis Engulfing Nature Matters for Business and the Economy

This is the report that WEF jointly produced with PwC. Launched in Jan. 2020, it is the first report in the NNE report series. It explains how nature-related risks matter to business, why they must be urgently mainstreamed into risk management strategies and why it is vital to prioritize the protection of nature's assets and services within the broader global economic growth agenda.

The report notes that "every industry sector has some degree of direct and indirect dependency on nature" and identify nature loss as a "fat-tail risk like the 2008 asset-price bubble." They point out that pursuing a "nature-positive way of doing business" can mitigate future economic and societal shocks.

The report discusses how risks to nature manifest as risks to business across all sectors. It highlights "that $44 trillion of economic value generation—more than half of the world's total GDP—is moderately or highly dependent on nature and its services and is therefore exposed to nature loss."

2.5 The Nature of Risk: A Framework for Understanding Nature-Related Risk to Business

Scientific consensus is building around risks to business from the loss and degradation of nature, or "nature-related risks." These risks are not adequately addressed by businesses, and to be addressed, they need to be considered together with climate-related risks. The two are inextricably interlinked because climate change drives change in nature, and change in nature drives climate change.

The terminology used in this report draws on both nature- and climate-related risk to facilitate a unified approach. This report and framework aim to catalyze the incorporation of nature-related risks into private sector decisions in a manner that facilitates sustainable development at all scales.

This WWF report launched in September 2019 includes:

(1) A literature summary of existing work on the topic that outlines how nature-related risk is not adequately accounted for by businesses.

(2) A synthesis framework for how nature-relate risk emerges that builds on the many existing frameworks and that brings together understanding of natural capital and climate-related risk.
(3) A typology based on analysis of existing literature which serves as a proxy for risks that are most widely acknowledged as high importance.
(4) A set of case studies—examples of businesses facing consequences due to nature-related risk.

2.6 Nature is Too Big to Fail—Biodiversity: The Next Frontier in Financial Risk Management

This is a report jointly launched by PwC Switzerland and WWF Switzerland in Jan. 2020. The report finds that the financial risks associated with the loss of biodiversity will become increasingly important in 2020—especially in the lead up to the United Nations Biodiversity Conference in October in Kunming (China).

As climate change and the loss of biodiversity mutually reinforce each other, decision-makers face a huge challenge to respond to this double crisis, as the risk of financial market instability significantly increases.

Loss of Biodiversity is an Unrecognized Environmental Risk

Climate change is a financial risk and recognized as such by a growing number of financial actors and regulators. A related but unrecognized environmental risk is the rapid loss of global biodiversity. Climate change further accelerates the extinction of species and leads to rapid changes in ecosystems. This in turn drastically limits natural carbon sequestration of ecosystems, which again worsens climate change. A negative spiraling loop, which has been until today almost completely ignored by decision-makers, the financial sector and their regulators.

"It is particularly dangerous for the financial sector to not account for biodiversity loss, as all economic sectors in which they invest, finance or insure depend on biodiversity. To avoid financial instability, we urge central banks and financial regulators to assess the financial risks stemming from environmental degradation more thoroughly, and to act accordingly," says Andreas Staubli, CEO of PwC Switzerland.

Thomas Vellacott, CEO of WWF Switzerland: "Biodiversity-related financial risks have not only been completely ignored by the financial sector but also by decision-makers globally. It is time to respond swiftly to the double crisis from biodiversity loss and climate change. Thus, humanity is in urgent need for a New Deal for People and Nature. All market, governmental and civil society actors are needed. Nature is too big to fail."

Four Biodiversity-Related Financial Risks Defined

The report suggests a typology of four financial biodiversity-related financial risks: physical, transition, litigation and systemic risks. The report further highlights what can be learned from the discussions around climate-related financial risks, provides a framework on how to integrate biodiversity losses into the classical risk framework of financial institutions and also suggests recommendations to financial regulators/central banks, financial market players and states/international organizations:

- States agree to an ambitious Global Biodiversity Framework at Kunming (China) in 2020 by bringing all financial flows in line with biodiversity conservation and restoration (Paris Agreement Obj. 2.1.c equivalent for biodiversity).
- The funding gap for biodiversity conservation and restoration of at least half a trillion US dollars per year needs to be closed by all actors rallying together.
- As biodiversity-related financial risks and the spiraling effect with climate change pose a systemic risk, all central banks and financial regulators need to emphasize the importance that the regulated entities regularly disclose their biodiversity-related financial risks. Furthermore, stress tests regarding biodiversity-related financial risks should be run regularly.
- A task force on Nature-Related Financial Disclosures should be created in 2020. It should drive standardized disclosure on nature-related risks, taking into consideration the physical, transition, litigation and systemic financial risks that stem from biodiversity loss.
- All financial actors should proactively manage biodiversity-related financial risks and seize and secure opportunities offered by ecosystem services (e.g., flood protection, pollination, clean water, fertile soils and adaptation to climate change).

2.7 Economic and Financial Systems and Tools to Develop Biodiversity Conservation

This report jointly published by WWF France AXA in May 2019, is on the financial risks associated with the loss of biodiversity, including *physical* risks (e.g., linked to supply shortage), *transition* risks (e.g., related to industrial or commercial developments) and *reputational* risks. This report reviews existing initiatives related to assessments of companies' impact on biodiversity, risks and opportunities for financial institutions and related reporting stakeholders. It identifies best practices and the most promising technical and political perspectives on these issues. It also proposes a roadmap to develop biodiversity finance commensurate with the current biodiversity crisis.

2.8 Studies in the Making

- Costs, Gaps, and Benefits to protecting 30% of the Planet, sponsored by Campaign for Nature
- Financing mechanisms to effectively finance the world's conservation needs, sponsored by Paulson Institute, The Nature Conservancy, and Cornell University
- The business and economic cases for nature, sponsored by WEF
- Independent global review on the economics of biodiversity and its relationship to economic growth, sponsored by UK government, Professor Sir Partha Dasgupta

- Biodiversity: Finance and the Economic and Business Case for Action, sponsored by OECD
- Natural capital valuation and analysis of protected areas in Africa, sponsored by German Federal Ministry for Economic Cooperation and Development (BMZ) and *Deutsche Gesellschaft für Internationale Zusammenarbeit* (GIZ English: German Corporation for International Cooperation)
- Update to the 'Little Biodiversity Finance Book: A simple guide to financing the Global Deal for Nature', sponsored by Global Canopy
- Resource mobilization and financial mechanisms for major CBD components: conservation, sustainable use and access/benefit sharing of genetic resources, sponsored by Convention on Biological Diversity
- Study and quantify how biodiversity lose may affect future economic growth, sponsored by the World Bank
- Living Planet Report 2020, WWF.

Annex 3: Biodiversity and Pandemic Risk Reduction
Contributed by Dr. Alice C. Hughes, a scientist at the Centre for Integrative Conservation, Xishuangbanna Tropical Botanical Garden.

Pandemics will continue to occur whilst we unsustainably use natural resources, thus better modes of governance for the use of wildlife and the maintenance of healthy ecosystems are needed. Linking ecological integrity to pandemic risk reduction is an important consideration for China's Ecological Civilization. No time can be more significant for doing so than now, given the unprecedented efforts and costs associated with the existing outbreaks. Also given China's hosting of the CBD COP 15 in October 2020, there are opportunities for promoting innovative approaches. While the Global Biodiversity Convention has recognized the need for linking health of people to planetary health, stronger actions must be put in place in line with the UN 2030 SDGs. Thus living in harmony has inherent benefits, in addition to being key to fulfilling the vision of eco-civilization and maintaining beautiful China, in addition to strictly adhering to international conventions including CMS, CBD and CITES.

Whereas domesticated animals may be kept in good conditions, well fed and frequently dosed with antibiotics and screened for infection, the same is not true for animals captured from the wild. Poor quality habitat and poor nutrition can both lead to immunosuppression in wildlife, and thus may be particularly likely in habitat patches surrounding human habitations (3). In addition whereas most livestock should be vaccinated, and screened and therefore have reduced exposure to pathogens, the same is not true for wild-caught animals, which especially in degraded habitats may also have larger ranges and higher levels of exposure to pathogens.

The ability of wild animals to carry diseases also varies dramatically by group. In recent years a number of pandemic diseases have emerged either from the direct consumption of wild animals or close contact between wild-caught and domestic animals for consumption, especially when hygiene standards are poor. Threat is at it's highest if animal products are eaten raw, thus the consumption of fresh blood or tissue has a very high chance of passing on diseases, including not only viruses such as corona but even prions and viriods which would be destroyed through the cooking of tissue.

Globally most pandemics in recent years started with the capture and normally consumption of wildlife, with bats and carnivores posing the greatest risk of sources or diseases with the potential to cross into humans, and should be consumed in no circumstances. Pandemics originating in these groups, or using them as an intermediate vector include SARs, MERs, Ebola, Nipah and now Covid-19 (among others). Of these Ebola is likely through direct consumption of bats, though some other mammals can also be carriers, Nipah originates in bat urine (normally through the consumption of toddy wine, which if left open bats can drink and urinate into), and the three coronaviruses; SARs, MERs and Covid-19; which likely originated in bats or civets and were transmitted into humans through civets or another intermediate host. Bats (especially Rhinolophids) and Civets show similar expression of the viral genes, and similar viral genomes, thus both have the potential to be a source for the viruses though the transmission route into humans is little known. Thus minimizing contact between people and these groups, and ensuring high-quality habitats to reduce the susceptibility, spread and any infection risk between wild animals and humans has multiple benefits in terms of enhanced service provision and decreased risks of diseases.

We recommend that a well-coordinated effort be initiated to break the link of zoonotic sources with human disease outbreaks. The effort should become a key part of China's shift to become an ecological civilization by 2035. It should be started immediately, fully implemented during the 14th FYP, and be continued with partners abroad and globally. These regulations include multilateral agreements on biosecurity and trade, and be a well-integrated component of the Belt and Road Initiative.

Ten core principals to ensure the protection of biodiversity and maintenance of biodiversity have been outlined in the Berlin Principles. In addition domestic (i.e., 2018 Law of the People's Republic of China on Wildlife Protection) and in international regulations need to be implemented based on common standards, definitions and defined and agreed upon reporting structures to provide the requisite information for monitoring and ensuring sustainable and safe trade as detailed below.

Recommendations which refer to "wild animals" or wildlife" applies to any mammal species that is not a captive bred ungulate or rabbit with small numbers of exceptions (detailed below) and not limited to the 342–408 animals currently recognized as nationally protected, or 981 provincially protected.

Ultimately reducing the risk of transmission of diseases from animals to humans has three major facets which act to reduce the risk of infection and spread of diseases in wild animals, reducing the risk of any diseases moving from wildlife into humans and the decreased risk of captive populations harbouring or passing on diseases. Definitions and further discussion behind these recommendations are provided after recommendations.

Reportage of violations of any provisions outlined below should be possible through a wechat mobile app to reduce the cost of reportage. Reduced ambiguity herein enables enforcement by making the sale and consumption of wild animals illegal (given that at least 70% of zoonoses are from wild animals), and standardizing these regulations across all provinces and counties of China maximizes safely and enforceability.

(1) Maintaining healthy native populations and minimizing risk of contagion

   (a) Natural areas should be redlined and a zero net loss in intact habitat aimed for to provide healthy habitats.
   (b) Prevent the destruction of caves, and decrease mining of karsts with known caves.
   (c) Hunting of wildlife should be limited to ungulates, based on licences and a quota, all reportage should be overseen by local police. Consumption for food of mammals other than ungulates or rabbits should be considered illegal.
   (d) Farming of rodents other than rabbits should also be limited except under very special circumstances due to disease risk. For other mammals only ungulates should be farmed for commercial purposes.
   (e) Wildmeat (deer, pig) if for sale at all should only be from licensed sellers and of species known to be unlikely to transmit diseases based on a quota and must be refrigerated and kept separate from other meat.
   (f) Prevent wild animals entering supply chains, wild-caught animals should at no time be openly for sale, or breeding stock (outside conservation programs within zoos or scientific institutes).
   (g) Wildlife markets should be closed to prevent contact between wildlife and humans. As this is challenging in border markets (particularly Mongla and to a lesser extent Botan on Myanmar and Lao borders) borders should be entirely closed to human entry.
   (h) Imported wildlife intercepted at international crossings should be repatriated or sent to centralized holding facilities where they can be screened for diseases and rehomed to appropriate long-term facilities, all imported specimens require quarantine in designated facilities.
   (i) Plant based traditional medicine should be further developed. Where unavoidable animal ingredients should be through licensed sellers subject to regular screenings, and pasteurized or treated through ultra-heat treatment.
   (j) The Catalogue of National Key Protected Wild Animals for Artificial Breeding should be re-examined to list species that can be bred in captivity and their purpose, and align with measures listed in part 2. International trade should also reflect the Agreement on the Application of Sanitary and Phytosanitary Measures to prevent international trade of wildlife for consumption unless treated or cured using heat or chemical treatment to prevent any disease risk, based on common standards and clearly noted.
   (k) Import of species should utilize a system such as LEMIS which clearly states the origin, source, purpose and recipient of any imported products from wild animals, and be in full compliance with CITES regulations.

(2) Preventing infection in captive animals

Outside zoos and licensed scientific facilities (including medical facilities), only livestock (ungulates, rabbits, chickens, etc.,) should be bred for consumption, or other consumables (i.e., leather), below regulations relate to the captive rearing of these animals, exceptions to this should be detailed in article 1i the Catalogue of National Key Protected Wild Animals for Artificial Breeding.

(a) Animal welfare ties directly to disease susceptibility and spread, thus minimum welfare standards should be applied to the keeping of captive animals. Foods provided for livestock animals should not be based on waste animal products, or meat. Administrative Measures for the Safe Production of Animal-derived Feed Products should be updated accordingly.

(b) Central databases captive mammals for all mammals over 3 kg. Individuals should be listed in the database, health checks and vaccination status noted, in addition to owner and previous owners.

(c) For non-domesticated animals bred in captive conditions, especially carnivores (tigers, bears) online registry should include an individual genetic barcode which can be used to verify identity and prevent wild-caught individuals being bought into the system (see 2b). This practice should be limited as much as possible and largely used for zoos and scientific institutes rather than commercial facilities, as the disease risk is higher, as is the motivation to launder animals into the system.

(d) Entities responsible for fulfilling these criteria are listed below and should be regarded as an update to information on National Key Protection of Wild Animals Domestication and Breeding Licences; licences can only be attained as detailed for ungulate species, and some birds (once a list of species has been developed). The domestication and wild capture of other species for commercial or consumption purposes is not permissible http://www.forestry.gov.cn/portal/main/govfile/13/govfile_2156.htm.

(e) Develop certification and quarantine facilities for imported animals. Wild-caught animals should only be kept as part of scientific studies or conservation programs, not for consumption or commercial programs.

(f) Prior to domestic transportation live animals (excepting personal pets) require medical check certification.

(3) Preventing transmission into humans

(a) Meat should not be sold in open conditions but only by licensed sellers in shops where it can be kept refrigerated and isolated. Blades should be sterilized between uses, and waste meat incinerated. No meat should be for sale in open conditions. All meat sources should be inventoried.

(b) No restaurant should be able to sell uncooked or uncured meat or blood

(c) Live animals sold in markets must have bedding changed and burned and the market washed and disinfected three times weekly.

(d)   Alternatives to current high status food items (i.e., traditional regional Chinese dishes/food varieties) should receive investment.
(e)   Invest in development of synthetic alternatives to animal based materials.

References

1. "Beijing Call for Biodiversity Conservation and Climate Change (06 Nov. 19)." France Diplomacy - Ministry for Europe and Foreign Affairs, www.diplomatie.gouv.fr/en/french-for eign-policy/climate-and-environment/news/article/beijing-call-for-biodiversity-conservation-and-climate-change-06-nov-19.
2. Convention on Biological Diversity: Zero Draft of the Post-2020 Global Biodiversity Framework (CBD/WG2020/2/3). (2020).
3. Convention on Biological Diversity: Synthesis of the Views of the Parties and Observers on the Scope and Content of the Post-2020 Global Biodiversity Framework (CBD/POST2020/PREP/1/INF/1). (2019).
4. Locke, H., Ellis, E.C., Venter, O., et al.: Three Global Conditions for Biodiversity Conservation and Sustainable Use: an Implementation Framework. National Science Review. 6(6), 1080–1082 (2019).
5. Organisation for Economic Co-operation and Development: Biodiversity: Finance and the Economic and Business Case for Action. (2019).
6. Convention on Biological Diversity: Report of the Open-Ended Working Group on the Post-2020 Global Biodiversity Framework on Its Second Meeting (CBD/WG2020/2/4), Rome (2020).
7. IPBES Expert Guest Article by Professors Josef Settele, Sandra Díaz, Eduardo Brondizio, and Dr. Peter Daszak. https://ipbes.net/covid19stimulus.
8. Woodley, S., Locke, H., Laffoley, D., et al.: A Review of Evidence for Area-Based Conservation Targets for the Post-2020 Global Biodiversity Framework. Parks. 25, 31-46 (2019).
9. Wilson, Edward O: Half-Earth: Our Planet's Fight for Life. Liveright Publishing Corporation, (2016).
10. Yang, Rui, et al. Cost-Effective Priorities for the Expansion of Global Terrestrial Protected Areas: Setting Post-2020 Global and National Targets. Science Advances (in Press). (2020).
11. Dudley, Nigel: Guidelines for Applying Protected Area Management Categories, Gland, Switzerland: IUCN, (2008).
12. Woodley, S., Locke, H., Laffoley, D., et al.: A Review of Evidence for Area-Based Conservation Targets for the Post-2020 Global Biodiversity Framework. Parks, 25, 31-46 (2019).
13. UNEP: Green Is Gold: The Strategy and Actions of China's Ecological Civilization. (2016).
14. Wang X.Y: A Study on Problems and Countermeasures of China's Ecological Civilization Construction. Ecological Economy, (2014).
15. Hansen M.H., Li H.T., Svarverud R.: Ecological civilization: Interpreting the Chinese past, projecting the global future. Global Environmental Change, (2018).
16. Gao J.X.: How China Will Protect One-Quarter of Its Land. Nature, (2019).
17. He P., Gao J.X., Zhang W.G., et al.: China Integrating Conservation Areas into Red Lines for Stricter and Unified Management. Land use policy, (2018); Jiang B., Bai Y., Wong C.P., et al.: China's Ecological Civilization Program–Implementing Ecological Redline Policy. Land use policy, (2019).
18. Liu Y.S., Liu J.L., Zhou Y.: Spatio-Temporal Patterns of Rural Poverty in China and Targeted Poverty Alleviation Strategies. Journal of Rural Studies, (2017).
19. Koch J.M., Hobbs R.J.: Synthesis: Is Alcoa Successfully Restoring a Jarrah Forest Ecosystem after Bauxite Mining in Western Australia? Restoration Ecology, (2007).

20. Banks-Leite C., Pardini R., Tambosi L.R., et al.: Using Ecological Thresholds to Evaluate the Costs and Benefits of Set-Asides in a Biodiversity Hotspot. Science, (2014).
21. Whisenant S.G.: Repairing Damaged Wildlands: a Process-Oriented Landscape-Scale Approach. (1999).
22. Clements W.H., Vieira N.K.M., Church S.E.: Quantifying Restoration Success and Recovery in a Metal-Polluted Stream: a 17-Year Assessment of Physicochemical and Biological Responses. Journal of Applied Ecology, (2010).
23. Buckley M.C., Crone E.E.: Negative Off-Site Impacts of Ecological Restoration: Understanding and Addressing the Conflict. Conservation Biology, (2008).
24. Xiao L.G., Zhao R.Q. China's New Era of Ecological Civilization. Science, (2017).
25. Yang B.W.: Research on Regulatory Framework of Agricultural and Forestry Carbon Sink Trading for Ecological Poverty Alleviation in China Under Policy Guidance. Agricultural Economics and Management, (2019).
26. Marton J.M., Fennessy M.S., Craft C.B.: USDA Conservation Practices Increase Carbon Storage and Water Quality Improvement Functions: An Example from Ohio. Restoration Ecology, (2014).
27. Xu W.H., Xiao Y., Zhang J.J., et al.: Strengthening Protected Areas for Biodiversity and Ecosystem Services in China. PNAS, (2017).

Chapter 3
Global Ocean Governance and Ecological Civilization: Building a Sustainable Ocean Economy for China

3.1 Background

3.1.1 Foreword

The ocean is fundamentally important for humankind. Simply stated, the ocean helps us breathe, regulates our global climate, and slows down the rate of global warming by absorbing 40% of anthropogenic carbon dioxide.

The ocean is also vital for the world's economic development. Three billion people globally rely directly on the ocean for their livelihoods. Ocean-based industries, such as fisheries, tourism, and maritime operations, are already today critical providers of employment and income. The ocean also holds the potential for the future development of new and expanded industries such as offshore renewable energies and marine biotechnologies.

However, a healthy ocean environment is a prerequisite for drawing on these direct and indirect benefits that the ocean provides—and the oceans and the ecosystem services they provide are under ever more serious threat than before.

This report looks at the opportunities that the ocean provides and the challenges it faces in continuing to provide for us these benefits.

China, like many other coastal nations, is facing the reality of seeing its own coastal seas declining in quality, caused by both terrestrial and marine development and activities, such as increasing discharge of terrestrial pollutants into the ocean, land reclamation, overfishing, pollutants from mariculture, and so on.

At the same time, global ocean conditions are being seriously affected by large-scale environmental pressures such as global warming, increased ocean acidification under a continuously higher atmospheric carbon dioxide level, microplastics pollution, and overexploitation of natural resources.

© The Author(s) 2022
China Council for International Cooperation on Environment
and Development (CCICED) Secretariat,
Green Consensus and High Quality Development,
https://doi.org/10.1007/978-981-16-4799-4_3

The nature of the ocean ecosystem is both fragile, highly dynamic, and globally interconnected. Therefore, there is good cause to manage and govern maritime ecosystems to ensure a healthy and sustainable ocean supporting prosperous societies now and in the future. There is a need and a call to manage and govern the ocean with an ecosystem-based integrated approach to strike a balance between protection and production.

The work of this Special Policy Study has clearly demonstrated that now is the time for China and the world at large to ensure that the ocean environment plays a critical role in the national and international efforts toward developing an ecological civilization securing our own future.

Clear and directed actions are needed to limit the threats and minimize the impacts to the oceans, and thereby lay the foundation for the oceans' ability to continue to serve as the basis of human life. Dedicated efforts are required to ensure further development of current and emerging industries happens in a sustainable manner. The principle of ecosystem-based integrated ocean management needs to weave through ocean management like a red thread in order to achieve these goals.

The work underlying this report has been extensive, both in time and in the matters covered. Around 50 experts from the Chinese and international community have been directly involved in the work of this policy study, giving their time and expertise within a wide range of relevant ocean-related topics. Without their commitment and energy, this work could not have been as comprehensive and overarching as this complex and important topic deserves. We are truly grateful and thankful for the contributions from each and all.

The recent COVID-19 pandemic has shown how vulnerable societies are. This study was almost completed before the outbreak of the pandemic and thus does not reflect the effects of it, but we are certain that the ocean also carries important capability to support society in such an undesired event.

We hope that this report, its findings and recommendations, can be a contribution to the further discussions and actions required both nationally and internationally, enabling the global community to truly integrate the ocean environment into governance discussions.

Jilan Su   Jan-Gunnar Winther

People ask: Why should I care about the ocean? Because the ocean is the cornerstone of earth's life support system, it shapes climate and weather. It holds most of life on earth. 97% of earth's water is there. It's the blue heart of the planet—we should take care of our heart. It's what makes life possible for us. We still have a really good chance to make things betterthan they are. They won't get better unless we take the action and inspire others to do the same thing. No one is without power. Everybody has the capacity to do something.

– Sylvia Earle

3.1.2 Introduction

The ocean covers more than 70% of the Earth's surface and is vital to the survival of life. It is fundamental for oxygen production, food sources, medicine, and many other products and essential ecological services. The oceans determine climate and weather, both on local, regional, and global scales. In addition to providing us with one of the most pleasing environments, the oceans are essential to energy, trade, transportation, and a number of other traditional and emerging industries.

There is currently a rising societal awareness and understanding of the overarching global importance of the ocean system as a living space and the basis for human civilization—as well as its highly dynamic and connected but fragile nature. In this context, all nations—individually and collectively—need to use and govern the ocean in a manner that will allow it to support society today and into the future.

Ensuring that the ocean system is integrated into and included in overarching societal strategies will be vital to China's ability to reach its goal of achieving a robust and viable ecological civilization. Three main themes merit attention, both to draw on opportunities and meet challenges: Environment, Industry, and Management. The three themes should not and cannot be seen as stand-alone pillars, but must be seen as three synergetic elements where each and all interact and impact each other in a number of ways and manners (Fig. 3.1).

The CCICED has explored the opportunities and challenges that the ocean provides within these three areas, through the establishment and work of an SPS for Ocean Governance and Ecological Civilization. This report constitutes the overarching findings of the SPS Ocean Governance.

The SPS has structured its work around the central theme and concept of ecosystem-based marine management. This concept addresses the impacts on marine ecosystems as a whole. Further, the SPS has focused its work on integrated ocean governance, which is an overarching and comprehensive tool for governing all human

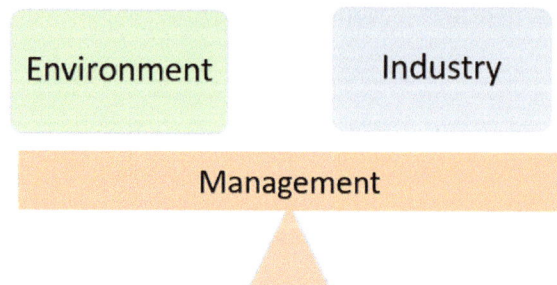

Fig. 3.1 Ensuring that the ocean system is integrated into and included in overarching societal strategies will be vital to China's ability to reach its goal of achieving a robust and viable ecological civilization. Three main themes merit attention, both to draw on opportunities and meet challenges: Environment, Industry, and Management

activities in the ocean space while at the same time considering environmental issues such as climate change, biodiversity, and pollution.

The Ocean Governance SPS has structured its work around this topic itself (ecosystem-based IOM and governance), as well as five additional interlinked themes:

- Marine living resources and biodiversity
- Marine pollution (plastics in particular)
- Green maritime operations
- Renewable energy systems
- Mineral resource extraction.

Climate change, technology, ocean economy, and gender have been a common thread through the various themes.

The Ocean Governance SPS has prepared reports for each of these six topics, and these reports serve as supporting documentation for this summary report. Below, we have included the summaries from each of the thematic reports to set the stage for this overarching summary report. Furthermore, the specific recommendations from the six task teams have been included as an appendix to this report.

Integrated and Ecosystem-Based Management

Summary of findings from the special study on integrated and ecosystem-based management

The oceans have always been key for the survival of humankind, and much of society's development has rested upon the qualities of the oceans. Marine ecosystems together form the largest aquatic system on the planet. The habitats of this vast system range from the productive nearshore regions to the barren ocean floor. The ocean has attracted multiple uses for centuries, including fisheries, shipping, and transportation, military, recreation, conservation, and, more recently, oil and gas extraction and mining. Ocean activities will be essential in meeting future global challenges, including food, energy, transportation, and climate regulation.

Considering the key role that the oceans play as the basis for society, humanity, and the global future, it is a given that maintaining the health of the oceans is essential. Today, the well-balanced species communities of the marine ecosystems are becoming increasingly unstable, putting the health of the whole ocean systems at stake. There are a number of pressures that affect the health of the oceans, such as pollution (including noise), biodiversity loss, invasive species, climate change, and overexploitation of resources. The need for more and better governance of ocean uses has been widely recognized. Globally, ecosystem-based IOM is accepted as the appropriate approach for ensuring the protection and the sustainable use of coasts and oceans, taking sufficiently into account knowledge and the particularities of the ecosystems to be managed.

In China, over half of the population now resides along the coast, with coastal provinces and metropolises contributing more than 60% of national GDP [1]. These critical regions also hold the key to future development of China's initiatives such as the Blue Economy and Belt and Road Initiative (BRI) and its interconnection to the rest of the world. The development of coastal regions is also an important aspect to consider in the context of the Chinese Government's vision of a Beautiful China and efforts to build an "ecological civilization" [2]. Challenges remain in implementing integrated ecosystem-based ocean management at national, provincial, and local levels. These include silo governance, differing national–regional–local frameworks, overlapping roles and functions in administration, lack of integrated land–sea management, lack of public–private partnerships, and lack of general (public) understanding of the importance of a holistic system.

China is well placed to develop and implement a fully integrated ecosystem-based ocean management system and take on international leadership in this field. There are ample opportunities for China to move in this direction given the well-developed management basis, strong political will, and the emerging consensus among general public and business sectors in conserving the ocean system.

The specific recommendations put forward for integrated and ecosystem-based ocean management are found in the appendix to this report.

Marine Biodiversity and Marine Living Resources

Summary of findings from the special study on marine biodiversity and marine living resources

Feeding more than 9 billion people by 2050 while protecting biodiversity and the natural systems on which life depends is one of the greatest planetary challenges we face today. As one of the world's top seafood-producing nations, China has a significant stake in solving this challenge.

The ocean shelters a vast array of species and nourishes billions of people who depend upon it for food and livelihoods. However, while the ocean's ability to produce food is enormous, it is limited. About one-third of fisheries for which data exist are overexploited or collapsed. As a major contributor to food safety, aquaculture can also have negative impacts: it can displace native coastal and marine ecosystems, require large inputs of wild fish for feed, introduce non-native species and diseases, and cause significant pollution. Sustainably managing living marine resources to optimize food production over the long term while minimizing damage to the ecosystem is not an easy task, although there are many proven solutions and new ones in development to achieve these goals. Climate change will likely compound the challenge. Warming and acidifying waters are altering the productivity of many marine species and driving others across borders, intensifying the struggle for resources among competing countries. More extreme storms, altered weather patterns, and disrupted water and nutrient cycles will increasingly stress coastal food production systems.

Rapid economic growth along China's coasts over the last 40 years has imposed significant costs on the coastal and marine environment. More than half the country's coastal wetlands, nearly 60% of mangroves and 80% of coral reefs have been lost to land reclamation, mariculture and pollution [1]. Significant portions of seagrass beds, tidal marshes, and tidal flats that once provided critical habitat for a diverse array of marine life have also been affected. China produces enormous quantities of wild and farmed seafood, but the rate of extraction and exploitation has outstripped the ability of marine ecosystems to absorb the impact, and most of the top marine predators have been eliminated or nearly so. Furthermore, China employs more people in the fishing and aquaculture sector than any other nation, making the social dimension of bringing these industries under control much more challenging. China has begun making significant progress in addressing the challenges. Bold, concrete actions, including strengthening the seasonal closings of nearly all domestic fisheries each year, protecting habitat and establishing marine reserves, imposing stricter standards for discharging pollutants into the sea, shutting down illegal mariculture operations, and establishing a total acreage limit for mariculture have been imposed.

More must be done if China is to re-establish healthy coastal and marine ecosystems that can sustainably provide the levels of nutritional and economic benefits historically enjoyed. Legal protections for living marine resources must be strengthened, monitoring expanded, and compliance improved, and more critical habitat restored and protected. Furthermore, because climate change is affecting living marine resources on a global scale, and because so many of these resources are shared, stronger regional and global governance is required to ensure living marine resources are managed sustainably at scale. Although these challenges are significant, solving them will produce enormous direct benefits for China by securing a large and sustainable supply of high-value seafood and livelihoods in its own waters. It will also create tremendous opportunities for China to demonstrate regional and global leadership among developing countries that lack the capacity to sustain the value of their living marine resources. These leadership opportunities are especially important and timely now as these countries begin to grapple with how to recover from the COVID-19 crisis and associated economic impacts.

The specific recommendations put forward for marine biodiversity and marine living resources are found in the appendix to this report.

Marine Pollution

Summary of findings from the special study on marine pollution

The ocean is vital to all life on Earth, providing many provisioning, regulating, and supporting services. Agricultural and industrial requirements for feeding, clothing, and housing the world's growing population and expanding economies have resulted in seriously degrading parts of the marine environment, especially near the coasts.

The lack of sewage and wastewater treatment and the release of pollutants from industrial, shipping, and agricultural activities are major threats to the ocean, particularly in terms of food security, safety, and maintenance of marine biodiversity. The ocean also suffers from the sewage, garbage, spilled oil, and industrial waste that we collectively allow to flow into the ocean every day. In addition to affecting marine and coastal ecosystems and biodiversity (as well as ecosystem services), such pollution has a direct connection to huge economic costs tied to marine fisheries, marine tourism, and human health and safety. Often, while production and emission to a large degree are land-based, the marine environment is, in fact, the end recipient. In addition to the well-known eutrophication effects from terrestrial nutrient input, the globally growing plastic pollution challenge is another prime example of such interactions. Continued growth in industrial production means that discharges and emissions likely will increase the inputs of heavy metals and other hazardous substances into the ocean. The use of the best practicable means to limit the creation of waste, discharges, and emissions can help control these problems.

In the last 40 years, China has formed a coastal ribbon of high economic development, which has brought with it population density and urbanization. This rapid development has subjected the coastal and marine ecosystems to tremendous ecological damage. More than 70% of nutrients discharged into the sea have land-based origins, and these and other sources of pollution being leached into the marine environment have led directly to a decline in marine water, sediment, and biological quality. Although laws and policies have been much improved in the last 10 years, there are still some significant gaps that prohibit China from fully implementing its obligations in the international conventions and protecting its marine environment and resources. The gaps exist in the lack of an integrated ecosystem-based view, lack of laws in protection of resources and ecosystem, lack of detailed implementation rules, and lack of a cross-sector implementation mechanism.

Recently, China stepped up efforts to promote ecological civilization. Pollution control is one of the three tough battles that the Chinese government must win. Innovations in cleaner production methods and means of reducing discharges and emissions would be important to keep pace with the growth in production, especially in areas of rapid industrial growth. There is a tremendous opportunity for China to promote the case for marine pollution governance. China's ecological civilization construction is a useful exploration and practice of sustainable development, providing an economic reference for other countries to deal with similar economic, environmental, and social challenges.

The specific recommendations put forward for marine pollution are found in the appendix to this report.

Green Maritime Operations

Summary of findings from the special study on green maritime operations

As a country trading across all regions around the globe, maritime operation is becoming a crucial foundation of China's socioeconomic growth. China ranks first in the world in port cargo throughput and has the largest ocean fleet and highest number of seafarers in the world. Its marine [3]. Marine fisheries and oil and gas exploration lead the world as well. With the sustainable development of this industry, marine environmental pollution caused by maritime operations has become a challenge that must be faced and solved. At the same time, it also brings great opportunities and challenges for China's maritime operation to adapt to the global trend of green development.

As the dynamic and vulnerable nature of the marine ecosystems has become increasingly understood and appreciated, higher standards for green maritime operations are required both domestically and internationally. While land-sourced pollution is an important concern and contributor to the degradation of the marine environment, the impacts on the marine environment from maritime transportation, oil and gas exploration, and fisheries—as well as port and ship infrastructure—also represent a challenge to be addressed. It is believed that pollution from ships is a large proportion of maritime pollution, with the main pollutants being oily water and domestic sewage. Internationally, the shipping industry is increasing its attention to these issues and moving toward modes of operation that reduce the ecological footprint. New international environmentally driven regulations for shipping are constantly complemented and improved. They are expected to bring significant benefits for human health and the environment, but at the same raise fresh challenges for the shipping industry.

China's booming ocean economy over the last 40 years has exacerbated the overall pressures on the marine environment. Ports claim large swathes of land from the sea. Land-sourced pollution further threatens ocean ecosystems, as does the inadequacy of environmental facilities at ports and on vessels and offshore oil platforms along with worsening air and water pollution. Ship-sourced GHG emissions remain significant. The massive scale of maritime transportation and coastal storage of hazardous chemicals such as petroleum products cause marine pollution risks to rise. At present, the Chinese government has made substantial efforts in promoting pollution prevention and control of ports and ships and has made some achievements, in particular through the adaptation and implementation of several laws and regulations related to maritime operations.

China's maritime operation needs to shoulder its share of responsibility in building an ecological civilization and tackling pollution prevention and control. In this context, there are more green goals and tasks for the shipping industry, marine fisheries, and oil/gas industry ahead. Government actions, such as subsidies and tighter supervision, have aided the industry's ability to follow up. Nevertheless, the lack of technology and environmental awareness is an impediment that calls for further measures. Internationally, there has been a significant advancement with regard to developing green maritime operational practices, such as emissions-control areas, green ports and fleets, emergency response, pollution prevention, and control for fishing vessels and harbours, which can provide valuable experience for China. China needs to increase its efforts to adapt to the greening of the global marine industry.

The specific recommendations put forward for green maritime operations are found in the appendix to this report.

Marine Renewable Energy

Summary of findings from the special study on marine renewable energy

Ocean renewable energy (ORE) is notable as an emerging sector of the maritime industry. China, the world's biggest energy consumer, is stepping up its push into renewable energy and proposing higher green power consumption targets, including in the ORE area. Achieving the needed renewable energy transition will not only mitigate climate change but also stimulate the economy, improve human welfare, and boost employment worldwide.

Different ORE technologies (wind, wave, current, tidal range, ocean thermal) are under consideration in different stages of development, and each will present its own unique challenges. ORE—specifically offshore wind—is seeing rapid growth in installed capacity and environmental; however, socioeconomic, and technical challenges need to be considered. Achieving a viable cost of electricity is a challenge to the offshore wind industry but provides an even bigger challenge to other ORE technologies. Understanding and assessing the environmental impact of ORE installations, operations, and decommissioning is substantially challenged due to, e.g., baseline data, socioeconomic and diverse developing technologies. Development of ORE affects or is also affected by numerous stakeholders. Understanding who the stakeholders are and how they are engaged in the process is necessary for improving the responsible development of ORE technologies. Key stakeholders include fishers, community members, regulators, developers, scientists, and tourists and so on that depend on the specific ORE project and the specific location. The seabed off China's east coast is characteristic of soft, silty soils which are unlike soil conditions in other countries contemplating ORE growth. This causes difficulty with regard to foundation type and installation techniques. Furthermore, the technical challenges for the offshore wind industry are much greater in other typhoon-prone regions, where the weather conditions can be quite impactful on turbine performance. China's current legal system of environmental consideration related to ORE activities is limited and further regulations need to be developed.

China is particularly active in developing offshore wind technologies, an area which is set to become an important sector for the global energy future, while also demonstrating wave and tidal energy technologies. The Chinese government has made a commitment that non-fossil fuel energy will account for 20% of energy production by 2030, and operational installed capacity of ORE (offshore wind) in 2019 reached 3.7 GW in total, with another 13 GW under construction and over 41 GW permitted. The development of ORE, including offshore wind, in China has reached a turning point in 2018, moving toward zero-subsidy. China's first auction for offshore wind projects in 2019 achieved a price of electricity at 0.75 Yuan/kWh, lower than the guide price of 0.8 Yuan/kWh. China has also become one of the few countries in the world that have mastered the technology of large-scale tidal current energy development and utilization.

ORE is a fast-growing ocean technology that is advancing the goals of the low-carbon and circular economy. Offshore wind technology only recently reached a policy turning point, while other ORE technologies are at an early stage of development. Nevertheless, there are encouraging signs that the investment cost of technologies and the price of electricity generated will decline further toward commercially viable ORE energy generation. Enhancing knowledge of the ORE technologies' potential impacts is crucial to inform future growth plans and as well as effectively licensing for ORE activities. Ongoing review of environmental impacts associated with the growing ORE sector and emerging ORE technologies will ensure that the best and most up-to-date information is available to decision makers, developers, and stakeholders. Furthermore, the opportunity of integrating emerging ORE technologies into military applications, electricity generation for remote communities, freshwater generation, or aquaculture applications, could be further opportunities. ORE technologies offer opportunities for China to develop a new industry, create jobs, and take advantage of opportunities within its competency to global markets.

The specific recommendations put forward for ocean renewable energy (ORE) are found in the appendix to this report.

Marine Mineral Extraction

Summary of findings from the special study on marine mineral extraction

The huge reserves of mineral resources (polymetallic sulphides, PMS; polymetallic nodules, PMN; cobalt-rich ferromanganese crusts, cobalt-rich crusts (CRCs); and rare earth elements [REE]) found on the deep seabed may be of great significance to the world's economic development as well as to the strategic reserve of mineral resources [4]. To date, 30 deep-sea mineral exploration contracts have been issued by the International Seabed Authority (ISA) for resources beyond national jurisdiction. Natural gas hydrates are widely distributed in most of the world's marine deep-water areas (~99%) and permafrost sedimentary environments (~1%) [5]. The amount of gas stored in the world's hydrate accumulations is significant, but estimates are highly uncertain.

Deep-sea mining (DSM) requires cutting, collecting, or dredging technology, and a rise-and-lift and return water system. Technology is currently being developed and tested. It is imperative that future exploitation of marine mineral resources can be carried out without causing significant environmental harm. There is, however, a challenge in that deep-sea environments in general are little-explored and poorly understood. The nascent DSM sector has reached a crucial juncture of transition from exploration to exploitation, and the regulatory framework is still not complete. Establishing such a new industry will require a joint effort and collaboration to succeed in establishing a sustainable industry and maintain the seabed's status as "the common heritage of mankind" [6]. In addition to a number of technical risks, NGH exploitation also comes with serious environmental risks such as the increased risk of seabed landslides and huge amounts of methane released into the water column and atmosphere.

China's law on Exploration for and Exploitation of Resources in the Deep Seabed Area was adopted on 26 February 2016 and entered into force on 1 May 2016. This is the first Chinese law specifically regulating relevant activities from China. Environmental protection principles and measures are well covered in the law, and stringent rules, standards, and effective measures regarding the protection of the environment are reflected and adopted. Chinese state-sponsored companies hold five of the exploration contracts related to PMS, PMN, and CRC. In 2020, China carried out a successful gas hydrate production test in the South China Sea.

There are ample opportunities for Chinese industry to engage at all levels of the DSM value chain, including research, exploration, exploitation, equipment manufacturing, technology design, and mineral processing and to promote DSM as part of the circular economy. In the collaborative work between operators and interested parties, there is a need to focus on reducing the environmental risk as well as sharing of data and experience to ensure the industry adapts best environmental practice and continuous improvement. There would be benefit in China reviewing and updating its Deep Seabed Area Law in order to comply with the new requirement of the exploitation regulatory framework developed by the ISA within the context of the domestic legal system, to deal more specifically with future exploitation activities, and consider developing additional regulations to supplement the ISA requirements, drawing on the concepts of sound environmental management. With its high level of engagement in DSM, China is well placed to continue to bring on initiatives to strengthen ISA as a regulator and actively engage with ISA.

The specific recommendations put forward for marine mineral resources **extraction** are found in the appendix to this report.

3.2 Major Research Results

3.2.1 Environment: The Ocean as the Basis for Life

The ocean and humans are inextricably interconnected. The choices we make and the actions we take in managing and governing the oceans consequently have profound and lasting impacts on human well-being and societal development.

The ocean is the largest ecosystem on Earth and covers 70% of the earth's surface. The world's oceans contain more than 97% of the world's water and are home to the greatest abundance of life on our planet. Marine biodiversity is remarkable at the species, genetic, and molecular levels. All the creatures that live in the ocean play an essential role in the trophic chain of the ecosystems.

Marine environments are normally classified as either pelagic (open water) or benthic (bottom), although they are closely interlinked in many ways. For example, pelagic plankton are an important source of food for animals on soft or rocky bottoms, because the upper part of the water column is where photosynthesis occurs. Food chains in the oceans are generally regulated by nutrient availability. These determine the abundance of phytoplankton, which in turn provide food for the primary consumers, such as protozoa and zooplankton that the higher-level consumers—fish, squid, and marine mammals—prey upon.

The distribution patterns of marine organisms have been and continue to be influenced by physical and biological processes over time, such as temperature, salinity, density, and current patterns. Cycles of plankton production vary around the world based on seasonal differences of light and temperature. Changes in production depend on the season, the proximity to fresh water, and the timing and location of upwelling, currents, and patterns of reproduction. Recent estimates of extant marine species range from ~300,000 to 2.2 million, revealing high levels of uncertainty in our knowledge of global marine biodiversity. Coastal and marine habitats in China are home to more than 20,000 species, including 3000 species of fish alone.

The importance of the ocean as a life support system for humankind relates to the air we breathe, the climate we have, and the food we consume.

The world's oceans generate more than 50% of the oxygen we (need to) breathe, and the photosynthesis of phytoplankton is the source of this oxygen. Through photosynthesis, these microscopic plants in the ocean take up carbon and release oxygen.

The oceans are key regulators of the global climate. The ocean absorbs the heat from the sun and transports warm water from the equator to the poles and cold water from the poles to the tropics—this continuous pumping of water frames regional climates around the world. Through the phytoplankton's photosynthesis processes, the ocean absorbs carbon and as such is the largest carbon sink we have. This means that the oceans play an important role in maintaining the balance of the overall carbon cycle, and thus also the stability or change in climate.

Oceans provide at least a sixth of the animal protein people eat and are the number one source of protein for more than a billion people. Fish, crustaceans, molluscs, algae and other sea plants are some of the food sources used around the world. With a growing global population, it is expected that the oceans will play an even more important role as a food source.

The marine ecosystem thus provides numerous benefits for people and society, including food, natural fibres, a steady supply of clean water, regulation of pests and diseases, medicinal substances, recreation, and protection from natural hazards such as floods. Environmental damage and degradation to marine ecosystems consequently may, therefore, have great social costs. Not only does it have economic costs such as those related to clean-up operations, reduced fishing catches, reduced coastal tourism, but also on general well-being due to negative impacts on, for example, recreational and aesthetic values of the marine environment. Aesthetic value, or beauty, is important to the relationship between humans and natural environments and is, therefore, a fundamental socioeconomic attribute of conservation alongside other ecosystem services.

Noting these important fundamental roles that the ocean system has for life, it is disturbing that the oceans are facing a constantly increasing number of threats, in particular habitat destruction, (coastal) pollution, overfishing, climate change, hypoxia, and ocean acidification.

China's marine environment has been deteriorating for many years. Major concerns regarding the coastal and marine environment in China include land reclamation and hardening of the coastlines of the Yellow and East China Sea, dramatic reduction in sediment discharge from the Huanghe and Changjiang River; navigation channel construction and deepening of ports; and increased nutrient and contaminant loading to the estuarine and coastal environment. Eutrophication related to the use of fertilizer is a prominent and overall "invisible" pollution source, which lead to a series of ecological effects such as harmful algal bloom, seasonal seawater acidification and hypoxia in the coastal oceans [7]. Pollutants from mariculture, agriculture and other land-based industries have eroded key habitats, including those further offshore that are buffered to some degree from alteration of the coastal zone.

Some marine ecosystems, most notably the central Bohai Sea, areas at offshore side of the Yangtze River Estuary and Pearl River Estuary, have been severely degraded and become seasonally hypoxic, while severe eutrophic pollution has occurred in other large bays or estuaries, such as Liaodong Bay, Bohai Bay, Laizhou Bay, Hangzhou Bay and Zhujiang River Estuary, compromising survival of fish populations and other living marine resources.

Many ongoing and emerging human activities have the potential to and are, in fact, destroying areas and ecosystems that marine plants and animals need to survive. The clearing of mangrove forests for shrimp production is one such activity, noting that approximately 240,000 ha of shrimp ponds have been built in the coastal areas of south-eastern China during the past 40 years, largely by destruction of mangroves and seagrass beds. Other activities that contribute to coastal degradation include land reclamation on coastal wetlands and destruction of the seabed by trawling. Furthermore, reclamation and eutrophication from both terrestrial input and mariculture are regarded as major causes for the decline of seagrasses, coral reefs, mangroves, salt marshes, and tidal flats.

Overfishing occurs when more fish are caught than are able to reproduce and to recruit. Normally, fishing occurs at a high level in the food chain, on the predators that live on the smaller organisms lower in the food chain. When essential predators are highly decimated or removed through overfishing, this impacts the populations of the other organisms in the food chain, often giving these the opportunity to increase. Affecting the top predators' role in balancing the ecosystem—from what they eat to how their bodies decompose—thus means important and potentially fatal ripple on effects for ocean ecosystems.

All the activities described above constitute threats to the ocean, causing degraded water quality, environmental degradation, decline in biodiversity, and the loss of ecosystem services, each of which may be difficult to account for in monetary terms but is significant nonetheless.

Overfishing is perhaps the most significant issue facing marine ecosystems overall. This is not to say that other issues are not important and in some locations more severe than overfishing; however, because fishing fleets are vast and widespread, and cause direct mortality to harvested organisms—often with incidental impacts on habitat—overfishing remains a major threat to ocean health at the global scale. The causes of overfishing are diverse and vary from fishery to fishery, though two of the most pervasive drivers include non-sustainable economic incentives that are not aligned with needed environmental outcomes, and fishers being disconnected from decision-making processes and therefore less likely to accept and comply with regulations.

Although the number of eggs a fish spawns usually reaches the level of one million, the fish eggs and larvae easily become prey for other animals. Therefore, the survival rates of fish eggs and juveniles (supplementary groups) are very low, possibly as low as 1 in 10,000 or 1 in 100,000. This is why larvae and juveniles of marine living resources need good habitats and shelters to grow properly and sustain their populations. The ecological function of coastal wetlands is very important for many coastal fish species. Numerous important fish habitats such as estuaries, seagrass beds, salt marshes, and tidal flats are distributed along the coasts of China. Besides being critical fish habitats, coastal wetlands also provide numerous ecosystem services, including water purification, which is indispensable for cleaning the coastal water by removing organic and inorganic nutrients, particulate matters and chemical pollutants.

Ocean ecosystems have no obvious physical boundaries, but rather are characterized by currents that transport nutrients and small marine organisms, by highly mobile species that migrate across entire ocean basins for feeding and reproduction. These horizontal and vertical movements connect the open ocean to coastal waters and the deep ocean and play an important role in maintaining healthy and productive ecosystems. Thus, transboundary connectivity is important in sustaining healthy populations of marine animals and functioning ecosystems which in turn provide a range of benefits for people and communities and society at large.

IPPC's Special Report on the Oceans and Cryosphere in a Changing Climate (SROCCC) records significant changes to the world's oceans as a result of climate change, noting that it is virtually certain that the global ocean has warmed unabated since 1970, and that since 1993, the rate of ocean warming has more than doubled.

Also, marine heatwaves have increased in frequency and intensity. SROCCC also documents that there are many marine species across various groups that have undergone shifts in geographical range and seasonal activities in response to the ongoing ocean warming and biogeochemical changes. This has resulted in shifts in species composition, abundance, and biomass production of ecosystems, from the equator to the poles. Coastal ecosystems, in particular, are affected by ocean warming, including intensified marine heatwaves, acidification, loss of oxygen, salinity intrusion, and sea-level rise.

By absorbing more CO_2, the ocean has undergone increasing acidification. Increasing anthropogenic nutrient discharge has resulted in widely distributed coastal hypoxia zones, often accompanied by severe acidification, called coastal acidification. Ocean acidification is expected to affect ocean species to varying degrees. Photosynthetic algae and seagrasses may benefit from higher CO_2 conditions in the ocean, as they require CO_2 to live, just like plants on land. On the other hand, studies have shown that lower environmental calcium carbonate saturation states can have a dramatic effect on some calcifying species, including oysters, clams, sea urchins, shallow water corals, deep sea corals, and calcareous plankton.

Impacts from human activities both on land and in the ocean are already observed on habitat area and biodiversity, as well as ecosystem functioning and services. Humankind can, through appropriate actions, contribute to slow down the ongoing changes. This will, however, require fundamental transformations in all aspects of society—how food is grown, land is used, goods are transported, and how the energy that support our lives and economies is produced. It will require a joint effort between governments, businesses, civil society, youth, and academia to make this shift.

3.2.2 Industry: Ocean Economy

Generally speaking, the entire global community depends in one way or the other on the services of the ocean, and it is estimated that globally 3 billion people rely directly on the ocean for their livelihoods, the vast majority of these in developing countries. Ocean-based industries, such as fisheries and tourism, are critical providers of employment and income. Expanding ocean-based sectors such as offshore renewable energies and marine biotechnologies can boost job creation, energy supply, food security, and infrastructure.

A healthy ocean environment is a prerequisite for drawing on the direct and indirect economic opportunities that the ocean provides, and thus investing in the ocean environment is also an investment in the ocean economies. It is important to ensure that ocean investors are made appropriately aware of how their investments affect the marine environment and of how a declining ocean environment, in turn, may affect the outcomes of their investments. Environment and industry cannot be considered as separate themes and management entities but as a synergetic system.

The importance of marine sectors for China cannot be overemphasized. Over half of the population now resides along the coast, responsible for 60% of national

GDP. More importantly, the recent rapid development of China's economy and social well-being was initiated in the coastal cities and their opening to the international community. Currently, the key ocean-based economies of China are fisheries, mariculture, ship-building/shipping, tourism, and recreation-based industries. Renewable ocean energy and seabed mining, as well as ocean-based biotechnology, are emerging as potential and likely future large-scale industries. In many areas, China is a world leader in the scale of the industry, and as such, contributes to setting global standards for the industries.

For example, China's aquaculture industry has grown markedly for more than six decades, and it is today the largest aquaculture producer in the world, accounting for around two-thirds of global production. China still leads the world in marine capture fisheries by far. In 2016, China's fishing fleets harvested over 15 million tons, nearly 2.5 times the catch of the next largest fishing nation. Approximately 90% of the catch is from domestic waters. Fisheries remain an important economic driver in China. Although mariculture production now far exceeds that of wild catch, China's position in the global seafood supply chain, its world-leading volume of wild fisheries catch, and the size of its fisheries labour force mean that management of wild fisheries remains a critical policy issue for social, economic, and environmental reasons. Approximately 19 million people are now employed directly or indirectly by fisheries and aquaculture in China [8].

As a shipping country, China ranks first in the world in port cargo throughput and has the largest ocean fleet and highest number of seafarers in the world. By the end of 2018, China's coastal ports had 5734 berths for production terminals, 2007 berths of 10,000 tons and above, with passenger throughput of 880 million people and cargo throughput of 9463 million tons [9]. In terms of container throughput, China's ports occupy seven positions in the world's top 10 ports.

Ocean energy and offshore wind energy are abundant, geographically diverse, and renewable. To extract these energy sources, there are six distinct technologies (i) offshore wind energy, (ii) wave energy, (iii) tidal current, (iv) tidal range, (v) salinity, and (vi) ocean thermal energy conversion. The variations in ocean resources and location will require different technological concepts and solutions. Many marine-based renewable energy technologies are at relatively early stages of development, and currently only contributes a tiny proportion (far less than 1%) of the global renewable energy production due to the technological difficulties as well as the high cost [10]. There is still little in the way of demonstrated effectiveness, cost, or environmental effects of large-scale ocean-based systems, and in particular assessments of the environmental impacts of installation, operation, and decommissioning of renewable ocean energy systems is facing critical challenges as there is a lack of baseline data as well as a diversity in the developing technologies that have their own specific environmental aspects [11].

Rising demand for minerals and metals, including for use in the high-technology sector, has led to a resurgence of interest in the exploration of mineral resources located on the seabed, in particular seafloor massive (polymetallic) sulphides around hydrothermal vents, CRCs on the flanks of seamounts, or fields of manganese (polymetallic) nodules on the abyssal plains. In addition to mineral deposits, there is interest in extracting methane from gas hydrates on continental slopes and rises. Seabed mining for minerals is still in the developing stage. The huge reserves of seabed mineral resources can be of great significance to the world's economic development, including China. However, there is a need to better understand the risks and potential impacts to ensure that any activity in this area is sustainable in the long term. Deep-sea environments are little-explored and poorly understood. Before initiating any large-scale DSM activity, it is necessary to understand the relevant ecosystems and the potential risks associated with the activity, both in general for an evolving industry and for specific exploration activities.

The creation of products and processes such as pharmaceuticals and cosmetics, food, chemicals, and biofuels from marine organisms through the application of marine biotechnology is a relatively young sector in the ocean economy, but which has the potential to contribute to economic and social prosperity by making use of recent advances in science and technology. Marine biotechnology is also playing an increasingly important role in the protection and management of the marine environment itself. Biotechnology, including marine biotechnology, has been one of China's strategic industrial sectors; however, this is field requires international cooperation in legally framing the industry and developing the technology to ensure economic prosperity in an environmentally sustainable manner.

In exploring and developing ongoing and potential future ocean-based industries, issues such as the environmental sustainability, the use and development of new technologies, social sustainability, and gender aspects are key factors to be considered.

Environmental sustainability is fundamentally important to consider in order to avoid putting the ocean's resources at risk, and thereby hampering the socioeconomic benefits they can deliver for future generations. The ocean is our greatest global common, and how we use it will determine much of the success toward achieving the SDGs. It is necessary to understand how various industrial activities may affect and degrade the environment, as well as invest in research on—and implementation of—technologies that minimize such impacts. The lack of sufficient scientific understanding of the vulnerability of the ocean ecosystems is an obstacle to understanding potential impacts, and efforts are required to decrease the knowledge gap. Emerging industries are in a unique position to define both national and global sustainable frameworks for the industry before it is fully developed.

Developing new technologies to support sustainable ocean economies is, in itself, an important contribution to the national and global economies. There are considerable potentials for pioneering technological innovations that could provide solutions for balancing the benefits of marine-based activities with a complex variety of risks that need to be carefully managed. In fisheries and mariculture for example, green technologies could include low-impact, fuel-efficient fishing methods, and innovative aquaculture production systems using environmentally friendly feeds; reduced energy use and greener refrigeration technologies; and improved waste management in fish handling, processing, and transportation. In the shipping arena, many national governments around the world have high aims for the greening of their national fleet and therefore formally encourage and support the introduction of incentive schemes aimed at moving ship owners to adopt ship designs and technologies that reduce fuel consumption and pollution. Developing green, smart ports is also essential. Such incentive schemes give the industry great autonomy in deciding how to meet the set of targets and provide it with incentives to search for cost-effective ways to meet them, yet contributes to a continuous greening effort.

The inclusion of women has been shown to increase the effectiveness of the Green Economy. According to the IMO, women today represent only 2% of the world's 1.2 million seafarers and 94% of these female seafarers work in the cruise industry. The number of female captains, officers, and general seafarers in China accounts for over 15% of the country's total maritime staff. Women account for 50% of the workforce in fisheries and aquaculture worldwide when accounting for the secondary industry [12]. Women account for 20% of the total professional fishery-related workforce in China. China has a long and strong commitment to gender equality (in its society). China was one of the first countries to ratify the UN Convention on the Elimination of All Forms of Discrimination Against Women (CEDAW) in 1980. Through this Convention, China, along with other nations, has agreed to take appropriate measures in all fields, including the economic fields, to ensure the full development and advancement of women on the basis of equality with men. UN SDG #5 calls for achieving gender equality and empowering all women and girls. China has committed itself to contribute to this goal through eradication of all forms of discrimination and prejudice against women and girls and through strengthening women's employment and entrepreneurship capabilities [13]. CEDAW calls for effective special measures in the fields of women's employment and participation aimed at accelerating de facto equality between men and women in line with the Convention. With its strong historical commitment and desire to move toward fulfilling UN SDG #5, China stands in a unique position to be a leader in contributing to changing the skewed gender balance in maritime operations, by committing to and implementing gender equalizing efforts in the further development of ocean industries in China.

> The ocean carries great potential for China's economic and social development. However, feeding more than 9 billion people by 2050 while protecting biodiversity and the natural systems on which life depends is one of the greatest planetary challenges we face today. With dedicated efforts to ensure further development of current and emerging industries in a sustainable manner, the ocean's potential can continue into the future. Recommended actions that the State Council could consider are summarized in Chap. 6.

3.2.3 Management: Balancing Environment and Economy

A healthy and sustainable ocean is essential for maintaining prosperous societies now and in the future. Recognizing the importance of the ocean for all aspects of human society and the habitability of the Earth, the need for more and better governance of ocean-related human activities has been widely accepted. There are ample opportunities for China to implement sustainable development approaches given the strong political will and the emerging consensus among the general public and the business sector in conserving the ocean system. This calls for ecosystem-based IOM to strike a balance between protection and production. IOM is an approach that brings together relevant actors from government, business, and civil society and across sectors of human activity. IOM is an important tool for ensuring the sustainable use of coasts and oceans and provides a framework for how the oceans could be managed under the premise of a knowledge-based and ecosystem-based approach. In short, IOM is a tool that enables society to optimize the benefits it derives from the ocean for the long term.

China issued the first National Marine Functional Zoning (MFZ) in 2002, which provided a basis for sea area management and preliminarily solved the problems of disorderly sea use. In the revised MFZ (2011–2020), China's coastal oceans are divided into eight functional zones, including agriculture and fishery, port and shipping, industrial and urbanization, minerals and energy, tourism and recreation, marine protected areas, special use, and reserved areas. An MFZ is the legal basis for rational development and utilization of marine resources and effective protection of the marine ecosystems. At the local level, Integrated Coastal Zone Management (ICZM) has been adopted in a few regions, with pilot projects successfully carried out over the past 30 years, to deal with certain cross-boundary issues. The Bay Chief System recently piloted in some regions is a coordinated mechanism led by governmental organizations with the participation of various departments and all sectors of the society. Its characteristics lie in the overall planning and coordinated management across departments to ensure that problems in bay environmental management are addressed more comprehensively. However, none of these practices (MFZ, ICZM or Bay Chief System) has an entirely "ecosystem-based approach," and challenges remain in implementing marine ecosystem-based integrated management at both national, regional and local levels, some of which are highlighted below.

Silo management is a common challenge within national and international governance, indicating management systems that are unable to operate with other systems and where management entities have different priorities, responsibilities, and visions. Sectoral silos need to be broken down to achieve IOM at national, regional, and local levels, as well as across borders. As a nation, China implements a top-down governance approach which includes three main governance levels (national, provincial, or metropolitan, and municipal or county). Despite the various strengths of this system, it also tends toward silo management. There needs to be a link between all levels to enhance the efficiency and effectiveness of implementation.

Traditionally, the focus has been on single pressures when developing management frameworks for the oceans. Single threats are indeed easier to study and understand, while multiple stressors are ubiquitous, and stressor interactions can lead to amplified effects. There is still very little understanding regarding how simultaneous stressors interact to affect species and ecosystems. Management rarely has the capacity to address more than one issue at a time. Greater reliance on the marine ecological sciences and application of a whole-system approach is needed.

The marine ecosystems are affected by human activities on both land and sea. However, standard approaches for management of the ocean often neglect to consider connections between ecosystems and thus are characterized by a sectoral approach to management. At present, China's coastal zone management has entered the stage where land and sea are coordinated. Land and its surrounding sea are an integrated system, so the marine ecosystem and land ecosystem should be considered as one in overall coastal planning. Challenges remain, however, and further measures to promote land–sea coordination are urgently needed.

The ecosystem-based management system can only be effective if it is understood by all concerned stakeholders. It is crucial to raise awareness of the multiple benefits provided by ecosystem-based approaches among all relevant policy sectors and stakeholders.

The recent COVID-19 pandemic has also shown how vulnerable societies are and could underscore the need to manage our ocean in a sustainable way. The ocean and its ecosystem services are important for people far beyond the coastal and marine regions themselves. Changes to the ocean itself, as well as changes to society that depend on the services of the ocean, may have detrimental effects, in particular in cases of unforeseen and sudden changes. In such a context, the resilience of the coastal and marine environment will be important. Societal resilience against climate-related, anomalous, and unprecedented disturbances of the ocean system can be strengthened by protecting resources, as well as other innovative approaches, such as a gender-based analysis.

Achieving a sustainable and integrated marine and coastal ecosystem management requires the involvement of women. Confronting gender inequality is essential to achieving the targets of the 2030 Agenda for Sustainable Development. It is important that women have an equal role in participating and managing ocean-related activities to enable this. Studies show that the participation of women in activities and their management often has a positive impact on issues related to the environment and sustainability.

Women and men use and manage marine and coastal ecosystems differently, have specific knowledge, capabilities, and needs related to this and are differently affected by changes in their environment due to climate change, pollution, and globalization. Gender mainstreaming is a critical and integral component because it is important to know how different groups of women and men use, manage, and conserve the marine and coastal environment, so that policies and projects can engage them equitably and effectively in sustainable management practices. Research has shown that involving women in decision making has positive impacts on social and environmental programs.

China would truly benefit by continuing to develop its efforts to manage its ocean interests (both nationally and globally) so as to strike a balance between protection and production and on the basis of the principle of IOM.

3.3 Recommended Actions

- *Bring the ocean environment clearly into the framework of "Beautiful China" the importance of the ocean environment as the basis of life in the 14th FYP, and as such consider strengthening policies and activities that support and sustain the biodiversity of the oceans—in particular, actions that*:
  - *Aim to* significantly reduce *land-based pollution (and pollution from maritime operations) reaching the sea, including promoting innovations in cleaner production methods and means of reducing discharges and emissions.*
  - *Avoid further marine habitat destruction and prohibit reclamation of coastal wetlands, and, during 2020–2030, re-establish degraded/destroyed coastal wetlands that once served as key habitats.*
  - *Through innovation, development and use of new technologies*, promote environmentally friendly mariculture and combat illegal fishing. *By 2025, full output control should be implemented for all fishery activities.*
  - *Implement the ecological* environment *damage compensation system and improve the public participation system for marine environmental protection. The Ministry of Ecology and Environment should carry out natural capital accounting for the ocean in order to undertake an evaluation of an official's ecosystem-based performance when carrying out his/her term assessment.*
- *Within the framework of the FYP, clearly flag a commitment to fulfil the Paris Agreement and actively use IPCC assessment reports and the IPCC special report on ocean and cryosphere in a changing climate in framing climate action policies. In particular, it is recommended that*:
  - *China considers how it can actively use the restoration of coastal wetlands in this context, noting the high capability of wetlands to sequester carbon.*[1]
  - *The development of a platform/framework for understanding and assessing the impacts of climate change on living marine resources and evaluate ways to mitigate the impacts.*

---

[1] Coastal wetlands are particularly effective at sequestering carbon. Although they occupy less area than woodlands, coastal wetlands take in carbon emissions and convert them into plant biomass more quickly, are better at trapping organic carbon from within their own ecosystem and other sources, and delay the decay of organic material (which releases carbon) for longer periods.

- *The ocean environment will, over the next decades, undergo substantial change, in particular due to climate change, as will the types and extent of ocean industries. The scale of these changes challenges current management regimes. There is, therefore, an essential and urgent need to understand these changes and develop ecosystem-based integrated management frameworks that capture this dynamic development in nature and ocean economies, e.g., climate change adaptation.*
- *Continue the efforts laid out in the current FYP with regard to strengthening the ocean economies, and, in doing so, in particular consider strengthening policies and activities within the ocean industries that:*
  - *Promote and embed the principle of circular and green economy in the ocean industries.*
  - *Enable restoration of important fishery habitats and implement enforceable measures on overfishing, including the two-way control of both fishing boat inputs and catch yield, based on the experience from the current pilot trial.*
  - *Enable the development of "green ports" and "green ships," including the fishing fleet/ports.*
  - *Deploy green technology and solutions along the BRI.*
- *China can take international leadership on issues and actions supporting sustainable ocean industries and promote international cooperation in matters relating to sustainable ocean management. These include: e.g.*
  - *Through the BRI, formulating guidelines for zero-environmental-impact DSM, developing green fishing vessels and green fishing ports, green mariculture and promoting green shipping in the Arctic, as well as promoting the concept of IOM as a management principle.*
  - *Participating actively and spearheading relevant discussions in key international processes and foras that provide a framework for sustainable activities, such as the Paris Agreement, the Biodiversity Convention, the ongoing discussions on Biodiversity Beyond National Jurisdiction, the International Maritime Organization, International Seabed Authorities, etc., and through this* inter alia *contribute to the timely development of appropriate legal and environmental framing of novel and emerging ocean industries, such as ORE (Ocean Renewable Energy), minerals, and biotechnology.*
- *Increase investment in maintaining an up-to-date knowledge base (science and technology) and data-sharing capability required for ocean management and governance and consider new and innovative approaches supporting such research and innovation, including: e.g.:*
  - *Expanding and implementing national and regional systematic programs for data and knowledge gathering and technology development, as well as innovative methods for disseminating data and knowledge.*
  - *Supporting and actively investing and engaging in the IOC Ocean Decade, acting as a spearhead for the international effort.*

- *Encourage, through relevant means, the use and development of green technologies supporting the ocean-based industries of China. More targeted government investment and preferential fiscal and tax policies in green technologies could help industries overcome financial obstacles which sometimes impede the creation of environmental technologies.*
- *Encourage and enforce the use of scientific knowledge and monitoring results relevant to the management of the ocean ecosystem and economies, in particular by providing mechanisms and opportunities to access to such a knowledge base. Consider establishing a formal mechanism at the national level, for example by a scientific advisory body, to underpin coordinated and holistic use of knowledge in instituting overarching policies on the development of ocean economy and the implementation of ecosystem-based IOM.*
- *Develop and provide organizational structures/bodies, guidance and legal frameworks that enable cross-boundary (administration and land–ocean connectivity) and cross-sectoral coordination and communication, both on and between national, provincial, and local levels. Specifically, it is recommended to establish a coordination mechanism across relevant government ministries to support the development of policies fostering and underpinning ecosystem-based IOM in China.*
- *Acknowledge the role of gender in sustainability focus in IOM and systematically work toward gender mainstreaming as an integral component in further developing IOM systems in China. Additional efforts are needed to understand the gender gap and improve women's educational, social, and economic opportunities and responsibilities in China's ocean economy. To support this, a clear, directed, and strategic gender program could be developed and implemented to enhance women's participation in all aspects of ocean economy, including industry, management and governance.*
- *Additional efforts are needed to understand what role the ocean can play in strengthening societal resilience in the event of anomalous and unprecedented disturbances such as the recent COVID-19 pandemic which has shown how vulnerable societies are. Management of the ocean system and services should take into account the ocean's ability to support society in such undesired events.*

3.4 Future Directions of Work

The ocean is expansive, and the related and relevant governance and management issues are complex and extensive. Thus, even if this work has been addressing a number of key aspects, we urge that ocean studies need to continue within the framework of CCICED to fully reflect the importance of the ocean to society.

Through our efforts, we have identified a number of areas that could merit further consideration to provide guidance to China's leadership.

Given that a new 5-year period of CCICED is starting already in 2021, we suggest *that CCICED in the period between AGM 2020 and AGM 2021 arrange and undertake an "ocean into the future" scoping workshop* where all members of the current ocean SPS and members from other relevant SPSs will be charged to discuss and suggest what and how ocean issues can and should be taken further in the next five-year period of CCICED, in essence pointing to the direction for establishing a roadmap for where China should be with regard to ocean issues in the future.

Topics that the workshop could consider include—but are not limited to—the following issues:

- Opportunities to promote Green BRI (e.g., eco-friendly fisheries, Arctic green shipping, mineral exploration framework);
- How to achieve 2030 SDGs through enhanced global ocean governance;
- Areas and issues where China's leadership is particularly relevant in the global context;
- Challenges and opportunities in technology supporting ocean economies;
- Management frameworks for the ocean economy industries;
- Ocean- and coastal-related tourism;
- Structures and mechanisms supporting innovation in developing new ocean economies;
- Opportunities to consider gender-related aspects of ocean activities and management;
- Coastal restoration supporting both protection and climate action (the ocean's answer to replanting of forests on land);
- Biodiversity in the ocean context, including issues relating to MPA planning, seen in the light of recommendations and findings from the coming/planned COP 15 of the CBD;
- Opportunities to promote land–ocean coordination through basin-wide integrated governance; and
- Identification and framing of key strategic areas upon which new ocean innovation and blue finance can be built.

Appendix: Specific Recommendations from the 6 Task Teams of SPS Ocean Governance

Integrated Ecosystem-Based Ocean Management

Recommendations from SPS TT 1

Overarching

Recommendation 1: Taking full potential of the new government structures (2018), develop and provide organizational structures/bodies, guidance and legal frameworks that enable cross-boundary (administration) and cross-sectoral coordination and communication (business) supporting ecosystem-based management, both on and between national, regional and local levels (across geographical scales, including land–ocean interactions). Specifically it is recommended to establish a coordination mechanism across relevant government ministries to support the development of policies fostering and underpinning ecosystem-based integrated ocean management in China.

Recommendation 2: An up-to-date and relevant knowledge basis is fundamental in order to undertake a fully ecosystem-based management of an ocean system that is in a constant flux, both due to natural variations, climate change and human interaction. It is suggested that:

A. Expand and implement national and regional systematic programs for data and knowledge gathering and innovative methods for disseminating data and knowledge, including but not limited to data and knowledge relating to coastal wetland ecosystems, nursery grounds, ecosystem services, sea level rise, phenological change.
B. Establish a formal mechanism on the national level, such as for example a scientific advisory body, to underpin coordinated and holistic use of knowledge supporting the development of overarching policies on ecosystem-based integrated ocean management.

Marine Spatial Planning

Recommendation 3: Marine Function Zones (MFZ) has been well established as key aspect for China's ocean and coastal management. Put increased emphasis on implementing an *ecosystem-based approach*, taking into account both the third (depth) and fourth (time) dimensions of the ecosystem dynamics, in the utilization of the well-developed spatial planning tool.

Recommendation 4: Integrate the ongoing MPA planning efforts into the broader marine spatial planning and ocean zoning efforts to strengthen the integrated and overarching management approach.

Land–Ocean Interaction

Recommendation 5: Ensure that the legal and administrative frameworks supporting IOM in China capture the connectivity and differences between land and ocean in an integrated and adaptive manner. Ensure the regulation and mandate from terrestrial and coastal management in same areas maintain consistent, while further improve cooperation and partnerships.

Recommendation 6: According to River Chief system, consider implementing a Bay Chief System nationwide, including providing an effective administrative support model through the establishment of a *Bay Chief Office* with comprehensive coordination capability.

Climate Change

Recommendation 7: Ensure that the significant ongoing climate changes affecting the entire ocean system is sufficiently considered in ocean governance and management through a dynamic and adaptive use of the best available projections of climate change.

Recommendation 8: Encourage and use knowledge, science and monitoring in the context of both global/regional climatic changes and localized stressors from the land and the coastal system as basis for ocean management and governance.

Sustainable Use of the Ocean

Recommendation 9: Develop and provide guidelines and principles to support the public–private partnership in the governance, management and financing for sustainable ocean-based economies. Learn from the global and international experience, such as the guidelines from UN Global Compact [14] and the High Level Panel for a Sustainable Ocean Economy.

Recommendation 10: Where appropriate, update current management and governance regimes relating to ocean economies to reflect the principles of knowledge-based, ecosystem-based and integrated approach.

Recommendation 11: China should design and implement a specific and strategic gender program targeting the enhancement of women's' participation in all aspects of ocean industry.

Marine Living Resources and Biodiversity

Recommendations from SPS TT 2

1. **Strengthen legal protections for coastal and marine ecosystems, while promoting sustainable production**. China should consider enacting a new aquaculture law that places limits on facilities' waste discharge and resource use and which should mandate stock reporting by all facilities, authorize routine onsite inspections and include other provisions that mitigate the industry's impacts on coastal and marine ecosystems. Ongoing efforts to shift toward output control in its capture fisheries should be integrated with rights-based approaches that allocate portions of the catch or local fishing areas to the fishing industry and communities. A Marine Habitat Conservation Law (MHCL) should also be enacted, to strengthen protections for coastal and marine habitats and encourage significant rehabilitation of lost ecosystem functions and resiliency.

2. **Implement a high-tech monitoring system to combat corrupt and illegal activities that undermine compliance and to improve marine science**. China's innovation in sensors, networking technologies and artificial intelligence can help create a transparent system that can operate across agencies, and even globally, to facilitate enforcement and promote compliance in protecting marine ecosystems. In addition to promoting compliance, a high-tech monitoring system will generate data for ecosystem understanding, emergency contingency actions, and climate change response and mitigation measures.

3. **Restore lost coastal and marine ecosystem functions needed to support fisheries production, biodiversity conservation and resilience to development, pollution and climate change**. Further actions than the ongoing redlining process need to be taken to restore lost habitat, including mangroves, seagrass beds, tidal marshes and flats, and coral reefs. If China's coastal and marine ecosystems are to withstand the impacts of pollution and climate change and continue to be a source of tremendous prosperity and food production, China should consider (i) establishing a national "marine ecological report card" on the health of China's coastal and marine ecosystems; and (ii) develop a national plan of action to restore lost ecosystem functions and services.

4. **Create a network of partnerships among countries along the Maritime Silk Road to promote sustainable marine governance and achieve the Sustainable Development Goals**. The Maritime Silk Road Initiative represents a historic opportunity for China to demonstrate leadership in global marine governance and advance the UN Sustainable Development Goals. Under this Initiative, China should consider creating a network of partnerships to encourage mutual learning and promote joint actions that promote a healthy ocean. Sustainability along the Marine Silk Road can also be promoted by information sharing and capacity building on developing and managing living marine resources sustainably. China's leadership could be further demonstrated by using the Maritime Silk Road Initiative to catalyze the development of regional and global approaches that can mitigate the impact of climate change on living marine resources.

5. **Assess the effects of climate change on living marine resources and evaluate ways to mitigate the impacts**. China could promote more research into the impact of climate change on capture fisheries and mariculture, and the natural ecosystem services upon which these industries depend. China may wish to consider ways to not only mitigate the effects of climate change, but effectively adapt to it.

Marine Pollution

Recommendations from SPS TT 3

I. **Establishing a holistic mechanism of land-sea coordination in joint marine pollution prevention and control**

Significantly enhance the land-sea ecological environment monitoring unity. In accordance with the principle of land-sea coordination and the unified plans, optimize the construction of a fully covered and refined marine ecological environment monitoring network, strengthen gridded and real-time monitoring, and develop the online monitoring for the primary rivers and outlets discharging and atmospheric deposition of pollutants into the sea. Establish a baseline survey/census system for marine pollution.

Enhance management and prevent land-based pollution from the agricultural, pharmaceutical sectors. Full consideration should be given to improving overall agricultural production capacity and to preventing and controlling rural pollution. Development of environmental protection facilities, such as those for handling rural wastewater and refuse, should be bolstered by subsidies from governments and village collectives, fee payments from residents, and the participation of non-government capital. A variety of assistive measures should be adopted to foster and develop market entities for the control of all types of agricultural pollution from non-point sources and for the handling of rural wastewater and refuse. Green production way in agriculture should be pursed to promote making full use of agricultural wastes. According to the market-based rules, a green finance system is encouraged to support the pilot of disposal and harmless treatment of livestock and poultry breeding. Comprehensive utilization of livestock and poultry manure might be gradually achieved on the spot. Subsidies for the production of organic fertilizers from the comprehensive utilization of livestock and poultry manure need to be increased, and simultaneously subsidies for chemical fertilizer to be reduced. The management of antibacterial drugs used for human and animals should be strengthened. Proper procedures should be introduced to restrict the use of chemicals such as antibiotics in accordance with the law, and prohibit the abuse of antibiotics.

Further improve China's marine environmental quality target system. China's marine environmental quality target system is mainly based on water quality targets, which are often expressed by the under-criteria rate of marine functional zoning or clean water (below the criteria of grade I, II). Suggest to further enrich the content of China's marine environmental quality target system, in addition to the water quality target, the spatial and temporal distribution characteristics of marine ecosystems need to be combined, further increase marine ecological protection target, such as the biodiversity, habitat suitability, ecosystem structure and function, etc., lay the foundation and direction for the marine ecological protection work. Strengthen the connection of sorting, indices selecting, and valuing of water quality standards between surface water and seawater, and introduce new indices such as total phosphorus, total nitrogen, and emerging pollutants. Advance seawater quality standards revision. Take a holistic approach for emissions control and water quality target management in the river basins and offshore areas.

Construct an integrated governance mechanism for the River Chief and Bay (Beach) Chief systems. In accordance with the holistic approach to conserving our mountains, rivers, forests, farmlands, lakes, and grasslands, strengthen coordination of the comprehensive management of rivers discharging into sea, bay and estuarine. Establish a joint-action mechanism between the River Chief System and Bay Chief System, set a regular consultation mechanism and emergency response mechanism, and enhance the capability of pollution prevention and control in a holistic approach for land and sea.

II.   **Strength lifecycle management for plastics, and formulates a national action plan for marine debris pollution prevention and control**

Strengthen the source control of plastics debris. Explore the waste reduction and harmless management pattern in line with national conditions, and effectively prevent the entry into marine environment of microplastics and plastic waste resulting from the manufacturing production and individual consumption process, severe weather events and natural disasters in coastal regions. Strengthen the management of plastic nurdle, and put on file and supervise of the process of "resin nurdles—plastic products—usage and circulating of commodity". Encourage extended producer responsibility (EPR) and related mechanisms. Promote EPR mechanisms to involve producers, importers and retailers in the establishment of resource-efficient product value chains from the design to the end-of-life treatment and in financing waste collection and treatment. Forbid to produce and sell personal care products containing plastic micro-beads. Introduce technologies in washing machines to better capture fibres from wash-loads in both domestic and commercial/industrial uses.

Support integrated sustainable waste management. Improve and developing national waste regulatory frameworks, including legal framework for EPR, and taking care for enforcement and governance. Support capacity development and infrastructure investments for improved waste management systems in cities and rural areas through existing instruments, and promote access to regular waste collection services and facilitate investments in waste management infrastructure in order to prevent plastic waste leakage into the sea. Establish sufficient waste reception facilities at harbors in coastal cities in order to allow ships to dispose of their waste in an environmentally sound manner.

Formulate a national action plan for marine debris pollution prevention and control. Promote the establishment of sound national regulatory frameworks on waste management. Construct an integrated coordination mechanism for marine debris prevention and control across sectors, regions and river basins. Encourage green development, speed up the research and application of innovative approach for substitute for plastic products and waste treatment, and urge the manufacturing and use of degradable plastic products and substitutes for plastic. Strengthen researches on sources, transport and fate of microplastics as well as the impact on marine ecological environment, and improve the scientific understanding of microplastics. Call on all relevant stakeholders to engage and encourage social organizations, communities and the public to reduce plastic waste generation, hold clean-up activities, significantly reduce the unnecessary use of single-use plastics, and live green-consumption lifestyle, with the aim to prevent and significantly reduce marine microplastic pollution.

III.   Develop a market system which allows economic levers to play a greater role in marine environmental governance and ecological conservation

Accelerate industrial innovative and green development and transformation in coastal areas. Promote industrial upgrading toward to emerging industries and modern service industries. Strengthen the construction of industrial zones, promote circular economy and green production, build ecological industrial zones, and enhance the integrated and recycling utilization of resources. Set the binding requirements including industrial structure and layout, resource and environmental capacity loads, and ecological red lines. Strengthen the management of project approval, enhance the market entry, compel industrial transformation and upgrading, and progressively fall into disuse lagged behind production capacity.

Improve the system for compensating marine ecology conservation efforts. Persist to the principle of "who benefits, who compensates", comprehensively use fiscal, taxation and market measures, adopt the form of incentive instead of subsidies, and establish a compensation mechanism for marine ecological conservation.

Strictly implement compensation systems for ecological and environmental damage. Tighten manufacturers' legal responsibilities for environmental protection, and significantly increase the cost of illegal activities. Improve legal provisions concerning marine environmental damage compensation, methods for appraising damage, and mechanisms for enforcing compensation. In accordance with the law, mete out penalties to those who violate environmental laws and regulations, determine compensation for ecological and environmental damage by the extent of damage and other factors, and pursue criminal liability when violations result in serious adverse consequences.

Establish a diversified funding mechanism. Integrate various types of marine environmental protection funds by central budget, increase financial support, and keep supporting the rural environmental governance and Blue Bay restoration actions. Bring into full play the initiative of local budget, enhance local financial support, make full use of market investment and financing mechanisms, and encourage and attract private, social, venture capital and other funds to gather in the area of marine environment protection.

Improve coastal wetland grading management system. Establish important coastal wetlands grading management systems at national and local levels, release in batches the national important coastal wetlands list, and identify the control proportion target of coastal wetlands at local level. Innovate the protection pattern, and establish the coastal wetland pilot national park.

Establish degraded coastal wetlands restoration system. In accordance with the natural attributes of marine ecosystems and the characteristics of coastal biota, carry out the coastal wetland restoration. Implement the restoration projects, including restoring the coastal aquaculture farms back to wetlands, culturing densely vegetation, conserving habitat, improve the community structure of wetland vegetation, and raise the biodiversity of wetland habitats. Expand the coastal wetland area and recover the ecological services of wetland, such as water purification, carbon sequestration. By 2020, the restored area of coastal wetland will be more than 20,000 ha.

IV.   **Strengthen cooperation and exchanges, and jointly address global marine pollution**

Strengthen research on emerging marine environmental issues of global concerns. Conduct survey and research on ocean acidification, plastics and microplastics, oxygen deficiency in hotspot areas, and comprehensively analyze the emerging marine environment issues of global concerns, particularly in the high seas and Polar Regions. Deeply participate in the designation of high seas protected areas, environmental impacts assessment of seabed development activities, and research on marine environmental protection in Polar Regions, and play our part in global marine environmental governance.

Establish Maritime Community with a Shared Future to jointly address marine pollution. With the aid of the twenty-first century Maritime Silk Road, carry out pragmatic and efficient cooperation and exchange under the framework of the Asian Infrastructure Investment Bank, China-Pacific Island Economic Development Cooperation Forum, China-ASEAN Maritime Cooperation, and Global Blue Economy Partnership Forum etc. Strengthen research on marine environmental issues of global concerns, build a broad blue partnership, jointly improve the ability to address and control marine pollution. Establish China-ASEAN Marine Environmental Protection Cooperation Mechanism, and promote international cooperation. Enhance capacity on pollution monitoring and governance through sharing knowledge making best use of other relevant efforts in the region such as PEMSEA, APEC, NOWPAP and COBSEA, GPML, GPNM, GWI and work together to build a community of shared future for mankind.

Green Maritime Operations

Recommendations from SPS TT 4

1.   **Establish emission control area under the framework of the MARPOL Convention**

Expand the geographical area of ECAs and tighten emission standards. Meanwhile, cooperate with neighbouring countries in the application of IMO-designated cross-national emission control areas.

2. Implement special protection for the Bohai sea

Further improve laws and regulations on environmental protection for the Bohai Sea, and introduce *the Bohai Sea Protection Law*. Establish a committee dedicated to the management of Bohai matters such as ecological protection, pollution prevention and sustainable utilization. Delineate a special control area around the Bohai Sea where tighter emission controls apply. Apply to IMO for the recognition of the Bohai Sea as a particularly sensitive sea area.

3. Carry out Green Port Action Plan

Promote integrated growth and synergetic planning between port and urban–rural development. Optimize cargo handling systems by improving railway connections, sea-rail and sea-sea transfers as well as landscape construction. Remediate old ports by upgrading the infrastructure for receiving and treating ship-sourced pollutants. Promote new and clean energies as well as operations geared towards *"zero emission"*.

4. Enhance vessel cleanliness

Establish lifecycle-based evaluation standards. Encourage Chinese ship owners to join international incentive programs and reduce emissions voluntarily. China should formulate its own incentive policies based on advanced experience at home and abroad, rewarding greener vessels through incentive programs. Scrap outdated ships and revise *the Old Vessel Management Regulations* to tighten the age threshold. Set up a funding pool to subsidize scrapped vessels, diesel engines, exhaust cleaning systems and emission control technologies. Draft an emission inventory for ship-sourced pollutants and evaluate the current port pollution, providing data support to precision pollution control of vessels.

5. Improve GHG reduction mechanisms

Establish a measurement, report and verification (MRV) mechanism for ship-sourced GHG emissions. Formulate standards and guidelines for emission verification as well as planning on vessel energy efficiency and statistical analyses. Incorporate coastal shipping into emissions trade system, which will lay a solid foundation for Chinese shipping to participate in international carbon trade systems.

6. Strengthen maritime risk control focusing on chemical emergency response and emergency response integration

Enhance risk prevention of both land-sourced and marine pollution. Improve law enforcement, reduce latent risks and nib accidents in the bud. In the coastal waters with high risk of hazardous chemicals, relying on the existing ship oil spill emergency equipment storehouse, supplement the emergency equipment and materials of hazardous chemicals, and continuously improve the capacity of hazardous chemical pollution accidents. Establish an inter-departmental emergency information system for more effective coordination, information exchange and decision making. Actively participate in international cooperation and accumulate more resources for China while contributing to global emergency response to pollution accidents at sea.

7.   **Improve environmental requirements for fishing vessels/ports**

Conduct specialized actions to improve the environmental quality of fishing vessels and harbours. Invest in environmental facilities and increase the protection level of fishing vessels and harbours. Tighten emission controls by extending the scope of atmospheric pollutants to fishing vessels, which shall apply the same standards as commercial vessels and harbours. Strengthen supervision of fishing vessels via safety inspection and screenings against potential hazards, further lowering risks of accidents.

8.   **Engage in and explore potential for Arctic green shipping**

Step up research cooperation with Russia and Nordic countries on the utilization of the Arctic shipping route. Advance research on Arctic navigation. Build north-eastern China into an important hub connecting China to Arctic routes, opening up new directions for deeper economic and trade cooperation between China and Europe via the *Ice Silk Road*.

Marine Renewable Energy
Recommendations from SPS TT 5
The following are the specific recommendations put forward for ocean renewable energy (ORE). First of all, they emphasize that an industrial supporting policy mechanism should be established and improved. Furthermore, the scale of ORE utilization should be promoted, whilst financial or venture capital communities as well as private capital should be encouraged by governmental policies. Finally, offshore wind should be accelerated whilst environmental and socio-economic impacts assessed; mechanisms to accelerate commercial realisation of other ORE technologies should be supported by the government.

Policy

- Industrial supporting policy mechanism should be established and improved.
- Scale of ORE utilization should be promoted.
- Enable RD&I to address challenges to reduce costs further to reach parity with other energy technologies.
- Enhance capacity to accelerate innovative and resilient technology development.
- Engage at an early stage with stakeholders include fishermen, community members, regulators, developers, scientists, and tourists.
- Integration emerging ORE technologies into wider applications such as military applications, electricity generation for remote communities, desalination, hydrogen production or aquaculture applications.

Market

- Financial or venture capital communities as well as private capital should be encouraged by governmental policies.
- Strengthening the global export and market opportunities.
- Grow ORE industry, create jobs and take advantage of opportunities within its competency to global markets.

Offshore wind

- Offshore wind should be accelerated whilst environmental and socio-economic impacts assessed.
- Increase Offshore Wind deployment addressing many strategically important goals such as decarbonisation, security of supply, and new business opportunities.

Marine Energy

- Tidal current energy research and development should be encouraged by government as expected to be next type of ORE.
- Mechanisms to accelerate commercial realisation of ORE technologies ORE technologies (wind, wave, current, tidal range, ocean thermal) should be supported by the government.

Mineral Resource Extraction—Deep-Sea Mining

Recommendations from SPS TT 6

1. Improvement of environmental management system

    - **Engage with development of environmental rules**: China should actively engage with International Seabed Authority (ISA) regarding development of Regulations, Standards and Guidelines, specifically towards environmental baseline, EIA, and EMMP development.
    - **Further improve national legislation**: China may review and update the Deep Seabed Area Law in order to comply with the new requirement of the exploitation regulatory framework developed by the ISA within the context of the domestic legal system, to deal more specifically with future exploitation activities, including financial terms, inspection and management, and indemnities to ensure the State is properly protected. Based on the assessment China may seek to develop additional regulations to supplement the ISA requirements, drawing on the concepts of sound environmental management.

2. Filling gaps in environmental understanding and technology

    - **Strengthen scientific understanding and develop key technologies**: China should aim to improve the understanding of, and better assess both the risks and opportunities associated with DSM as well as exploitation of

NGH. This includes (but is not limited to) (1) strengthening environmental data collection in important marine areas to improve the understanding of deep-sea ecosystems; (2) developing environmentally critical technologies concerning environmental monitoring, EIA, safe operations and environmental restoration; (3) actively promoting the development of environmentally friendly solutions to key technical problems for exploration, exploitation and transportation of deep-sea mineral resources and natural gas hydrates.

- **Improve the understanding for NGH**: China should aim to improve the understanding of, and better assess both the risks and opportunities associated with NGH exploitation.

3. Expanding value chain and promoting circular economy

- **Expand value chain**: China should seek opportunities for Chinese industry to engage at all levels of the DSM value chain, including research, exploration, exploitation, equipment manufacturing, technology design and mineral processing.
- **Promote circular economy**: China's DSM policies should proactively support the intentions described in SDG #12, where the ambition of creating a circular economy is embedded in the design from the beginning of the design and concept phase and that "all" collected materials are fully utilized while waste streams are minimised. In addition, should NGH exploitation be deemed environmentally and economically feasible, China should promote the development of carbon capture and storage to accompany the development of hydrate extraction technologies that enable NGH to become a "bridge fuel" towards a low-carbon future.

4. Creation of cooperative and transparent mechanisms and platforms

- **Enhance data sharing**: Seabed mineral contractors should be encouraged to share widely through globally and publicly accessible databases all environmental data acquired through DSM research programmes. China should play a leading role in establishing good practice for quality control, data sharing, and transparency.
- **Conduct cooperation**: China should strengthen international cooperation, especially bilateral and multilateral cooperation and exchanges, including jointly contributing to the development of cooperation mechanisms and platforms, jointly building open markets, and jointly promoting marine technology exchanges.

5. Enhancement of leadership towards the ISA and active support of the UN SDGs

- **Support the UN SDGs**: China should actively relate to the UN SDGs when further maturing the business case for DSM, such as contribution towards # 14—life below water and #5—gender equality in education and training for DSM professionals within geology, engineering and environmental technology.

- **Enhance of leadership**: China should continue to initiatives to strengthen ISA as a regulator, and actively engage with ISA, such as to take opportunities for convening group discussions as well as take active leadership both within thematic groups and in its geographic group (Asia–Pacific); to show the leadership around a good model for State sponsorship, to establish a network for consultations and informing of its national positions on DSM.
- **Support REMP process**: China should support a standardised, transparent and consultative REMP process at the ISA. This should include the establishment of a network of biologically representative, fully protected no-mining zones.

References

1. Ma, Z., Melville, D.S., Liu, J., Chen, Y., Yang, H., Ren, W., Zhang, Z., Piersma, T. and Li, B. 2014. Rethinking China's new great wall. Science, 346(6212), 912–914.
2. Xi, JP, 2014, The governance of China. Beijing: Foreign Languages Press.
3. MOT (Ministry of Transport), 2019a, Report on China's Shipping Development 2018, Beijing: China Communications Press.
4. Sharma, R., 2017, Deep Sea Mining: Resource Potential, Technical and Environmental Considerations. Springer International Publishing.
5. Klauda, J. B. & Sandler, S. I., 2005, Global Distribution of Methane Hydrate in Ocean Sediment, Energy and Fuels, 19(2): 459–470.
6. Gerber, L. J. & Grogan, R. L., 2020, Challenges of Operationalising Good Industry Practice and Best Environmental Practice in Deep Seabed Mining Regulation, Marine Policy, 114: 103257, doi:https://doi.org/10.1016/j.marpol.2018.09.002.
7. Rabalais, N.N., Diaz, R.J., Levin, L.A., Turner, R.E., Gilbert, D., Zhang, J., 2010. Dynamics and distribution of natural and human-caused hypoxia. Biogeosciences 7 (2), 585–619.
8. Bureau of Fisheries (BOF), Ministry of Agriculture and Rural Affairs (1949–2020). China Fisheries Statistical Yearbook. China Agriculture Press, Beijing.
9. MOT, 2019b, Statistical Bulletin on the Development of the Transport Industry 2018, http://xxgk.mot.gov.cn/jigou/zhghs/201904/t20190412_3186720.html.
10. University of Edinburgh's Policy and Innovation Group, Energy Systems Catapult, 2020. Wave and Tidal Energy: The Potential Economic Value. https://periscope-network.eu/analyst/wave-andtidal-energy-potential-economic-value.
11. Smart G, Noonan M, 2018. Tidal stream and wave energy cost reduction and industrial benefit. Offshore Renewable Energy Catapult. https://s3-eu-west-1.amazonaws.com/media.newore.cat apul/app/uploads/2018/05/04120736/Tidal Stream and WaveEnergy-Cost-Reduction-and-Ind-Benefit-FINAL-v03.02.pdf.
12. Information office of MOT, 2019, Report on Chinese crew development 2018, https://www.msa.gov.cn/public/documents/document/mdk1/mdm5/~edisp/20190626095039643.pdf.
13. Government of the People's Republic of China, 2016. China's National plan on implementation of the 2030 agenda for sustainable development.
14. "Homepage: UN Global Compact." Homepage I UN Global Compact, www.unglobalcompact.org/.

Chapter 4
Green Urbanization Strategy and Pathways Towards Regional Integrated Development

Accelerating China's Green Urbanization Based on Ecological Civilization[*]

4.1 Why Is the Green Urbanization Transition So Critical

4.1.1 The Basic Tasks for China's Urbanization

An important driving force for the rapid development of China's economy is rapid urbanization. In 1949, only 10.6% of China's population lived in cities. In 2019, China's urbanization level reached 60.6% [2]. According to the experience of industrialized countries, it is estimated that by 2035, about 70% of China's population will live in urban areas. In 2050, this proportion will rise to around 80% (Fig. 4.1). The National Population Development Plan (2016–2030) predicts that China's total population is expected to reach 1.45 billion by 2030, which will gradually decline afterwards. The urbanization rate will be 70% at that time. The United Nations Population Division also predicts that China's population will peak around 2030, and would decrease to 1.4 billion by 2050 [3]. This means that China's urbanization level still has about 20% points to increase, and the newly added urban population is about 200 million people.[1]

Therefore, China's green urbanization faces two basic tasks: first, how will the 300 million people be urbanized in a green way; second, how will the existing cities

---

[*]Part of this thematic policy study report has been published. Refer to: Zhang [1].

Please pay attention that the main body of this chapter has been published in *Chinese Journal of Urban and Environmental Studies* https://doi.org/10.1142/S2345748121500019

[1] However, there is a dispute over the prediction of China's new urbanized population in the future due to great differences in the population forecast by 2050. The emphasis of this report is not to study the newly urbanized population but to illustrate the importance of urbanization.

© The Author(s) 2022
China Council for International Cooperation on Environment and Development (CCICED) Secretariat,
Green Consensus and High Quality Development,
https://doi.org/10.1007/978-981-16-4799-4_4

Fig. 4.1 Rapid growth of urbanization in China 1970–2050. *Data Source* Development Research Center of the State Council (DRC) Green Team

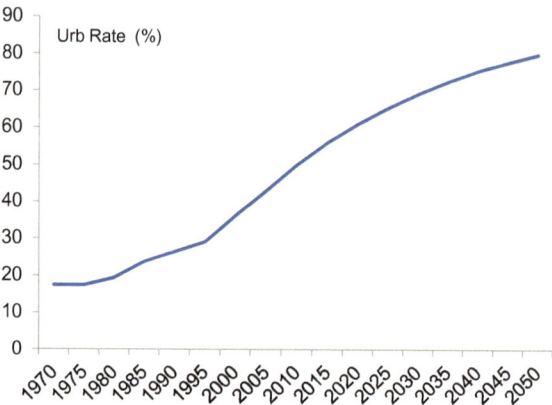

formed in the era of traditional industrialization become sustainable through a green transformation that injects vigour into the economy.

To answer these two questions, it is better to understand the nature of urbanization. The agglomeration process of population and economic activities in cities, i.e., urbanization, has greatly accelerated the process of industrialization. Human society has thus formed a modern social structure based on industrial civilization and the basic urban–rural economic geography of "urban–industry and rural–agriculture." Therefore, urbanization has long been regarded as an important driver for economic growth. However, there are problems with modern cities, and environmental pollution is one of the most severe problems. That is why green urbanization is so currently important.

4.1.2 The Basic Characteristics and Consequences of Traditional Urbanization

Historically, it took a long time for cities to emerge, but large-scale urbanization is a product of the Industrial Revolution.

The urbanization that took place in the traditional industrial era has two basic characteristics:

First, from the perspective of economic development, the function of the city is mainly to promote the production and consumption of industrial wealth, i.e., to promote the process of industrialization. Correspondingly, the function of urban infrastructure is also largely to facilitate the production of industrial products. More generally, the economic development process based on traditional industrialization is a process of urbanization that transfers a large amount of agricultural labour to urban manufacturing, forming a pattern of urban–rural economic geography of "urban–industry; rural–agriculture."

Second, the organization of the city itself is mainly based on the centralized logic of traditional industrialization. The city's design philosophy relies too much on industrial technology, rather than relying on ecological ideas to make nature work for the benefit of humanity. For example, heating, energy, construction, water treatment, etc., are often costly. Fully tapping natural forces will reduce the costs of the city and increase its efficiency (see The Nature Conservancy, "Valuing Nature's Role").

This urbanization model, while greatly promoted industrialization, inevitably brings unsustainable consequences to the environment and the regional economy.

First, it results in serious environmental consequences, including air pollution, water pollution, noise pollution, and solid waste pollution, etc. This is because the traditional industrialization model centred on the production and consumption of material wealth must be based on material consumerism (e.g., encouraging over-consumption, planned obsolescence, instant products, etc.), resulting in "excessive resource use, severe environmental damage, high-carbon emissions." If economic growth still heavily depends on the material wealth, urbanization based on this will inevitably become a major source of environmental problems.

Second, the transformation of traditional agriculture into industrialized and chemical agriculture with the logic of urban industrialization has brought about serious rural ecological and environmental consequences, including environmental pollution (industrial pollution, chemical agriculture, aquaculture pollution, domestic pollution), ecological consequences caused by overmining, ecological chain destruction, monoculture, and chemical agriculture.

Third, the consequences of urban–rural and regional imbalance. In the process of industrialization and urbanization, the population will migrate from rural areas that have no industrial advantages to urban or coastal areas, which will bring irreversible impact to the former and inevitably lead to urban–rural and regional disparities.

Fourth, social and cultural costs. On one hand, there appears to be modern social diseases in big cities, where high income and low well-being have become a prominent problem. At the same time, it is difficult for migrant workers to be truly integrated into the city. On the other hand, urban and rural problems have become two sides of one coin. The original rural social fabric has been impacted by large-scale urbanization. The problem of "agriculture, rural areas, and farmers" has become serious, and a large number of hollow villages, left-behind children etc. have appeared. To this end, the 19th National Congress of the Communist Party of China has made "rural revitalization" a major strategy.

As the basis of the traditional urbanization model, the traditional growth model has significantly enhanced human well-being, while also affecting people's well-being through two channels. First, ecological damage and environmental pollution will reduce people's quality of life and well-being. Environmental problems such as air pollution, food safety, water contamination, noise, extreme weather, and biodiversity loss have penetrated all aspects of people's lives, seriously affecting their quality of life, health and safety (for example, [4]). Second, economic growth centred on the production and consumption of material wealth has failed to simultaneously improve people's quality of life and happiness. Numerous studies have shown that in many countries, including China, economic development under the traditional industrialization model does not continuously improve the level of happiness in the way people think [5–9]. When basic material needs are met, the further expansion of material wealth, although it will bring bright GDP figures, will have little effect on improving people's well-being. Meanwhile, the so-called modern way of living compatible with the industrial model brings about the disease of affluence.

In short, the traditional industrialization model, which is the basis of the existing urbanization, has brought about high material productivity, but it is unsustainable with high costs. Since these high costs are not reflected in the internal cost of enterprises but reflected as social costs, hidden costs, long-term costs and opportunity costs, they are easily ignored. At the same time, the well-being this growth model brings is also relatively low, while improving well-being is the ultimate purpose of economic growth. With the transformation of this unsustainable growth model to sustainability, the corresponding urbanization model must also be redefined on the basis of ecological civilization.

4.2 Green Urbanization: An Analytical Framework

4.2.1 The Theories Regarding Green Urbanization

There are two popular theories regarding green urbanization or sustainable urban development [10], but it seems hard for them to solve the fundamental problems of urbanization because they fail to think outside the traditional industrial box.

The first is to understand development issues and urbanization based on traditional industrialization thinking. In this model, cities represent opportunities, and economic development is the process of continuous population transfer to cities. Population concentration is conducive to economies of scale and technological innovation, so the larger the city is, the better it would be. Unsustainable issues such as the environment in cities can be addressed through technological advances and better urban design. Quite a few mainstream urban thinkers, especially economists, can be classified as such [11–15]. Some scholars believe that although many urban problems arise because of their large scale, the resolution of these problems also depends on the size of the city [16]. This thinking does not think (or fails to realize) that behind the problem of urban unsustainability, the essence is the unsustainable development model. As Einstein pointed out, "We cannot solve our problems with the same thinking we used when we created them." Another related idea is to advocate the path of small and medium-sized towns. This kind of thinking makes natural sense. The biggest problem here is that whether the emphasis is on the path of large cities or the paths of small and medium-sized towns, it is a pseudo-problem, because the fundamental driving force behind the size of the city is market forces rather than administrative planning. No force can design a city in advance, whatever its size.

The second is to emphasize the ecological and environmental capacity of green urbanization, and to emphasize that the development or scale of a city should be planned scientifically in accordance with local resources and environmental capacity. This statement is widely accepted and it seems reasonable because it is impossible for any city to exceed its environmental capacity—which seems self-evident. However, when the development content carried by a city and its organization changes, its corresponding environmental capacity will also change. The same environmental capacity can accommodate different city sizes. This urbanization approach is essentially the same as the first one that emphasizes technology, because given the constant development content, economic development must rely on technological breakthroughs; otherwise, environmental capacity will restrict economic development.

These popular ideas are, to a large extent, discussing green urbanization in the framework of traditional industrialization. Since modern economic activity mainly occurs in cities, most of the environmental problems also originate in cities. In this way, people naturally treat green urbanization as an urban issue rather than a development issue, and take existing towns as a starting point for discussion.

However, the environmental problem of a city is, in its nature, an issue of the development model, rather than the problem of the city itself. When the content and methods of economic development—which are the basis of urbanization—face a profound transformation because they are not sustainable, the corresponding urbanization model must also undergo a profound transformation. When people discuss ecological civilization, they often talk about the so-called "Green Industrial Civilization," that is, achieving the goal of sustainable development through green technological innovation without changing the traditional industrialization model [17]. However, Green Industrial Civilization is not an ecological civilization, and the two are essentially different [18].

The existing urbanization model, whether it is the economic content carried by the city or the specific organizational form of the city itself, is largely based on the logic of traditional industrialization. This development model has brought great progress to humankind, but also brought serious unsustainable problems.

Urbanization is a spatial manifestation of economic development. When the technical conditions and content of economic development change, the spatial form they require will change accordingly. Therefore, urbanization does not always increase productivity. Although the emergence of cities has a history of thousands of years, the phenomenon of large-scale urbanization in the modern sense is based on the industrialization model formed after the Industrial Revolution. In the agricultural era, cities were more used as political, religious, military, and other non-economic centres. Because agricultural activities depend on land, large-scale urbanization in the agricultural era cannot increase but reduce productivity.

Therefore, when thinking about green urbanization, we should proceed from the starting point of why cities exist, rather than starting from existing cities and towns. The environmental problem of a city is fundamentally an issue of development model, not just a problem of the city itself.

This means that on the basis of ecological civilization, the existing urbanization must be reshaped, and green urbanization promotes China's economic transformation and high-quality development.

4.2.2 Analytical Framework

Thinking about green transformation of urbanization must begin with the question of why there is a city. Before answering this question, we must first understand the mechanism of economic growth and how urbanization promotes economic growth.

The driver of economic growth is the improvement of the division of labour, and the division of labour is limited by the extent of the market [19]. There is a trade-off here, that is, a higher specialization and division of labour mean higher productivity, but the division of labour necessarily requires trade, which incurs transaction costs. If the transaction costs are too high to exceed the benefits of specialization and the division of labour, it would be difficult for the division of labour to occur and for the economy to grow [20, 21].

Therefore, how to increase transaction efficiency is the key to promoting economic growth. Urbanization is crucial for increasing transaction efficiency. In addition to the improvement of (i) hardware infrastructure such as road transportation and communication and (ii) the soft aspects of institutional and mechanism design (including effective government, property rights system, enterprise system, patent system, etc.), the geographic agglomeration of economic activities in urban areas—i.e., urbanization—plays a crucial role in increasing transaction efficiency.

Because an industrial chain is concentrated in the city, it is easier to develop the division of labour and collaboration compared to being scattered in rural areas, thus driving economic growth. The other benefits of the city include: First, the agglomeration of the population in the city can expand the market, which creates conditions for the increase of the division of labour. Second, the centralization of urban facilitates the provision of infrastructure and government public services. The concentration of public facilities such as water, electricity, gas, and communications would, compared to decentralized provision, greatly improve the efficiency of use and reduce construction costs. Third, the concentration of population in cities facilitates the exchange of ideas and is conducive to the creation and diffusion of innovation and new knowledge. In addition to the perspective of division of labour, there are many other perspectives in urban research [12, 22–25].

Therefore, there are three key factors determining the urbanization model: The first is the change in transaction efficiency; the second is the change in the provision of public facilities and public services; the third is the change in the content of development, i.e., the content of production, consumption, and trading. **Among them, the development content shifts from the industrial wealth characterized by "high resource consumption, high environmental damage and high carbon emissions" to high-quality service industries that rely more on intangible resources such as**

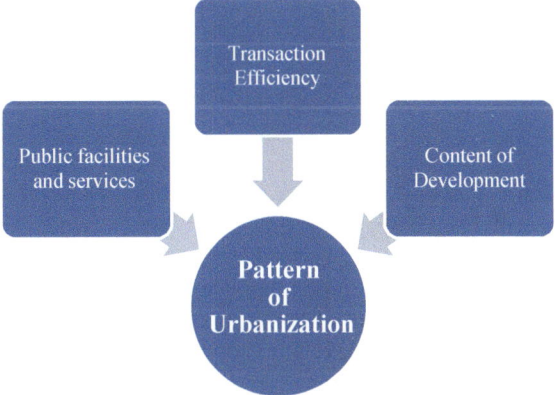

Fig. 4.2 Three defining factors of urbanization: an analytical framework. *Source* Author diagram

knowledge, eco-environment, and culture. When the three defining factors undergo profound changes, the requirements for spatial agglomeration of economic development will change, and the content and forms of urbanization will also change accordingly. A core objective of this research is to investigate the changes of these defining factors in the digital and green development era and its implications for China's urbanization, and how the government should formulate corresponding green urbanization strategies (Fig. 4.2).

4.2.3 The Emergence of Urban Clusters

Now that the agglomeration of population and economic activity is so important to economic growth, according to this logic, should the entire population gather in a super-large city? The answer is no. Driven by market forces, a hierarchical structure of large, medium-sized, and small cities will be formed, and then several central cities will be formed in different regions, which together constitute several urban clusters and metropolitan areas.

Why is there a hierarchy of large, medium-sized and small cities? Although big cities improve productivity, they also have disadvantages, including high prices and various "urban diseases" in terms of urban pollution, traffic congestion, high housing prices, crime, high mental stress, etc. Therefore, the real utility in big cities is not as high as their nominal income suggests. For example, a 10,000-yuan income in a big city does not mean that its real utility is twice that of a small city of 5000 yuan, because a large part of the income is used to pay for various costs including transport, high rent, and so on. If further taking non-monetary factors such as pollution and pressure into account, the real utility of big cities and small cities should be quite similar. This is why, under strictly market forces, different people will choose different cities to work and live, thus forming a hierarchical structure of large, medium-sized and small cities [25].

How do the urban clusters appear then? The reason is that different regions forming their central cites can minimize the spatial cost of the overall economy. In particular, a vast country with a dense population like China will certainly form a number of regional central cities and metropolitan areas, and within each of them will form a hierarchical city structure. The phenomenon that a large part of a country's population is concentrated in a mega city is more likely to occur in territorially small countries. The total transaction costs of the population to agglomerate in different central cities are often lower than that of the entire population agglomerating in a single large city. In addition to cost, how cities are geographically distributed also depends on the benefits of agglomeration to production, which is affected by land area, population size and initial distribution, industrial structure, natural endowments, transportation, climate, culture, institution, and etc. Such factors affect the cost and benefits of agglomeration and thus affect the geographical pattern of urbanization.

4.3 The Future Green Urbanization Model in China

4.3.1 The Key Factors Defining Urbanization Are Changing

As human society enters the digital and green era from the traditional industrial era, the three key factors that determine the urbanization model are undergoing dramatic changes. These changes are particularly dramatic in China. This means that China's future urbanization model will undergo profound changes.

First, there has been a dramatic increase in transaction efficiency. With the advent of the Internet, the digital age, and the rapid transportation system, the traditional conceptualization of space and time is undergoing major changes. Many economic activities do not necessarily have to rely on the large-scale physical concentration of production factors and markets as in the industrial era, and no longer have to be undertaken in the city or at a fixed location.

Second, technological changes have made it possible for some public facilities and services that originally relied on physical concentration to be provided in a decentralized manner. For example, heating, sewage treatment, distributed energy, garbage disposal, etc., can be transferred from centralized supply to distributed supply in many cases. This means that in some small towns and villages, high-quality living can be achieved at a low cost. In the digital age, many government services are also accessible through digital platforms.

Third, and more importantly, the economic development driver changes. As discussed above, the traditional industrialization model will inevitably lead to an unsustainable environment. One of the important tasks of green urbanization is to change the supply that meets people's new demands. Among them, **increasing demand for the emerging services that meet people's expectations of a better life is the direction of green development and is the new economic foundation of green urbanization**. Although urban agglomeration is still very important, economic development requires less physical concentration than it once did. In particular, rural areas and small towns excel in good environment and culture. As a result, many new economic activities may emerge in the countryside, and the urban–rural relationship will be redefined.

4.3.2 The Implications of Green Urbanization

It is important to point out that although the above changes in the three key factors have made many economic activities less dependent on the physical concentration of the factors of production as in the past, this does not necessarily mean "the decline of the city," nor does it mean that a large number of economic activities will leave the city. It means the traditional urban and rural concepts need to be redefined through which new sources of economic growth would be emerging.

- **The profound change of economic activities carried by the city**. People's demands for a good life are not just reflected in material wealth. As demands upgrade, the content of economic development expands from traditional material wealth to emerging services. Many economic activities that did not exist under traditional development will appear. For example, the large population of existing cities could be an advantage for developing a culturally creative and experience economy, thus transforming the development content; in addition to producing agricultural products, rural areas could also represent a new type of geospatial space that can accommodate many new economic activities, including economies of experience, ecological tourism, education, health, etc.
- **The change of city's own organization and geospatial layout.** For example, the way of urban life will change a lot; the centralized energy supply may be partially replaced by distributed supply, and urban infrastructure will be based more on ecological principles.

The above changes have the dual effects of increasing agglomeration and decentralization of economic activities. Whether the urban areas will become more agglomerated or decentralized depends on which effects of the above three defining factors become dominant.

4.3.3 Spatial Distribution of Future Urbanization

As for the trend of spatial distribution of urbanization in the future, it seems that a consensus has yet to be reached in the academic community. There are two different visions regarding future urban forms. One is support for the decentralization trend. Henderson et al. show evidence that Chinese cities are experiencing a decentralization trend with the emergence of high-speed rail [26]. Another is that the Internet and convenient transportation will accelerate the concentration of population to large cities, such as [13]. These two different conclusions may be due to different urban theories and different definitions. Therefore, it would be more effective to measure the real situation of urbanization through big data on population and economic activities distribution than traditional statistical data.

For China's future urbanization strategy, it is very important to clarify the relationship between city size and economic development. Though city scale is emphasized in much of the literature, in the theory of economic growth, population size is not always conducive to economic growth. For example, in Solow's growth theory, endogenous growth theory, and Lewis' surplus labour theory, population size has a negative, positive, or neutral effect on economic growth. The new economic geography, represented by Krugman and Fujita [12], emphasizes the benefits of population size for economic growth. However, as [27] pointed out, the "extent of market" emphasized in Smith's theorem is not "mass production" and population size. Zhang and Zhao [28] show that the economies of scale of enterprises in the Fujita–Krugman urbanization model are not in line with reality. Some empirical studies that emphasized the size of the city show there is a strong correlation between the size and per capita GDP [11]. However, the conclusion may not be that simple, and we explained how the urban hierarchy structure is created earlier. Because large cities have increased market size and a higher level of division of labour, their nominal GDP is usually higher than that of small and medium-sized cities, but the GDP of large cities contains more transaction costs including commuting costs, high house prices, congestion, etc., and so the net utility is not necessarily higher. The regression analysis on urban population and GDP can always lead to the conclusion that the larger the city is, the higher the GDP will be. This conclusion may be misleading from academic and policy perspectives.

In reality, we can find a large number of examples of "small but advanced cities," and "large-scale but poor cities." In Europe, more than half of the population lives in small and medium-sized cities with a population of 5000–100,000 [29]. At the same time, the size of the urban population is not equal to prosperity. 22 out of the 29 megacities in the world with more than 10 million people are in Africa, Asia and Latin America, and these super-large cities have not prospered. In China, the development of many cities no longer depends on population growth, and there is an inverted U-shaped relationship between population and urban economic growth [30].

4.3.4   The Evolution of China's Urbanization

The actual level of urbanization in China is higher than statistics suggest. The urban area is defined as an area with a population density of more than 1000 people per km^2. According to a study of DRC Big Data Lab for Macroeconomy with Baidu HUIYAN Population Big Data, China's actual urbanization level in 2015 was 62.2%, 6.1% points higher than the traditional statistics [31] (Fig. 4.3).

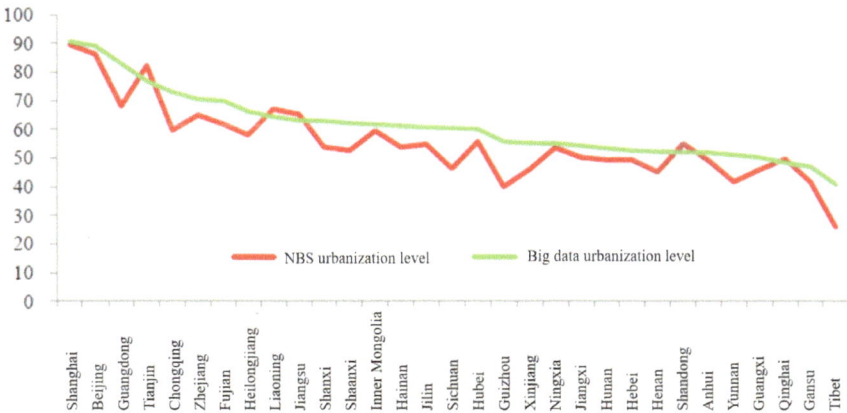

Fig. 4.3 Comparison of provincial urbanization level: Big data-based and statistics. *Source* Chen and Shi [31]

Second, China's cities have entered a high-quality development stage from quantity expansion. Economic development and population in some cities have started to show an inverted U-shaped relationship [32]. Over the past two years, the net inflow of daily migrants in some of China's most attractive cities has not changed significantly. The resident population in a few megacities has declined. As the regional economy becomes more balanced, an increasing number of people tend to return to their hometowns to work or start a business.

Third, the spatial pattern of the city is undergoing major changes: the rise of urban clusters and metropolitan areas will dominate the future economic development in China. According to the calculations of the SPS team, the proportion of GDP, population and land area in 2017 in China's 20 urban clusters accounted for 90.87%, 73.63%, and 32.67% of the nation, respectively. Lan Zongmin's research, based on Baidu migration data, mobile phone density data and nighttime lighting data, showed that the divergence of urban clusters is obvious, and the spatial scope of the planned urban clusters is generally smaller than that measured with big data [33].

This means that in the future, the main spatial scope of green urbanization will occur mainly in the existing urban clusters and at the county level. At the same time, the content and form of urbanization are undergoing profound changes.

4.4 The Impact of Green Urbanization on Regional Integrated Development

Green urbanization in the digital age will profoundly change China's regional economic structure. In the era of traditional agriculture, economic development was highly dependent on natural conditions, so a population distribution pattern was formed based on natural and geographical conditions. In China, there is the so-called Hu Huanyong Line. In 1935, geographer Hu Huanyong proposed the "Aihui–Tengchong Line, and internationally as the Hu line" in the dissertation "Distribution of Chinese Population," and found that the population west of this line represents 6% of China's total population and the population east of this line about 94% of China's total population. This line was later called "Hu Huanyong Line" by academics. According to the "Urban Development Trends" data of the East China Normal University Urban Development Institute, previous census data show that in the 80 years since 1935, China's population distribution pattern has remained almost unchanged (quoted from "Baidu Encyclopedia").

This population distribution pattern also provides different conditions for industrialization in the North and South regions. Overall, this pattern of population and economic development has been further strengthened in the industrial era. However, because industrial production can be largely freed from the constraints of natural geographical conditions, and the accumulation of population, that is, the process of urbanization, has greatly accelerated the process of industrialization, and urbanization has increased regional economic differentiation. Industrialization requires support from convenient transportation and markets. Many areas that thrived in the agricultural age no longer have advantages in the industrial age. A large number of agricultural populations and populations in areas without industrial advantages have flowed on a large scale to large and medium-sized coastal cities with industrial advantages. Therefore, development based on the industrialization model will inevitably bring regional disparities in urban and rural areas along with regional economic differentiation. What's more serious is that this traditional industrialization model not only brings about differentiated development levels, but also the systematic destruction of the socioeconomic ecosystems of backward areas and villages. This is why many rural areas declined during the industrialization process.

As human society enters the era of mobile Internet and ecological civilization, the development paradigm formed in the traditional industrial era is undergoing profound changes, including the development concept, development content, and resource concept, and the spatial meaning of economic activities is also undergoing profound changes. This is expected to fundamentally change this urban–rural imbalance and regional imbalance. This means that although the spatial differences in the sense of natural geography will persist for a long time, the gap in spatial development in the sense of economic geography may break through from a larger space and time range. Therefore, it offers the western regions a good opportunity to follow a path of ecological civilization with a new paradigm shift in the digital era.

Specifically, it is green urbanization that is changing the spatial pattern of the regional economy. The combination of "green" and "urbanization" has special significance. Meanwhile, urbanization can drive economic growth by reshaping the spatial pattern where population and economic activities concentrate. Green is an important factor to meet the needs of a better life. Its corresponding resource concept goes beyond the concept of material resources underlying traditional industrial civilization and is closely related to the ecological environment and culture which is the endowment of the backward regions. Therefore, from the perspective of green development, the endowment concept of the regional economy will be redefined. This will bring new opportunities to the backward areas that lack development advantages in the industrial age.

4.5 Green Urbanization: Case Studies

There are many good practices in China and around the world in promoting green urbanization. How to select valuable cases for research is the first problem to be solved by case studies. Einstein has famously said that "It is the theory which determines what we can observe." In the face of the complicated real world, we cannot simply collect random cases for research, but we must step out of the traditional industrialization model and identify valuable cases based on new theories and concepts. Specifically, we hope to discover and even create valuable cases by participating in local experiments under the new green urbanization theory framework. The difficulty is that due to path dependence, the realization of the theoretical vision often faces the "chicken and egg" dilemma. When there is not enough evidence for green urbanization, governments often do not take strong action to avoid the risk of failure. The evidence is less likely to occur without sufficient action. Therefore, we cannot judge the feasibility of green urbanization based on whether there are enough successful cases of green urbanization. The principle of evidence-based decision making is not always reliable for the transformation of the development paradigm. In many cases, the foresight and action of decision-makers are often more critical.

4.5.1 Cases in China

Case 1. Shenzhen New Energy Transportation (National Climate Assessment Report 2019)

Shenzhen is the first batch of demonstration cities for the application and promotion of new energy vehicles in China and has become the city with the highest number of new energy vehicles in the world. As of July 31, 2018, the number of motor vehicles in Shenzhen was 3.3307 million, of which the number of new energy vehicles reached

187,100, accounting for 5.6% of the total. By the end of 2018, Shenzhen had realized electrification of all buses and taxis. The value of this case is that many obstacles to the promotion of new energy vehicles can be solved through a series of policies and mechanism designs.

The measures taken by Shenzhen are as follows. First is to resolve financial pressure. In order to solve the problems of the promotion of new energy vehicles in the face of their relatively high purchase prices—along with the mismatch between power battery life and vehicle service life—Shenzhen has adopted a "financial lease, vehicle-electricity separation, and combination of charging and maintenance" model. Second is to innovate the promotion and application model, prioritize public transport, and initially realize the combination of asset light-weighting, rent-purchase, mileage guarantee, hire payment in installments, self-charging, and benefit sharing. Third is to focus on providing financial support to charging facilities. On the basis of a network of charging facilities combining fast and slow charging, it has continuously innovated and diversified charging methods. Fourth is to promote the development of the new energy vehicle industry, a number of industry leaders such as BYD, Wuzhou Long, and Waterma have emerged, forming the most complete new energy vehicle industry chain in China.

Case 2: Shenzhen's Coordinated Governance of Air Quality and Carbon Emissions (National Climate Assessment Report 2019)
Protecting the blue sky and addressing climate change are two major challenges facing China, and emissions of air pollutants and carbon emissions have the "same roots" to a certain extent, which can be attributed to the burning of fossil fuels and production of steel and cement. Therefore, there is a synergy in reducing air pollutants and carbon emissions. From the perspective of emission sources, the focus of Shenzhen's air pollution prevention and carbon emissions is on transportation. At present, Shenzhen has implemented measures in the field of transportation. According to a 2015 study by Peking University Shenzhen Graduate School and Shenzhen Comprehensive Transportation Operation Command Center, the proportion of new energy vehicles is one of the main controlling factors for the development of urban low-carbon transportation. The value of this case is that the benefits of reducing carbon emissions are not just global, but are felt at the local level, so reducing emissions can be a self-serving behaviour.

Case 3: Zero-Waste Pilot Cities
In January 2019, State Council issued the Zero-Waste City Construction Pilot Program. Zero waste does not mean that there is no solid waste generated, nor does it mean that solid waste can be fully utilized as a resource, but aims to ultimately achieve the goal of minimizing the amount of solid waste generated in the whole city, maximizing the use of resources, and ensuring the safety of disposal. At the present stage, we should make overall plans for solid waste management in economic and social development through piloting, vigorously promoting source reduction, resource-based utilization and harmless disposal, resolutely curbing illegal transfer and dumping, exploring and establishing a quantitative index system, summarizing

the pilot experience, and forming a replicable and applicable construction model. On April 30, 2019, the Ministry of Ecology and Environment announced 11 zero-waste pilot cities, including Shenzhen, Baotou City in Inner Mongolia Autonomous Region, Tongling City in Anhui Province, Weihai City in Shandong Province, Chongqing Municipality (main city area), Shaoxing City in Zhejiang Province, Sanya City in Hainan Province, Xuchang City in Henan Province, Xuzhou City in Jiangsu Province, Panjin City in Liaoning Province, Xining City in Qinghai Province.

The green urbanization SPS studies the mechanism behind the zero-waste city and reveals its policy implications. Research shows that, although technically all garbage and waste can be called "gold in the wrong place," effectively converting it into gold depends on whether it is cost-effective. Being technically feasible doesn't necessarily mean economically feasible. However, there are many ways to promote the consistency of technical effectiveness and economic effectiveness. For example, we can strengthen the requirements for waste disposal (the "polluter pays" principle), and make related manufacturers concentrate geographically as much as possible. For each pilot city, there are specific constraints, and different measures need to be taken.

Case 4: Chengdu Private Car Carbon Emission Trading System (National Climate Assessment Report 2019)
Chengdu Rong e-Travel carbon project aims to encourage private car owners to stop driving to reduce emissions. By establishing a quantitative methodology for carbon emission reduction, we can quantify the actual contribution of citizens who stop driving vehicles to carbon emission reduction and provide a scientific basis for the "carbon assets" rights and interests of carbon participants. As of October 2018, the number of registered users of "Rong e-Travel" reached 2.03 million, and 16,000 private car owners have voluntarily stopped driving to reduce emissions. Each private car stopped driving on average 14 days, and the total emission of major pollutants has been reduced by about 13 tons. The value of this case lies in that an effective incentive mechanism plays a great role in promoting green consumption mode.

Case 5: Old City Culture Revitalization
The case is about the "four seasons market" in Dali, Yunnan Province. This case will activate an old vegetable market that can't realize its value under traditional industrial thinking, and produce good economic and social benefits. Because traditional industrialization is based on material wealth, and the industrial production process needs more input of material resources, intangible culture is not only difficult to find a role in production processes, but is also destroyed in the industrialization model. For example, in industrial thinking, the function of the old vegetable market is to sell vegetables. In the process of urban reconstruction, such places are often demolished. However, once we jump out of this traditional thinking, we can see that the old vegetable market has great historical value and cultural value besides selling vegetables. Through the activation of entrepreneurs, designers, and artists, these cultural values can become new products and services.

However, unlike tangible industrial products, culture is often difficult to trade, and not easy to commercialize. If it can't be commercialized, it usually depends on government investment, and the government usually has a lot of rigid expenditure, which makes it difficult to take into account such projects that are difficult to bring direct financial revenue. At this time, the new business model is very important for cultural development. One potential business model is that an enterprise take responsibility for the cultural development of a specific region. These cultural services inspire direct transactions, and the development of the region will generate added-value for enterprises in the region, and then the investment enterprise will share part of the benefits from these enterprises and get the return on investment.

The value of this case lies in that it redefines the traditional concept of resources from a new perspective, recognizes the value of culture, promotes its value through creative design and effective business model. This case provides a useful exploration of how to promote those cities formed in the era of traditional industry.

Case 6: Rural Revitalization
Cities and the countryside are two sides of the same coin. The problems of cities will be reflected in the countryside. To discuss the problem of green urbanization, we must discuss the problem of rural development at the same time. Under traditional industrialization thinking, development is defined as the process of industrialization, urbanization and agricultural modernization. In order to produce industrial wealth more efficiently, population and industrial activities need to gather in cities, while rural areas are narrowly positioned as the supply base of surplus labour, agricultural products and raw materials, forming the basic urban–rural division pattern of "urban industry; rural agriculture." Industrial production activities are based on economies of scale, while agriculture is transformed into so-called industrialized agriculture, mono-agriculture, and chemical agriculture. The process of economic development has become a process of transferring a large percentage of the agricultural population to cities and towns, while other functions of agriculture and villages have not been fully recognized and developed. This kind of development mode has brought a lot of material wealth but has also led to unsustainable well-being, serious urban–rural imbalance, regional imbalances and other problems.

In order to promote the construction of ecological civilization in China, since 2016, the Green Development Research Team of the Development Research Center of the State Council has helped Shishou City in Hubei Province to carry out a green development pilot project. Instead of following the old path of treatment after pollution, it has redefined the countryside with new the development concept and realized leapfrog development through green transformation. In the new development concept and digital era, the countryside is no longer just the traditional definition of the "three rural" concept (rural, farmers, agriculture), but a new geographical space that can carry all kinds of modern civilization and green economic activities. The redefinition of the countryside brings infinite possibilities.

The pilot demonstration mainly focused on four aspects. (1) The chemical agriculture system has been transformed into ecological agriculture on a large scale, and pesticides and chemical fertilizers were no longer used, so as to enhance the value of ecological environment. Their integrated rice–frog–duck farming method has developed the village into the largest rice–duck–frog base in China. The income of villagers has increased significantly, and the rural environment has also improved. (2) Local culture activation: fully explore rich local culture and activate it in modern forms. (3) With the ecological concept, the rural residential areas have been transformed into high-quality ecological communities. (4) Through the new green infrastructure and Internet, a large number of green economic activities that are pro-environment and pro-culture are generated, and the sound ecological environment and rich local culture are turning into invaluable assets, realizing synergies between environmental protection and economic development." After four years of experiment and demonstration, the region has achieved remarkable progress and explored a new model of rural green development. Officials from developing countries keep coming to study its experience, and some foreign universities even take it as an overseas summer camp base. Shishou has, in this way, become an international knowledge centre for green development.

4.5.2 International Case: Valuing the Role of Nature in Urbanization

So far, the key measures to achieve sustainable cities have focused on how to minimize the environmental hazards that cities can cause. However, these practices still maintain the dichotomy between nature and city. For example, traditional twentieth century environmental protection tools are nature reserves and national parks. An example of the twenty-first century version of the reserve is China's ambitious national ecological functional area and ecological red lining. These plans extend conservation and restoration efforts to key areas of ecosystem service delivery. However, in fact, these areas are mainly located in rural, mountainous, or sparsely populated areas, which are different from densely populated urban areas. These exclusive nature reserves, of course, are also an important part of the ecological civilization for urban development. However, the benefits of biodiversity and ecosystem conservation have been delivered to cities. We need to integrate nature into urban planning and a value system that covers wastelands to urban cores.

Some cases about how to play the role of nature in urbanization can be found in the Appendix.

- Greening case of Melbourne metropolis in Australia
- Greening of the Bay Area of California
- Bringing life back to Qingxichuan in South Korea
- Mangrove conservation in the Philippines
- Rain and flood management case in sponge city of China
- Providing incentives for urban stormwater management facilities
- Urban stormwater credit transaction case
- China Water Conservation Fund
- How nature becomes an economic driver.

4.6   Strategic Approach to Green Urbanization and Policy Recommendations

4.6.1   Strategic Approach

The overall approach is to reshape China's urbanization based on the ecological civilization, rather than relying on quantitative urban expansion. Green urbanization should act as a driver for green transition and high-quality development in China, and green urbanization strategy should be an important part of the 14th Five-Year Plan.

4.6.1.1   Three Major Components of Green Urbanization

Component 1: Reshaping Existing Cities, That Is, Transforming Cities According to the Requirements of New Production and Lifestyle in the Digital Green Era

The first is to promote the green new economy. The advantages of existing cities for green transformation lie in demand and supply. In terms of market demand, the existing population size provides huge market for the new service economy; on the supply side, relying on its intangible endowments such as high-quality talents, urban culture, and history, a large number of experience economy and creative economy could be formed. At the same time, it is of great potential to upgrade the traditional sectors by applying new business models and Internet technologies, and China has lots of successful cases, including transformation of old neighbourhoods, old industrial parks, and old malls into creative and experience economic zones, and successful transformation of resource-exhausted cities.

The second is the green transformation of urban infrastructure. Renovating existing urban infrastructure based on the concept of ecological civilization will reduce urban costs and improve efficiency. For example, research by the Nature Conservancy shows that valuing the role of nature brings better results. When ecosystem functions and services are included in a cost–benefit analysis, hybrid infrastructure combining nature-based and traditional infrastructure can provide the most cost-effective protection from sea-level rise, storm surges, and coastal flooding. The traditional flood infrastructure has higher costs and could miss opportunities for generating additional economic benefits and providing ecosystem services that could otherwise be provided by recreation, carbon capture, and habitat (TNC, "Urban Coastal Resilience: Valuing Nature's Role").

Component 2: New Urbanization, That Is, Urbanizing New Population in a Green Way

In the future, 300 million people shall be urbanized in a new green concept and model. A large number of these people will be transferred to existing towns, while some will be urbanized locally in the county area to form new characteristic towns. The future between cities and villages is more of a difference in physical form than the difference between modernity and economic development level. Due to the new opportunities in the countryside and the substantial improvement in the quality of rural life, a large number of new urban–rural commuters is likely to emerge. The traditional statistical definition of urbanization also needs to be changed accordingly.

There are many good cases and studies in this regard in China. For example, the Rocky Mountain Institute is promoting "near-zero emission demonstration zones" in some parts of China. It is based on the concept of integrated governance while promoting economic growth, minimizing pollution, garbage and carbon dioxide emissions. The demonstration takes an integrated concept to solve environmental problems, considering the protection of air, water, soil and ecosystem as a whole. It provides an integrated solution from perspectives of the ecosystem, production process, full value chain, etc.

Component 3: A New Definition of the Countryside
The city and the countryside are two sides of one coin. When the content and mode of economic development change, the definition of the village and the urban–rural relationship will change accordingly. Under the traditional concept of development, it is a process in which agricultural labour is transferred to cities for manufacturing on a massive scale—i.e., industrialization and urbanization—while agriculture and rural areas are restructured from the perspective of industrialization, becoming a base of labour, food, and raw materials for use by urban industries. The mode of agricultural production is also transformed into monoculture and chemical agriculture in accordance with the logic of industrialization, which brings serious ecological and environmental consequences. This traditional rural definition from the perspective of industrialization not only limits the economic development potential of the rural areas, but also sacrifices many valuable rural cultural and ecological resources. In fact, the countryside is a versatile new geospatial space that can accommodate a large number of new economic activities. In this regard, China also has many successful cases. For example, the DRC Green Team helps underdeveloped regions achieve leapfrog development through green transformation under the framework of "redefining countryside."

4.6.1.2 *Two Strategic Focuses of Green Urbanization: Green Urban Clusters + County-Level Urbanization*

The two strategic focuses of China's green urbanization are: first, the green transformation of urban clusters and metropolitan areas and second, county-level urbanization. The reasons are as follows.

First, the economy and population of the 20 urban clusters currently account for an absolute proportion in the country. In 2017, China's 20 urban clusters accounted for 90.87%, 73.63% and 32.67% of the national GDP, population, and land area, respectively. It could be said that the success of the green transition of the whole country hinges on the green transition of its city clusters (Table 4.1; Fig. 4.4).

Second, since the urban clusters include three major components of green urbanization, namely existing cities, population to be urbanized, and rural areas, they can

Table 4.1 Share of GDP, population, and land area of urban clusters to the nation's total

	GDP (100 million)	Population (10 thousand)	Land area (km²)
Urban clusters (A)	743,771	10,2351	314,7710
National 2017 (B)	818,461	139,008	9,634,057
Share (A/B) (%)	90.87	73.63	32.67

Source Made by author according to statistics

Fig. 4.4 Share of GDP, population and land area of each urban cluster to the nation's total. *Source* Made by author according to statistics

receive both advantages of the urban and rural areas. Focusing on the city clusters and metropolitan areas can activate the advantages of both urban and rural areas and the potential market. Based on their ecological resources, villages located in urban clusters and metropolitan areas could provide new green supplies to the surrounding cities.

Third, revitalizing the county economy is a major task of rural revitalization in China. In addition to moving to the capital city of the county, a large number of people will be urbanized in the form of characteristic towns, so as to take both advantages of the urban and rural areas.

4.6.1.3 Shift from Functional Cities to Pro-nature Cities

The pro-nature urban model is not to divide urban land into different uses, but to integrate the land use so as to incorporate the natural characteristics into the city. In addition, pro-nature cities do not regard nature as an externality but introduce the value of nature into urban planning and decision making while making wise choices to promote public interests. Pro-nature cities can also increase the ecological service supply with the market economy. The challenge is to change the social value system so that natural capital is no longer independent of existing financial systems and land-use planning decisions.

4.6.2 Policy Recommendations

The first recommendation is to achieve a breakthrough in understanding. First, we should fully realize the impact of the current digital era and green development concept on the urbanization model, and we can't use the old urbanization thinking for green urbanization planning. Second, green urbanization is not just about architecture, planning and green technology, but relates more to development content and development mode. Third, the decisive role of the market in urban planning should be further explored, and the government has a better role to play.

Second is to promote the free flow of urban and rural elements, and the effect of urban planning on rural areas must be taken into account. Cities and the countryside are two sides of the same problem. When it comes to green urbanization planning and policies, we should take an integrated approach and consider its impact on the rural economy, ecology, society and culture. Meanwhile, we should encourage urban talents to move to the countryside and gradually give urban residents access to the lease and use of rural homesteads.

Third is to accelerate the promotion of green technologies. This would begin with the removal of the institutional barriers that hinder the promotion of new green technologies. The biggest obstacle to green urbanization is not a lack of good green technologies—it is the difficulty of implementing them even if they are cost-effective and technically feasible. The next is attaching greater importance to the promotion of low-cost green applicable technologies, such as small constructed wetland sewage treatment systems and passive buildings. The last is to achieve breakthroughs in some green technologies with huge potential and less difficulty, and resolve the application problems. Energy saving by means of room air conditioners is a possible breakthrough.

Fourth is to shift from a functional city to a pro-nature city model. To this end, efforts need to be made in the following four directions (see the Appendix for details). First, we should protect the important urban biodiversity and natural habitats, and to overcome the traditional planning that separates cities from nature. Second, we should integrate biodiversity and ecosystem services into urban planning and design cities with integrated land use, including natural infrastructure essential for human well-being. Third, we should introduce policies and incentives to give value to ecosystem services, and regard ecosystem services as a key part of the market, rather than as externalities. Fourth, we should take the opportunity to provide citizens with ecosystem services as the key to realizing sustainable development.

Appendix 1: Valuing the Role of Nature in Urbanization and Regional Development

TNC Input to SPS 2-1 Green Urbanization Strategy and Pathways Towards Regional Integrated Development

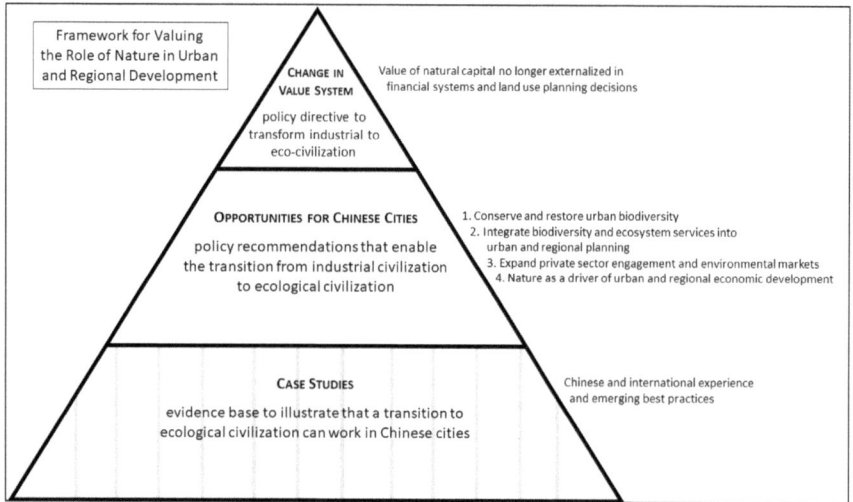

(1) **Introduction**

Ecological civilization (eco-civilization) is a set of values and development concepts enshrined in the constitution of the People's Republic of China in 2018 and now a key driver in the country's transition to higher quality economic and social development. In an unprecedented fashion, this concept connects the primacy of ecological health with traditional development elements. Explicitly recognizing that people rely on a healthy planet for their economic and social progress, eco-civilization is being used by China to provide a coherent conceptual framework for adjustments to development that meets twenty-first century challenges [34]. With this recent commitment to an eco-civilization development pathway, enormous strides are being made to increase the sustainability of Chinese urban systems. However, to fully transition from our current industrial civilization pathway to eco-civilization, a new paradigm of urban development is needed [35].

(2) **Status and Trends**

Under industrial civilization, the dominant urban model was what has been called the functional city (Fig. 4.5). In the functional city, nature is thought of as something separate and independent from cities. The functional city was designed with segregated human land uses in discrete zones, with no place for natural features within urban bounds. Moreover, the numerous ways that urban residents and firms depend on nature for their well-being were ignored in decision making and markets. For instance, the clean drinking water of urban residents depends, in part, on natural vegetation preventing erosion into the city's reservoir. The amount of tree canopy cover there is in a region affects its air quality. Parks and other urban public space contain natural features like lawns, forests, and lakes that are an amenity to people who recreate there. Most ecosystem services that cities rely on are common or public goods. In the functional city model, ecosystem services are considered externalities to markets and decision making. This means they will be degraded and underprovided relative to the true needs of society.

(3) Progress to Date

Steps to make cities more sustainable to date have focused on minimizing the environmental harm cities can do. However, they have retained the sharp dichotomy between nature and cities. For example, the traditional twentieth century tool of conservationists is the protected area, parks and reserves containing important biodiversity set aside from human development. An example of the twenty-first century version of the protected area is China's ambitious programs of National Ecological Function Zoning and ecological red lines, which extend protection and restoration efforts to areas of key ecosystem service provision. However, in practice these zones are primarily in rural, mountainous or sparsely populated zones, conceived of as distinct from populous, urban areas. These exclusive or near-exclusive nature zones are important components of an eco-civilization approach to urban development, but critical elements of biodiversity and ecosystem benefits occur outside these zones. We need to integrate nature into urban planning and value systems along the full gradient from wild lands to the urban core.

(4) Challenges

To continue China's transformation into an ecological civilization, we need to move from the functional city model to the biophilic city model (Fig. 4.5). Instead of segregated land use within the city, this model has integrated land uses, which explicitly incorporated natural features within the city. Moreover, instead of treating nature as an externality, the biophilic city brings the value of nature into urban planning and decision making, making wiser choices that promote the public good. Biophilic cities also bring value to ecosystem services in their markets, harnessing the power of market economies to efficiently enhance ecosystem service provision.

The challenge or overall policy directive is to change the societal value structure whereby natural capital is no longer externalized in financial systems and land-use planning decisions. To help achieve this transition, we propose the following four policy recommendations:

1. Conserve important urban biodiversity and natural habitat within urban areas, overcoming the traditional dichotomy in planning between urban and rural areas.

2. Integrate biodiversity and ecosystem services into urban planning, designing cities with integrated land uses that include natural infrastructure crucial for human well-being.
3. Create policy incentives that give value to ecosystem services, treating ecosystem services as a key part of markets rather than as an externality.
4. Use the need to increase ecosystem service provision for urban residents as a sustainable pathway for future economic development in urban, county, and rural areas.

Each of these recommendations is expanded on below.

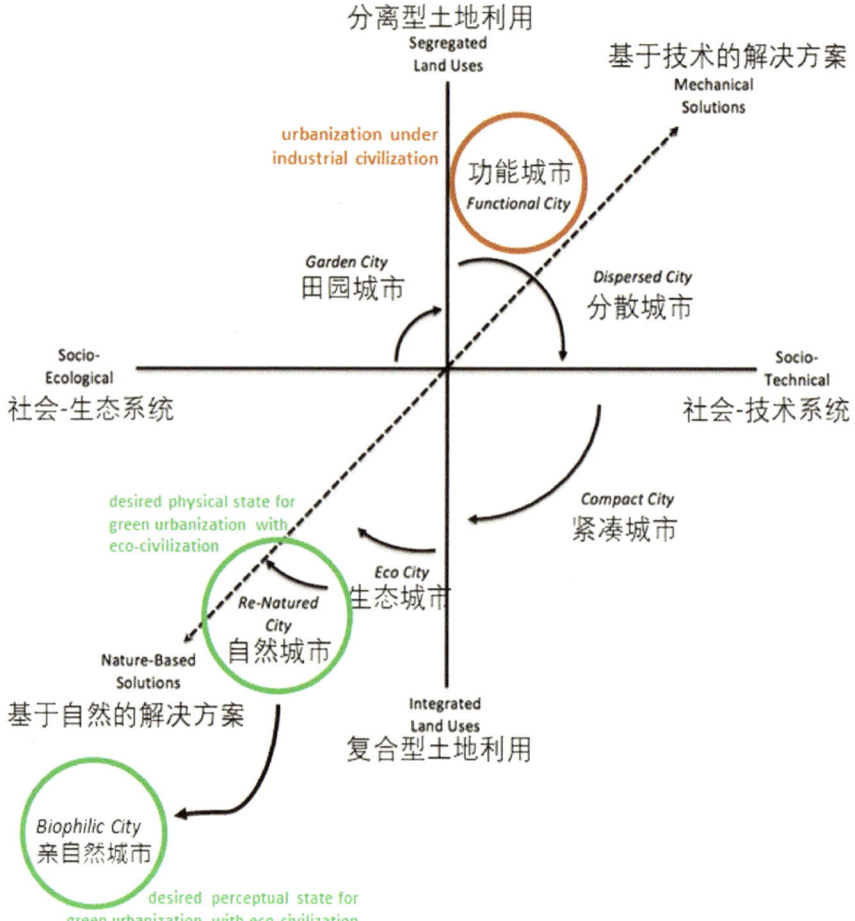

Fig. 4.5 Conceptual framework for urban transitions from industrial to ecological civilization and internalizing the value of natural capital into financial and land-use decisions. *Source* Modified from Scott and Lennon [36]

Opportunities for China: Policy Recommendations for Valuing the Role of Nature in Urbanization and Regional Development

Policy Recommendation 1: Conserve and Restore Urban Biodiversity

- Recognize urbanization as a key driver of biodiversity loss nationally as well as the importance of urban biodiversity to human well-being. Take global leadership in developing laws, policies, best practices, and measures of success to prevent further loss due to ill-planned urban growth and strengthen the efforts to better protect and restore urban biodiversity in existing cities.

Governance and Implementation

As a party to the Convention on Biological Diversity (CBD), China has committed to protecting biodiversity both for its own value and for its instrumental value, that is, helping solve problems across society, including in and around cities. While China remains one of the most biodiverse countries on Earth, it has experienced significant loss over the centuries. The last 40 years has seen significant loss due to urbanization [36]. Recognizing the significance of biodiversity loss due to urbanization and building on earlier policy directives from the Ministry of Housing and Urban–Rural Development, the China National Biodiversity Conservation Strategy and Action Plan (2011–2030) proposed a planning and demonstration project for biodiversity conservation in urban development. Yet urbanization remains a critical driver of biodiversity loss as reported in China's Sixth National Report to the CBD Secretariat in 2019.

Beyond treaty commitments and moral motives, biodiversity is also important for its instrumental value, that is, helping solve problems in human communities. Long-term resilience of society, generally, and cities, particularly, require diversity and the conditions where all elements can be present. Sustaining biological diversity is embodied in eco-civilization as a core principle. Preventing loss of natural diversity (and restoring elements already lost) is essential so that we do not foreclose on options to solve challenges faced by future peoples. Historically, most people think there is no biodiversity worth conserving in urban areas, yet we now know that urban nature is an indispensable component of overall biodiversity. Urban areas are important refuges for wildlife, are evolutionary hotspots for novel elements of biodiversity, are key to habitat connectivity across landscapes, and also because nature in urban has much closer interactions with people therefore directly affect people's lives.

China has recognized that the nature can help solve many urban problems today and created special programs for targeted solutions (i.e., sponge city, national forest city, etc.). But we cannot predict all the challenges cities will face in 50 or 100 years, therefore need all the parts that can help us innovate and create new solutions. Transition from industrial to eco-civilization will not be successful without this.

As China prepares to host the CBD 15th Convention of the Parties in October 2020, we recommend:

- China should encourage cities to participate in global networks that are creating science-based metrics for evaluating urban systems (municipalities, city clusters, regional rural–urban landscapes) as to their measurable contributions in achieving biodiversity targets of the China Biodiversity Strategy and Action Plan and to the CBD, generally.

- Develop and adaptively strengthen laws and policies for biodiversity conservation in urban systems based on global experience and emerging best practices in China, including standards for (1) cataloging urban biodiversity, (2) preventing loss of extant diversity, (3) strategic restoration of degraded habitats and extirpated species, and (4) conservation programs that engage stakeholders from all sectors.
- Pilot urban biodiversity improvement programs with clearly defined goals in range of pioneer cities, including tier 1–3 municipalities and city clusters that encompass their regional rural–urban landscapes. An ideal candidate as a demonstration pilot would be the Guangdong-Hongkong-Macau Greater Bay Area. This city cluster and surrounding rural counties lie in the Southeast China Subtropical Broadleaf Evergreen Forest Ecoregion, which globally is projected to have among the greatest loss of biodiversity due to urbanization processes.

Case Studies (Chinese and International Experience and Emerging Best Practices)

- Greenprinting a Metropolis with Living Melbourne, Australia
- Greenprinting the Bay Area of California, USA
- Bringing Life Back to Cheonggyecheon Stream, Korea.

Policy Recommendation 2: Integrate Biodiversity and Ecosystem Services into Urban Planning and Design

- China should approach cities (including municipalities, city clusters and regional rural–urban landscapes) as a system rather than the implementation of narrow, specific programs on ad hoc basis. To that end, incorporate biodiversity and ecosystem services into China's new urban spatial planning framework to help better demarcate urban growth boundaries, coordinate the various biodiversity and ecosystem service agencies/programs, and plan for optimizing the value of natural capital along the full land-use gradient.

Governance and Implementation

Seeking synergies across all sectors of society lies at the heart of eco-civilization. For example, protecting or restoring natural habitats in cities will help biodiversity conservation as well as provide ecological services of value to urban settlements, such as abating urban heat, dealing with urban flooding and storm water pollutions, as well as providing recreational green space. Finding synergies is a means to reduce costs while getting better returns on eco-civilization investments. Addressing these synergies can also be a means of improving legal frameworks, financial incentives, and institutional arrangements.

Current urban planning approaches evolved under the industrial civilization model whereby spatial design applied a segregated land-use approach as seen under the lens of socio-technological problem solving (Fig. 4.5). The services nature provides people were externalized from this value system. To make the transition to an eco-civilization we must consider the intersections between ecosystem processes and spatial planning frameworks and consider the city as an integrated socio-ecological system. In other words, internalize the value of nature into urban planning and design. Currently China is designing a new land-use spatial planning system, which already puts a lot of emphasis on ecosystem services. China should keep on this rule within the spatial planning process from national level to city level and help better demarcate urban growth boundaries, especially in newly built cities or towns, to prevent further biodiversity loss due to ill-planned urbanization.

China has made a commitments and established programs to protect biodiversity and maintain ecosystem services. Unfortunately, and reflecting the current state of affairs globally, the approach in China is most typically sectoral isolation instead of synergist integration and codesign. Lost are the obvious advantages gained by treating the multiple benefits of urban nature together instead of as single benefits isolated from each other and planned for separately in space and time. For example, several national programs in China focus on one or a small number of ecosystem benefits such as sponge city, national urban forest city, garden city, and eco-city. There may also be municipal biodiversity conservation plans.

China needs to codify in law and regulation the integration of all overlapping biodiversity and ecosystem service programs in their urban operations and spatial planning. This will result in better outcomes for both biodiversity and people at a spatial scale that matters to the whole rural–urban system and greater economic efficiency in delivering public services.

Case Studies (Chinese and International Experience and Emerging Best Practices)

- Greenprinting the Bay Area of California, USA
- Greenprinting a Metropolis with Living Melbourne, Australia
- Valuing Protective Services of Mangroves in the Philippines
- Investing in Natural Solutions for Stormwater Management with Sponge Cities, China.

Policy Recommendation 3: Expand Private Sector Engagement and Environmental Markets

- Through public policy, create incentives that give value to key ecosystem services by establishing new environmental markets that encourage investment by the private sector in natural assets to improve infrastructure and city services. Key to a successful market design is to drive project-based and technology innovation and lowest cost solutions.

Governance and Implementation

As the value of urban natural assets is internalized within the financial and spatial planning systems of regional development under eco-civilization, there may be greater opportunities to engage the private sector through expansion or creation of market-based approaches to deploying natural infrastructure in regional development. Markets allow us to look at a landscape as a whole and make development and mitigation decisions holistically. Investment in urbanization that puts stressors on the environment, should have a corresponding investment in mitigation that corrects, or ideally improves overall conditions.

Increasingly private business and private property are shaping the landscapes of Chinese cities. While a single development site might not cause the collapse of a local ecosystem, unmitigated development without an eye towards its role in the larger ecosystem can be problematic. Engagement with the private sector around hosting and investing in mitigation, through the power of markets can help drive investment in low cost solutions. For example, multiple development sites, whether as sources of supply or demand, can be linked into a collective economic force for investment in ecosystem management at a scale and location where it matters most. The goal being to improve the overall landscape, while also giving flexibility to developers to invest in projects that reduce their compliance costs.

This expands existing ecological compensation and payment for ecosystem services programs in China to incentivize private capital into the urban greening.

Implementation recommendations that apply:

- Strengthen ecosystem impact mitigation requirements for all new developments based on science-driven data analysis. Advise on underlying regulatory requirements to create/support market, where applicable.
- Research on tradable ecosystem services important to China cities/region (e.g., stormwater pollution; pluvial, coastal, and/or fluvial flood mitigation; heat stress from lack of vegetation; air quality; biodiversity).
- Create crediting methodology for ecosystem impacts most important to city/region (e.g. gallons of water managed, habitat managed, pollutants managed).
- Allow for developments to trade among one another or into select regional mitigation banks.
- Create select financial subsidies or benefits to encourage market participation.
- Create guidance on administering market program for whatever unit of government (city-scale, regional, etc.) would be responsible for tracking and clearing trades.
- Work with a couple of select regions/developers to do demonstration projects that provide indicative information on project economics, permitting and regulatory challenges/costs, and legal arrangements between credit/offset buyers and sellers.

Case Studies (Chinese and International Experience and Emerging Best Practices)

- Incentivizing Distributed Green Infrastructure for Urban Stormwater Management
- Market-Based Solutions to Landscape-Scale Water Quality Mitigation with Stormwater Retention Credit Trading

- Water Funds for Source Water Protection in China
- Valuing Protective Services of Mangroves in the Philippines

Policy Recommendation 4: Nature as a Driver of Urban and Regional Economic Development

- Design policies to capitalize on natural assets and ensure their long-term provision as part of an economic development strategy for urban areas, counties, and rural landscapes. The national land-use planning is an ideal platform to identify areas suited for this strategy.

Governance and Implementation

Examples from China and around the world have shown the success in using nature as a direct driver of economic development at the same time as being pathway to realizing green urbanization and eco-civilization. For urban areas, it has been shown that the environment is a key indicator of quality of life, which is crucial to attract knowledge workers. Nature has also proven to increase property values and consumer spending. Agritourism in county areas and ecotourism in rural areas have also shown to be powerhouses for economic development. A major driver for agritourism is enjoying nature, which makes protecting and restoring nature important for both agritourism and ecotourism. For county and rural areas, agritourism and ecotourism are service industries that do not require drastic increase in educational level while creating large employment with good wages.

Some specific recommendations:

- For urban areas that already have components of attracting knowledge workers such as convenient transportation and high-quality higher education institutes, set a minimum target for nature that the areas should have to ensure they can attract knowledge workers in the long term. Elevate the level of funding for the natural environment to the same level of importance as other man-made hard and soft infrastructure, such as transportation and education. Include and enhanced/restored natural environment as part of the strategy for urban revitalization.
- For counties and rural landscapes, agritourism is suitable for areas that are identified as prime farmland and are also close to urban centres. In these areas, protect farmland and restore the surrounding natural environment. For areas identified as suitable for ecotourism, protect and restore natural assets, develop long-term management framework and funding schemes for restoring, maintaining, and expanding the natural assets.
- For both county suitable for agritourism and rural areas suitable for ecotourism, elevate agritourism and ecotourism as a key poverty reduction-economic development-employment creation strategy. Invest in supporting infrastructure such as transportation, health care, and hospitality facilities as well as soft infrastructure such as education and training for local residents to equip them to take advantage of agritourism, and visitors will have easy access to the natural assets

and the supporting infrastructure to enjoy them. Ensure the visitation does not damage or deplete agriculture assets and natural environment through limitation on land use, visitation numbers, and strict requirements for resource usage (such as water in dry areas), sanitation, and waste management in hospitality facilities.

Case Studies (Chinese and International Experience and Emerging Best Practices)

- Nature as an Economic Driver in Urban and Rural Environment
- Bringing Life Back to Cheonggyecheon Stream

Appendix 2: From Industrial Civilization to Ecological Civilization: Changing Implications to Gender Equality

Gender equality is a prominent problem facing the world today. How to improve women's social status and give full play to their unique role in all aspects is an important issue in both China and the world.

Different development paradigms have internal relations with gender issues. The role of gender in society changes with the content and mode of development. The transformation from the matrilineal society to patriarchy society and the strengthening of the latter could be attributed to the changes in the content and mode of economic development.

Before the large-scale industrialization, women played a leading role in many aspects, such as agricultural production, family education and management.[2]

After the industrial revolution, human beings established the industrialization mode based on the creation of large-scale material wealth, and the productivity leaped greatly, which promoted the unprecedented progress of human civilization. Accordingly, the fact that industrial "violence" (by means of tools) conquers the nature improved men's status. On this basis, the whole social division of labour and organizational structure would inevitably bring about gender inequality.

The traditional industrialization mode has damaged ecological environment and caused social problems. It failed to bring women's advantages into play, but made them, especially those in rural areas, bear more consequences. In the process of rapid industrialization and urbanization, a large number of young and middle-aged labour migrate to urban factories to work, leading to hollow villages and left-behind women and children which is a severe social problem. The environmental pollution caused by chemical agriculture also puts women in a relatively disadvantaged position. Therefore, if the mode of development behind gender inequality is not changed, it will be quite difficult to solve gender problems.

---

[2] In the long-term collection activities, women understood the law of crop growth and developed agriculture. At the same time, women were responsible for managing residence and raising children. These activities established the leading role of women in the social and economic activities at that time.

The Chinese government takes ecological civilization and green development as its national strategy, which brings new opportunities to better address gender equality. The important content of the new development concept of ecological civilization is to transform the beautiful natural ecological environment and rich local culture into wealth and realize Green is Gold. At the same time, the 19th National Congress of the Communist Party of China put forward the strategy of rural revitalization. Because of the natural connection between women and nature and local culture, they can play a unique role in the green revitalization of rural areas, thus greatly contributing to gender equality, green development and rural revitalization at the same time.

References

1. Zhang, Y. (2020). Promoting China's green urbanization based on ecological civilization— CCICED Report of 'Green Urbanization Strategy and Pathways towards Regional Integrated Development'. Chinese Journal of Population Resources and Environment, 30(10), 19–27.
2. National Bureau of Statistics. China Statistical Yearbook 2020 [M]. National Bureau of Statistics Press, (2020). (in Chinese).
3. United Nations, Department of Economic and Social Affairs, Population Division. World Population Prospects 2019[R]. Online Edition. Rev. 1. (2019).
4. Yang, J. D., and Zhang, Y.R.: Happiness and Air Pollution[J]. China Economist, 10(5), (2015).
5. Easterlin, R. A., Morgan, R., Switek, M., & Wang, F.: China's life satisfaction, 1990–2010[J]. Proceedings of the National Academy of Sciences of the United States of America, 109(25), 9775–9780 (2012).
6. Jackson, T.: Prosperity Without Growth: Foundations for the Economy of Tomorrow. Taylor & Francis, (2016).
7. Ng, Y. K.: From preference to happiness: Towards a more complete welfare economics[J]. Social Choice & Welfare, 20, 307–350 (2003).
8. Scitovsky T.: The Joyless Economy: The Psychology of Human Satisfaction [M]. Oxford University Press USA, Revised edition (1992).
9. Skidelsky, E., & Skidelsky, R.: How much is enough?: money and the good life[M]. Penguin UK (2012).
10. Zhang, Y.: Redefining Urbanization. Working paper, (2020).
11. Bettencourt, L. M. A.: The Origins of Scaling in Cities[J]. Science, 340, 1438 (2013). DOI: https://doi.org/10.1126/science.1235823.
12. Fujita, M., Krugman, P.: When is the economy monocentric? von Thunen and Chamberlin unified[J]. Regional, Science & Urban Economics 25(4), 505–528 (1995).
13. Glaeser, E.: Triumph of the city: How our greatest invention makes us richer, smarter, greener, healthier, and happier[M]. Penguin (2011).
14. Lobo, J. et al.: Urban Science: Integrated Theory from the First Cities to Sustainable Metropolises"[J]. Mansueto Institute for Urban Innovation Research Paper Series, (2020).
15. Romer, P.: The City as Unit of Analysis (2013). https://paulromer.net/the-city-as-unit-of-analysis/.
16. Lu, M.. Cities, Regions, and National Development: Current Topics and Future of Spatial Political Economics [J]. Economics (Quarterly), 16(4): 1499-1532(2017). (in Chinese)
17. Acemoglu, D., P. Aghion, L. Bursztyn, & D. Hemous: The environment and directed technical change[J]. American Economic Review, 102, 131–166 (2012).
18. Zhang, Y.: Differences between Ecological Civilization and Green Civilization[M]//A Beautiful China: 70 Experts on Ecological Civilization at 70 years of PRC.China Environmental Press Co., (2019a). (in Chinese)

19. Smith, A.: An inquiry into the nature and causes of the wealth of nations[M]. W. Strahan and T. Cadell, London (1776).
20. Bettencourt, L. M. A.: Impact of Changing Technology on the Evolution of Complex Informational Networks[J]. Proceedings of the IEEE, 102(12), (2014).
21. Yang, X. K.: Economics: New classical versus neoclassical frameworks[M]. Blackwell, New York (2001).
22. Fujita, M.: Urban Economic Theory: Land Use and City Size[M]. Cambridge University Press, New York (1989).
23. Henderson, J. V.: The Sizes and Types of Cities. American Economic Review, 64, 640–657 (1974).
24. Yang, X.: Development, Structure Change, and Urbanization[J]. Journal of Development Economics, 34, 199–222 (1991).
25. Yang, X. and Rice, R.: An Equilibrium Model Endogenizing the Emergence of a Dual Structure between the Urban and Rural Sectors[J]. Journal of Urban Economics, 25, 346–368 (1994).
26. Baum-Snow, N., Brandt, L., Henderson, J. V., Turner, M. A., & Zhang, Q.: Roads, railroads and decentralization of Chinese cities[J]. Review of Economics and Statistics, 99 (3), 435–448 (2017).
27. Young, A.: Increasing returns and economic progress[J]. The Economic Journal, 38, 527–542 (1928).
28. Zhang, Y. & Zhao, X.:Testing the scale effect predicted by the Fujita–Krugman urbanization model"[J]. Journal of Economic Behavior & Organization, 55, 207–222 (2004) .
29. European Commission: Cities of tomorrow: Challenges, visions, ways forward[R]. EC, Brussels (2011).
30. Zhuo, X.: A Re-examination of New Trends in Agriculture Population[J]. Shandong Economic Strategy Research, 10, 45–46 (2019). (in Chinese)
31. Chen, C., Shi, G.: Urban Population Ratio in China from a Big Data Perspective[M]//Demographic Migration and Changing Cities: Urbanization in China from the Lens of Big Data. China Developmental Press, 2019. (in Chinese)
32. Chen, C., Wei, D..: Analysis of Urban Developmental Potential with Big Data[M]//Demographic Migration and Changing Cities: Urbanization in China from the Lens of Big Data. China Developmental Press, 2019. (in Chinese).
33. Lan, Z.:Identification of City-Clusters and their Spatial Characteristics with Big Data[M]//Demographic Migration and Changing Cities: Urbanization in China from the Lens of Big Data. China Developmental Press, (2019). (in Chinese).
34. Hanson, A.: Ecological Civilization in the People's Republic of China: Values, Action, and Future Needs[R]. ADB East Asia Working Paper Series No. 21, (2019).
35. Zhang, Y.: Why ecological civilization is different from green industrial civilization [R]. Draft manuscript, (2019b).
36. Scott, M., & Lennon, M.: Nature-based solutions for the contemporary city[J]. Planning Theory and Practice, 17, 267–300 (2016).

Chapter 5
Ecological Compensation and Green Development Institutional Reform in the Yangtze River Economic Belt

5.1 Introduction

In 2018, China Council for International Cooperation on Environment and Development (CCICED) launched a Special Policy Study (SPS) project called Ecological Compensation and Green Development Institutional Reform in the Yangtze River Economic Belt (YREB), which aimed to solve the problem of how to construct ecological compensation mechanisms and implement green development system reform in the context of ecological protection and sustainable development. Through two years of project research, the research team has conducted in-depth studies on ecological compensation and green development system reform in the YREB and submitted specific and operable policy recommendations to promote ecological compensation and green development in the region. In 2020, the project will continue to deepen the research on reform of the YREB and green development. It will examine how to solve the lack of ecological compensation standard and green financial innovation by focusing on innovative research into natural ecological capital accounting (NECA) and an eco-financing mechanism (EFM), and their application in policy practices. It will provide scientific support for the strategic development and decision-making mechanism of YREB as well as in the Yellow River basin.

Based on the review of local NECA practices and the current situation and trends in ecological investment and financing policies, the overall goal of this project is to identify the current challenges of applying NECA and summarize the best national and international practical cases, experiences, and enlightenment that could inspire China in current and future practices. Through the design of short-, medium- and long-term policy implementation roadmaps, the project will eventually propose policy recommendations for incorporating NECA and EFM with high-quality development and high-standard protection for both YREB and Yellow River basin.

© The Author(s) 2022
China Council for International Cooperation on Environment
and Development (CCICED) Secretariat,
Green Consensus and High Quality Development,
https://doi.org/10.1007/978-981-16-4799-4_5

5.2 Status and Trends

5.2.1 Methodology of Ecological Capital Accounting

Ecological capital includes natural capital as "stock" capital and ecosystem services as "flow" capital. At present, NECA is mainly reflected in the accounting of ecosystem services in order to estimate "flow" capital value. The objective of ecosystem services accounting is to assess the benefits for human welfare and sustainable socioeconomic development. These ecosystem services are subdivided internationally into provisioning, regulating, and cultural services.

Provisioning refers to the production of services like food and water. Regulating refers to the control of climate and disease and includes climate regulation, water conservation, soil conservation, flood storage, pollutant degradation, carbon sequestration, oxygen release, and pest control. Cultural services include spiritual and recreational benefits.

In 1967, Krutilla introduced existence value into the mainstream economic literature for the first time, laying a theoretical foundation for the subsequent quantitative assessment of the value of ecological capital [1]. NECA includes physical and monetary accounting, which mainly accounts for the direct use value, indirect use-value, and non-use value of ecological capital. Among them, the physical amount of ecological capital is quantified by various approaches, including statistical methods and remote sensing analysis methods, such as the CASA model, USLE model, InVEST model, etc. The monetary value can be calculated by a variety of methods, including adopted market value method, expense expenditure method, travel cost method, recovery and protection cost method, shadow price method, opportunity cost method, contingent valuation method, and other environmental–economics methods. Due to different conversion standards, the value of ecological capital is not consistently evaluated by different valuation methods. In general, the direct use value of ecological capital uses the market value method, the indirect use value uses the alternative market method, and the non-use value applies the contingent valuation method.

In practice, the value of ecosystem services can be calculated by complementary, bottom-up and top-down equivalent factor methods. The bottom-up valuation method is based on the quantity and the unit price of ecosystem services that are aggregated to estimate the total value of ecosystem services. This method originated from the article, "The Value of the World's Ecosystem Services and Natural Capital," published by Costanza et al. in *Nature* in 1997 [2], and applied in China by Wenhua Li [3], Zhiyun Ouyang [4, 5], and other national scholars. It is very pertinent for assessing local conditions. According to the ecological characteristics of a specific region at a certain time, the ecosystem services are summarized one by one, and the corresponding value is estimated according to the unit price of the service for that year. However, due to the different types of ecological assets, specific accounting indicators, and valuation methods, the comparability of accounting results remains uncertain.

Due to the difficulty and uncertainty in the implementation of the bottom-up approach, some scholars have used the equivalent factor method to simplify how the value of ecosystem services is calculated using a top-down perspective. Based on the research of Costanza et al. [2], Gaodi Xie conducted several questionnaire surveys of more than 700 professionals with ecological backgrounds in China and published the *Equivalent Factor Table of Value of Terrestrial Ecosystem Services in China* [6]. This method estimates the value equivalent of various services provided by different types of ecosystems based on quantifiable criteria and then evaluates ecosystem services according to their regional distribution. It has the advantage of using unified methods and consistent standards. It is intuitive and easy to use with fewer data requirements and provides a comprehensive evaluation and easily compared results. However, it provides a rather macroscopic average value that cannot fully reflect the specific ecosystem characteristics of different regions.

Currently, the literature on understanding the gendered implications of ecological capital accounting is limited, but it is growing and will be factored into the overall approach recommended through this SPS.

5.2.2 Ecological Investment and Financing Policies

As the core of ecological protection, policies for ecological investment and financing determine how funds for ecological protection are collected and spent. They also reflect the various relationships among ecological economies. Therefore, the design of ecological investment and financing policies should not only consider the entities and channels of investment and financing but also reflect the economic, social and eco-environmental benefits they provide [7]. "Green finance" refers to *financial services for project investment, financing, operating and risk management that support projects involving environmental improvement, and tackling climate change, promoting conservation and the efficient use of resources, energy conservation, clean energy, green transportation, green building,* etc. [8]. The scope of projects supported by ecological investment and financing specifically in this study is smaller than that of "green finance," but the means of investment and financing are similar.

It is generally believed that, internationally, in the design of ecological investment and financing policies, considerations should be given to strengthening the government's leading role in the EFM, developing the EFM in the form of public–private partnerships under the guidance of the government. This approach effectively reduces the risk of private investment, improving the efficiency of ecological investment and reducing the cost of ecological investment [9–11]. Chinese scholars believe that China's current ecological investment demand and financing needs are huge; however, the government's devotion to ecological environmental protection is so far insufficient. China should draw lessons from international experience, accelerate the development of a diversified pattern on investment and financing in the field of ecological, environmental protection, expand investment and financing channels for ecological and environmental protection, and formulate relevant policies to ensure investment and financing channels are more feasible [12, 13].

Several national and international research projects involving the quantitative analysis of ecological investment and financing policy design have been carried out. Subhrenduk [14] adopted the market value approach to calculate the value of ecological services provided by the upper stream of the river basin to the lower stream, served as references for the ecological compensation standards. The Organisation for Economic Co-operation and Development (OECD), together with the Danish Environmental Protection Agency and the Danish consulting firm COWI, applied computerized decision support tools to develop environmental financing strategies for countries in Eastern Europe, the Caucasus, and Central Asia (OECD, 2003). By building a system dynamics model for the study of the effects of multi-channel financing, using water quality changes across river basins as a criterion, Mingkai Zhang [15] simulated the effects of multi-channel funding for a watershed ecological compensation fund. He indicated that a single source of financing could not achieve the effectiveness of watershed eco-compensation, but diversified financing channels could have a better effect. Accordingly, China should accelerate the establishment of a leading platform for trading ecosystem services, similar to those for emissions trading, to attract the participation of various forms of finance.

5.3 Progress to Date

5.3.1 Progress on NECA

In 1997, Costanza et al. evaluated global ecosystem services for the first time and proposed a classification of ecosystem services that included 17 indicators. In 2001, the UN-initiated Millennium Ecosystem Assessment (MA) [34] classified ecosystem services into four functional categories: provisioning services, regulating services, cultural services, and supporting services. Since then, The Economics of Ecosystems and Biodiversity (TEEB) [35], supported by the UN Environment Programme, examined the economic valuation of ecosystems. The Final Ecosystem Goods and Services Classification System (FEGS-CS) of the United Nations Statistics Division and the National Ecosystem Services Classification System (NESCS) has developed new accounting systems based on the MA accounting framework.

Building on international accounting experience, Chinese scholars have also actively explored this issue. Zhiyun Ouyang, Gaodi Xie, Bojie Fu, and other scholars have helped to construct China's evaluation index system of ecosystem services. The former State Forestry Bureau and State Oceanic Administration have issued guidelines including the LY/T 1721-2008 [37] *Specifications for Assessment of Forest Ecosystem Services in China*, the GB/T 28058-2011 *Technical Directives for Marine Ecological Capital Assessment*, the LY/T 2006–2012 [36] *Assessment Criteria of*

Desert Ecosystem Services in China, and the LY/T 2735-2016 *Norm of Techniques for Valuation of Forest Resources Assets*, to promote ecosystem services assessment processes covering forest, ocean, wetland, desert, etc.

Wetlands are the most complicated ecosystems from the perspective of NECA. They involve the mechanisms and processes of hydrology; the material cycle of carbon, oxygen, nitrogen, phosphorus, and other elements; biodiversity maintenance; water flow regulation; water and soil conservation; and cultural landscape. Their ecological processes, functions, and services contain a non-linear relationship with social welfare. Ecosystem services present a high degree of spatial heterogeneity and dynamism. The international experiences of ecological services focus on water resources, water conservation, and disaster mitigation based on integrated river basin management, land and water resources use, and economic policy design [16, 17] to analyze relationships between economic development and ecological services, upstream protection and downstream development, agricultural development, and land-use scenarios with associated impacts on ecosystems [18–21].

5.3.2 Practice of PES

The results of NECA have been studied and applied in the design of ecological compensation policies. At present, ecological compensation has been extended from a single element, such as for river basins, forests, grasslands, and wetlands, to comprehensive coverage, such as for red lines for ecological protection, key ecological functional zones (KEFZ), and national parks. The most direct purpose of ecological compensation is to protect the ecosystem in order to achieve the goal of sustainable provision of ecosystem services. Therefore, the services provided by ecosystems are an important scientific basis for the design of an ecological compensation system [4, 5].

As an important tool to adjust the environmental and economic behaviours of relevant stakeholders, achieve the sustainable development goal of the river basin, and effectively protect aquatic ecosystems, payment for ecosystem services (PES) in river basins has been implemented internationally for a long time. Some PES has focused on improving ecological services: the compensation standard is tied to the ecological function of water resources and costs of restoration [19, 22, 23] or the trade of ecological services based on a voluntary framework [15, 22]. This approach mainly involves the provision of ecosystem services and products in a river basin being compensated through market mechanisms or government funds. Eco-compensation and ecological restoration practices have been carried out in the Danube River basin in Europe, the Nile River basin in Africa, the Mississippi River basin in North America, the Amazon River basin in South America, and other major transnational and inter-state basins.

China also attaches great importance to the development of an eco-compensation mechanism in river basins. In 2016, China issued the *Guiding Opinions on Accelerating the Establishment of River Basin Upstream and Downstream Lateral Ecological Compensation Mechanism*. It proposed that eco-compensation would be accelerated by applying water quality and water quantity benchmarks to a trans-boundary section of a river basin as a compensation benchmark and that this would promote the development of ecological civilization. However, due to the imperfect way in which the law was drafted, the lack of specific, operational mechanisms, and the diverse methods available, eco-compensation practices in river basins need to be further explored and supported.

Eco-compensation in the Xin'an River basin is the first, and most successful, inter-provincial pilot in China so far. The compensation agreement was signed in 2012, and a three-year eco-compensation program was initiated. By the end of 2017, two rounds of pilot practices came to a successful conclusion. The water quality in the river basin was stable and improved, and economic development was maintained at a relatively fast pace with high quality. Public awareness of ecological civilization and participation in ecological and environmental protection had increased significantly. The linkage and coordination mechanism between the upper and lower streams of the river basin has been continuously improved, and the pilot targets have been fundamentally achieved.

5.3.3 Application of NECA in Spatial Planning

Internationally, a great number of research projects have been carried out using the value of ecosystem services as a reference for regional land-use scenario selection. Stephen et al. [24] quantitatively evaluated the change in ecological capital value and habitat caused by land-use change in Minnesota, United States, from 1992 to 2001 using environmental–economic accounting methods. They also simulated other land-use-change scenarios and comparatively analyzed the results. Research showed that landowners could only obtain the highest economic benefits when the farmland vastly expands, while other social benefits and ecological capital values were relatively lower. The results also showed that long-term land-use planning should consider not only direct economic returns but also the social and natural benefits included. Erik et al. [25] also used the InVEST model to evaluate and predict the change of ecological capital value in the Willamette River basin of Oregon, United States, when agricultural land takes up different proportions. They were able to provide valuable suggestions for land-use planning and decision-making in this area.

Zheng et al. [26] analyzed the relationships between provisioning and regulating ecosystem services and biodiversity protection. Taking the Hainan Ecological Function Protection Area as an example, they calculated changes in ecological capital value due to the large amount of rubber forest planting in this area from 1998 to 2017. The study showed that a 70% increase in the rubber planting area over the past 20 years had caused a decline in natural forests, reduced regulating ecosystem services, and destroyed large areas of wildlife habitat. The results showed that the damage costs and ecological benefits of ecological capital should be considered comprehensively in long-term land planning to maintain the balance between industrial crop supply and natural ecosystem health.

In May 2019, the Central Committee of the Communist Party of China (CPC) and the State Council jointly issued *Opinions on Establishing and Supervising the Implementation of the Territorial Spatial Planning System (Opinions)*. This marked the formation of the top-level design of China's spatial planning system and a milestone in the process of China's historical planning. *Opinions* pointed out that the interrelationship between natural resources and economic development should be taken into account by the regulatory system and that the relationship between natural resources and economic development should be fully analyzed. In terms of the spatial governance system, it is necessary to build a protection system for territorial space to balance the relationship between ecological space, energy security, and food safety, and to achieve better preservation and appreciation of ecological capital through the protection of ecological space.

5.3.4 Ecological Investment and Financing Based on NECA

In addition to investments directly made by government financial funds, ecological investment and financing policies in the rest of the world have focused more on attracting private finance to participate in ecological environment protection through various means, including funds, bonds, and trusts. Government funds also tend to channel investment through funds and other financial means. Some developed countries have started to implement charging policies based on ecosystem services to increase ecological protection funds. At the same time, they also pay great attention to ecological investment and financing policies based on ecological and environmental benefits. For example, Land Retirement for Natural Resource Conservation in the United States compensates farmers that return farmland to forest, grassland, or farrow land. This project has achieved good results. One of the reasons for its success is the Environmental Benefits Index (EBI) introduced by the United States Department of Agriculture to comprehensively assess the environmental benefits of the rehabilitated land. The compensation standard has continuously improved based on the actual conditions.

China's ecological investment and financing policy system has been continuously improved in recent years. Especially since the 18th National Congress of the Communist Party of China in 2012, momentum has accelerated. The *Decision of the Central Committee of the Communist Party of China on Some Major Issues Concerning Comprehensively Deepening the Reform* adopted at the Third Plenary Session of the 18th Central Committee calls for the establishment of a market-based mechanism to attract non-governmental investment for ecological and environmental protection.

In September 2015, the Central Committee of the CPC and the State Council issued the *Integrated Reform Plan Promoting Ecological Progress*, defining the "Establishment of a Green Financial System" for the first time. The *13th Five-Year Plan Outline* clearly proposes the goal of "Building a Green Financial System," and this has become a national strategy. In July 2016, the Central Committee of the CPC and the State Council formulated and promulgated *Opinions on Deepening the Reform of the Investment and Financing System*, which established a series of specific arrangements for the government to deepen the reform of the investment and financing system, specifically for ecological and environmental protection and restoration. In August 2016, the People's Bank of China and other ministries jointly issued *Guidelines for Establishing the Green Financial System*, which marks China as the first country in the world to build a systematic green financial policy framework. Relevant ministries continue to innovate in the areas of fiscal, monetary, and regulatory policies related to ecological investment and financing, from the perspectives of green credit, green bonds, public–private partnerships, ecological compensation, etc. In addition, some local governments have also actively reformed policies and measures to innovate in the field of ecological investment and financing. For example, Huzhou city in Zhejiang province and Jiangsu province have introduced regulations to provide incentives in the form of reduced interest for green credit and green bonds.

5.4   Challenges

5.4.1   A Lack of Standardized NECA Framework and Methods

The connotation and concept of ecological capital are not unified. As a research field, NECA has a short history. Different scholars have diverse understandings of the connotation and the definition of related concepts. The concept and connotation of ecological capital are similar to the terms of ecological assets, ecological products, and ecosystem services. There are also overlaps and crossovers in the definition of assets or capital categories. The inconsistency of terms and concepts makes it difficult to compare results, despite having a large number of achievements in the academic field. The non-standardized accounting results also mean that they are difficult to promote and apply at the policy level, which directly affects the application of research results by decision-makers.

A lack of standardized accounting framework and methods. Although NECA has been carried out in the MA, the United Nations Experimental Ecological Account (EEA) and TEEB study, there are differences in accounting frameworks and methods used in different studies. One of the most important issues where there is a lack of consensus on the "Four Categories" approach proposed by the MA. This distinguishes four types of ecosystem services: supporting, provisioning, regulating, and cultural services. In its technical guidelines, the EEA advocates the use of a "Three Categories" approach that includes provisioning, regulating, and cultural services. At present, there is no consensus in national and international studies on using either of these approaches for accounting indicators. The indicators of both supporting services and regulating services are also not unified.

5.4.2 Economic Development and Planning Decisions not Fully Reflecting the Value of Ecological Capital

For quite a while, China's planning system has been extensive and complex. Since the implementation of integrated urban planning in a single scheme, the core of the planning system has become spatial planning with the goal of optimizing the allocation of territorial space and promoting social and economic development. However, at present, China's planning and decision-making system lacks the quantitative evaluation of the costs and benefits in order to achieve this objective. The planning and decision-making process also lacks guidance on the concept of sustainable development, which leads to unclear economic accounting of planning goals. As a result, the hidden costs and benefits related to ecological capital cannot be fully considered in planning and decision making. The lack of quantitative evaluation methods leads to the questioning of the scientificity and sustainability of planning decisions.

Meanwhile, the objective indicator system of national spatial planning has also lacked consideration of ecological capital. The indicators related to ecological capital are not included in the planning indicator system as binding indicators. Different types of ecological capital and the services they provide cannot be fully reflected in the planning process. The value of ecological capital cannot play a leading role in planning decisions, nor in the process of planning. The long-term absence of ecological capital value in planning has affected the balanced development of the overall layout of China's "Five-in-One" national strategy. The outcome of ecological civilization development, reflected in the increased value of ecological capital, cannot be fully achieved in the planning process.

5.4.3 NECA not Fully Working for the EFM

The pricing mechanism and trading rules of eco-products have not been established yet. Although China has made some progress in developing the pricing mechanism of eco-products, a unified standard procedure has not yet been formed. This leads to the subjectivity and arbitrariness of some NECA, further affecting the authority of assessment results. At the same time, the regulations regarding the eco-product market, which includes an eco-products value realization mechanism, market permits and allocation, as well as the management of relevant stakeholders, are lacking standards, which impedes the implementation of a trading mechanism for eco-products. There are also technical obstacles to incorporating the results of NECA into the financial and investment system. Furthermore, it is difficult to carry out financing operations by using mortgage loans, green bonds, and green funds, which is unfavourable for green development.

The evaluation system and tools of green financial products based on NECA are in great demand. The consideration of ecological capital is missing during the decision-making process for financing projects, which still treats the natural environment as an unvalued factor of production without considering the "capital" attributes of the natural environment. As a result, it becomes difficult for investors to identify green investment projects or to design projects to protect or enhance natural capital. Moreover, it fails to stop investors from providing credits or funds to enterprises that seriously pollute the environment and have no possibility of reaching emission standards. Without such consideration, it is also difficult to help other green projects, including environmental remediation, contaminated site restoration, and ecological conservation, to obtain investment. This makes it difficult for the green industry to develop through the capital market, which further seriously restricts the development of green investment and EFM.

5.5 Chinese Experience and Emerging Best Practices

5.5.1 Ecological Bank: Wuyishan City

Located in the northwest of Fujian province, Wuyi Mountain is one of China's 11 key areas for biodiversity conservation in features of global significance and the only one in the southeast region. Wuyi Mountain was listed as a world heritage site by UNESCO in December 1999. Within national KEFZs, Wuyishan City undertakes important ecological functions, such as water flow regulation, mass flow regulation, and biodiversity maintenance, which are crucial to the ecological security of Fujian province. In 2017, Wuyishan City piloted the NECA tasks designated under the *Implementation Plan of the National Ecological Civilization Pilot Zone (Fujian)*. The accounting results indicated that, in 2010 and 2015, the total value of ecosystem services in Wuyishan City was 183.09 billion RMB and 221.99 billion RMB, respectively. The value of ecosystem services achieved 27.8 and 16.0 times the city's GDP, depending on the accounting methods adopted.

In 2017, Nanping where Wuyishan City is located, released The *Pilot Implementation Plan of "Eco-Bank."* The Wufu township of Wuyishan City was set as the location of the pilot. Through the exploration and innovation of natural resources management, evaluation, circulation, and transaction, a typical path was formed to transform the advantages of ecological resources into economic benefits. The operation process of the "eco-bank" model includes three stages: acquisition and storage of resources, consolidation of assets, and introduction of invested capital. Among them, resource acquisition and storage includes two modes: actual acquisition and storage and pre-storage. The "actual acquisition and storage" refers to the collection and storage of scattered and fragmented resources into the "eco-bank" operating platform or the natural resources bureau of Wuyishan City, the government of the Wufu township, and the villages. Natural resources are collected by means of resource purchasing, circulation, leasing, mortgage loan of the right to use, share cooperation and trusteeship, etc. "Pre-storage" refers to the use of the Wufu Township Ecological Resources Registration Card issued by the "eco-bank" to register their private ecological resource development expectations, including information on the acquisition and storage method, purchase and storage price, purchase and storage period, and purchase and storage use of these resources at the "eco-bank" operation company. This is run through the town government's service centre, which will be included in an Ecological Resources Map and the eco-bank can only be used when it conforms to the control requirements of city and town planning.

The Eco-Bank of Nanping City is not a financial institution, but a natural resources operation and management platform. By making an Ecological Resources Map, the fragmented ecological resources are centrally acquired, stored, renovated, and converted into contiguous high-quality and efficient resource packages and entrusted to operators to realize the transformation of resources into assets and, finally, capital under the premise of ecological protection. The ultimate purpose of the eco-bank is to realize the transformation of resources into assets and capital and to allocate high-quality and environmentally friendly green industries through the connection of green industries to the capital market.

5.5.2 Voluntary Carbon Emission Reduction and Public Welfare Tree Planting: The Ant Forest

The Ant Forest, launched in August 2016, is an application on the Alipay platform developed by Ant Financial of the Alibaba Group. All registered users will get a virtual tree, which can be cultivated by obtaining energy values through personal daily low-carbon practices. When the virtual tree is fully grown, Ant Financial will plant a real tree in the real world. By August 2019, 500 million users in Ant Forest had reduced carbon emissions by 7.92 million tonnes and planted 122 million real trees, creating a green area the size of 1.5 Singapores, according to the *Report on Public Low-Carbon Lifestyle in the Context of Internet Platforms*. At present, Ant Forest's protected areas are mainly concentrated in Heshun of Shanxi Province, Pingwu of Sichuan Province, Yanghu of Huangshan City, Deqin of Yunnan Province, Wangqing of Jilin Province, Yangxian of Shaanxi Province, and Kuubuqi of Inner Mongolia, among others.

Alipay incentivizes consumers to use its services by rewarding their behaviour with a tree-planting scheme. Users will generate a green energy value representing

Table 5.1 Partial green energy value comparison

Green behaviours	Green energy values (carbon dioxide emission reduction)
Online Ttcket purchase (including movie tickets and performance tickets)	180 g
Walk steps	296 g(Maximum)
Offline payments	5 g per purchase
Living consumption (Including water, electricity, gas and etc.)	262 g per purchase
Online train ticket purchase	277 g per purchase
Subway	52 g per ride
Bus	80 g per ride
Second-hand goods trade on idle fish	790 g(Maximum)
Choosing no cutlery for takeout orders from Ele.me	16 g

Note At present, more than 10 kinds of saplings can be planted by Alipay, such as ammodendron, camphor pine, populus euphratica, and salix. The amount of green energy required to plant a certain type of tree ranges from 16,390 to 22,400 g

their carbon dioxide emission reduction after they practice low-carbon behaviours such as offline payment via Alipay, online payment for living consumption, online ticket purchase, appointment booking, applying for electronic invoicing, and taking a walk. However, green energy is not generated immediately. It can only be generated the day after the user completes the green behaviour. The green energy will automatically disappear if it is not collected by the user within three days (Table 5.1).

Ant Financial has three expansion plans regarding the Ant Forest. The first plan involves improving and standardizing the calculation method of carbon dioxide emission reduction through either cooperating with Beijing Environment Exchange and other institutions to standardize a personal carbon footprint algorithm or working with the UN Environment Programme to transform the calculation method to international standards, which will be promoted and applied to other payment platforms to jointly practice carbon emission reduction. The second step is to set up an open green platform to realize personally centred carbon emission calculation methods for enterprises and non-governmental organizations (NGOs) and turn emission reduction into projects related to environmental protection, such as tree planting and water conservation. The final plan is to build a multi-purpose green finance platform to help small and medium-sized enterprises to enter the carbon trading market and reward their carbon reduction activities; assist small and medium-sized enterprises and individuals in conducting offline transactions outside of the market; establish a certification system for green products; and develop green financial tools to support the green investment and financing industry.

5.5.3   Forest Coupon System: Chongqing City

In recent years, Chongqing has maintained ecological and green development priorities, vigorously implementing ecological protection and restoration initiatives. According to the *Action Plan for Implementing Ecological Priority Green Development in Chongqing (2018–2020)*, a total of 17 million acres of forest will be planted between 2018 and 2020, and the city's forest coverage rate will increase to 55% by 2022. In October 2018, the General Office of the Chongqing Government issued the *Work Plan for Implementing Horizontal Ecological Compensation to Improve Forest Coverage in Chongqing (trial)*, exploring a PES mechanism based on forest coverage. In order to achieve the goal of afforestation, Chongqing started the construction of a forest coupon trading mechanism on the basis of a land coupon trading pilot combined with PES.

Chongqing divided its 38 districts and counties into four categories and assigned target tasks accordingly. Among them, the forest coverage rate of major grain-producing counties and major vegetable oil-producing areas (except the national KEFZ) is 45%. After deducting the transaction index, the forest coverage rate of districts and counties that are eligible to sell the forest area index shall be not less than 60%. The Chongqing Forestry Bureau regulates the overall transaction price, stipulating that the subsidy for afforestation per acre should not be less than 1,000 RMB, which can be paid in a lump sum or in installments before 2022. At the same time, the corresponding area of forest management and protection expenses, which shall be no less than 100 RMB per acre per year, should be paid as well for at least 15 years of management and protection. In March 2019, the Jiangbei District of Chongqing and Youyang Tujia and Miao Autonomous County signed the first PES agreement based on forest coverage.

Under the principle of occupation and compensation balance, the forest coupon trading system encourages landowners to voluntarily develop construction land or unused land into qualified forest land, which will be issued an evaluation report by third-party professional institutions. The local government will eventually issue a forest coupon according to the quality and area of forest land. Through the trading platform built by the government, the coupon is transferred to the occupier of the forestland with compensation in the way of market bidding. Only after the occupier obtains enough forest coupons can it purchase the right to use the corresponding amount of land for development and operation.

5.5.4   Green Investment and Financing Policy Practice: Quzhou City of Zhejiang Province

Since the successful approval of the national green finance reform and innovation pilot zone in 2017, Quzhou City of Zhejiang province has established a green standard evaluation system and a factor priority guarantee supply mechanism, actively guiding financial capital to upgrade traditional industries.

A green finance standard system has been developed that suggests the following steps. First, take "finance supports for the green transformation of traditional industries" as the guideline, compile the evaluation methods for green projects and green enterprises, and guide social capital to increase support for the transformation and upgrading of traditional industries, the happiness industries of a beautiful economy, and the smart industries of the digital economy. Second, raise the standardization of comprehensive liability insurance services for production safety and environmental pollution to provincial standards. Third, be proactive in carrying out a quality assessment of special statistical data on green loans to achieve the unification of local standards for green credit. Fourth, take the lead in establishing green banking system standards for local legal entities. This Quzhou model produces a group of green financial products.

The first step to creating green credit in the Quzhou model is to develop green credit products such as pledge loans for intangible assets, environmental rights and interests and receivables, seamless loan renewal, investment-loan linkages, debt-to-equity swaps and debt-to-equity combination. The second step is to innovate the Quzhou model of green bonds by actively exploring the combination model of government industrial funds and private convertible bonds and then issue the first private green entrepreneurship and innovation convertible bonds in China. Then, actively issue green financial bonds to support local green projects. Quzhou City launched a special program called One Point Carbon Turns Green to Gold, which innovates an ecological compensation and forestry development mechanism through forestry carbon sink trading. The third step of the Quzhou model is developing green insurance. China's first comprehensive liability insurance for production safety and environmental pollution established the comprehensive insurance service mechanism of "insurance + process management." It developed China's first joint mechanism of pig insurance and harmless treatment to solve the environmental pollution of livestock from the source and pioneered comprehensive liability insurance for electric bicycles. Fourth, explore the "giant model" of green finance to support the transformation and upgrading of traditional enterprises and accelerate the green transformation of traditional chemical industries.

The Quzhou model also developed a green finance approval process system. Quzhou focuses on rebuilding the green credit approval process in the city's agricultural credit system and plays an exemplary leading role. At the same time, it encourages and guides commercial banks to set up separate green finance departments, independent green credit approval channels, separate green credit scales, and a green credit assessment and incentive system, all of which have received extensive responses and achieved positive results.

5.5.5 Green Finance Standard: Gui'an New Area of Guizhou Province

Gui'an New Area of Guizhou province is the first approved green financial reform and innovation experimental zone in Southwest China. In recent years, it has actively explored innovative green financial standards and assessment systems and mechanisms, issued a Guizhou Province Green Finance Projects Standard and Evaluation Method (trial), and clarified green financial project assessment criteria and procedures. Guizhou Province Green Finance Projects Standard and Evaluation Method (trial) is composed of two parts: Guiding Standards for Key Green Finance Support Industries in Guizhou Province (trial) and Evaluation Methods for Major Green Projects Supported by Green Finance in Guizhou Province (trial).

On the basis of the *Green Industry Guidance Catalogue* (2019 edition) issued by seven ministries and commissions, including the National Development and Reform Commission, Guizhou Province Green Finance Projects Standard and Evaluation Method (trial) has been improved and enriched from the aspects of financialization, localization, practicality, and internationalization. From the perspective of industry selection, according to the industrial characteristics of Guizhou province, key green industries are selected from the *Green Industry Guidance Catalogue* (2019 edition), and green industries with the characteristic of Guizhou are appropriately increased, such as ecological tourism, biodiversity protection, the green data centre, maintenance and management of arable land and soil, fertilizer and water speed measuring technology development and application, green transportation, green public transportation, etc. In terms of standard form, it gives priority to quantity and takes quality as complementary; in terms of presentation, the "index system method" is adopted to clearly characterize the green financial standards and characteristics of various industries. For example, the green data centre adopts the standards promulgated by the Ministry of Industry and Information Technology that the energy use efficiency of new large and super-large data centres has significantly improved through electricity transformation. The Hong Kong Quality Assurance Agency Green Finance Standard, standards related to the Equator Principles, and a number of internationally recognized industry standards have been added to attract international green finance funds.

According to the investment amount of green technology, the level of technology, the comprehensive eco-environmental benefits, and the replicable and popularizing value of green financial innovation, etc., Guizhou Province Green Finance Projects Standard and Evaluation Method (trial) lists projects that are of great significance to the protection of the ecological environment and the innovation of green finance that can be replicated and promoted as major green projects. Mainly major green projects will be evaluated, supported, and promoted, with supporting policies including but not limited to financial incentives, policy support, priority review, and financial support.

At present, the green finance project bank of Gui'an New Area has absorbed more than 1,000 projects in Guizhou Province and Southwest China, with the project financing demand reaching more than 400 billion RMB. It has successfully launched the distributed energy project supported by green asset securitization called Two Lakes and One River, the Guiyang metro line S1 project, and other projects, in cooperation with financial institutions.

5.5.6 Ecological and Environmental Damage Compensation: China

Since 2018, when China launched a pilot reform of the compensation system for ecological and environmental damage, the principle of Taking Responsibility for Damage and Valuing the Environment has been implemented. After a two-year nationwide trial, it has effectively promoted the restoration of the damaged ecological environment and paid compensation in cash for the damage that could not be repaired. While solving the dilemma of the public becoming victims of pollution caused by enterprises while the government is paying the bill, it accumulated funds for compensation for ecological environment damage. So far, 942 cases of ecological and environmental damage compensation have been handled, involving compensation of about 2.5 billion RMB. More than 12.09 million cubic metres of soil, 19.98 million square metres of forest land, 6.05 million square metres of grassland, 42.23 million cubic metres of surface water, and 460,000 cubic metres of underground water have been repaired. About 227.92 million tonnes of solid waste has been cleaned up. There are some typical cases selected:

(1)   Hazardous Waste Dumping in Haide, Jiangsu Province

In 2014, a company repeatedly entrusted unqualified personnel to dispose of hazardous waste, resulting in more than 100 tonnes of waste lye being dumped into the Yangtzerriver, causing serious water and environmental pollution. The parties bore criminal responsibility and were sentenced and fined between 10,000 RMB and 2.58 million RMB. After the criminal case was concluded, the People's Government of Jiangsu Province filed a claim with the people's court as the obligee of compensation, and the enterprise undertook fines of 54 million RMB to repair the damaged ecological environment.

(2)   Ecological Environmental Damage Compensation in the Dabusu Nature Reserve

In 2005, a company started an oil recovery project in the core and buffer areas of Dabusu Nature Reserve without approval. Due to long-term oil exploitation, the soil and vegetation in the reserve area have been damaged. In 2018, the People's Government of Jilin Province, as the obligee of compensation, made a compensation request with the company for ecological and environmental damage. Based on the environmental damage assessment and suggestions for restoration, the two sides reached an agreement on ecological and environmental damage compensation. Jilin Oilfield Company entrusted a third party to prepare the restoration plan for the ecological and environmental damage caused by illegal mining and organized the restoration work by itself. After the completion of the restoration, the environmental authority evaluated the result of restoration. In addition, the company paid 2,303,600 RMB for the loss of service function during the ecological and environmental restoration. The damages are transferred to the financial account designated by the compensation right holder and used as the non-tax income of the provincial government, which is handed over to the Provincial Treasury and managed according to a provincial financial budget.

(3) An ecological restoring park instead of atmospheric emission violation in Shaoxing City, Zhejiang province

An enterprise in Zhejiang province interfered with automatic monitoring data and illegally discharged air pollutants. After negotiation, the enterprise not only paid 1.1 million RMB for air pollution damage compensation but also voluntarily added 1.76 million RMB to build an ecological park in the village where the damage occurred. In this case, the polluters improved the ecological environment by building the ecological park, which enhanced the surrounding villagers' sense of a gain in environmental improvement.

In summary, for the first case, in which damage to the ecosystem cannot be fixed, compensation by money is the only solution. In the second case, the damage is partially repaired and additionally compensated with money. The third case adopts an alternative restoration plan to achieve eco-environmental protection. It should be noted that, at present, the Ministry of Finance plans to integrate eco-environmental damage compensation and fines from environmental public interest litigation into the fiscal budget management at all levels but cannot allocate the fund to address eco-environmental restoration needs specifically. How to efficiently utilize this fund is an urgent problem and remains unsolved.

5.6 International Experience and Emerging Best Practices

5.6.1 Gender as a Factor in Ecosystem Services: Nepal and Kenya

Nepal: Ecosystem services protections for women and vulnerable groups need to be carefully designed so as not to create additional burdens. Chaudhary et al. [27] examine provisions that are intended to ensure social equity but find that high-income groups are still able to disproportionately access the benefits of ecosystem services. In particular, the authors note that these provisions place additional burdens on the groups they sought to help. The policy of imposing fines on those who do not participate in community forestry meetings is intended to encourage the participation of marginalized groups. However, these groups have both the least capacity to attend meetings and the most difficulty paying fines.

Kenya: A randomized trial demonstrates that, for PES schemes to be equitable, they must consider the relative status of women in a given context and overcome cultural and economic barriers. Examining a randomized trial that utilized auction-based contracts, Andeltová et al. [28] find that women tend to be relatively more risk averse than men and theorized that this is likely due to women's significantly lower income in this context. The authors also observed that the trees planted by woman have lower survival rate than men, despite women working harder, and argued that this is due to inequality in reciprocal labour between men and women. The authors argue that targeting women in PES schemes can improve gender equity by granting them greater access to decision-making, training, and cash. Andeltová et al. [28] also argue that the participation of women in these schemes can significantly improve the schemes' effectiveness.

PES can economically empower women if deployed in a context-sensitive approach. In an examination of agroforestry schemes with PES in sub-Saharan Africa, Benjamin et al. [29] find that female participation "reduced profit inefficiency," which contributes to economic empowerment. The authors suggest that agroforestry schemes with PES have the potential to empower women in sub-Saharan Africa if they target poor female smallholders.

Access to education is gendered and plays a central role in determining who benefits from ecosystem services. In a study of landowners in Brazil, Lima and Bastos [30] found that years of formal education has a significant impact on whether an individual perceives the value of ecosystem services that are considered to be more difficult to observe (e.g., pollination and pest regulation).

Accounting for gender in ecosystem services is likely to result in the prioritization of different services, which in turn could lead to different outcomes for livelihoods. In a study of nine Indigenous communities in the Colombian Amazon, Cruz-Garcia et al. [31] compared the ways in which men and women value ecosystem services. In the study, services such as land for agricultural fields and the provision of fish and medicinal plants were equally important to both men and women. However, while women tended to consider wild fruits and resources to make handicrafts as important, men more frequently mentioned timber, materials for making tools, and coca leaves. In their examination of the Fijian men's and women's perspectives on the use, benefit, and value of mangrove ecosystems, Pearson et al. [32] found that women and men in Fiji valued the ecosystem services provided by mangroves differently, based largely on the traditional gender-specific tasks to which they were assigned. The authors call for a gender-sensitive valuation framework of ecosystem services to ensure that the decision-making process is inclusive.

5.6.2 Natural Capital Management Experiences: United Kingdom

Implementing a natural capital approach: The work of the Natural Capital Committee. Over the last decade, the United Kingdom (U.K.) government has introduced a set of ambitious policies to protect and enhance natural capital. This has put the U.K. at the cutting edge of efforts to advance the conceptual, empirical, and practical policy approaches needed to better manage natural capital. However, despite these good efforts, results have fallen short of ambitions. Some limited progress has been achieved in some areas, but, in many others, the state of the environment has continued to deteriorate. Ecosystems remain fragmented. The difficulties inherent in developing a robust evidence base to guide policy has been a major obstacle. Equally, policy implementation has been impeded by the failure of the U.K. government to provide the legal, institutional, and financial means to achieve its ambitious policy objectives.

In 2012, the U.K. government established the Natural Capital Committee (NCC). During its first phase of work, the NCC identified the key elements that it considered necessary to support the elaboration of a coherent strategy for protecting and enhancing natural capital. It also elaborated key policy principles to guide the development of a natural capital approach and recommended that pilot projects be implemented in order to gain experience in applying a natural capital approach in different areas and circumstances. The NCC considered that there were three basic foundations on which a better strategy for managing natural capital should be based: measurement, accounting, and valuation. In addition, the NCC argued that a framework was needed to identify and prioritize actions. The framework should address not only actions to protect nature in the future but also actions to reverse historic losses in natural capital. In 2017, the NCC produced a workbook to help decision-makers to implement a natural capital approach. It synthesized some of the NCC's main findings and referenced tools and sources that could assist planners, communities, and landowners to make place-based decisions to protect and enhance natural capital.

Natural capital accounts. In 2011, the government committed Defra and the Office of National Statistics (ONS) to incorporating natural capital into the U.K. Environmental Accounts by 2020.

Measuring natural capital. In 2014, the NCC proposed a framework for defining and measuring natural capital.

Prioritization framework. In its 2015 report, the NCC considered accurate measurement, accounting, and robust economic valuation to be essential for implementing a natural capital approach. It argued that a commitment to reverse historic losses in natural capital was also required. To this end, the NCC proposed a framework for identifying and prioritizing actions organized around three questions: How much natural capital is needed and what targets should be adopted? Which natural assets and benefits need urgent action? How should priority action be determined?

Natural capital valuation. In 2017, the NCC published a guide to natural capital valuation. The NCC considered that the failure to adequately value the full economic costs and benefits of natural capital assets in public and private decision making was a key factor in their deterioration.

Financing the protection and enhancement of natural capital. While the provision of mechanisms to finance investments in natural capital is important, government policies and institutional arrangements can provide important incentives (and disincentives) for such investments. In the U.K., some of the main policy approaches that are in operation or under consideration include:

- Replacing European Union subsidies for agriculture, which have often intensified pressures on natural capital with a system targeting public goods and promoting a better market for environmental goods and services.
- Establishing an "environmental (or biodiversity) net gain" principle for development projects.
- Integrating nature-based approaches into flood management schemes.
- Establishing a domestic carbon offset program that helps to create a market for carbon sequestration, e.g., tree planting.
- Product certification schemes.

During preparations of the 25 Year Environmental Plan [33], Defra compiled information on the main sources of finance for natural capital. The 2019 *State of Nature* report also examined financing for conservation. Defra found that most investments in the U.K. had been provided as subsidies or grants from the European Union, central and local authorities, philanthropic organizations, and the national lottery. Central government spending was estimated at £805 million in 2015–16. The 2019 *State of Nature* report's estimate was lower; it concluded that public sector spending on biodiversity in 2017–18 was £456 million, a decline of about a third from a high point in 2008–09.

Defra estimated that spending by non-governmental environmental organizations with a nature or biodiversity objective was £236 million in 2014–15, and support from the national lottery was about £100 million per year. The 2019 *State of Nature* report's estimate was similar: NGO expenditure on biodiversity and conservation amounted to £239 million in 2017–18, an increase of about one quarter over five years. This report also stressed the importance of volunteer work, estimated at 7.5 million hours to support the preparation of the *State of Nature* report. The authors concluded that "although financial investment is crucial, as are government policy and legislation,….the most successful conservation action arises from partnerships, across governments, charities, business, landowners and individuals working together."

5.7 Accelerating and Scaling up Solutions

5.7.1 Design Unified Standards for Eco-Products Valuation

To establish a unified classification and pricing standard for eco-products, several steps must be taken, including:

- Further strengthening the study on the classification of eco-products.
- Harmonizing the existing classification system of eco-products adopted by the Ministry of Ecology and Environment, the Ministry of Natural Resources, and the National Development and Reform Commission with corresponding international guidelines to identify and analyze the concept, connotation, and scope of eco-products.
- Preparing a classification guide of eco-products in different regions using the main ecosystem types in different regions of China in order to guide the classification and indexation of eco-products.
- Carrying out an analysis of international and national NECA results, including international and national market prices for the same eco-products produced in different areas, to try to establish a price mechanism model of eco-products that could lay the foundation for forming the pricing standard of eco-products.

It is critical to factor in the valuation effects based on gender and vulnerable populations.

Gradually improve the ecological capital accounting methods system. At present, the framework and methods of NECA are not unified and the technical parameters needed to characterize regional ecological systems are lacking. Therefore, we recommend the adoption of two approaches for valuing ecosystem services: a bottom-up valuation of ecosystem services and a top-down equivalent factor method based on macro measurement results. The aim is to establish the assessment framework, accounting method, and technical parameter system of NECA at the regional, watershed, and national levels for different purposes. There is also a need to clarify the application of different accounting methods, to guide various regions to carry out eco-product valuation within a unified framework, and to provide technical support for the standardization of eco-product value.

5.7.2 Establish and Unify the Concept and Rules of an Ecological Financing and Investment Mechanism (EFM)

Further define and unify the concept and connotation of ecological investment and financing. In order to distinguish eco-financing from green financing, and to accurately reflect the status of EFM in China's ecological environment protection system and its driving effect on economic growth, we need to clearly define the concept of EFM needs. Responsibility for the indicators related to EFM was allocated to relevant governmental departments, which would jointly promote the design, revision, implementation, dissemination, supervision, and management of standards in key areas of EFM, such as securities, insurance, environmental rights and interests, and trading, to ensure the unity and universality of the standards.

Establish determination and evaluation rules for ecological investment and financing. Based on the valuation results of ecological capital and eco-products, work should be carried out to explore and formulate a universal and unified catalogue of ecological investment and financing sectors and specific projects. Regional authorities should be encouraged to produce regional standards such as Evaluation Criteria for Ecological Investment and Financing Projects, Evaluation Criteria for Ecological Investment and Financing Enterprises, and Specifications for the Construction of Ecological Finance Franchise Institutions, as well as other guidelines that could be widely applicable in banks and other financial institutions dealing with bonds, credit, stocks and funds. The catalogue of ecological investment and financing sectors and projects should comply with the overall requirements of national ecological civilization development as well as development trends of the green industry system. They should clearly specify the scope and key areas of ecological investment and financing projects.

5.7.3 Carry Out Planning and Project Cost-Effectiveness Analysis Based on NECA

Establish a planning methodology that comprehensively considers the protection and enhancement of ecological capital and the related economic benefits. The ecological capital background of the targeted region or watershed should be mapped, and the major ecosystem services and associated economic benefits they provide should be evaluated. This information should be integrated into the objective indicator system of spatial planning, with an awareness of the gendered characteristics of the value of resources. Research should be carried out on the valuation of eco-product provisioning, ecosystem regulation, and culture services under different planning scenarios, as well as on the associated economic cost–benefit analysis during the planning stage. A cost-effectiveness analysis should be used during planning to ensure that the approach chosen protects or enhances ecological capital at the least cost.

Require cost-effectiveness analysis for ecological investment and financing projects. This should apply to the different types of ecological investment and financing projects, such as the ecological restoration and environmental treatment of mountains, rivers, forests, farmlands, lakes, and grasslands, mines, and river basins.

The external impacts on and benefits of ecological capital should be comprehensively considered within the project life cycle. Research should be carried out on the ecological and environmental life cycle impacts of ecological investment and financing projects, including the preparation of an inventory of impacts and a scenario analysis. A breakthrough is needed to establish the framework and technical methods of NECA for the management and ecological restoration of river basins. Fundamental data should be collected as well as the technical parameters of water flow regulation, climate regulation, water and soil conservation, biodiversity conservation, and other ecosystem services involved in different types of watershed and wetland construction projects. Technical guidelines should be prepared for the cost-effectiveness analysis of projects for the ecological restoration of river basins, and support should be provided for optimizing project decision making.

5.7.4 Strengthen the Design of Ecological Investment and Financing Policies

Promote the integration of ecological capital into the design of ecological investment and financing policies. When designing ecological investment and financing policies and mechanisms, we should consider not only the goal of economic benefit as traditionally defined but also the goal of ecological protection and the implications based on the gendered use of resources. We should comprehensively evaluate the cost of supplying eco-products and the benefits of providing those services at different scales by using NECA as a quantitative assessment tool for investment and financing decision making. We should determine whether the design of ecological investment and financing policies and mechanisms are sustainable by comparing the changes in the stock of ecological capital and other capitals. We should continuously revise ecological investment and financing policies to improve the reliability of decision making based on the evaluation of these changes.

Establish a performance evaluation system of ecological investment and financing projects based on NECA. Taking an ecological and environmental performance perspective, NECA should be combined with the existing project performance evaluation system. First, the existing performance evaluation should be appropriately adjusted based on changes in the provision of ecosystem services. It should be applied flexibly based on activity type, and the implications should be based on the gendered use of resources. How project funds are used should objectively and fairly evaluate the effectiveness, efficiency, and benefits. The evaluation should strengthen ecological protection and restoration, improve the quantity and quality of ecological capital, and enhance the capacity to supply eco-products through the means of PES, property transaction, ecological industrialization management, etc.

5.8 Governance and Implementation

In future development, the application of NECA and ecological investment and financing policy should be divided into different stages according to the extent of urgency and difficulty.

Short-term plan (2020–2023): First, raise public awareness. Through the media, make the public generally aware that ecological capital is not only valuable but also closely related to the interests of everyone. Foster awareness that protecting ecological capital could increase its value and the return on investment. Second, standardize NECA. The frameworks for accounting, applying, investing, financing, and trading ecological capital should be established under a unified methodology, indicator, and valuation system. In order to better promote NECA and its application in various policies and planning, the standardization of NECA should be studied to make the accounting results comparable at both regional and time-frame levels. Third, further improve capacity building. Officials of the relevant government departments at all levels should be trained in and certified to use NECA to implement ecological investment and financing policies. Local governments should be encouraged to set up ecological cloud information platforms, share basic accounting data, and design automatic NECA systems to ensure the sustainable advancement of working practices.

Medium-term plan (2024–2026): First, promote pilot projects. Apply and disseminate the standardized framework and methods of pilot studies to more regions, so as to form a mature NECA method and pricing system. Second, promote the application of NECA in the policy fields of ecological investment and financing and spatial planning. Improve the design of relevant ecological investment and eco-products' financing policies by establishing an eco-products trading market and its operational mechanism. Carry out an analysis of national spatial planning and decision-making based on NECA to strengthen the optimal allocation of land and resources. Establish regulatory mechanisms for NECA and EFM to support comprehensive decision-making by governments at all levels. Third, strengthen the application of NECA at a strategic level. Promote NECA and EFM innovation to support national strategic plans such as the Belt and Road Initiative, development of the Yellow River basin and YREB, etc. Provide methods and policy guidance for national strategic development regions such as Beijing-Tianjin-Hebei and Great Bay Area.

Long-term plan (2027–2035): Form a complete NECA and ecological investment and financing system. Adapt top-down government institutional reform by continuously improving ecological capital management. Establish working mechanisms for coordinating socioeconomic development with the comprehensive management of ecological capital and ecological environment quality. Mainstream NECA and ecological investment and financing policies into decision-making processes at all government levels (Fig. 5.1).

Fig. 5.1 Governance and implementation roadmap (2020–2050)

5.9 Recommendations

5.9.1 Accelerate the Promotion of NECA and Its Policy Application: Unifying Values Through Standardization

At present, the basic concepts of ecological capital and eco-financing have not reached a mutual agreement. The accounting framework and accounting methods carried out around the world have not been unified. The unstandardized accounting results are making it difficult to apply the results at the policy level due to a lack of comparability. Take the lab-initiative as the carrier, promote the standardization of the accounting framework and methods, integrate the mechanisms and process parameters of different regions and different types of ecosystems, and form an integrated regional solution consisting of basic data, quantitative accounting, scenario simulation and policy innovation. This approach also provides the support for eco-product transactions and eco-financing mechanisms at the national level.

5.9.1.1 Establish a Normative and Standardized NECA System to Provide a Basis for Value Realization

Form a NECA framework system. On the basis of NECA research experiences so far, an international NECA committee jointly organized by international organizations such as the United Nations Statistical Division, the World Bank, and

ADB etc., is proposed to: (i) provide basic information and agreement on essential concepts and connotations such as stock accounting and flow accounting; (ii) propose accounting frameworks for provisioning services, regulating services, cultural services, and/or supporting services in different application scenarios; and (iii) explain and standardize the applicability and basic data requirements of different accounting methods.

Publish technical guidelines on NECA. To comprehensively summarize the specific practices of NECA in different regions, form a standardized and unified technical guideline for NECA that clearly defines the principles for indicator selection for different application scenarios and makes uniform stipulations on professional terms, accounting indicators, applicable models, data sources, and technical parameters, etc. The standardization of NECA methods provides an approach to the comparability of accounting results between different regions and the applicability of cost-effectiveness decisions. It will also serve to standardize cross-regional transactions of eco-products.

Promote the development of an ecological resources monitoring network. Establish an ecological resource monitoring network that coordinates national technical standards, regional and river basin technical supervision, local promotion and implementation, and community participation. Promote the standardization of a monitoring network that combines multi-source remote sensing and ground observation of ecological resources to form a comprehensive survey and monitoring coverage on forest, grassland, wetland, farmland, oceans, minerals, water resources, and other important ecological elements. This will finalize data sharing, verification, and a working mechanism for cross-industries, cross-regions, cross-sectors, and the various stakeholders that participated.

5.9.1.2 Establish a Lab Initiative for NECA and Policy Planning to Provide Tools for Accurate Policy-Making

Develop a standardized NECA platform. Summarize various types of statistics, surveys, inventory, remote sensing and monitoring data from the ecological resource monitoring network. Propose principles on data selection and cleaning as well as the selection standards for accounting model tools to establish a fundamental database for NECA. The database will cover the structure, quantity, quality, and spatial distribution of land, forests, grasslands, water resources, and biological resources, etc. Integrate NECA model tools and the technical and value parameters required for the calculation of different types of ecological services to form a NECA model tool base that enables the ecological capital accounting functions on both physical and monetary perspectives.

Develop a decision-making consultation platform based on NECA. Clarify the simulation model, methods, and data requirements of eco-financing policy plans such as bonds, funds, eco-product trading, PES, spatial planning, effects and post-evaluation analysis, etc. Develop a policy and planning decision-making methods base; each of these methods needs to factor in gender and vulnerable populations. Propose technical guidelines for cost-effectiveness analysis at the policy, planning, and project levels. Clarify the ecological services, ecological service accounting methods, applicable parameters, and economic benefit accounting methods for different types of projects. Develop corresponding cost-effectiveness analysis tools to form a comprehensive platform for policy and planning simulations and ecological capital institutional innovation.

Establish a national NECA and policy planning simulation lab. With NECA as the core, set up a national-level NECA and policy planning simulation lab that is cross-sectoral, cross-disciplinary, and cross-domain. The lab should have clearly functional divisions between its institutions and initiate working mechanisms, including data sharing, information disclosure and dissemination, and multiple participation. The lab aims to provide institutional and practical support for the formulation of regional, river basin, and national eco-financing policies and for the implementation of cross-regional and cross-domain strategic planning.

5.9.1.3 Establish an NECA Mechanism and a Policy Application that Guarantees Transformation "From Green to Gold" and Includes Gender Implications

Draft and issue *Guiding Opinions on Accelerating the NECA and its Policy Application.* In order to ensure the mainstreaming of NECA into the management and decision-making processes of governments at all levels, it is necessary to clearly normalize and standardize NECA and promote the application of NECA in the fields of eco-financing and spatial planning as both short-term and long-term objectives. Refine target indicators and define completion requirements at each stage, and form the guarantee system for a sound NECA and eco-financing policy development, which provides a basis for securing "lucid waters and lush mountains" as invaluable assets.

Establish a working mechanism to promote NECA and its policy application. Secure an organizational guarantee that the State Council should take the lead in establishing a high-level ecological capital working group with the participation of relevant ministries and agencies, such as the National Development and Reform Commission; the Ministry of Finance; the Ministry of Natural Resources; the Ministry of Ecology and Environment; the Ministry of Water Resources; the Ministry of Agriculture; the Ministry of Housing and Urban, Rural Development; the Ministry of Culture and Tourism; the National Forestry and Grass Administration; and the Chinese Academy of Sciences. The working group should designate the responsibilities of relevant departments, improve the working mechanism, arrange work tasks associated with the NECA system and its policy application, and provide sufficient funds.

Deploy the key tasks. Summarize the experiences of pilot projects such as development of Ecological Civilization Experimental Areas, accelerate the establishment of NECA technical methods and specifications, and form standardized accounting guidelines. Accelerate the lab-initiative for NECA and policy simulation by clarifying the work tasks and deadlines of different departments and institutions. Strengthen the verification and evaluation of the accounting results to improve the scientificity and applicability of the accounting results in policy formulation and decision-making, including how ecosystem services are realized via gender and vulnerable populations. Design policies for realizing eco-products' value and strengthen the application of NECA results in the field of eco-financing policies. Formalize and improve related policies for eco-product transactions, ecological banks, PES, and ecological funds to promote the transformation of ecological environment governance from cost attributes to value attributes. Start the pilot projects in Ecological Civilization Experimental Areas to establish the projects' evaluation mechanisms by applying NECA in spatial planning and eco-financing fields. Strengthen capacity building to guarantee ecological capital accounting work.

5.9.2   Promote Eco-Products Pricing and Trading in Stages: Realizing Value Through Transactions

Eco-products are different from general products in that their production has the dual attributes of natural reproduction and social reproduction. Their value therefore has the dual attributes of market transactions and compensation. The products have incomplete competition and public goods attributes. They feature higher consumption opportunity costs and complicated consumption willingness. To truly turn the "lucid waters and lush mountains" into "invaluable assets," it is necessary to realize the circulation chain of "ecological resources to ecological assets, then to ecological capital and finally to liquid ecological capital" in stages. To do this, the key is establishing an eco-product pricing and trading system.

5.9.2.1   Form a "Three-in-One" Pricing Mechanism for Eco-Products and Establish a Benchmark for the Circulation of Eco-Products

Determine the price of tradable eco-products through the market. The supply value of eco-products can usually be fully reflected through market transactions. The development of green agriculture and green tourism industries will enhance the intrinsic value of eco-products. To explore the "ecological + " model, establish ecological+, brand+, and Internet+ mechanisms to improve the output efficiency of eco-products. To strictly manage an eco-products certification mechanism, establish a unified eco-products standard, certification, and labelling system to guide various social capital investments relying on market mechanisms. Provide a large number of differentiated eco-products to meet consumer demands at different levels.

The government takes the lead in enhancing the value of regulating services and supporting KEFZ services. For eco-products with outstanding ecological functions, non-competitive, non-exclusive and public product attributes—such as national parks, nature reserves, ecological conservation areas, water conservation areas, wild forest areas, ecological restoration management areas, etc.—their function and service value have to be reflected through PES and/or government investment. Through the secondary distribution of financial transfer payments, one can mobilize the enthusiasm of eco-product producers; reasonably allocate the rights and interests of eco product producers, investors, and beneficiaries; and improve the long-term mechanism of differentiated and coordinated development of different main functional areas.

Comprehensive pricing of eco-environmental resources is explored through government-led trading of eco-environmental resources. On the basis of pollutant emission trading and carbon trading pilots, combined with the standardization process of NECA, scientifically determine the types of rights and interests of eco-environmental products that could enter the market. Ensure ownership rights by establishing property rights, and promote the appreciation and circulation of ecological property rights. Explore policy innovations such as bidding and auctioning, mortgage loans, green securities, and eco-financing to promote ownership circulation and ecological assets value increase.

5.9.2.2 Implement Eco-Product Trading Pilots and Explore Eco-Products Trading Mechanisms

Based on water rights trading, explore the mechanism for eco-products trading. Promote water rights trading pilots under the premise of total water use control. Through the registration of water resources use rights, water users are given the right to use and gain water resources in accordance with the law. These water rights need to recognize gendered implications, both in terms of whose labour is involved and who receives the benefits. Based on the regulatory operation and supervision of the primary water rights and the water rights transaction platform, use market mechanisms and information technology to promote water rights transactions across river basins, regions, industries, and different water users. Through the establishment of water rights systems and water rights transactions, promote the optimal allocation of water resources and efficient use in accordance with market competition and market rules.

Carry out ecological capital and eco-products trading pilots. Select and determine pilot areas. Check and ratify the supply of eco-products, such as water conservation, carbon sequestration, and pollution purification, and then determine the quantity, quality, and spatial layout of eco-products. Formulate the principles of the ecological rights of the primary market and the trading rules of the secondary market in the pilots. Study the system of empowerment and consumer payment of ecological assets and eco-products, and formulate transaction processes. Explore to build an ecological assets and eco-products trading platform, and allow ecological assets and eco-products to be exchanged with development rights quotas, such as energy rights, water rights, and pollution rights. Try to develop land development rights and ecological rights transactions in the downstream of the river basin. Establish a diversified, market-based OCB mechanism based on the ecological environment capacity, with the participation of government, enterprises, organizations, and communities.

Establish a trading platform for ecological capital and eco-products. Based on the technical support of the National NECA and Policy Planning Simulation Laboratory, establish a national eco-products trading platform and a unified pricing standard for pollution rights and ecological rights. Start rights transactions in virtual markets to form a unified, open, and competitive ecological capital and eco-product market system to promote ecological, environmental and resource transactions across regions and categories. Improve the guarantee system for ecological capital trading platforms, and build a property rights system for ecological capital that is suitable for market transactions. Form a public trading mechanism with standardized procedures.

5.9.2.3   Accelerate a Transaction-Based PES System and Establish a Diversified PES Mechanism

Increase PES programs for regions with prominent ecological functions but relative poverty. To continuously improve PES standards, expand the scope of national KEFZ payments transfer and change the direction of investment from ecological protection to a combination of ecological protection and the improvement of local livelihood to improve the vital capacity of ecological protection areas. Strengthen benefits evaluation after PES implementation. Encourage the enthusiasm and proactiveness of ecological protectors, and establish a PES mechanism that is compatible, harmonious, and promoted along with local economic development.

Speed up PES along the river basin. In accordance with the principles of "government procurement, market competition, contract management, and rewards & punishments," establish a PES mechanism based on "beneficiary purchasing" through "open tendering + price negotiation + agreement signing." Issue guidelines and conduct pilot projects for PES along river basins. Establish PES standards for both upstream and downstream of the river basin from water resources, water environment, and water ecology perspectives. Establish joint meetings for the protection and governance of the river basin by linking districts and counties from upstream and downstream together. Set reward and punishment mechanisms for PES in order to form a dynamic and scientific PES mechanism with clear subjects, objects, standards, and with various forms.

Establish a trading mechanism for PES based on OCB. With the objective of increasing the total value of ecosystem services, encourage landowners to voluntarily develop the constructed or unused land into ecological land in accordance with regulations; then it can be served as the transaction subject to trade in a PES market. Drawing on ecosystem provisioning and regulating services aspects, formulate transaction standards based on OCB principals. Issue management measures for supervising ecological land transactions, improve supporting systems for ecological land transactions, and conduct trading trials to promote value realization of ecological capitals.

5.9.2.4 Carry Out PES Pilot Projects and Trading in the Yangtze River Basin and Strengthen Biodiversity Protection

Develop PES practices based on water flow regulation services. Provide an accounting of ecological service values such as water regulation, conservation of soil and water and biodiversity conservation in the upstream of the Yangtze River Basin. Based on the accounting results, determine the PES standards between downstream and upstream of the Yangtze River Basin, and promote the implementation of a PES mechanism based on water resource supply and water quality. Increase PES for KEFZ in the upstream of the Yangtze River, and rationally determine the proportion of central fiscal transfer and local payments in the midstream and downstream. Strengthen the promotion and dissemination of PES experiences from the Xin'an River, the Chishui River and rivers across Chongqing, to formulate the *Measures for Reward and Punishment on Ecological Quality of Cross-Border Sections of the Yangtze River Basin, and PES Management*. Establish a reward and punishment mechanism for mobilizing the proactiveness of PES, and improve the joint alliance of local governments from both upstream and downstream to co-govern the river basins. Unified monitoring and supervision mechanisms will build a long-term mechanism for river basin protection to ensure continuous improvement and the stability of water environment quality.

Explore the establishment of a "forest coupon" trading centre based on an OCB mechanism. In accordance with the basic principles of "total control, quota management, land conservation, reasonable land supply, and balance between occupation and replenishment," draw on Chongqing's experience through National Bureau of Statistics and manage industrial and mining wasteland in the Yangtze River basin to effectively replenish the amount of forest land to ensure a dynamic and balanced growth of forest land resources along the Yangtze River basin. Set the Yangtze River basin's forestry red line to increase year by year, and establish a Yangtze River basin forest coupon transaction centre. Incentivize PES between occupied forest land and newly added forest land to ensure a steady increase in terms of volume and area of forest land.

Increase the protection of and investment in biodiversity in the Yangtze River basin. With the construction of a large number of reservoirs and economic development, a large number of rare and endemic fish-specific spawning grounds and habitats in the upstream of the Yangtze River no longer exist; as a result, the flowing water habitat suitable for spawning of some fish will disappear. Innovate existing regulations on local ecological and environmental damage compensation funds, and promote a joint fund from 11 provinces and cities along the Yangtze River to protect biodiversity and carry out eco-financing policy innovations by initiating the Yangtze River Delta integrated eco-environment protection fund and habitat bank as pilots. Establish a Yangtze River basin projects inventory for ecological protection and eco-financing that includes projects related forestry cultivation, wetland protection, ecological resettlement, etc. Guide various institutional investors to invest in green financial products.

5.9.3   Strengthen the Application of NECA in Spatial Planning: Optimize Value Through Planning

China's planning system has long faced problems of systematic incompatibility, content conflicts, inadequate coordination, and a lack of comprehensive evaluation of ecological capital changes and economic and social benefits caused by land-use changes. China is currently carrying out the pilot and reform work of "multi-planning integration," with spatial planning as the core. This will provide an opportunity for ecological capital value to be incorporated into planning decisions, as well as establishing a planning method system based on the integrated considerations of ecological capital and economic benefits and improving the evaluation mechanism of spatial planning. Through scientific planning, the lucid waters and lush mountains will be transformed into invaluable assets to make China a contributor and leader in the construction of the global ecological civilization.

5.9.3.1 Include Natural Ecology in the Management of Compulsory Indicators and Strengthen the Role of Spatial Planning in Optimizing and Allocating Natural Ecology

Strengthen the binding forces of natural and ecology indicators in planning. Quantitative and qualitative indicators of the five important ecosystem types—forests, grasslands, wetlands, arable land, and oceans—are clearly defined as mandatory indicators and included in the target indicator system for spatial planning. With the overall goal of ensuring that the value of ecological capital does not decrease, propose key planning tasks from multiple aspects, such as establishing an economic and ecological win–win situation, maintaining the ecological capital stock, providing a reasonable flow of ecological capital, strengthening ecological environment construction, and preventing and controlling environmental pollution. Coordinate land and space layout, economic and industrial layout, and urban infrastructure layout through reasonable index settings.

Enhance coordinating analysis between natural ecosystems and economic development. Encourage regions where conducting NECA to incorporate indicators of conversion efficiency and output intensity between ecological capital and production capital into the index system. Carry out land resources and environmental capacity evaluation as well as land-use suitability evaluation to promote the continuous adaptation and coordination of the three industrial economies and ecological environment and social development.

Promote a post-evaluation and guarantee system. On the basis of the existing Natural Assets Balance Matrix System, establish an ecological capital assessment system and a dynamic evaluation system of "planning, preparation, implementation, evaluation and plan revision." Coordinate the nationwide results of surveys and evaluations of water, forests, grasslands, wetlands, farmland, and marine minerals to form a unified standard database for spatial planning. In parallel with the preparation of planning, develop a "one-map" supervision information system that comprehensively adopts Big Data, remote sensing, and network informing means to carry out the dynamic monitoring of the planning implementation, and strengthen the supervision of the planning.

5.9.3.2 Carry Out Spatial Planning Based on NECA to Improve the Scientificity and Rationality of Spatial Planning

Establish a technical standard for spatial planning based on NECA. Create a differentiated spatial planning target index system based on the balanced growth of ecological capital and socioeconomics. Through the realization of the value quantification of ecological capital, a spatial planning method system that consists of scenario cost-effectiveness analysis, input–output analysis, and econometric analysis will comprehensively improve the scientific rationality of planning strategic decisions.

Promote cost–benefit/effectiveness analysis for both the project and planning levels. Build a framework system for ecological capital and socioeconomic impact assessment for spatial planning with specific assessment methods for different types of planning. Propose an ecological capital and socioeconomic impact assessment index system and methods throughout the entire planning and implementation life cycle. Focus on cost–benefit and/or cost-effectiveness evaluation methods of spatial planning and project implementation. Define assessment frameworks for different types of tasks and projects. Compile a *Technical Guide for Cost-Effective Evaluation of Spatial Planning* to regulate evaluation indicators, evaluation methods, technical parameters, value parameters, and the application of results. Ensure the accuracy of ecological capital assessment results and provide methodological guidance for funding needs analysis and the project prioritization of ecological protection projects.

Strengthen risk assessment for spatial planning. Establish a risk assessment system for spatial planning from perspectives on population and industry layout, ecological capacity, biodiversity change, and early-warning risk response for the eco-environment. From the perspective of global climate change, major natural disasters, major security events, major environmental events, and extreme events such as major public health events, assess the scientificity, rationality, and effectiveness of spatial planning in terms of protecting public health, maintaining biodiversity stability, and safeguarding national ecological security. Steadily achieve urban–rural spatial planning in all factors with overall coverage and full implementation.

5.9.3.3 Insist on "All Ecological Elements" Management and Handle Global Environmental Issues Effectively

Achieve socioeconomic and eco-environmental "win–win" via spatial planning. Government departments at all levels must adhere to the principle concept of a win–win situation for ecological protection and socioeconomic development when preparing and implementing spatial planning, taking spatial planning as the basis for implementing various development, protection, and construction activities. Fully figure out the bases of various ecological environment factors in the region, and comprehensively evaluate the ecological environment impacts of implementing planned tasks. Promote high-quality economic development and high-level protection of the ecological environment.

Coordinate regional development and ecological balance via spatial planning. Adhere to the thinking of ecological environment as bottom line, carry out NECA, break the administrative regional boundaries of spatial planning, coordinate regional ecological protection and economic development, and ensure the synchronization of regional development. Adhere to the horizontal linkage of planning and implementation; break boundaries between upstream and downstream of the river, land, and ocean; promote the restoration and protection of landscapes, forests, lakes, grasses, and marine ecosystems; and maintain the integrity of the ecosystem.

Promote solutions on global environmental issue via spatial planning. Uphold the basic strategy of harmonious coexistence between man and nature, and comprehensively consider the relationship between economic development and global environmental issues, such as biodiversity conservation and climate change. Coordinate environmental governance, ecological protection, and climate change. Through the establishment and improvement of market-based PES and eco-financing policy mechanisms, maintain the healthy and sustainable development of global biodiversity and ecosystems.

5.9.4 Establish an NECA and Eco-Products Guarantee System for Value Realization: Solidify Value Through Institutional Arrangements

Strengthen the system design in terms of improving the legal system, the institutional mechanism, and scientific and technological research and development to ensure the innovation and promotion of ecological capital accounting and ecological investment and financing policies. Explore the establishment of a system for determining the property rights of natural resource assets and a trading system for ecological products to improve the spatial planning and ecological compensation system. Through fore-running demonstration and international cooperation, further promote the implementation of ecological product value pilots, fulfill the concept of ecological civilization development proposed by President Xi Jinping, meet the growing needs of the people for satisfied ecological environment, and promote the country's high-quality leapfrog development.

5.9.4.1 Establish an Ecological Asset System with Explicit Property Rights and Guarantee the Establishment of Ecological Investment and Financing and Trading Systems

Improve an ecological capital property rights system. With the State's improving natural resource property rights and the establishment of usage regulation system, promote the verification, registration, and certification of the rights of a spectrum of ecological capital, including water flows, trees, mountains, grasslands, barren land, and tidal flats. Clarify the ownership entity of ecological resources, regulate the right to use ecological resource capital, protect the right to receive ecological resource capital income, activate the right to transfer ecological resource capital, and rationalize the right to supervise ecological resource capital. Establish a natural ecology capital property rights system with clear ownership, well-defined rights and responsibilities, and effective supervision.

Clarify ecological capital property rights owners. In accordance with the rule of property rights and the different types of ecological resources, initiate the implementation of the split-operation mechanism of rights, including ecological capital ownership, management rights, contracting rights, etc. Appropriately expand the property rights of a range of ecological capital. Clarify the rights, responsibilities, and interests of the subject of property rights in the possession, use, income, and disposal of ecological capital. Strengthen the supervision of the exercise of property rights over natural resources capital.

5.9.4.2   Strengthen Regulations and Technical Support, and Promote the Implementation of Planning Policies and Systems Focusing on NECA

Establish and improve legal systems to support payment for eco-system use. Fix the rights and obligations for relevant stakeholders, trading of eco-products, compensation for and investment in NECA, spatial planning and eco-financing mechanism in the form of laws and regulations to ensure that the goal of maintaining and increasing ecological capital value is included in the design and implementation of any plans and policies.

Accelerate the establishment of a performance evaluation mechanism based on ecological capital. Develop an integrated regional development index that comprehensively considers economic development and the status of ecological assets as an indicator to reflect the extent of ecological civilization at the regional and river basin levels. Establish distinct assessment index systems for different natural backgrounds and socioeconomic development levels, and clarify the assessment mechanism, assessment subject, assessment object, and application method of the assessment results.

Carry out institutional reform for government agencies. Further clarify the responsibilities of relevant departments, such as Development and Reform, Natural Resources, Ecology and Environment, Agriculture and Rural Development, Water Resources, and Urban Development, in the management of ecological capital. Establish a decision-making mechanism that harmonizes ecological capital, ecological environment quality, and socioeconomic development.

Establish key R&D projects to improve fundamental capacity building. Implement the key R&D plan for the realization of ecological capital value and eco-financing innovation. Develop an ecological resource monitoring network. Start the R&D of NECA and the policy planning simulation lab. Carry out research on an ecological capital value realization road map and analysis on eco-financing to solve technical bottlenecks in terms of tool platforms, technical methods, market mechanisms, institutional policies, and assessment systems that constrain the application of ecological capital in the field of planning and policy.

5.9.4.3 Strengthening International Cooperation and Capacity Building, Lay a Solid Foundation

Strengthen dissemination and training. Carry out basic theoretical training on NECA, spatial planning, and eco-financing policies. Enhance the promotion and interpretation of the practical experiences in Quzhou of Zhengjiang, Wuyishan of Fujian, and Gui'an District of Guizhou. Improve the working capacity of local authorities.

Implement pilot and demonstration works. Set up policy mechanism innovation pilots in the YREB and the Yellow River basin for NECA, spatial planning, eco-products trading, and market-based PES. Innovate a road map on ecological value realization and transformation, and summarize experiences to form practical advantages.

Enhance global cooperation. Carry out international cooperation on NECA tool development, market cultivation of eco-products, strategic development planning for ecological capital, and eco-financing policy innovation.

5.9.5 Design Eco-Financing Policies for Yellow River Basin: Preserve and Increase Value Through Investment

At present, the market value realization and distribution mechanism of ecological capital has not been established. The integration of NECA results into traditional financial investment systems is facing technical and institutional obstacles. Eco-financing projects lack a set of scientific and applicable evaluation indicators, evaluation standards, and evaluation methods. The Yellow River is an important ecological barrier and an important economic zone in northern China. The ecological security of the Yellow River basin is related to the prosperity of the country and the nation's revival. Broaden the application of financial instruments in the area of ecological protection of river basins. Promote the interconnection between the ecological asset property market and the traditional financial market. Through the piloting of ecological investment and financing policies in the YREB and the Yellow River basin, a number of replicable, scalable, and applicable river basin ecological investment and financing policy models shall be formed to promote a virtuous circle between economic and social development that recognizes the gendered implications, resources, and environment.

5.9.5.1 Coordinate the Economic Development and Eco-Environmental Protection of the Yellow River Basin and Secure an Ecological Barrier

Accelerate the formulation and implementation of the Yellow River Ecological Protection and High-Quality Development Plan. From the perspective of spatial development strategy, clearly identify the "three zones" (urban zone, agricultural zone, ecological zone) and the "three lines" (red line for ecological protection, red line for permanent farmland, and urban development boundary) of the Yellow River basin, and define the ecological spatial layout, ecological function, and ecological protection goals of the upstream, midstream, and downstream, as well as key tasks for ecological protection and high-quality development of the Yellow River in the short and medium terms.

Formulate and implement control measures according to local conditions that incorporate impacts on gender and vulnerable populations. Carry out a background survey on ecological resources in the Yellow River basin. Investigate the industrial development, energy structure, transportation structure, and land-use status. Scientifically assess the ecological damage and the status of environmental pollution. Focus on sorting out ecological degradation, soil erosion, water pollution landscape fragmentation, animal and plant habitat destruction, and other related ecological protection and environmental governance issues. Fully consider the differences of ecological management requirements in the upstream, midstream, and downstream and estuaries of the Yellow River. Implement comprehensive protection and restoration management in stages and regions.

Gradually build institutional mechanisms that promote the integration of protection and development. Improve the river basin management system, establish and improve a cross-regional management coordination mechanism, and develop a decision-making mechanism based on "government guidance, market leadership, and social participation." Use incentive instruments such as an environmental tax, environmental liability insurance, green credit, green bonds, and eco-product trading means to take comparative advantage of the river basin. Promote the rational flow of various factors for securing high-quality development of urban agglomeration regions along the Yellow River basin.

5.9.5.2 Promote Successful Experiences in the YREB for Coordinating the Development of the Upstream, Midstream, and Downstream of the Yellow River

Explore key role of NECA in eco-financing policies for river basins. Establish the project's economic evaluation system, including the evaluation of ecological capital. Conduct a monetized economic analysis on the eco-environmental impact of investment projects. Fully consider the potential impact on the eco-environment during economic development and planning decision-making process. Consider putting potential eco-environment costs, benefits, risks, and returns into investment and financing decisions. Promote economic, ecological, and social sustainable development through the guidance of economic resources.

Explore the ecological capital value formation mechanism. Establish an evaluation method for ecological asset value accounting, and scientifically evaluate the potential value of various types of ecological assets. Strengthen the eco-environment protection payment transfer to areas or river basin that experience poverty but are rich with ecological function, taking into account how those payments have differential impacts based on gender. Explore the establishment of a financial transfer payment mechanism based on the quality and value of ecological assets to determine the PES quota. Improve the trading mechanism of ecological assets in the river basin and promote the linking of initial quotas for energy use rights, carbon emission rights, and pollution rights with ecological asset value accounting.

Expand financing channels. Establish a set of green project identification and evaluation standards, build a project inventory for ecological protection and green development in the Yellow River basin, and evaluate the effectiveness of green projects. Encourage, guide, and attract public–private partnership projects to participate in watershed ecological protection and green development. Encourage financial institutions to carry out green crediting, such as forestry rights loans, public ecological welfare forestry rights pledge loans, and pollution rights loans. Eligible local corporate banks and enterprises are encouraged to deeply connect with the green bond market by issuing green finance bonds and green corporate bonds. Explore the establishment of a river basin green development fund to guide social capital to increase investment in green industries and promote the development of energy conservation and low-carbon industries.

5.9.5.3 Innovate the Eco-Financing Policy of the Yellow River Basin, and Alleviate Poverty Through Value-Added from Ecological Capital

Establish an inventory of ecological assets and NECA in the Yellow River basin. Organize nine provinces and regions along the Yellow River Basin to carry out Natural Ecological Assets Inventory and NECA according to ecosystem elements. Find out the stock and flow of ecological assets. Consider the differences between the ecological background and the economic level of the upstream, midstream, and downstream of the Yellow River basin, trading ecological assets and eco-products in the river basin, determining initial quotas and integrated the pricing of ecological assets and eco-products, and innovating eco-financing policies to provide technical support.

Protect KEFZ in the upstream of the Yellow River and improve the design of PES policies. Focusing on Sanjiangyuan, Qilian Mountains, and south of Gansu, etc., as water conservation areas of the upstream Yellow River, establish special funds for ecological protection, restoration, and construction. Draw on the Xin'an River's PES experience to comprehensively consider the water amount and water quality as well as the changes in ecological assets and liabilities to enlarge PES for KEFZ in the upstream of the Yellow River.

Build a green financing reform and innovation pilot area in Gansu's Lanzhou New District in the midstream of the Yellow River basin and establish an eco-financing market. Explore the establishment of an eco-financing centre in the Yellow River basin. Develop ecological banks, ecological capital trusts, ecological assets, eco-product trading platforms, and third-party payment pilots to use ecological financial means to support projects such as soil and water conservation, ecological restoration, and biodiversity protection in the river basin to promote the preservation and value-added of ecological capital. Explore the mechanism of cross-regional, eco-environmental damage compensation and PES. Establish an eco-environmental damage compensation fund and a high environmental risk projects' trust fund for off-site transactions to achieve comprehensive eco-environmental governance and protection across regions.

Develop diversified blue financial products downstream. In response to the development and protection of the Yellow River Delta wetlands, innovatively introduce green credit means, such as mortgages for the use of near-ocean areas, mortgages for the use of beaches, and loans for supporting waste circulation. Focus on the finance development of the marine industrial chain. Develop regional industrial clusters, a business district, and a supply chain to create a sustainable "blue finance" service model. Promote the ecosystem health of the Yellow River, and provide fund and policy guarantees for high-quality development in the Yellow River basin.

References

1. Krutilla. Conservation Reconsidered. Environmental Resources and Applied welfare Economics: Essays in Honor of John V. Krutilla [M]. Washington DC: Resources for the Future: 1988: 10.
2. Costanza R, D'arge R, De Groot R, et al. The value of the world's ecosystem services and natural capital [J]. Nature,1997,387:253–260.
3. Li W, Zhang B, Xie G. Research on Ecosystem Services in China: Progress and Perspectives [J]. Journal of natural resources, 2009, 24(1): 1–10.
4. Ouyang Z Y, Zheng H, Yue P. Establishment of ecological compensation mechanisms in China: perspectives and strategies [J]. Acta Ecologica Sinica, 2013, 33(3): 0686–0692.
5. Ouyang Z, Zhu C, Yang G, et al. Gross ecosystem product: concept, accounting framework and case study. Acta Ecologica Sinica, 2013, 33 (21): 6747–6761.
6. Xie G, Zhang C, Zhang L, et al. Improvement of the Evaluation Method for Ecosystem Service Value Based on Per Unit Area [J]. Journal of natural resources, 30(8). 1243–1254.
7. Zhu X. The system reform of investment and finance on the Chinese ecosystem environment construction [J]. Social Sciences in Ningxia, 2006, 135(02): 31-38.

8. The People's Bank of China, Ministry of Finance of the People's Republic of China, National Development and Reform Commission, et al. Guidance on building a green financial system [R/OL].

9. Easterly, William. 2001. "Think Again: Debt Relief", Foreign Policy, November/December: 20–26.

10. Esmon, M. J. and Uphoff, N. T., 1984. "Local Organizations: Intermediaries in Rural Development". Oornell University Press.

11. Shin, Myoung-Ho., 2001, "Financing Development Projects:Public-Private Partnerships". Paper read at OECD/DAC Tidewater Meeting in Penha Longa, Portugal.

12. Chen P, Lu Y, Chen H, et al. Study on the Investment and Financing Channels for Environmental Protection [J]. Ecological Economy, 2015, 31(07): 148–151.

13. Wang J, Ge C, Yang J. Environmental Financing Strategy [M]. China Environmental Science Press, 2003.

14. Subhrenduk, Pattanayak. Valuing watershed services: concepts and empirics from Southeast Asia [J]. Agricultural, Ecosystems & Environment, 2004, 18(12):171–184.

15. Zhang, M., J. Zhou, and R. Zhou. 2018. Interval Multi-Attribute Decision of Watershed Ecological Compensation Schemes Based on Projection Pursuit Cluster. Water 10:1–12.

16. Wang, X., S. Peng, H. Ling, H. Xu, and T. Ma. 2019. Do Ecosystem Service Value Increase and Environmental Quality Improve due to Large-Scale Ecological Water Conveyance in an Arid Region of China? Sustainability 11.

17. Zhao, X., Y. He, C. Yu, D. Xu, and W. Zou. 2019. Assessment of Ecosystem Services Value in a National Park Pilot. Sustainability 11.

18. Aneseyee, A. B., T. Soromessa, and E. Elias. 2019. The effect of land use/land cover changes on ecosystem services valuation of Winike watershed, Omo Gibe basin, Ethiopia. Human and Ecological Risk Assessment.

19. Jiang, Z., X. Sun, F. Liu, R. Shan, and W. Zhang. 2019b. Spatio-temporal variation of land use and ecosystem service values and their impact factors in an urbanized agricultural basin since the reform and opening of China. Environmental Monitoring and Assessment 191.

20. Solomon, N., A. C. Segnon, and E. Birhane. 2019. Ecosystem Service Values Changes in Response to Land-Use/Land-Cover Dynamics in Dry Afromontane Forest in Northern Ethiopia. International Journal of Environmental Research and Public Health 16.

21. Wang, Y., and J. Pan. 2019. Building ecological security patterns based on ecosystem services value reconstruction in an arid inland basin: A case study in Ganzhou District, NW China. Journal of Cleaner Production 241.

22. Gao, X., J. Shen, W. He, F. Sun, Z. Zhang, X. Zhang, C. Zhang, Y. Kong, M. An, L. Yuan, and X. Xu. 2019b. Changes in Ecosystem Services Value and Establishment of Watershed Ecological Compensation Standards. International Journal of Environmental Research and Public Health 16:1–30.

23. Yan, D., Y. Fu, B. Liu, and J. Sha. 2018. Theoretical Study of Watershed Eco-Compensation Standards. Pages 1–6 in K. Wang, editor. 5th Annual International Conference on Material Science and Environmental Engineering.

24. Stephen P, Erik N, Pennington, et al. The Impact of Land Use Change on Ecosystem Services, Biodiversity and Returns to Landowners: A Case Study in the State of Minnesota [J]. Environmental and Resource Economics , 2011, 48(2): 219–242.

25. Erik N, Guillermo M, James R, et al. Modeling multiple ecosystem services, biodiversity conservation, commodity production, and tradeoffs at landscape scales [J] . Frontiers in Ecology and Environment, 2009, 7(1):4–11.

26. Zheng H, Wang L, Peng W, et al. Realizing the values of natural capital for inclusive, sustainable development: Informing China's new ecological development strategy [J]. PNAS, 2019, 116 (17): 8623-8628.

27. Chaudhary, S. et al. "Environmental justice and ecosystem services: A disaggregated analysis of community access to forest benefits in Nepal." Ecosystem Services. Vol. 29. 2018.

28. Andeltová, L. et al. "Gender aspects in action- and outcome-based payments for ecosystem services—A tree planting field trial in Kenya." Ecosystem Services. Vol. 35. 2019.

29. Benjamin, E. et al. "Does an agroforestry scheme with payment for ecosystem services (PES) economically empower women in sub-Saharan Africa?" Ecosystem Services. Vol. 31. 2018.
30. Pereira Lima, F. and R. Pereira Bastos. "Perceiving the invisible: Formal education affects the perception of ecosystem services provided by native areas." Ecosystem Services. Vol. 40. 2019.
31. Cruz-Garcia, G. et al. "He says, she says: Ecosystem services and gender among indigenous communities in the Colombian Amazon." Ecosystem Services. Vol. 37. 2019.
32. Pearson, J. et al. "Gender-specific perspectives of mangrove ecosystem services: Case study from Bua Province, Fiji Islands." Ecosystem Services. Vol. 38. 2019.
33. Defra (2018), A Green Future: Our 25 Year Plan to Improve the Environment. https://www.gov.uk/government/publications/25-year-environment-plan.
34. Millennium Ecosystem Assessment. Ecosystems and human wellbeing: synthesis [M]. Washington DC: Island Press, 2005.
35. TEEB (edited by Kumar P) . The economics of ecosystems and biodiversity: ecological and economic foundations[M]. London: Earthscan Ltd. 2010.
36. National Forestry and Grassland Administration. LY/T 2006—2012 Assessment criteria of desert ecosystem services in China[S]. Beijing: Standards Press of China, 2012.
37. National Forestry and Grassland Administration. LYT 1721-2008 Specifications for assessment of forest ecosystem services in China [S]. Beijing: Standards Press of China, 2008.

Chapter 6
Green Transition and Sustainable Social Governance

6.1 Green Consumption, Transition and High-Quality Development

As the scale of consumption in China is continuing to expand rapidly, its structure is shifting from being subsistence-based to being like that of a well-off country. Its contribution to economic growth is soaring, and it is becoming an important engine for economic development. As shown by the China Council for International Cooperation on Environment and Development (CCICED) (2019) study [1], a green transition in the consumption sector will lead to and enforce the greening of production, facilitate green production and green lifestyles, mobilize the general public to actively practice green concepts, and improve the governance system for a green social transition. It will play a decisive role in the overall green transition and realization of high-quality development in China.

6.1.1 Status and Trends of Green Consumption in China

6.1.1.1 Overall Status of Consumption in China

In recent years, consumption in China has steadily and rapidly grown. In 2019, the total retail sales of consumer goods amounted to 41.2 trillion yuan, nearly doubling the volume of 2012, which was 21 trillion yuan. The corresponding growth rate reached 8.05%, which was twice as much as the 2018 level of 4.02% and 2% higher than the 2019 GDP growth rate. According to data released by the National Bureau of Statistics of China, the contribution of final consumption expenditure to GDP growth

© The Author(s) 2022
China Council for International Cooperation on Environment
and Development (CCICED) Secretariat,
Green Consensus and High Quality Development,
https://doi.org/10.1007/978-981-16-4799-4_6

accounted for 57.8% for the whole year, 26.6 percentage points higher than the contribution of total capital formation. At the same time, consumption has moved to a new stage in which the spending power of residents is rising rapidly, the trend toward upgraded consumption has become more explicit, the demand for mid- and high-end goods and services has risen, and the consumption of services has expanded. In 2019, per capita consumer spending on services for the whole country accounted for 45.9% of per capita consumption expenditure, an increase of 1.7 percentage points over the previous year; the Engel coefficient was 28.2%, showing a drop of 0.2 percentage points. In 2019, the per capita consumption expenditure of the country exceeded 20,000 yuan for the first time, reaching 21,559 yuan, with a real increase of 5.5%. Consumption expenditures by rural households grew faster than by urban households. The nominal growth rate and real growth rate of rural household consumption expenditures were 2.4% and 1.9%, respectively, higher than those for urban households. The level of urbanization has further increased, with the urbanization rate of the permanent population surpassing 60% for the first time as of the end of 2019, creating a huge space for investment growth and consumption expansion.

6.1.1.2 Progress in Green Consumption Policies in China

In recent years, China has put forward hundreds of concepts, guidelines, and specific policies related to the promotion of green lifestyles for citizens. It has witnessed promising achievements in greening household consumption in areas such as clothing, food, residences, and mobility. However, in general, an effective policy framework has not yet taken shape due to a lack of systematic planning and top-level design. Existing policy focuses more on resource and energy conservation rather than on ecology and environmental protection. Responsibilities for green consumption promotion are not clearly defined and are scattered across many government departments, with neither synergies created, nor an important place given to green consumption on key policy agendas. Without a systematic design and the integration of relevant policies, the environmental benefits and economic effects linked to green consumption might be greatly impeded. China now has strong political will, an increasingly mature social foundation, and a good practical basis for promoting the green transformation of consumption. The timing and conditions are right for incorporating green consumption into the national 14th FYP.

6.1.1.3 Trends and Features of Green Consumption in China

As noted by the *Survey Report on the Status of General Public's Green Consumption in China* (2019), the concept of green consumption is becoming increasingly popular in the public's understanding of daily consumption. Around 83.34% of the respondents expressed support for green consumption behaviour, and 46.75% were "highly supportive." In addition, there has been a constant increase in both businesses' willingness to procure, utilize, and sell environmentally friendly green products and consumers' willingness to buy safe, reliable, and green products. There is also much closer attention from the general public to green food and green home improvement. Consumers are not only willing to buy high-quality green products but also concerned about the impacts of production methods on the eco-environment.

According to the *Report on Green Consumption Trend and Development 2019* released by the JD Big Data Research Institute, the number of types and categories for "green consumption" commodities exceeds 100 million. The year-on-year increase in the total sales volume of green consumption related products on JD.com was 18% higher than the overall sales increase of all products on the platform in 2019. The top five categories with the largest sales volume included grain, oil and seasoning, facial care, children's clothes and shoes, furniture, and car trim. With regard to the share of commodities for "green consumption" in markets at all levels, the proportions of "green consumption" commodities were relatively higher in second-tier and third-tier markets, while the total green consumption volume ranked top in the first-tier market. Judging from the two-year change in the proportion of "green consumption" commodities in markets at all levels, the sales volume in new markets grew with the fastest speed.

In the context of occupation, gender and age, medical practitioners/public organization employees, females, and people above the age of 46 were the groups that were paying the closest attention to "green consumption." The share of green consumption held by these groups was 7.4%, 11.5%, and 24.8% higher than the overall sales of all products on JD.com, respectively. According to the online survey results, 80% of the household consumption decisions were made by women. Women are important decision-makers for consumption, and female consumers have become the pioneers and main force of "green consumption" (Fig. 6.1).

6.1.2   Green Consumption is the Key to Push Forward a Green Economic Transition

The green transition of the economy has been primarily driven by the transformation of two sectors—production and consumption. Built on the results of research conducted in 2019, the SPS team has improved the design of the existing green transition index and indicator system and conducted further empirical evaluation on the role and trends of green consumption in the green economic transition. The results are as follows:

6.1.2.1   There Has Been a Shift Towards a Green Transition year by year; However, the Growth Trend Has Gradually Flattened

From 2004 to 2008, green transition index values increased substantially on a yearly basis, clearly indicating a growth in the green transition change rate. From 2009 to

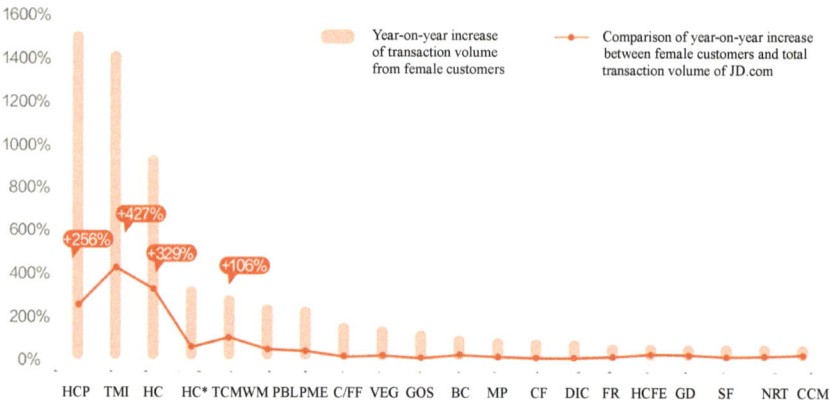

Fig. 6.1 Top 20 categories by increase of transaction volume from female customers vs. sales volume of all products on JD.com. (*Source* JD Big Data Research Institute, Female Consumption Trends Report 2020). Notes: HCP: Health care and protection (including mask); TMI: Toys and musical instruments; HC: Home care; HC*: House cleaning (including disinfectant); TCMWM: Traditional Chinese medicine and western medicine; PBL: Pork, beef, and lamb; PME: Poultry meat and eggs; C/FF: Chilled/Frozen food; VEG: Vegetable; GOS: Grain, oil and seasoning; BC: Body care; MP: Milk powder; CF: Complementary food; FR: Fruit; HCFE: Health care and fitness equipment; GD: Gaming device; SF: Seafood; NRT: Nutrients (including Vitamins); CCM: Chess, cards, and mahjong

2015, however, the momentum behind the green transition slowed, and the speed of the green transition decreased. Since 2016, a slight decline has been detected in green transition index values, revealing a declining trend in green transition (Fig. 6.2).

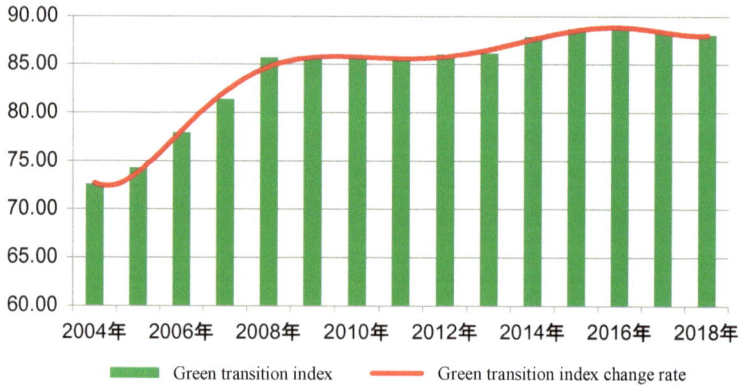

Fig. 6.2 Trend of the green transition index change (2004–2018)

6.1.2.2   The Green Transition Index Related to Lifestyles Has Dropped, Suggesting that This is a Bottleneck Area Restricting Overall Progress on a Green Economic Transition

The green transition index in the production sector has been on an upward trend since 2004. In contrast, the green transition index in relation to lifestyles has shown a sharp downward trend since 2008. The green transition index value from 2016 to 2018 was even lower than that of 2004. This has contributed to a slowdown in the overall green transition of the economy (Fig. 6.3).

The slowdown of the green transition in the lifestyle area can be primarily attributed to the steady growth in consumption of resources and energy, the increase of pollutant emissions from domestic sources, and the reinforced adverse impact of lifestyle on eco-environmental quality. From 2004 to 2018, with the improvement of living standards, there was an increase in household energy consumption. The per capita domestic energy consumption increased from 191 kgce in 2004 to 441 kgce in 2018. Both the emission of major domestic pollutants (including carbon dioxide and domestic wastewater) and domestic waste collection volume have shown a clear year-on-year upward trend.

As revealed by the above results, the positive environmental effects and benefits created by the improvement of resource and environment efficiency in the production sector in China are not yet able to compensate and offset the negative environmental impacts brought about by the expansion of the country's consumption sector. The green transition process in domestic consumption is sluggish and deteriorating, resulting in delays to both the pace and depth of the overall green transition of development in China. From a resource and environmental performance perspective, there is huge room for advancing China's green transition in both the production and

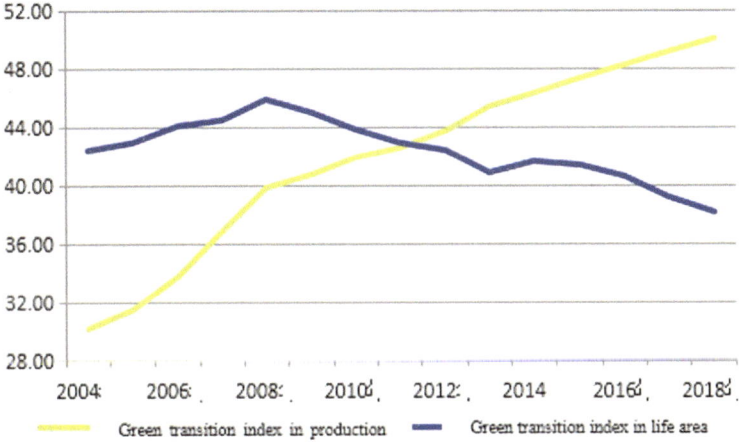

Fig. 6.3   Green Transition index trends in production and lifestyle sectors (2004–2018)

consumption sectors, especially related to lifestyles. Urgent measures are needed to green the lifestyles and the consumption behaviour of the public.

6.1.3   Advancing Green Consumption is an Important Option to Push Forward High-Quality Development

6.1.3.1   Demand and Supply Are Closely Related Aspects of Economic Activity, and High-Quality Development on the Supply Side Depends on Optimization and Upgrading on the Demand Side

The booming development of green consumption can create an ample market space for building a green economic system. In 2019, the Engel coefficient of China's household consumption dropped to 28.2%, with a decrease of 3 percentage points from the level of 31.2% in 2013. In the future, as the demographic structure changes and the urbanization rate increases, the joint effects of pro-consumption factors such as employment, income, and social security will lead to further decreases in the Engel coefficient. It is projected that the Engel coefficient will continue a decline to 20% by 2035, reaching the well-off line of 20% to 30% set by the United Nations (UN). In the meantime, the household consumption model in China will further shift from a material-oriented and subsistence model to a well-off, service-oriented model. An increase can be expected in the share of per capita consumer spending on services such as transportation and communication, education, culture and entertainment, as well as health care. Such important changes and trends in consumption scale, structure, and preferences in China will inevitably induce corresponding adjustments on the supply side for production and services provision. In this process, should the general public be positively guided towards green consumption, it will effectively boost the development of green product manufacturing, energy conversation, and environmental protection industries, and act as a new driving force for green growth. Green products and industries for energy conversation and environmental protection have long industrial chains, high levels of inter-connectivity, and a strong capacity for job creation; thus, they can also pull the development of other related industries as they themselves develop. Green consumption's direct growth and indirect pulling effects will exert positive effects that help drive economic growth.

6.1.3.2   Advancing Green Consumption is an Important Direction for Fostering New Momentum and Achieving Stable Economic Growth

Generally speaking, the production of green products involves longer industrial chains and higher-quality processes compared to traditional products. Gradually expanding the consumption of green products will have a stronger pulling effect on the economy. However, in the short term, there might be some adverse impacts that require attention and evaluation, such as relatively high product prices, which may take up larger proportions of total spending and affect the budget available for other goods and services. Therefore, the SPS team built a large-scale dynamic computable general equilibrium (CGE) model to analyze the multiple scenarios for replacing traditional consumption with green consumption. Under the simplified assumption that people's consumption preferences remain unchanged, the substitution of green consumption for traditional consumption will have limited negative impacts on the economy in the short term; but in the medium to long term, it will bring about a continuous increase in positive economic growth.

Figure 6.4 illustrates consumption trends for 2020–2025 under a baseline scenario and under a green consumption model. It is assumed that, starting from 2020, consumer products worth over 400 billion yuan (around 1% of total household consumption volume) will be replaced by green products in the areas of food, automobiles, buildings, household electric appliances, and household items. Should that happen, in the short term, the GDP will drop slightly by 0.06% (around 6.1 billion yuan) relative to the baseline scenario, indicating a short-term, small, and manageable negative impact. The cause for this decline lies in the relatively high price of green products, which leads to an increase in the composite average price for household consumption by 0.11% relative to the baseline scenario. This increase is significantly higher than the GDP deflator (only a 0.02% increase). Given the balanced

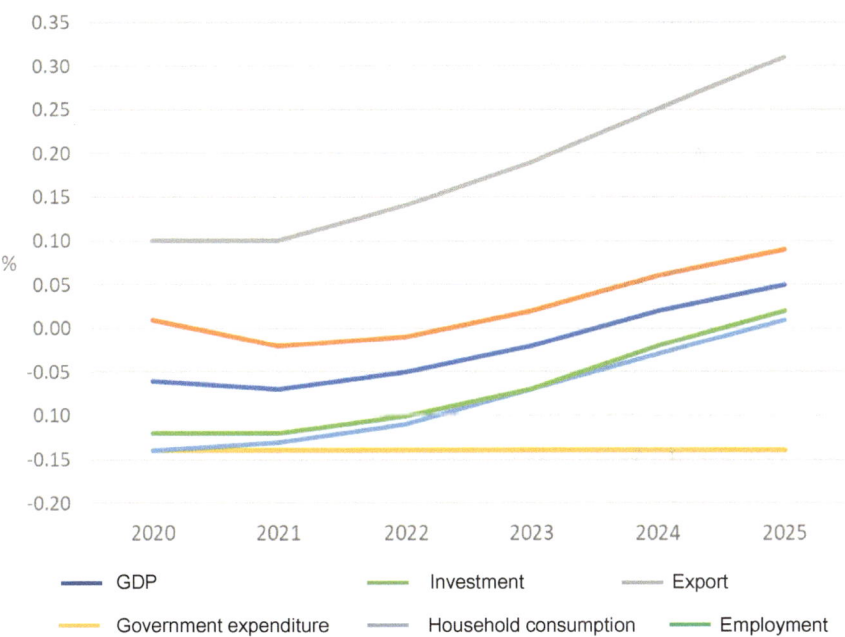

Fig. 6.4   Macro impact of 1% of household consumption volume being replaced by green consumer goods

relationship between GDP by output approach and GDP by expenditure approach in macroeconomics, the price difference will cause short-term growth of -0.12% in employment on the production side and a -0.06% drop in GDP.

However, in the medium to long term, as investments in the production of green products become profitable, there will be a gradual turn to positive GDP growth relative to the baseline scenario. It is projected that the GDP will increase by 0.05% relative to the baseline scenario in 2025 (equivalent to an increase of 73 billion yuan relative to the baseline scenario in 2025), with investment rising by 0.09% (equivalent to an increase of 49 billion yuan), household consumption by 0.01% (equivalent to an increase of 6 billion yuan), leveraged exports by 0.3% (equivalent to an increase of 59 billion yuan), and the corresponding employment by 0.02% (equivalent to an increase of 600,000 jobs).

Assuming a replacement of traditional products worth 2% of the total household consumption volume with green products, a similar trend is obtained, i.e., there is greater economic loss relative to the baseline scenario in the short term (-0.25%), but a larger positive economic growth gained in the medium to long term (0.14%).

The above scenarios, moreover, do not take into account the fact that green products will provide consumers with better enjoyment, a sense of honour and gain, and bigger positive environmental effects than traditional products. For the middle and upper classes who are economically well off, an enhanced supply of green products will not drive out traditional consumer goods. They will instead stimulate more consumer desire and expand consumption scale. At present, the CGE model has not yet quantified such a scenario, but a qualitative judgment shows that, under this quite realistic assumption, the gains from such a scenario might significantly offset the short-term negative economic impacts projected by the previous scenarios, even resulting in positive macro-economic growth in the short term. In this case, there could simultaneously be positive growth in the medium to long term.

6.1.3.3 Replacing Traditional Consumption With Green Consumption Will Effectively Help Optimize and Upgrade the Industrial Structure, and the Manufacturing Industry of Green Products Will Achieve Sustainable and Rapid Growth

According to the industry output results obtained by the above CGE model (Table 6.1), should green consumption substitution be implemented in 2020, the output value of green consumer goods will grow rapidly and constantly in both the short term (2020) and the medium to long term (2025), driving industry-wide overall growth (offsetting negative impacts) with new momentum for green growth. The food manufacturing industry will foster the strongest green growth momentum while electric vehicle manufacturing in automobile manufacturing and green wholesale and retail in service industries will have the largest increment. All of these industries should be treated as priority industries for promoting green consumption.

Table 6.1 Impact of substitution of green consumer goods on value-added of major sectors (0.1 billion yuan, current price of 2017)

Value-added of major sectors	2020	2025
Food	−1571	−1744
Green food	1600	1856
Building	−3	30
Green building	5	13
Household appliances	−9	4
Green household appliances	5	8
Automobile	−49	−54
Electric vehicle	61	75
Wholesale and retail	−217	−112
Green wholesale and retail	200	247

6.1.3.4 The Substitution of Green Consumption for Traditional Consumption also Has Obvious Resource and Environmental Effects

As clean energy such as electricity and natural gas is more frequently used in the production and use of green consumer goods, the corresponding consumption of coal and oil will reduce, which will spur the green transition in the energy sector. According to energy consumption results from model estimations, under the baseline scenario (without consideration of the COVID-19 impact), China would consume 4.95 billion tonnes of standard coal equivalent in 2020, of which, 4.05 billion tonnes of coal, 630 million tonnes of oil, and 320 billion cubic metres of natural gas. Should products worth 1% of household consumption volume be replaced by green consumer goods, the total energy consumption would reduce by 0.05%, out of which coal demand will reduce by 0.07%, oil by 0.08%, natural gas by 0.06%, and non-fossil power generation would increase by 0.05%. It is estimated that the emissions of carbon dioxide would be reduced by 7 million tonnes, those of sulphur dioxide by 56,000 tonnes, and those of nitrogen oxides by 31,000 tonnes.

6.1.4   Changes in Consumption in China and the Corresponding Long-Term Impacts of the COVID-19 Pandemic

6.1.4.1   Consumption Volume Has Dropped Drastically Due to COVID-19

According to data released by the National Bureau of Statistics of China, China's total retail sales of consumer goods from January to March 2020 reached 7.858 trillion yuan, showing a year-on-year nominal decrease of 19.0%. Retail sales of consumer goods other than automobiles were 7.2254 trillion yuan, a decrease of 17.7% (Fig. 6.5).

Retail sales of consumer goods in urban areas were 6.7855 trillion yuan, a year-on-year decrease of 19.1%, and in rural areas, 1.0725 trillion yuan, a decrease of 17.7%.

Retail sales of commodities amounted to 7.2553 trillion yuan, a year-on-year decrease of 15.8%; sales revenue from the catering industry was 602.6 billion yuan, a decrease of 44.3%.

Supermarket retail sales in units above a designated size (i.e., normally those with total annual sales greater than 5 million yuan) increased by 1.9% on a year-on-year basis, while those of department stores, specialty stores and exclusive stores decreased by 34.9%, 24.7%, and 28.7%, respectively.

Nationwide online retail sales amounted to 2.2169 trillion yuan, showing a year-on-year decrease of 0.8%, 2.2% less than the decrease rate in the January–February period. Online retail sales of physical commodities amounted to 1.8536 trillion yuan, with an increase of 5.9%, accounting for 23.6% of total retail sales of consumer

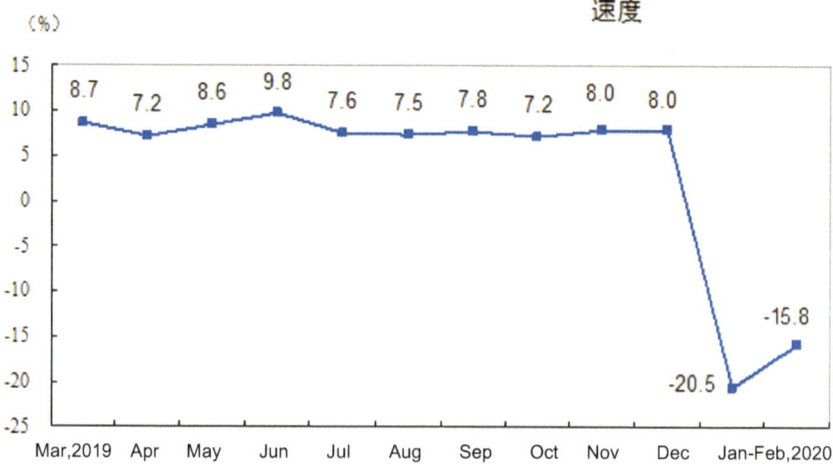

Fig. 6.5   Year-on-year change in the rate of total retail sales of consumer goods by month

goods. In terms of online retail sales of physical commodities, the sale of food and daily utensils increased by 32.7% and 10.0%, respectively, while the sale of clothing-related commodities fell by 15.1%.

6.1.4.2 The Long-Term Stable Growth of Consumption Will Remain Unchanged

The Covid-19 pandemic has had a significant short-term impact on consumer markets, resulting in a sharp decrease in retail commodity sales. Retailers and restaurant businesses experienced an accelerated transition to and growth of online sales. Self-service retail formats such as supermarkets have seen a slight increase in sales; sales of community retail stores have declined but less than the overall level; the retail sales of food commodities (such as grain and oil) grew rapidly; and the sales of protective equipment increased significantly.

However, as China's consumer market is large in scale, filled with potential, and resilient, the promising trend of long-term steady growth has not changed. Affected by short-term external factors, consumer demand has been suppressed temporarily, yet residents' willingness and capacity to consume still exists. The short-term fluctuations will not alter the promising long-term development trend. The long-term stable growth of the consumer market and the momentum for its accelerated transformation and upgrading have remained unchanged. Once the pandemic's influence resides, and with the gradual release of restrictive consumption and the continuous optimization of the market supply structure, China's consumer market will maintain stable growth.

The COVID-19 pandemic has resulted in huge, profound, and far-reaching economic and social impacts on China and the world, sounding an alarm bell warning people of the impacts of their lifestyles and consumption patterns. First, it has shown that it is important to establish ecological values and build a lifestyle of harmony between man and nature. A disharmonious relationship between man and nature may cause immeasurable and tremendous damage to human society. It is necessary to build and foster an ecological value wherein man respects nature and nature's laws and acts to protect it. Second, it has shown the importance of promoting green lifestyles and green consumption. It is crucial to end our harmful lifestyles and replace them with simple and moderate lifestyles that strive for quality of life improvements through reducing resource consumption and pollutant discharges to the maximum extent possible. It is crucial to strongly oppose overconsumption, advocate green and moderate consumption, and prohibit illegal hunting, trading, and abusive use of wildlife. Third, we must accelerate the pace of the green transition in the consumption sector. This means employing administrative, economic, legal, and policy instruments to both the supply and the demand sides to enhance the effective supply of green products, foster green consumer markets, advance the greening of lifestyles and consumption patterns, and reduce the resource and environmental impacts of lifestyle-based consumption.

6.2 Green Consumption Case Studies

6.2.1 Green Building: Green Renovation of Residential Buildings

6.2.1.1 Status and Problems

According to data released by the National Bureau of Statistics of China in 2018, residence-related expenditure accounted for 23.5% of household consumption in China, second only to food (28.4%), and building energy consumption made up over one third of total energy consumption of the whole society. Advancing the greening of residential buildings has become an important component of promoting the green consumer market. China's existing buildings have nearly 60 billion square meters of floor area, more than 95% of which belong to high-energy-consumption buildings, with per unit energy consumption 2–3 times higher than that in developed countries with the same climate conditions. The recovery rate of construction waste in China is barely 5%, far below the level of 90% in developed countries. The huge consumption and waste of energy and resources directly and adversely affect the high-quality development of the national economy and energy.

In 2019, the National Development and Reform Commission and the Ministry of Housing and Urban–Rural Development jointly conducted a survey. According to the survey results, there were 170,000 old residential communities across the nation, with a total residential area of more than 4 billion square metres, which were home to over 100 million people. It is projected that the actual growth rate of real estate investment in the next decade may fall back to zero at the end of the 14th FYP period and gradually go into a negative growth rate while the growth rate of residential areas in aging communities will speed up significantly. In recent years, China has carried out a number of old community renovation projects in many areas at various levels. The following problems and weaknesses have been observed in the process. First, more focus has been placed on individual renovation projects with limited attention to the overall design. Simple construction projects are highlighted, yet people's most needed services and functions are missing. Follow-up scientific management is not being used. Second, renovations are mostly arranged by the government. There are no adequate channels for private capital and the general public to participate in the process. Private enterprises have insufficient motivation to participate. Third, national-level attention and support is not sufficient, since supporting policy measures such as financial subsidies and tax incentives are relatively small, making it difficult to drive investment into the green building sector.

6.2.1.2 Multiple Benefits Brought by the Green Renovation of Old Communities and Green Buildings

Renovation of old communities includes renewal and on-site reconstruction. In the process of pushing renovation efforts forward, it is essential to fully mobilize local government, residents, and enterprises and evoke their sense of initiative and ownership. The decisive role of the market in renovation and renewal projects should be brought into full play. Emphasis needs to be placed on overall planning, long-term effects, and classified measures for different targets, details, and initiating pilots so as to fully release the triple dividend in people's welfare, the economy, and the environment that can be brought about by renovation and renewal efforts.

Green buildings are an important starting point for smart city development. A fundamental goal of building a smart city is to let citizens fully experience how resource-efficient, convenient, easy, and comfortable life can be in a smart city. Green buildings serve as an entry point where the new concept of the "intensive, intelligent, green, low-carbon" approach emphasized by an Ecological Civilization is integrated into urbanization. Green building is also the dominant future trend in the construction sector and symbolizes a typical feature of a smart city. Green building development should respond to the requirements of smart cities. The construction industry needs to follow relevant planning and construction standards and technical specifications, draw on the Internet of Things, and make coordinated use of the new generation of information technology to comprehensively meet the needs of energy conservation, emission reduction, and comfortable living. Green building will contribute greatly to the planning and development of people-oriented, highly-efficient, and sustainable smart cities.

Since the promulgation of the *Notice on Conducting National Smart City Pilot* by the Ministry of Housing and Urban–Rural Development in November 2012, nearly 600 smart city pilots have been established nationwide, and more than 500 cities have clearly proposed relevant plans for building smart cities. Such specifications as the *Smart City—Top-level Design Guide* (GB/T 36333-2018) and the *Technical Outline for Smart City Spatio-temporal Big Data and Cloud Platform Construction (2019 Edition)* have been released one after another to promote the construction of smart cities.

At present, the development of a green building industry in China is in a "naive stage." Due to relatively high construction costs, green building-related products are not competitive in the market, and it is difficult to form a scale effect. The SPS team probed into the issue from the two perspectives of the green renovation of old communities and new green buildings and quantitatively analyzed the economic benefits, social welfare benefits, and environmental benefits of green building development in the short term with the help of a CGE model. A comprehensive review of results from the CGE model analysis and exciting research in the field reached the following findings:

(1) **Economic benefits**: Should a moderate increase in green building investment take place, there will be a significant positive effect on economic growth in the

short term. The four leading economic indicators of GDP, investment, export, and government consumption exhibit an upward trend in the short term, with the rate of change increasing year by year.

(2) **Social welfare benefits**: The moderate shift of investment from traditional buildings to green buildings will have a positive effect on promoting the healthy development of the real estate industry, expanding employment opportunities, and protecting people's livelihoods. But at the same time, it is also necessary to guard against the negative effects caused by the expansion of investment expectations or excessive investment.

(3) **Environmental benefits**: The statistical results show that more than 50% of the raw materials obtained by mankind from nature are used to build various types of buildings and their ancillary equipment. The construction and use of these buildings have consumed about 50% of global energy. The share of building- and construction-related pollution such as air pollution, light pollution, and electromagnetic pollution, account for 34% of the total environmental pollution. Construction waste makes up to 40% of the total waste generated by human activities. On the one hand, green buildings can save energy consumption by using green building materials; on the other hand, energy will be saved when people are using green buildings. As revealed by related studies, green buildings can save about 30% of energy compared with conventional buildings. In addition, green buildings also have remarkable performance in water and land savings. In conclusion, the resources, energy, and environmental benefits of green buildings are considerable.

6.2.1.3   Preliminary Conclusions and Recommendations Pertinent to the Green Renovation of Residential Buildings in China

(1)   It Is Necessary to Attach Great Importance to Green Building Development

Against the background of economic recovery after the COVID-19 pandemic, high-quality development, and the building of an Ecological Civilization, it is a necessary choice for China to accelerate research on new development models of energy conservation and emission reduction, break through the bottleneck of traditional industrial development, and seek new driving forces for economic development. Of the three major energy-consuming sectors (i.e., the construction, transportation, and secondary industries), the construction industry has the largest potential for energy conservation. Therefore, it is necessary to strongly support the promotion and application of green buildings and facilitate the healthy development of the green building industry so as to achieve the goal of high-quality economic and energy development and satisfy people's desire for a better life. The construction of new buildings should follow green building standards, while close attention needs to be directed to the green renovation of old buildings. Green renovation should be treated as the entry and focal point of the first round of efforts at change.

(2) **Establish and Improve Governance Mechanisms for Renovation and Renewal of old Communities**

Define the rights and obligations of the government, enterprises, and residents, and clarify the roles and functions of various stakeholders in the process of renovation and renewal. Formulate management regulations for the definition and adjustment of ownership involved in renovation and renewal. Establish a handover system for construction institutions and management institutions and a corresponding judicial mediation mechanism. Develop top-level design and guidance for the planning and adjustment of the renovation and renewal of old communities to maximize their quality. Establish a mechanism for mutual consultation, joint construction, and sharing. Encourage the communities to set up multistakeholder consultations and push forward planning, design, construction, and acceptance in an orderly manner. Establish innovative and supportive fiscal and taxation policies, and provide discounted interest loans to support in-situ reconstruction of old communities in the same area. Innovate investment and financing mechanisms, promote cooperation between the government and private capital, facilitate the use of fiscal funds in leveraging social funds, and encourage financial institutions to reinforce financial support. Implement an incentive and linkage mechanism for neighbouring households, and execute differentiated preferential policies so that the more households that complete the renovation and renewal rapidly, the more benefits each household receives.

(3) **Improve the Standard System and Supervision System for Green Renovation**

Improve renovation and renewal standards and promote green renovation. Lift the restrictions on green space, sunshine and other indicators and compile a list of green renovations. Support energy-saving renovation to meet the standards of new buildings in a one-step process. Encourage the application/construction of structural performance testing and reinforcement, heating measurements, rainwater collection, elevator installation, external wall insulation, integrated renewable energy systems, greywater recycling, parking lots, etc. Vigorously promote the recycling and recovery of construction and demolition waste, formulate technical standards for the recycling and recovery of construction waste, and improve pollutant emission control standards and supervision systems. Give priority to supporting in-situ reconstruction, strengthen research to coordinate the overall planning of old community renovation and renewal, and release the potential of new communities. Old communities with little renovation value should be reconstructed in situ. Make full use of underground and aboveground space, facilitate lightweight upgrades of buildings, in order to maximize floor area (i.e., enhance the land-use efficiency). For those communities with sufficient capacity, sufficient space should be reserved for installing elevators. Optimize management and supervision processes and simplify approval, complaint, and supervision procedures.

(4) **Use Intelligent Technological Means to Greatly Improve the Quality of Green Renovation**

Significantly improve the performance of buildings, emphasize safety and health performance and adaptive design for the elderly during the construction process, and create a more comfortable working and living environment through means of intelligent technology. Use intelligent technology-based construction modes to link information on people and things together in the construction process to achieve interconnectivity and information sharing. Integrate the application of information surveying and mapping, digital construction, standardized design, factory production, assembly, construction, integrated decoration, information management, and intelligent applications. Streamline various links in the value chain, including investment and financing, planning and design, production and transportation, construction, as well as operation and management. All of these measures will contribute to and guarantee the attainment of energy and material savings, cleanliness, safety, high quality, and high efficiency as well as the overall large benefits of construction activities.

6.2.2 Green Consumption and Production in the Automobile Industry

6.2.2.1 Status and Problems

The automobile industry has become the strategic pillar industry of China's economy. For 10 consecutive years since 2009, the volume of car sales in China has ranked the top in the world, reaching nearly 30% of global car sales in recent years. As shown by statistics from the China Association of Automobile Manufacturers, the share of employees in automobile-related industries has surpassed 10% in urban employment for many years in a row, and the number of employees exceeds one sixth of the total employed population in China. Each additional employee in the automobile industry can leverage an increase of 10 employees in related sectors. According to statistics, in 2017, the consumption of gasoline and diesel in the transportation sector in China accounted for 46% and 66% of national total consumption, respectively. In 2018, automobile-emitted NO_x amounted to 43.6% of total national NO_x emissions, yet the contribution of the automobile sector to China's total NO_x emissions reduction was less than 20%.

As shown by relevant research at home and abroad, if non-fossil energy is used for power generation and hydrogen production, the promotion of electric passenger vehicles can effectively mitigate climate change; future vehicle efficiency improvements are expected to reduce emissions from internal combustion engine vehicles (ICEVs) to ~$450gCO_2e$/mi (grams of carbon dioxide equivalents per mile), and those of hybrid electric vehicles (HEVs), fuel cell electric vehicles (FCEVs), and battery electric vehicles (BEVs) to 300–$350gCO_2e$/mi. In the entire life cycle, the environmental cost of HEVs and BEVs is lower than that of ICEVs, and the environmental cost of BEVs is only 36.04% of that of ICEVs. The total energy consumption of HEVs and BEVs, respectively, equals 59.92% and 52.20% of ICEVs' total energy consumption. Compared with ICEVs, BEVs and HEVs have lower energy consumption during the use phase. Vigorously developing new energy vehicles will generate outstanding energy-saving and emission-reducing effects, and thus should be an important means of achieving green consumption in the automobile industry.

China depends on a multi-faceted approach to promote the green transformation of the automobile industry from the perspectives of consumption, production, transportation, and energy policies. To promote green consumption related to automobiles, the Chinese government has successively introduced a range of policies, including purchase subsidies, tax incentives, accelerated construction of charging infrastructure, financial loan support, and promoting accessible transportation. The design of the tax system takes into consideration the role of taxes in guiding energy savings and emissions reductions. Such policy guidance is reflected in the setting of tax rates for passenger vehicle excise taxes and vehicle and vessel taxes, as well as the reform of the refined oil excise tax. Positive results have been obtained in the implementation of relevant policies. In terms of green production, China has taken active actions to reduce hydrofluorocarbon (HFC) refrigerants. It attaches great importance to the development of the remanufacturing industry, with management regulations and policy measures developed for the recycling and use of used parts, market entry, production authorization, taxation, pilot and demonstration projects, quality management, marketing, and incentives. China has established framework legislation for the recycling and reuse of new energy vehicle batteries, making use of an extended producer responsibility system as a fundamental guiding principle.

However, there are also problems and weaknesses in these policies. In terms of policies addressing the consumption of new energy vehicles, the first problem is the unbalanced structure of the tax regime. The tax burden on a purchase is relatively high while that on utilization is low, which is not conducive to the economical use of vehicles. The second problem is that the current tax system is not directly linked to energy-efficiency indicators. The emission index does not directly reflect the energy efficiency of automotive products. Third, the new energy vehicle subsidy policy places more emphasis on purchase than on utilization. The construction of related infrastructure is lagging behind. Fourth, the new energy vehicle transportation policy is only implemented in a limited number of cities; policy coverage is not sufficiently wide. In the field of automobile production, there are also problems such as the lack of a policy system for HFC emission reductions from air conditioning refrigerants, limited development of a parts remanufacturing industry, and inadequate legislation and standards for battery recycling, all of which have seriously inhibited green consumption in the automobile industry.

6.2.2.2   Evaluation of New Energy Vehicles

China has already listed the new energy vehicle industry as one of its strategic emerging industries and established a comprehensive industrial policy system covering industry guidance, research and development (R&D) support, production supervision, purchase incentives, and use incentives. Driven by industrial policies, the new energy vehicle industry has achieved remarkable results, which are highly reflected in three dimensions, including the gradual growth in the market share of new energy passenger vehicles, the continuous enrichment of vehicle types, and the constant improvement of key technologies. However, the industrial issues caused by industrial policies are also outstanding. The coexistence of achievements and problems has triggered debates in the industrial and academic circles on the implementation effect of industrial policies for new energy vehicles. It is necessary to systematically evaluate this issue.

The SPS team sorted and deconstructed China's policies on new energy vehicles issued from 2009 to 2017 by their roles and targeted sections and then divided them into four categories. The first category is policy instruments for the R&D section, including financial support at the national or local level to encourage new energy vehicle R&D. The second category is policy tools for production, including the creation of a separate set of qualifications for investing in new energy vehicle-related projects and the setting of new energy vehicle production ratios as a requirement. The third category includes policy measures dedicated to purchasing, mainly made up of direct incentive policies such as the direct fiscal incentives and tax preferences on product purchases, as well as indirect incentive policies, such as restrictions on the purchase of ICEVs in private areas. The fourth category is policy tools targeting product use, consisting of various fiscal and tax incentive policies that reduce the costs of vehicle use as well as non-financial and tax policies, such as non-restrictions to road access and preferential parking.

The SPS team employed an improved analytic hierarchy process (AHP) model to analyze the contribution of industrial policies. As shown by the results, China's policies promoting the new energy vehicle industry are generally effective. Of the measures, purchase subsidies have made the highest comprehensive contribution to the development of the new energy vehicle industry, with a ratio close to 50%. They also have had the most significant effects on promoting technology progress, cost reduction, and market growth. The contributions of the preferential tax policy, product access regulations for enterprises, national R&D plans, and transportation support policies decreased in descending order. In addition, from the perspective of individual indicators, preferential tax policies and transportation support policies have made a larger contribution to technological progress, R&D support and preferential tax policies to cost reduction, and transportation support policies and preferential tax policies to market growth. At the same time, the evaluation has unveiled flaws in the existing policy system. First, the comprehensive contribution of the purchase subsidy policies is excessively large, resulting in a high degree of dependence on subsidies by industries, enterprises, and consumers. This could result in drastic drops in the market should the subsidies be phased out. Second, support from national R&D plans needs to be strengthened to further enhance their contribution to technological progress.

6.2.2.3 Conclusions and Recommendations

First, establish and improve a sound policy system covering the entire automobile industry chain to support green consumption and production. In the automobile production process, the development and use of non-HFCS alternatives and alternative technologies should be encouraged. For automobile purchases, tax reform should be promoted to enhance the effect of taxes in leading energy conservation and emission reduction, reduce the cost of purchasing green car products, and encourage green consumption. With regard to the actual use of vehicles, efforts are needed to make green car products easier to use and reduce corresponding costs. In terms of vehicle scrappage and recycling, it is important to push forward the improvement of power battery recycling policies and standards, improve policies related to the remanufacturing industry, and enable the integrated development of the remanufacturing and insurance industries so as to facilitate the better development of the remanufacturing industry.

Second, in order to guide the green consumption of automobile products, it is necessary to give further play to the role of fiscal and taxation policies in leading the trend to energy conservation and emission reductions. With reference to international experiences, and in consideration of China's industrial development and tax system status, priority should be given to implementing a green tax system based on passenger vehicle fuel consumption indicators. For passenger vehicles that meet the fuel consumption targets in advance, a certain range of tax preferences should be given in the purchase tax and consumption tax according to the degree that they outperform targets. For passenger vehicles that fail to meet targets and standards, the tax should be increased according to the degree of underperformance in meeting targets.

Third, specific plans are needed for implementing a green tax scheme for automobiles in 2021–2035, based on the current tax system and preferential tax policies. From 2021 to 2025, gradually phase out the current purchase tax exemption policy for new energy vehicles, and initiate preliminary research. From 2026 onward, implement tax preferential policies based on fuel consumption, and set up a dynamic policy adjustment mechanism to respond to changes in fuel consumption regulations. From 2031 to 2035, raise the threshold for preferential policies and introduce a punitive tax system. Two policy scenarios, one based on the introduction of a strong policy (greater tax preferences) and one employing a weak policy (smaller tax preferences and greater punitive tax), have been proposed, and their respective policy effects are analyzed. The calculation results show that the implementation of a green tax system can effectively adjust the structure of the automobile market and increase the market share of energy savings and new energy vehicles while having a significant positive effect on energy conservation and emissions reductions (Table 6.2).

Fourth, in order to encourage green production in the automotive industry, automotive air-conditioning refrigerants with low global warming potential (GWP) values such as 2,3,3,3-tetrafluoropropene (HFO-R1234yf) can be selected to replace existing HFC refrigerants so as to slow down global warming. Further steps include improving

Table 6.2 Policy effectives under different scenarios

Scenarios	Share of energy-saving vehicles and new energy vehicles (NEVs)			Fuel conservation (10,000 tonnes)			Pollutant emission reduction (10,000 tonnes)		
	2025	2030	2035	2025	2030	2035	2025	2030	2035
Strong policy	20%	52%	62%	440	690	824	3.1	9.1	12.8
Weak policy	16%	42%	56%	402	582	766	2.1	5.4	10.6
No policy	12%	34%	47%	375	576	730	1.4	4.2	8.6

Table 6.3 Framework of green consumption index and indicator system in China

I. Overall index	Current value
1. Growth rate of per capita domestic energy consumption (%)	6
2. Growth rate of per capita domestic CO_2 emission (%)	6
3. Per capita daily domestic water consumption (l/person)	179.7
4. Output value of major green products (100 million yuan)	–
5. Proportion of green government procurement (%)	Around 90
II. Domain specific indicators	Current value
6. Clothing: Recycling rate of waste and old textiles (%)	Around 30
7. Food: Rate of food waste (%)	12
8. Housing: Proportion of green buildings in new buildings in urban areas (%)	50
9. Housing: Energy consumption per unit of building area of public institutions (%)	Cumulative decline of 10% in five years
10. Household appliances: Recycling rate of urban–rural domestic waste (%)	Around 15
11. Mobility: Proportion of urban green mobility (%)	Around 70
12. Mobility: Share of new-energy vehicle in total sales volume of automobile of the year (%)	
13. Tourism: Proportion of green hotel and restaurant (%)	

Explanatory Notes: (1) The values for the growth rate of per capita domestic energy consumption and growth rate of per capita domestic CO_2 emission indicate the annual average growth rates from 2016 to 2018; (2) The values for per capita daily domestic water consumption, proportion of green government procurement, recycling rate of waste and old textiles, rate of food waste, and recycling rate of urban–rural domestic waste use 2018 data; (3) The values for the proportion of green buildings in new buildings in urban areas, energy consumption per unit of building area of public institutions, and proportion of urban green mobility are the projected values for 2020; (4) The indicator for urban green mobility refers to the proportion of green mobility in the central urban area of large and medium cities; (5) The output value of major green products includes certified energy-saving and water-saving products, green labeling products, green and organic food, etc.; (6) The proportion of green government procurement refers to the share of green products among products of the same category purchased by the government; and (7) This indicator system uses nationwide statistics, and does not distinguish between urban and rural areas

the legislative system for auto parts remanufacturing, promoting the integrated development of the remanufacturing industry and the insurance industry, and cultivating and expanding diversified marketing and promotion approaches for remanufactured parts. Also important will be facilitating the deployment of industry standards and promoting sustainable development through power battery recycling, and further improving regulations and systems for an integrated and efficient resource utilization industry.

6.2.3   Green Power Market and Reform

6.2.3.1   Status of Green Power Development in China

Green power generally refers to the power generated by renewable energy-based power generation projects. By the end of 2019, the installed capacity of renewable energy-based power generation in China reached 794 million kWh, showing a year-on-year increase of 9%. The installed capacity accounted for 39.5% of the entire installed capacity, showing a year-on-year rise of 1.1%. The level of renewable energy utilization in China is also rising. In 2019, renewable energy power generation amounted to 2.04 trillion kWh, a year-on-year increase of approximately 176.1 billion kWh. The power generated from renewable energy sources made up 27.9% of the total power generation, exhibiting a year-on-year increase of 1.2 percentage points. It is anticipated that during the 14th FYP period (2021–2025), newly installed wind power capacity will reach 120 to 200 kWh, and the installed capacity of photovoltaics (PV) will be around 200 to 300 kWh. Estimations also show that, by the end of the 14th FYP period, the share of renewable energy-based power generation will exceed one third of China's total power generation and approach 40%.

Distributed power generation has developed rapidly in China in recent years, with distributed PV power generation for enterprises and households dominating, and distributed wind energy development for use in industrial parks and rural collectives expanding. By the end of 2019, the cumulative grid-connected capacity of distributed PV power generation in China reached 62.63 million kW, accounting for 31% of the total installed capacity of photovoltaic power generation. With the drop in cost in wind and PV power generation, the innovation in and maturation of distributed generation business models, and the successive introduction of supporting policies in various regions, China's market for distributed generation from renewable energy will continue to expand.

In 2019, the actual consumption of power from all renewable energy sources, including hydropower, was 2.0141 trillion kWh, accounting for 27.9% of the total electricity consumption in China, a year-on-year increase of 1.4%. The national consumption of electricity from non-hydro renewable sources amounted to 738.8 billion kWh in 2019, accounting for 10.2% of the total electricity consumption, a year-on-year increase of 1%.

6.2.3.2   Progress and Challenges Related to the Reform

There are three major ways enterprises consume green power in China at present. The first is when enterprises invest in renewable energy-based power generation projects on their own or through a third-party developer. The second is when enterprise consumers purchase green power directly from power generation companies. The third is when power users purchase green power certificates.

With the resumption of the power sector reform, the access conditions for power users wanting to participate in market-oriented transactions have been gradually relaxed. The changes in the power pricing mechanism, with the liberalization of trading methods and varieties, have created conditions for power users to participate in market-based transactions. However, China's green power market is still in an initial stage of development; the promotion of green electricity consumption needs to be a critical component of substantive power market reform. The power market reform related to green power consumption mainly involves two levels: one is the reform of policies and institutional arrangements for renewable energy power generation (including an on-grid price formation mechanism, a guaranteed purchase system, a green certificate system, and a market-based trading mechanism); the other is the overall reform of the power market, with particular focus on user-side reform to allow power market access to facilitate the participation of various power users, and reform of the transaction mechanism. To date, the following reforms have been achieved: first, the green power policy has gradually shifted from the practice of "purchase with guaranteed price and volume" to market-oriented practices; second, the reform of the power market has shifted from system design to implementation; and third, green power trading has shifted from inter-grid trading to full-scope trading.

The establishment of China's green power market is still in process. Limited procurement channels and unclear trading mechanisms are currently the biggest obstacles for companies to achieve their renewable energy consumption targets. At present, the most mature path for enterprise users is investment in distributed PV power generation projects, either by enterprises themselves or through third-party developers, yet the scale of such practices is limited. The most expensive path is purchasing green power certificates for renewable energy; however, the price for green certificates is falling. The path that has been attracting the most attention is the use of power purchasing agreements (PPAs) for the procurement of renewable energy power, a system for dealing with high market entry barriers and not clearly defined trading rules. The use of virtual power purchasing agreements (VPPA) is still limited and is dependent on the establishment of a power spot market in China.

6.2.3.3 Thinking and Recommendations for Reform

In order to boost the development of green power markets, further improve the market base, and unlock the demand for green power in various industries, the following recommendations are made.

First, accelerate the establishment of a green power market system. Promote the use of PPAs and VPPAs to further define specific rules and regulations for various power sources, including renewable energy, to be engaged in market-based transactions. Effectively mobilize all market participants and spur their sense of ownership.

Second, cut down improper administrative interventions from local governments, liberalize power generation plans, use plans and users' choice. Liberalize power development plans and respect users' choices within provinces and remove interprovincial barriers and restrictions on market players' participating in cross-provincial and cross-regional and inter-market transactions.

Third, further liberalize and protect power users' choices. Facilitate the market-oriented transactions between power users and clean energy-based power generation companies, such as hydropower, wind power, and solar power. Lift restrictions on inner-provincial and inter-provincial power purchases by power grid companies, power users, and power sales companies, and give priority to cross-provincial transactions of renewable energy in the transmission network.

Fourth, improve the policy and market environment for various users to jointly develop and use distributed renewable energy-generated power. Government and power grid companies should continue to deepen reforms, streamlining administration, decentralizing governance, strengthening regulation, and optimizing services. Cultivate innovative business models to introduce new business entities, such as virtual power plants and integrators, and approve their participation in distributed and wholesale power markets.

Fifth, gradually expand direct trading of renewable energy-based power pilots. Provide capacity-building support for power generation enterprises. Encourage power generation enterprises to carry out mid- and long-term power transactions with such users in close proximity as industrial enterprises and data centres that have a relatively large electric load and continuous stable use of electricity. Reduce the transmission and distribution prices for direct transactions in close proximity, and reduce the corresponding policy cross-subsidies.

Sixth, recognize the environmental attributes of renewable energy certificates to enhance enterprises' confidence in the trading market. Expand the scope of certificates to cover all kinds of renewable energy power generation projects. Expand the supply of affordable green certificates through grid parity projects, which will help decouple the price of certificates from the intensity of subsidies. Support the purchase of unbundled renewable energy certificates.

Seventh, establish a communication platform that includes various stakeholders to strengthen communication and cooperation. Build a case-sharing platform that will facilitate the exchange and learning of the best practices of green power purchasing in China, and collect information on the renewable energy procurement needs of various enterprises. Promote exchanges and cooperation among governments, power generation and consumption enterprises, industry associations, research institutions, and international organizations.

6.2.4 Green Logistics

6.2.4.1 Current State and Problem

By the end of 2018, the total number of express deliveries in China had reached 50.71 billion, which surpassed the total volume of developed countries and economies, including the United States, Japan, and Europe. In 2018 alone, 50 billion waybills, 24.5 billion plastic bags, 5.7 billion envelopes, 14.3 billion packaging boxes, 5.3 billion woven bags, and 43 billion metres of tape were consumed in the express logistics industry. Management of express delivery waste, including landfill and incineration, cost 1.4 billion yuan. In megacities, the increase of express delivery packages accounted for 93% of the total increase in domestic garbage, reaching 85% to 90% in some large cities. In addition, logistics transportation in China mainly depended on vehicles using traditional fuels. With almost 20 million vehicles for logistics consuming gasoline and diesel, a large amount of pollutants were emitted, imposing immense burdens and pressures on resources and the environment.

The Chinese government attaches high importance to the development of green logistics. In 2009, the State Council released the Plan for Adjusting and Vitalizing Logistics, proposing to encourage and support energy conservation and emissions cuts in the logistics industry and to develop green logistics. Afterward, China released relevant policy documents to advocate for the development of green logistics at the national, ministerial, and local levels in areas of transportation, storage, packaging, flow processing, and recycling, among other logistics links.

Their analyses into relevant policies on green logistics led the research team to identify several problems in current policies: (1) Legislation on green logistics lags behind. Although existing laws on environmental protection and resources regulate green logistics, there is no systematic special planning for green logistics. The duties, rights, and responsibilities of relevant entities are not clearly defined, and effective restraint mechanisms have not been established. (2) Green logistics have been incorporated into national strategies, but policies are relatively general and lack clear and definite targets as well as comprehensive supporting mechanisms. (3) Practices and measures related to green logistics focused more on green packaging and recycling of waste and old products, with weak practical support from the national level. (4) Decentralization of the green logistics authority led to insufficient follow-up of green logistics policy assessments and a lack of dynamic follow-up evaluations. (5) Green logistics pilot programs are important instruments for promoting the green transition of logistics. There have been positive outcomes achieved with the pilot program on recyclable bags, and pilot projects on green purchasing are actively promoted. However, it is important to proactively track and evaluate the effects of pilot programs, promoting and up- and out-scaling the successes.

6.2.4.2   China's Practices on Green Logistics

Currently, e-commerce and logistics companies are actively promoting cloud-based warehouse management, smart sorting and route planning, packaging algorithms, electronic waybills, environmentally friendly bags, green packaging boxes (such as use of recycled paper and printing using environmentally certified printing ink), sharing delivery boxes, logistics vehicles fuelled by new energies, solar energy logistics parks, and technologies free of oil and ink, etc., to meet the goal of realizing green logistics and cutting carbon emissions. A summary of the efforts of 10 companies[1] involved in green storage, green transportation and delivery, green packaging, green flow processing, recycling of waste and old products, and green information provision identified the following characteristics.

First, priority measures for the green logistics industry are gradually shifting from reducing consumable items and packing materials to green packaging, storage, transportation, recycling, and green information processing. Especially after 2016, there have been rapid developments in this direction. Multiple measures have been adopted to facilitate green logistics, with storage, transportation and delivery, package and information processing becoming the main focus. The recovery of waste and old products is attracting attention in the development of green logistics, while the field of green flow processing needs further development.

Second, the adoption of new technologies is an important factor in facilitating green logistics, as well as an indispensable measure for e-commerce and logistics companies to realize a green transition. It plays a key role in promoting green storage, green transportation, green packaging, and green recycling.

Third, some measures are difficult to implement. For instance, green storage (i.e., storage facilities characterized by low environmental impacts, low storage damage and loss, as well as saving transportation costs) requires a lot of financial support. In addition, new energy vehicles and unmanned aerial vehicles (UAVs) face a series of issues like cost and traffic conflicts. The high cost of green packaging makes it difficult to be widely adopted. The foundation of green recycling infrastructure is weak, heavily relying on consumers and deliverymen, with the overall recovery rate of packaging being less than 20%. The delivery end still composes a weak link in package recycling.

Fourth, green logistics need scientific methods of assessment. However, guidelines for assessing green logistics have not been established. An inventory of green logistics technologies to be promoted needs to be formulated to provide a reference for all logistics companies and vendors. The management of the green logistics supply chain is underdeveloped. Currently, green logistics measures are mainly targeted at logistics companies, while those for vendors, suppliers, and consumers are relatively weak, with insufficient participation from consumers.

---

[1] The 10 companies are: JD Logistics, SF Express, Suning, Meituan, Cainiao, STO Express, ZTO Express, YTO Express, Best Express [formerly known as Huitong], and Yunda Express).

Case 1. Green packaging

Green packaging activities in China's main e-commerce and logistics companies include:

(1) Ecological design of packaging: in 2013, SF Express formed a research team on packaging; then, in 2016, it established the Packaging Laboratory of SF Technology. In 2016, JD and Tung Kong Inc. jointly initiated JD Packaging Laboratory. In 2019, Suning.com established Green Packaging Laboratory. These actions all aimed to promote green packaging. Related efforts include environmentally friendly design, such as disposable plastic bags promoted by Cainiao of Alibaba Group and bio-based packaging bags promoted by Best Express. These designs reduce raw materials, for example, by reducing the weight of packaging boxes (JD) and reducing the thickness of tape (SF). Designs can also extend the usage of packaging, such as Feng BOX from SF, Qingliu Box from JD, and the sharing of delivery boxes by Suning. Environmental design of packages reduces the consumption of materials, lowers costs, and decreases waste.

(2) Greening the use of packaging: Cainiao built the first "Green Warehouse" in the world for all sorts of products, delivering products to consumers in recycled boxes. The whole process requires no secondary packaging, that is, it uses zero tape, zero packing, and zero new paper boxes. On Alibaba platforms, consumers can purchase commodities with green packages and receive "green energy" scores on Ant Forest, which encourages retailers to use green packaging. The "Qingliu Initiative" by JD promoted simplified printing of transportation package boxes and the adoption of packaging for direct delivery and circulating boxes. Meituan initiated the first alliance of recovering food delivery boxes in the catering industry to recycle take-out boxes.

(3) There are mainly two types of reuse for recovered packages. The first type is targeted at internal recovery within logistics companies without the involvement of consumers. The second type is recovery targeted at consumers, conducted through three modes: door-to-door recovery by deliverymen, establishing recovery stations, and setting up recovery boxes. By June 2019, JD recovered more than 5.4 million paper boxes. The proportion of old paper boxes directly recovered and reused in the delivery service by retail stores of Alibaba LST reached 30%.

Case 2. Green transportation and delivery

According to the Report on the State and Trend of New Energy Vehicles Development in Express Delivery in China (2018), as of June 2018, 12,988 new energy vehicles had been put into operation for express deliveries in the 31 provinces (or districts, municipalities) of China, a fourfold increase compared to 2016. Of these, 82% are micro-small cars, and 84% are rentable. In terms of city distribution of new energy vehicles, the most new energy vehicles are in Shenzhen, followed by Tianjin, Beijing, and Shanghai. Currently, JD has replaced all of its logistics vehicles with electric vehicles in Beijing and planned to replace all delivery vehicles in JD with new energy vehicles in five years. In terms of city delivery, delivery by UAVs and robots is emerging, mainly targeted at the hub to consumer transportation route to solve the last-mile delivery challenge. SF Express, Cainiao, JD, Suning, ZTO, and YTO and other logistics companies have adopted UAV delivery. SF, Cainiao, JD, Suning, STO, ZTO, YTO, Best Express, and Yunda Express have established cloud platforms that are responsible for delivery schedules and management.

6.2.4.3   Conclusion and Suggestions

First, establish special planning for promoting green logistics at the national level, and guide and supervise e-commerce companies to develop green logistics. Clarify the main responsible entities for green storage, green packaging, green transportation and delivery, as well as reverse logistics recovery systems; set up targets and goals for mid- to long-term assessments; specify the responsibilities of all parties, including the government, industry and consumers; and promote green development of e-commerce logistics.

Second, establish a mechanism for evaluating green logistics technologies to assess new measures to promote green logistics and facilitate the implementation of excellent green logistics measures. Issue and periodically update a green logistics technology inventory for the reference of and adoption by logistic companies in order to facilitate the successful implementation of new green logistics measures.

Third, encourage the development of green packaging in the industry, promoting a green transition in logistics packaging. Incorporate the green packaging industry in the *Guideline Inventory of Green Industries*, and facilitate the development of packaging reuse and biologically disposable packaging. In addition, it is important to evaluate green packaging, apply green purchasing requirements to logistics packaging, promote standard packaging, establish unified reverse recovery systems for logistics packaging, break barriers between companies and facilitate the reuse of logistics packaging.

Fourth, deepen pilot work on green logistics and improve the exemplary role of pilot projects. While drawing on the experiences related to logistics transportation gained during the fight against Covid-19, expand the scope of green pilot projects, combine green logistics and urban governance, address the challenge of broken chains witnessed in single logistics companies so as to assure the flow of supplies and smooth transportation channels, and connect upstream and downstream elements of the logistics supply chain to realize efficient collection, distribution, storage, transportation and delivery of all kinds of production and living materials.

Fifth, improve the environmental awareness of consumers to promote package recovery at the consumer end. Promulgate measures that could motivate consumers to recover logistics packaging and promote or innovate favourable mechanisms to encourage consumers to undertake environmentally responsible behaviours when choosing a package or recovering packages, such as the green energy system of Ant Forest, deposit for package, or the financial rewards for recovering packages to facilitate the recycling of logistics packaging.

6.2.5 Digital Platforms for a Low-Carbon Lifestyle

6.2.5.1 The Current State of and Challenges Facing Low-Carbon Lifestyle Programs and Platforms

In recent years, many kinds of low-carbon lifestyle programs and platforms have been experimented with, including Ant Forest, Tanpuhui, the Zero Carbon Group app, and the Lvdoya app. These have proven to be effective, innovative mechanisms for guiding transitions towards a low-carbon lifestyle. The Ant Forest digital platform built by business and the Tanpuhui platform (literally meaning "low carbon benefits all") built by the government are representative cases.

There are still many difficulties and challenges inhibiting the comprehensive use of these kinds of digital platforms, which aim to guide the public to live a low-carbon life nationwide. First, simply depending on companies to operate these platforms is unsustainable, as they require special policy support. Yet currently, the policy base underpinning a transition towards a low-carbon life is relatively weak. The guiding role of government needs to be further enhanced. In an effort to protect the privacy of individual data and the security of data providers related to emission reductions, low-carbon lifestyle platforms are not able to acquire sizable and effective data on emission reductions. The motivation for companies to participate in carbon reductions is insufficient, and it is difficult to attract business companies to cooperate in the promotion of digital platforms. Second, there is no unified standard for assessment or centralized supervision of such platforms. As a result, different platforms might calculate and double-count emission reductions. The methodology or algorithm used by different platforms in measuring individuals' voluntary behaviours in reducing carbon emissions can be quite different, leading to huge gaps in the assessment results of reduced carbon emissions. This could lead to the situation that users might question the seriousness, scientific evidence, or effectiveness of emission reduction data. Without unified supervision and monitoring at the national level, the carbon reduction numbers from users could be double-counted by authorized platforms.

Case 3. Low-carbon Military World Games

Wuhan Municipal Government initiated a Low-carbon Military World Games Program at the 7th Military World Games in 2019. This program quantified and summarized citizens' emission cuts through their green and low-carbon behaviour and their contribution to a carbon–neutral Military World Games. This Low-carbon Military World Games program was connected to the platforms of Wuhan City Pass, Hello Global and Bank of Communications, among others. After collecting users' low-carbon behaviour data and measuring it in carbon emission cut equivalencies, the program issued carbon credits.

This Low-carbon Military World Games' program was officially launched on June 18, 2019. After four months of operation, it had created favourable social and emission reduction benefits. First, the program issued users with electronic Carbon Offset Certificates of Honor for the Military World Games, which enhanced their sense of honour as low-carbon citizens. Carbon credits could be redeemed for Military Games gifts, which motivated citizens to practice low-carbon lifestyles. Companies and vendors were attracted to the platform, as it could help them establish green branding. In addition, the program ran for 122 days, with a total of 1,747,089 visits. It certified 75,700 users, reducing carbon emissions by a total of 92.61 metric tonnes. There were 100,739 instances of low-carbon green consumption behaviour, with carbon dioxide emission cuts equivalent to 22.08 metric tonnes. It is estimated that the emission of carbon dioxide for athletes taking shuttle buses between the Games Village and venues was about 80 to 100 metric t CO_2e. Carbon offsetting was thus successfully promoted.

6.2.5.2   Corresponding Recommendations

In order to fully benefit from digital platforms aimed at furthering green consumption among the public and guiding people to practice low-carbon lifestyles, the following recommendations are proposed, drawing on experiences from the operation mode of the Low-carbon Military World Games program.

First, establish a national digital platform for a low-carbon lifestyle. Based on the operation mode of the Low-carbon Military World Games Program, gradually attract organizers of large sports games and international and domestic meetings to use the platform. Construct an ecosystem for carbon neutralization and build a digital platform for a low-carbon lifestyle that is nationally influential and makes use of a comprehensive application standard.

Second, have the government play an exemplary role and set up a normalized carbon offset mechanism. Further refine the implementation plan for the *Guidelines of Realizing Carbon Neutral Large Events (trial)*, give full play to the exemplary role of government, and require hosts to use the low-carbon lifestyle digital platform to achieve carbon neutrality when activities (sports games, meetings, etc.) organized by the government would emit over 1 tonne of carbon.

Third, issue an *Annual White Book of Enterprises Supporting Carbon Neutrality*, and include emission reductions into the social credit system. Include participation in carbon offset efforts as one of the indicators for the annual assessment of the performance of state-owned enterprises and multinational enterprises, and provide examples of leading enterprises actively participating in carbon reduction and performing climate responsibilities. Meanwhile, the government could formulate favourable policies for companies or individuals that make contributions to realizing carbon neutrality at large events, sports games, or meetings.

Fourth, set up a special carbon-neutrality fund to provide financial safeguards for realizing carbon neutrality at large events (e.g., sports, conferences). Ecological and environmental authorities should establish the special carbon-neutrality fund. Hosts of large events should contribute a specific portion of advertisement benefits to this special fund in order to guarantee the daily management and operation of the fund. Investments from private and public welfare capital should also be encouraged.

Fifth, initiating green consumption coupons will help develop new consumption growth areas. The government should promote a green consumption coupon program system, formulate relevant policies for green consumption coupons, and make green coupons available to individual consumers making use of a digital platform to stimulate the consumption of green products.

6.2.6 Other Cases Promoting Green Consumption

Case 4. Green financing aid for green consumption

There are two main steps to be taken so that green financing can support green consumption. First, increase the accessibility of financial resources for green consumption to help consumers that favour green consumption receive financial support, and give full play to the leveraging role of finance in consumption. Second, use green financial instruments to reduce green consumption costs to make green consumption products more price competitive, promote the flow of social resources into green consumption industry chains, facilitate companies to produce green products, and realize the green and sustainable development of the economy. China has formulated a multi-level system of service providers that finance consumption and has gradually formed a system of consumption financing mainly composed of commercial banks, consumption financing companies, and financial platforms for Internet-based consumption. According to the Report of the Development of Consumption Financing in China, in the five years from 2014 to 2018, the financial credit volume for consumption on the Internet expanded from 20 billion yuan to 7.8 trillion, representing a 400-fold increase.

Specific measures include the green buildings mortgage services provided by Industrial Bank and Maanshan Rural Commercial Bank, loan businesses for green car consumption provided by China CITIC Bank, the green energy-efficiency loan business of Maanshan Rural Commercial Bank, and the green credit cards issued by China Construction Bank, Industrial Bank, China Everbright Bank, Agricultural Bank of China, Ping An Bank, etc.

Case 5. Sustainable food supply chain and consumption system

In 2017, the Report on China Sustainable Consumption jointly released by the United Nations Environment Programme and the China Chain-Store & Franchise Association (CCFA) showed that over 70% of consumers in the urban areas in China are aware of sustainable consumption to some degree. Almost half expressed a willingness to pay more for sustainable products up to a rate of less than 10%. However, the lack of sustainable consumption brands is restraining the further development of sustainable consumption. In 2018, the World Wildlife Fund (WWF) released Seafood Consumption Guidelines for sustainable aquatic products. Through assessing the sustainability of seafood products, the guidelines provide references and a feasible instrument for green consumption choices.

It is estimated that about 54% of food waste in the world takes place upstream in the value chain, during the production and post-harvest treatment and storage of food. The remaining 46% takes place downstream during the processing, circulation, and consumption of food. WWF promoted industry initiatives and pilot work in the restaurants and cold-chain logistics industries. In 2018, WWF started pilot work at 5-star hotels in Changxing County and promoted tools and training videos for reducing food waste in kitchens. In 2019, WWF and the Cold-chain Logistics Committee of China Federation of Logistics and Purchasing (CFLP) jointly launched the Initiative of Sustainable Aquatic Products Cold-chain of China, advocating for cold-chain companies to reduce resource waste and cut greenhouse gas (GHG) emissions during transportation, and working to jointly contribute to mitigating global warming.

WWF is committed to promoting sustainable food production, processing, and circulation systems globally. It advocates for the concept and practice of sustainable food consumption to improve efficiency and productivity, reduce waste and change consumption patterns, and ensure that human beings receive adequate food and nutrition while fully maintaining and protecting our natural resources. WWF carried out work under the following three goals:

(1)   By 2030, realize sustainable management of 50% of agriculture and aquaculture with no land cultivated for food at the expense of natural habitats;

(2) Reduce global food waste per capita by half and reduce post-harvest food loss;
(3) Have 50% of food consumption meet the dietary guidelines of the United Nations Health Organization and Food and Agriculture Organization of the United Nations in target countries.

Seven European countries signed the Amsterdam Declaration in 2015, committing to support the measures of the private sector in resisting deforestation in the supply chain. In Europe, 74% of palm oil imported for food production needs to meet the certified sustainable standards established by the Sustainable Palm Oil Roundtable Initiative (RSPO).

China has also initiated a series of industry actions for realizing a sustainable supply chain. In 2017, WWF, China Meat Association, and 64 companies jointly released the Chinese Sustainable Meat Declaration, aiming to create a sustainable meat industry and enterprise supply chain. The eight commitments of the Declaration covered zero deforestation and improving efficiency, among other aspects. In 2018, WWF, China Chamber of Commerce of I/E of Foodstuffs, Native Produce and Animal By-products (CFNA) and RSPO jointly launched the China Sustainable Palm Oil Initiative to promote sustainable palm oil and have it become a mainstream commodity in the Chinese market.

Case 6. Walmart Project Gigaton

Walmart launched Project Gigaton in the United States in 2017, a major initiative to engage suppliers, non-governmental organizations, and other stakeholders in climate action. Project Gigaton aims to avoid one billion tonnes (a gigaton) of GHGs from the global value chain by 2030 by engaging suppliers in target-setting and initiatives in six areas: energy use, sustainable agriculture, waste, deforestation, packaging, and product use. The Project Gigaton platform includes a variety of tools, including calculators for setting and reporting goals, best practices workshops, and links to additional resources and initiatives to make progress.

To date, Project Gigaton is one of the largest private sector consortia for climate action. Since its launch, more than 2,300 Walmart suppliers from 50 countries have signed up to participate and reported cumulative avoided emissions have reached more than 230 million tonnes (MMT) of GHG emissions (calculated in accordance with Walmart's Project Gigaton Methodology).

Walmart launched Project Gigaton in China in 2018, setting a sub-target of 50 MMT by 2030. To date, suppliers have reported over 5 MMT toward this target. Among these suppliers is China's Technical Consumer Products Inc., which supplies lightbulbs to Walmart stores in China and globally. Taking its

Project Gigaton commitment even further through product innovation, Technical Consumer Products introduced new energy-efficient bulbs at its Shanghai plant that are currently available around the world and in more than 400 Walmart stores in China. The redesigned light bulbs consume 36% less energy than their predecessors. The amount of energy that could be saved from 2018 sales alone is enough electricity to power 2,768,000 Chinese households for a year.

Moreover, in 2016, Walmart launched a Mill Sustainability Program to support suppliers and their mill partners in improving their manufacturing practices to help reduce environmental impact. Since then, participation has increased to more than 65% of apparel and soft home sales in U.S. Walmart stores in 2020 being sourced from suppliers working with mills that have completed the Sustainable Apparel Coalition's Higg Index Facility Environmental Module (FEM).

The Higg FEM Index is an industry-accepted tool that uses a cross-functional approach, allowing facilities to work internally to track their environmental impact, set goals, and improve their overall environmental performance. Of the 334 mills that completed last year's Higg FEM and shared their results with Walmart, over 54% of the facilities were in China. The total GHG emissions directly related to Higg reporting mills was more than 4.7 MMT CO_2e, where over 1.9 MMT CO_2e of those emissions were produced in China.

6.3   International Experiences with Sustainable Consumption Policy

6.3.1   Sustainable Consumption: Different Concepts, Different Implications

Green consumption, sustainable consumption, and sustainable lifestyles are three closely related terms often used interchangeably in discourse on demand-side sustainability. However, they have different conceptual approaches and emphases; each concept is tied to particular policy objectives, implementation strategies, and implications [2].

- *Green consumption* is closely tied to the greening of products and services in the marketplace and facilitating consumer access to them; it seeks to improve the quality of economic growth.
- *Sustainable consumption* is a broader concept that goes beyond the marketplace, although it is still very highly aligned with materialism and product/service consumption. The application of sustainable consumption highlights the need for

eco-efficiency (getting more from less) while also being attentive to the increased utility of consumption.

- A yet broader concept is *sustainable lifestyles*, which extends beyond material consumption and markets to reflect the more intangible aspects of everyday life, such as the values and social norms that shape our daily practices. Applying the sustainable lifestyles concept necessitates not only accounting for the environmental impacts of everyday living, it also acknowledges the fulfillment of needs through non-market options and includes practical instruments to support well-being and equity for all. An even more comprehensive term, in keeping with a sufficiency approach, is *sustainable living*.

Which term is selected depends on the level of ambition and how expansive a policy approach should be. This paper employs the concept of sustainable consumption, although elements of sustainable lifestyles and green consumption are included as well. While acknowledging that domestic, public, and private sector consumption are integral to sustainable consumption, this paper focuses on government policy to promote sustainable consumption, and specifically domestic consumption—by individuals, households, and community groups. It does this through a comparison of sustainable consumption perspectives, policies and programs, and experiences in the European Union (EU), Germany, Japan, and Sweden. It does not analyse government and public consumption (e.g., green procurement), nor does it analyse the numerous initiatives by businesses (e.g., sustainability reporting or greening of value chains[2]) and civil society organizations (e.g., sustainability campaigns).

6.3.2 A Comparison of Government Approaches to Sustainable Consumption Policy

Governments have an important role to play in setting the framing conditions around how production and consumption are practiced. They can establish visions and guidelines for a sustainable society, and promote change among households and organizations through various incentive structures and regulatory measures. Key approaches governments have used include treating sustainable consumption policy as (i) a part of an overall development strategy; (ii) a standalone strategy or action plan; (iii) a part of a sectoral policy, issue strategy, or program; or (iv) a part of the mandate of a public agency or organization. Several country governments combine more than one of the above approaches. Examples are discussed below.

Sustainable consumption integrated into an overall (sustainable) development strategy. In this case, the government identifies priorities for sustainable consumption and, instead of preparing a separate strategy, integrates those priorities in a broader strategy that guides government planning and operations. Examples include when

---

[2] China Green Transition Outlook 2020–2050 Project http://www.cciced.net/cciceden/POLICY/rr/prr/2016/201612/P020161214521503400553.pdf).

sustainable consumption is built into national vision documents, national (sustainable) development strategies, national green growth or green economy strategies, and national Sustainable Development Goal (SDG) implementation plans. The *EU Circular Economy Strategy* [3] and the attendant Action Plan [4] are high-level examples of how sustainable consumption can be reflected in a sustainable development strategy. Japan's *Basic Act for Establishing a Sound Material-Cycle Society* underpins its sustainable consumption and broader sustainability initiatives. An important strategy choice in Sweden is to integrate sustainable consumption into its overall framework for reaching the environmental objectives—a system to guide the overall efforts to safeguard the environment. Efforts towards resource-efficient material cycles and more sustainable consumption have been identified as essential ingredients for obtaining environmental objectives [5]. Germany's National Sustainable Development Strategy has aligned with Agenda 2030 in order to support the implementation of the 17 SDGs.

Incorporating sustainable consumption into the national development framework has the advantage that consumer behaviour is not addressed in isolation from the broader development trajectory. Since lifestyles and consumption touch on a variety of soft (e.g., education, health) and hard (e.g., industry, infrastructure) issues, a coherent and concerted approach is needed, as is often the case with overall national strategies. The risk, however, is losing focus, especially if there is competition for resources and political attention among priority issues, leading to sustainable consumption being subordinated to more short-term politically charged issues. This would postpone the need to address rising consumerism or inequality and could entrench unsustainable consumption problems, making it even more difficult to address if, and when, it eventually receives attention. There can also be missed opportunities to address unsustainable practices from a demand-side approach, including solutions that reduce harmful consumption altogether.

Sweden's Sustainable Consumption Strategy (Strategi för hållbar konsumtion) and Germany's National Programme for Sustainable Consumption (Nationales Programm für Nachhaltigen Konsum) are two unique examples of *dedicated national strategies for sustainable consumption*. The two countries also co-lead two of the six programs under the United Nations' 10-Year Framework of Programmes on Sustainable Consumption and Production. Sweden co-leads with Japan on Sustainable Lifestyles and Education, and Germany co-leads with Indonesia and Consumers International on Consumer Information. This paper will go into more detail about these national strategies.

The most widely adopted approach is to integrate *sustainable consumption as part of a sectoral policy, issue strategy, or program*. Sustainable consumption is tied to sectoral policies, including for energy, water, transportation, health, and housing and infrastructure. An example is the Swedish National strategy and Action Plan on sustainable food [6]. Sometimes sustainable consumption is embedded in programs established in response to societal issues driven by citizens or in which citizens are affected. Examples include national programs or strategies for poverty reduction and national programs on social health and obesity.

Although not officially government policy, the Science Council of Japan has proposed a Roadmap to Healthy Low-Carbon Lifestyles, Cities and Buildings. It is aimed at ensuring a high quality of life for citizens as Japan faces a super-aged society of elderly citizens. It targets infrastructure in cities and buildings, making sure they are suitable for the demographic and also deliver a low-carbon footprint and high environmental performance. The recommended policies are divided into four parts: increasing motivation for new, healthy, low-carbon lifestyles and behavioural changes; designing healthy low-carbon cities and traffic systems for a mature society; accelerating low-carbon housing and buildings, health measures, and energy generation; and applying Japan's low-carbon cities, buildings, and traffic systems strategically across Asia [7].

Such sectoral or issue-based programs run the risk of being time-bound and can disappear when governments change. Thus, while effective during the operational term of a priority government program, approaches should be institutionalized to achieve long-term stability.

Sustainable consumption through public agencies/institutions or civil society organizations. The rise of consumer organizations, especially in Europe and North America, has paralleled the rise in public concern about the market's willingness to prioritize consumer interests over profits. Organizations such as Test Achats in France, Which UK! in the United Kingdom, Consumentenbond in the Netherlands, Stiftung Warentest in Germany, the Swedish Consumers' Association in Sweden, and Consumer Co-operatives in Japan are examples of transitional consumer organizations that, in conducting products tests and ensuring producer responsibility, are shifting concerns from product price, quality, and respect for consumer rights to broadened mandates that include responsible/sustainable consumption.

6.3.3 Supranational Level: The EU Policy Approach to Sustainable Consumption

The EU internal market is built on common law for all products and its trade within the EU. The EU is actively moving to direct member states' economies to limit resource use and waste, develop new industries, promote green jobs, redesign urban structures, and change societal behaviour by promoting and enabling sustainable consumption. The importance of promoting a sustainable consumption policy gradually took root in the EU, moving from an early focus on recycling and waste minimization to increased attention on sustainable product design and information availability for consumers about the energy use and environmental impacts of products. The EU's renewed Sustainable Development Strategy [8] (2006) provided an important impetus that led to the development of a range of initiatives and instruments. These include the Ecodesign Directive [9] (2009), which set ecodesign requirements for energy-consuming products, the Ecolabelling Directive, and the Energy Labelling Framework Regulation, which aimed to ensure consumers are provided with information on the energy and environmental performance of products. Substantial improvements were incentivized through these regulations, but numerous gaps remained. Product legislation tended to address only specific aspects of a product's life cycle, and many of the environmental impacts of products were not addressed.

With heightening concern in the EU about global warming, pollution, inefficient material use, natural resource depletion, and import dependency for energy and natural resources, in 2008, the European Commission drew up the Sustainable Consumption and Production and Sustainable Industrial Policy (SCP/SIP) Action Plan [10]. The SCP/SIP Action Plan sought to develop a more comprehensive approach, expanding the coverage of the Ecodesign Directive to cover all energy-related products, setting environmental benchmarks for products, and conducting periodic reviews; establishing a harmonized base for public procurement by EU institutions and Member State authorities; and, importantly, promoting smarter consumption. A range of actions were launched to encourage retailers and manufacturers to green their supply chains and to raise consumer awareness and involvement.

The newest and most important policy development is the European Green Deal, which addresses the areas of clean energy, sustainable industry, building renovation, sustainable mobility, food production and consumption, and biodiversity protection. It aims for climate neutrality and zero pollution by 2050. It also aims to ensure European global leadership in this realm and to provide a model for other countries, including China, to consider. In March 2020, a new Circular Economy Action Plan [11] was adopted as one of the main components of the European Green Deal. Going significantly beyond the 2015 Circular Economy Package, it aims to make sustainable products the norm in the EU, empower consumers and public buyers to consume sustainably, and achieve a system where no waste is produced. It focuses on the value chains in areas where circularity potential is high: electronics, batteries and vehicles, packaging, plastics, textiles, construction and buildings, food, water and nutrients. It further aims to make circularity implementable for people, regions, and cities. Product durability, reusability, and recyclability, energy and resource efficiency, and recycled material content are to be enhanced. Single-use is to be restricted and premature obsolescence of products countered while product-as-a-service is to be incentivized. Various measures to empower consumers are central to the plan, including enhanced information about product lifespan and the availability of repair services, and the establishment of minimum requirements for sustainability labels and information tools. A "right to repair" is being considered, as are requirements for companies to substantiate their environmental claims using Product and Organisation Environmental Footprint methods. The Monitoring Framework for the Circular Economy is to be updated. Furthermore, a Circular Economy Stakeholder Platform offers the public an opportunity to share their ideas about good practices, publications, events, and networks related to sustainable consumption, production, waste management, and innovation [12].

6.3.4 National Government Strategies on Sustainable Consumption

The following sections provide background details on three case study countries—Germany, Sweden and Japan—highlighting the unique characteristics of each country's approach and presenting options that may be of interest to China as it designs its own sustainable consumption policies.

Consumer protection, consumer rights, consumer safety, and consumer information have long been important for industrialized countries.[3] More recently, the sustainability of consumer behaviour has become a critical area of concern. In industrialized countries, upwards of 50% of environmental impacts (including GHG emissions, resource use, pollution, noise, and loss of biodiversity) can be linked to available domestic consumption options and practices. However, political measures introduced to address the impact of consumption have typically done so with a focus on industrial sectors, without taking into account the conditions and drivers of demand-side behaviour.

Germany

In February 2016, just shortly after the SDGs were agreed upon, Germany became the first nation to adopt a standalone National Programme on Sustainable Consumption [13]. The strategy was developed in the Federal Ministry for the Environment and negotiated within the framework of a formal inter-ministerial working group on sustainable consumption. The composition of this inter-ministerial working group—led by the Federal Ministries for the Environment, Nature Conservation and Nuclear Safety; Justice and Consumer Protection; and Agriculture and Nutrition—reflects the cross-sectional approach and transversal nature of sustainable consumption.

Germany's strategy outlines five foundational principles: making sustainable consumption feasible (by increasing the ability of consumers to make decisions and take action); taking sustainable consumption out of the niche and into the mainstream (e.g., by creating protected spaces and promoting new initiatives, with incentives for the use of particular technologies or for enabling sustainable behaviour); ensuring all sections of the population are included (by tailoring the approaches to specific target groups); looking at products and services from a life-cycle perspective; and shifting the focus from products to systems and from consumers to users.

Despite the recognition of the impact of harmful consumption, efforts to develop a strategy in Germany entailed debates on how to avoid consumer scapegoatism [14]—a situation where the burden is shifted to consumers without a critical analysis of their capacities or drivers of their behaviours. The strategy therefore sought to understand aspects of the supply side that shape consumer behaviour. Supply-side instruments like the Ecodesign Directive, producer responsibility schemes, and warranty regulations are explicitly part of the government approach because they

---

[3] See, for example, the German Consumer Information Act.

heavily influence consumption patterns and can help reduce impacts through, for example, lower-energy consumption or the greater durability of products.

Sweden

Sweden's *Strategy for Sustainable Consumption* aims to make it easier for citizens and consumers to act sustainably. This is a government strategy to promote sustainable household consumption and was introduced in order to address the increasing environmental impact following Swedish consumption. Implementation of the strategy is led by the Ministry of Finance and is carried out mainly by the environment and industry ministries [15].

The strategy was adopted in the fall of 2016. It addresses what the State, together with municipalities, the business sector, academia, and civil society, can do in order to make it easier for citizens to shift to more sustainable consumer behaviours and lifestyles. The strategy has seven strategic focus areas: increasing knowledge and deepening cooperation; encouraging sustainable ways of consuming; streamlining resource use; improving information on companies' sustainability efforts; phasing out harmful chemicals; improving security for all consumers; and putting a special sectoral focus on food, transport, and housing. The Ministry of Finance is responsible for the implementation. The Forum on Eco-smart Consumption, a national dialogue and a digital platform, is one of the implementation mechanisms supporting knowledge building and communication about sustainable consumption. The Swedish Consumer Agency has been assigned by the government to operate the forum, supported by an inter-agency reference group consisting of the Swedish Environment Protection Agency (EPA), the Swedish Chemical Inspectorate, the Swedish Energy Agency and the Swedish Food Agency. An advisory board with representatives from industry, academia, etc. has also been appointed to guide the implementation.

The strategy is currently being implemented. Some examples of government policy initiatives are new education materials to facilitate knowledge building on consumption's environmental impacts; economic incentives to increase repair and maintenance of selected product groups; stricter requirements on consumer information in the business and financial sector; debt counselling services to support families and individuals to get their financial situation in order; measures to stop companies from engaging in unlawful marketing practices; annual national workshops on sustainable living; and the introduction of well-being indicators.

Japan

While Japan does not have a dedicated sustainable consumption strategy, it has a broad sustainability policy framework and programs dedicated to consumption and lifestyles. The Japanese approach is highly influenced by its history of development and waste management. In the 1950s, Japan's economy changed significantly in both size and structure. The scale grew drastically, restructuring was driven by heavy and chemical industries, and the concentration of the population in cities intensified. Both municipal and industrial waste rapidly increased. Inadequate waste disposal and illegal dumping became rampant. Developing a waste management policy, improving sanitation, and prevention of environmental pollution became important. These developments fed into the government's sustainability strategy, with its dedication to building a "sound material cycle society." Given the topical issue of the coronavirus pandemic and that suggested solutions are linked to sanitation, Japan's history of linking waste management systems to sanitation and public health provide indications of how sustainable lifestyles can also be addressed through public health policy and infrastructure.

In a more recent iteration, Japan established The Basic Act on Establishing a Sound Material-Cycle Society. The act establishes basic principles centered on cyclical utilization of resources and disposal of waste, including clear priorities and hierarchical measures for the 3Rs (reduce, reuse, recycle). The act requires the government to develop and renew a Fundamental Plan for Establishing a Sound Material-Cycle Society every five years. With this plan, Japan is building on cultural traditions embedded in programs and projects, such as sustainable packaging and recycling through its 3R campaign, promoting household waste separation and energy efficiency at home. Given how integral consumption is to lifestyles, culture, and tradition provide important levers. The Japanese language has a word that embraces the concept of sustainability: *mottainai*. *Mottainai* is used to indicate the wastefulness associated with discarding something that still has value. The idea was picked up in a comic strip by the same name in an effort to help educate the manga-loving population.

One example of the above is the nationwide "Cool Choice" program, launched in 2005 in order to encourage people to choose decarbonized products, services, and lifestyles, such as through the use of public transportation and energy-efficient appliances. The government also launched a "Cool Biz" campaign, encouraging sustainable and temperature-appropriate clothing in the work environment. In a country known for its formal working environment, in the summer, people are encouraged to wear light, casual clothing, meaning no neckties for men and no suit jackets. By dressing more "coolly," in addition to feeling more comfortable at work, this campaign reduces the need to set very low temperatures for air conditioners in offices. Results show strong savings: approximately 6.95 million people and 100,000 companies have adopted "Cool Choice" practices. While the decline by 10% in the carbon dioxide emissions from the household sector between 2013 and 2017 cannot be tied exclusively to this program—energy-saving initiatives after the Fukushima nuclear disaster also contributed—the "Cool Biz" campaign has certainly played a significant role.

6.3.5 Determining Priority Areas: European, Swedish, and German Experiences

In over 50 years, consumption policy has evolved from addressing the consequences of end-of-pipe issues (such as waste and local contamination) to taking on a broader systemic lens (such as shaping social norms and values that inform the economic system within which production and consumption activities occur [16]). In the late 1960s and 1970s, industrial manufacturing caused serious environmental problems due to air and water pollution and from poor waste management. Government policies were mainly reactive, with a heavy focus on public health and the emergence of consumer protection legislation. By the 1980s, the more preventive approach of cleaner production was adopted, and, in the 1990s, this approach was re-emphasized in eco-efficiency and product-oriented approaches. Demand-side policies emphasized increased efficiency in material and energy use, supported by ecolabels and better household waste management—known as the 3R approach. Towards the turn of the century, government policy began to acknowledge the negative impacts of harmful overconsumption and the role of social inequities in driving unsustainable patterns. Contemporary European policies tend to be a mix of eco-efficiency and inclusive characteristics of social welfare, as encapsulated in the "leave-no-one-behind" motto of the SDGs.

Priority sectors and themes for sustainable consumption policy in Europe have been heavily driven by research. The Seventh Framework Programme for Research and Technological Development (FP7) [17] was the EU's main instrument for funding research in Europe and ran from 2007 to 2013. It was designed to respond to employment, competitiveness, and quality of life in Europe. The successor program, Horizon 2020 [18], supported research and innovation with nearly €80 billion of funding for the period from 2014 to 2020. Under these schemes, several research projects were funded which focused on consumption, lifestyles, and policy analysis, in order to support EU- and national-level government strategies and approaches—for example, baseline and needs assessments; assessment of risks and uncertainties; life-cycle assessment; material flow analysis; cost–benefit analysis; environmental and social impact assessments, etc. The scale of funding and the rigour of research undertaken has established a clearer understanding of areas where consumption and lifestyles have the highest impact on the environment, and, therefore, should constitute priority areas for policy. These include food systems, mobility, housing, consumer goods, leisure and tourism, and cross-cutting areas such as energy, water, and waste. Although emphases might differ from one country to the next, these major areas often remain as top areas of focus and have come to shape priorities for sustainable consumption policy.

Despite the scientific clarity, addressing consumption also needs buy-in from citizens. As such, the policy-framing phase is crucial; the governments of Sweden and Germany have relied heavily on public consultations and deliberative processes, inviting non-governmental organizations, businesses, and local communities into carefully designed public citizen consultation processes. Analyses of these processes show that, once the public understands the critical nature of the issue, proposed policies are widely accepted. In fact, most post-consultation commentary suggests that citizens who acquire an awareness of the impact of sustainable consumption tend to propose more ambitious actions than governments eventually reflect in policy.

In Sweden, consumption-based GHG emissions have been an important (leading) criterion used to assess the impact of consumption. The main reasons are that climate impact is of high priority, and that methodology and data are accessible. Major efforts

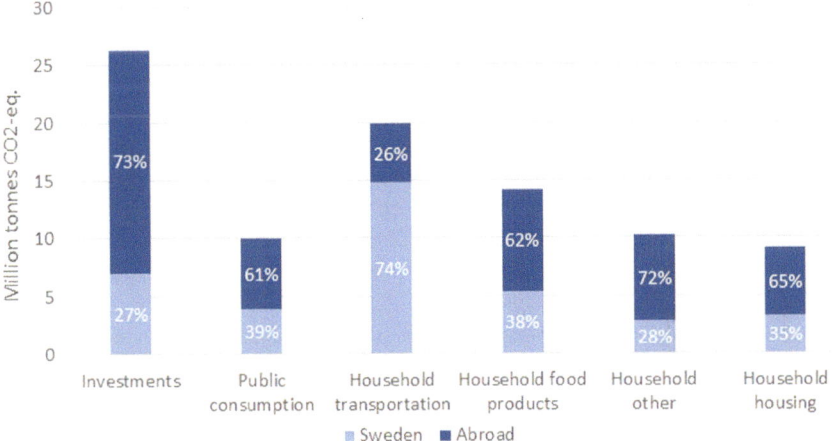

Fig. 6.6 Swedish consumption-based emissions per consumption category within Sweden and outside Swedish borders (2017). *Source* Sweden Statistics (SCB)

have been made to develop consumption-based indicators and to collect data on GHG emissions along product life cycles. Household consumption represents a dominant part of the total consumption-based GHG emissions, with three areas standing out: food, transportation, and housing. These three areas are therefore of high priority for sustainable consumption policy.

According to the most recent annual assessment, total GHG emissions from Swedish consumption in 2017 were 90 million tonnes of GHG-equivalents—about 9 tonnes per capita. Of this total, 58% arose outside of Swedish borders (see Fig. 6.6). Emissions abroad are generated from, among others, air flights, the import of palm oil, electronics, and textiles. Swedish household consumption accounts for 60% of the total GHG emissions, while emissions from the public sector account for 11%. For household consumption, shares of emissions from food, transportation, and housing are, respectively, 15%, 20%, and 10%, while the remaining 29% is attributed to investments.

For Germany, the National Programme for Sustainable Consumption places its emphasis on six key areas of private consumption: mobility, food, home, workplace and office, clothing, and leisure and tourism. Mobility accounts for 26% of GHG-related emissions, with food at 13% and housing at 36%.

6.3.6 Policy Instruments

Common to most EU Member States, EU legislative frameworks and policy instruments that are highly relevant for sustainable consumption and production are transposed to the national level. Beyond the EU common SCP policy agenda, there are additional national governmental initiatives in place to facilitate the shift to sustainable consumption. Examples of national government policy instruments and measures in place addressing key Swedish Environmental Quality Objectives at the macro (sector) and micro (product supply and demand) levels include a carbon tax and subsidies; flight tax; bonus-malus system on new cars; electric car subsidy; subsidies on solar panels, investment support programs ("Climate step" and "Industrial stride," waste prevention in a circular economy): 12–25% reduced value-added tax on repair of bikes, shoes, leather goods, clothing and household linen; tax reduction for repair and maintenance of white goods (large electrical goods such as fridges and washing machines); and a financial support program called Sustainable Use of Plastics.

The Swedish government is also supporting stakeholder collaborations. Some current examples are national and regional workshops on sustainable consumption and lifestyles, a multistakeholder dialogue on the textile chain, and a national forum on eco-smart consumption (described in more detail in Sect. 3.4.2). Examples of instruments mentioned here have been reported in accordance with Agenda 2030 indicator 12.1.1 (Number of countries with SCP national action plans or SCP mainstreamed as a priority or a target into national policies [19]).

Sweden has implemented a large number (over 100) of environmental consumer policy instruments designed to directly influence consumer behaviour (demand-side) in an environmentally sustainable direction. Among these instruments, 32 have been evaluated [20]. They include a mixture of administrative, economic, and informative instruments (economic is dominating) addressing housing, travel, and food. A majority address carbon dioxide emissions and other air pollutants, and a few are cross-objective (e.g., labelling schemes).

The German National Policy for Sustainable Consumption includes more than 170 measures, which are either "hard" instruments, directly addressing one of the six priority policy areas discussed above or "soft" cross-cutting instruments.

6.3.7   Governance and Institutional Arrangements

6.3.7.1   Institutional Arrangements

Both Sweden and Germany have made the implementation of the sustainable consumption strategy a multi-agency, inter-ministerial responsibility. In Sweden, since 2014, 18 strategically relevant national and regional agencies have been cooperating at the highest management level (Director General) in the Environmental Objective Council [21]. Annual and in-depth assessments of environmental progress are made. Based on these, the council aims to contribute to solving conflicts between environmental and other goals in society and will present proposals to the government. Each year (2016–2019), the council has presented around 20–30 measures to be undertaken in collaboration by relevant national agencies in order to speed up the work towards meeting environmental objectives. So far, approximately 30% of the collaborative measures have been targeting the area "economy, growth and consumption." In February 2020, the council presented its fifth list of inter-agency collaborative measures. Several tasks will be undertaken in the period 2020–22 in order to strengthen Swedish environmental policy towards the objectives. Policy instruments for sustainable consumption is one out of seven priority areas for 2020–22 [22].

In 2016, when the Swedish sustainable consumption strategy was adopted, the government allocated 43 million Swedish kronor annually until 2020, and thereafter 9 million kronor per year to strengthen the Swedish Consumer Agency's work in environmentally sustainable consumption. The Consumer Agency was assigned by the government to establish a new National Forum on Eco-smart Consumption [23]. The vision is to make the sustainable choice the standard choice by sharing ideas, knowledge, and solutions (see Sect. 3.4.2 above).

For the implementation of the German National Program for Sustainable Consumption, a National Competence Centre on Sustainable Consumption has been established at the Federal Environment Agency (UBA). It also involves other agencies, such as Deutsche Gesellschaft für Internationale Zusammenarbeit (GIZ) and the Federal Agency for Agriculture and Nutrition. The centre organizes the implementation process of the program and ensures the involvement of relevant stakeholders in a National Network on Sustainable Consumption. It develops a reliable knowledge base on sustainable consumption and provides information to the public, organizing workshops and conferences.

An inter-ministerial working group was established to support the implementation of the program, bringing together representatives from all government departments concerned with sustainable consumption. It is headed by three federal ministries: the Federal Ministry for the Environment, Nature Protection, and Nuclear Safety (BMU); the Federal Ministry of Justice and Consumer Protection (BMJV); and the Federal Ministry of Agriculture and Nutrition (BMEL).

The program will create a public platform designed both to expand ideas about instruments and approaches that have already proved successful and to instigate new ones. This will ensure a continued assessment of the diversity of approaches available in the field of sustainable consumption and also encourage as many actors as possible to participate. A change in consumption patterns towards greater sustainability can only be achieved with the participation of all parts of society and a range of policy approaches undertaken in an integrated fashion.

6.3.7.2 Monitoring

The regular monitoring of the implementation of the German National Program for Sustainable Consumption is aligned with the required assessments of progress related to the SDGs. In line with international indicators intended to measure SDG 12 on sustainable consumption and production patterns, consumption-related indicators and targets were developed for the German National Sustainable Development Strategy. These indicators monitor the market share of sustainable products in 19 product groups, including food, paper, textiles, cars, and electric appliances. A target of a 34% market share by 2030 has been set for these product groups; GHG emissions per capita tied to consumption are to continuously decline; the share of Blue Angell-labelled paper in public procurement at the federal level is to reach 95% by 2020; and GHG emissions from public vehicles per km are also targeted for continuous decline.

For Sweden, since 2008, the country has been articulating the relationship between national-level consumption and impacts on climate by analyzing emissions along whole product life cycles. To improve its methods, the Swedish EPA recently funded the PRINCE project, which sought to develop state-of-the-art indicators for the environmental impacts of Swedish consumption [24]. The project was led by Sweden Statistics, with participants from academia [25]. Based on the research results, two new consumption-based indicators were introduced in the environmental quality objective monitoring system: consumption-based GHG emissions per consumption category and consumption-based emissions in Sweden and abroad [26]. Last year, the Swedish EPA proposed a set of consumption-based indicators to monitor the GHG emission trends from selected consumer categories (passenger travel, flights, food, housing construction and housing, textiles) to the government [27].

GHG emissions from Swedish households have decreased by 14% from 2008 to 2017, despite the (25%) increase in consumption volume. About two thirds of this decrease is estimated to be from the effect of increased eco-efficiency in imported goods as well as nationally produced goods. One third of the decrease is estimated to be related to a shift in Swedish consumption patterns. Consumption-based emissions, however, are still far above sustainable levels [28]. The government is probably going to explore the introduction of national consumption-based emission goals in 2020.

The Consumer Agency in Sweden developed a method in 2009/2010 to examine consumers' experiences related to different goods and service markets, such as those for dairy products and meat. The agency has used the results from its research to identify markets that are problematic for consumers. The results of this research have been published annually in the *Consumer Report*, which was released in six editions between 2013 and 2018 [29].

Sweden also monitors efforts in sustainable consumption from a gender perspective. Since 2015, the Swedish Consumer Agency has, on behalf of the government, mapped out what opportunities are available to consumers to act in environmentally sustainable ways. To understand this, consumer experiences are obtained by means of self-administered questionnaires. These are then supplemented with calculations from Statistics Sweden, showing the climate impact of consumption in general as well as in various markets. Similar calculations are made for GHG and other atmospheric emissions. The survey is designed in a way that provides disaggregated data for men and women and in different age groups. Results of the 2018 survey show that women see greater opportunities to make more environmentally sustainable choices. The consumer category that finds it most difficult to make environmentally sustainable choices are middle-aged men between 3 and 64 years, whilst women aged 65–75 find it the easiest to make more sustainable choices. Those with little or no interest in environmental issues see fewer opportunities to make more sustainable choices. There is a significant difference between women's and men's attitudes towards environmental issues. As much as 25% more women than men consider it important to consider how their consumption impacts the environment. Women have a better understanding of environmental choices and more often use ecolabels and other information to help them decide.

6.3.7.3 Gender Perspectives

Gender parity in commissions is an increasingly important topic in Germany. The German Council for Sustainable Development, an advisory body to the German government, has a predominance of female members in 2020 (9 women, 5 men). Sustainable consumption and resource management fall within its mandate. Gender is not explicitly addressed in the National Programme on Sustainable Consumption but is an implicit part. To become explicit, there are currently research projects to identify gender aspects of sustainable consumption, meant to enhance people's participation in the program [30, 31].

Gender equality is central to the Swedish government's priorities in decision-making. Women and men must have the same power to shape society and their own lives. The government's most important tool for achieving this, as a strategy to reach the goals declared for the Swedish gender equality policy dating back to 1994, is gender mainstreaming. According to the strategy, gender equality work must be integrated into the regular operations and not merely be dealt with as a separate, parallel track. The Swedish government has commissioned the Swedish Gender Equality Agency to support 58 government agencies, among others, the Consumer Agency, with the work of integrating a gender perspective in all their operations. The government ordinance, which governs the work of the Consumer Agency, instructs that the agency should integrate issues of sustainable development and work to achieve the Swedish environmental objectives, and the agency should also integrate a gender perspective.

For its work on gender integration, the Consumer Agency has set itself an objective to acquire knowledge showing different possibilities for women and men as consumers, and for the agency's activities to be based on this knowledge. The agency has, for example, organized workshops on "Women, Men, And Environment—Why Gender Perspective Is Relevant for Sustainable Consumption." In 2017, the agency assigned researchers to compile a report providing an overview of consumer behaviour and gender aspects. The report began with the question: "How do conditions differ for men and women as consumers to make active choices in the marketplace and in everyday consumer life? [32]" The report systematically identifies the latest research in relation to several market contexts or focus areas identified through discussions between the Consumer Agency and the Swedish Environmental Protection Agency. The report shows, for instance, that, in decisions concerning financial services, men and women exhibit different behaviours. Women tend to focus on minimizing potential losses from investments, whereas men express a conviction to maximize profit, socially and financially.

6.3.8 Conclusions and Recommendations

The increased role of consumption in determining the long-term availability of economic resources, in defining public attitudes of citizens, and in creating impacts on the environment requires that sustainable consumption commands priority attention in any future-proof national sustainability strategy. This is especially the case for China, given the combination of the size and the rate of growth of its economy. It is also an opportunity for the country to show global leadership, given that most strategies to address sustainable consumption are still evolving, and to achieve the ultimate objective of delivering on well-being for all while limiting ecological impacts within sustainable limits. Recommendations based on international experiences are made with the hope that they can inform and be adapted to the development of a China strategy.

(1)   **Embed the concept of sustainable consumption into the 14th FYP.** Attention to the importance of sustainable consumption can be accompanied by major goals to reduce the ecological and societal impacts of excess consumption. As well as clear direction for demand-side action, the plan should include actions geared at changing provisioning systems that determine consumer choice, as well as the physical architecture and socioeconomic systems that lock in everyday living.

(2)   **Develop a dedicated Chinese sustainable consumption strategy and related action plan** to further China's long-term ambitions to develop as an ecological civilization, achieve a well-off society, and address climate change and resource depletion concerns. The Swedish and German sustainable consumption strategies, as well as the recent European Green Deal, provide useful examples. The action plan should develop specific policy instruments to address different areas of consumption, mixing educational campaigns, information systems, incentive structures, and regulatory approaches. The plan can follow a life-cycle approach and address key areas where Chinese consumption has the highest impact on the environment.

(3)   **Construct an integrated indicator system that can comprehensively reflect the status and level of sustainable consumption by private consumers.** This will support the monitoring of sustainable consumption objectives in the Sustainable Consumption Plan. Consider including the proposed indicators within China's existing national system of statistics categories for green households or consumer goods and services.

(4)   **Develop explicit well-being indicators to be used in monitoring and reporting on how economic development is responding to the needs of all.** Monitoring of consumption should not be limited to only environmental concerns but also reflect people's growing demand for a better quality of life and a new effort by the government to drive high-quality development. Such a set of indicators could complement China's Green GDP efforts. As an

example, the Swedish government introduced 15 national well0being indicators [33]. Since 2017, these indicators have been reported in the government's proposed annual spring budget.

(5) **Develop clear definitions and technical criteria for ecolabels and minimum sustainability standards, including low- or net-zero levels for key sectors**, including housing, mobility, consumer goods, and food, especially in relation to resource use (including materials, energy, water, land) and waste (and pollution). One lesson from Germany is that there is a need for a coherent approach: instruments should be implemented in a coherent manner, acting institutions should use the same definitions and targets, and related information should be clear and accessible. The certification and standards should affect the design and manufacture of consumer products and also use and post-use phases. There are also opportunities to adopt solutions that catalyze sustainable consumption across sectors. For example, initiatives that enable sharing, reuse and repair can support goods, housing, and mobility sectors.

(6) **Set up a well-resourced and empowered coordinating agency to ensure the implementation and monitoring of sustainable consumption policy.** A successful coordinating agency should be delegated sufficient and capable staff, have financial and other resources, be given the authority to ensure implementation by related stakeholders, and have representation from not only the ministry responsible for the environment but also other government ministries (including economy, finance, housing, transport, etc.).

(7) **Create a dedicated funding stream to sustain the operations of the sustainable consumption coordinating agency.** To fund the agency, sustainable consumption programs, and implementation of the strategy, revenue could be generated from instruments such as a carbon tax (e.g., in Japan) or differentiated vehicle taxes (e.g., in Germany). As well as providing revenue, experience from Sweden shows that a congestion tax, green car premiums, and differentiated vehicle taxes have the highest effects on consumer choices among evaluated instruments designed to directly influence consumer behaviour.

(8) **Create an Ombudsman for Sustainable Consumption, Youth and Future Lifestyles.** In 2007, the Hungarian Parliament created a special Ombudsman for Future Generations, whose responsibilities include intervening where state policies would lead to overconsumption and thus endanger prospects for the society of tomorrow [34]. Other countries, such as the United Kingdom, have argued for similar foresight [35]. Such an Ombudsman would need to work together with a coordinating agency responsible for realizing the Chinese national sustainable consumption strategy as well as the elements nested within the FYP.

(9) **Establish a China Panel for Wellbeing Society and Future Sustainable Living** and task it with examining trends, anticipating future directions, and continuously advising the government and ombudsman on what actions are needed to ensure long-term sustainable ways of living for Chinese

society. This would go beyond directly addressing resources and consumption and look at critical aspects of everyday living that affect the choices and patterns of people. For example, it could make concrete recommendations for **programs on a future-proof digital society that examines how information communication technology, digitalization, and digital tools could be harnessed to reduce consumption impact** (e.g., digitalization and traceability in food systems and production value chains; contracted labour practices such as telecommuting and work-life balance; and opportunities for servicing instead of product ownership systems). It could also enhance the rights of consumers through the strengthening of consumer organizations. This is one area where China could leapfrog over the more traditional sustainable consumption policy approaches of industrialized countries and introduce innovative instruments.

(10) **Launch several highly communicative national programs that showcase the benefits of sustainable consumption to the public** while also helping to make a transition to a sustainable society. Research shows that some of the most effective programs are linked to key turning points and events in life, such as marriage, birth, or graduation from school. These are milestones in peoples' lives where they tend to re-examine lifestyles and practices and redefine aspirations as they switch from one stage to the next. Programs targeting these key turning points could be combined with **nationwide awareness-raising campaigns on the impacts of unsustainable consumption, opportunities for transitioning, and benefits of sustainable living**.

6.4 Overall Roadmap for Boosting Green Consumption in China During the 14th FYP

The 14th FYP period is a critical stage for China as it moves to high-quality growth and furthers achievement of the ambitious goal of building a beautiful China. Consumption is a major engine for economic growth and an important driving force for high-quality development. As shown by both domestic practice and international experience, vigorous promotion of green consumption has a highly significant positive effect on transforming the mode of development, changing lifestyles, and improving ecological and environmental quality. For a long time, the policy focus for China's green economic transition has been placed on green production on the supply side. In recent years, emphasis has gradually shifted to the green transition in the consumption sector. The European Union, Germany, Sweden, Japan and other pioneer economies have all attached great importance to green consumption and formulated corresponding national strategies and action plans on sustainable consumption and green consumption. Considering the goals and requirements of future high-quality development, in the 14th FYP period, China should take existing progress and practice as the foundation, fully learn from international experience, and enhance emphasis on green consumption on the demand side. It is recommended to establish a systematic national green consumption strategy and action plan, including goals and indicators, priority areas, key tasks, and policy measures, and push a green transition and the upgrading of consumption on the demand side to push forward the green transition in the production sector on the supply side, so as to advance the overall green transition of the economy and society as well as high-quality development.

6.4.1 Set Goals and Indicators for Green Consumption

In recent years, the Chinese government has formulated a number of sectoral policies on green consumption, such as the "Notice on Guiding Opinions on Promoting Green Consumption" and "Opinions on Accelerating the Establishment of Legal and Policy Framework for Green Production and Consumption" by the National Development and Reform Commission (NDRC) and other line departments, articulating the basic directions and key priority areas for China to push forward green consumption. However, in general, there has been no targeted and systematic establishment of strategic goals and specific monitoring and evaluation indicators that are clearly defined. It is necessary to deliberate and establish long-term strategic goals and specific target indicators for promoting green consumption as a national top-level design.

6.4.1.1 Identify Strategic Goals for Promoting Green Consumption

With regard to international experience, the European Union (EU), Germany, Sweden, and Japan have developed long-term strategic goals for sustainable consumption or green consumption. The European Green Deal sets the goal of climate neutrality by 2050 for major sectors with consumption included. Germany emphasizes the necessity of mainstreaming sustainable consumption. Sweden is committed to facilitating the easy shift to more sustainable consumption behaviors and lifestyles for its citizens. Japan is dedicated to building a Sound Material-Cycle Society by pushing forward resource utilization and proper waste disposal.

Taking into account the existing progress of green consumption policies and practice, and the requirements proposed by future green economic transition and high-quality development, the following long-term strategic goals for China to accelerate green consumption are recommended: adhere to the philosophy of Ecological Civilization, substantially improve the level of green consumption, and speed up the formation of green production patterns to foster novel internal impetus for eco-environmental quality improvement and high-quality development. Specific targets may include the following aspects:

(1) **A significant rise in the awareness of green consumption in the whole of society**: Society-wide visibility and awareness raising efforts and education

actions on green consumption are carried out in depth. Green consumption has become an integral part of various environment-related thematic campaigns and awareness raising activities, and is gradually incorporated into the products pertinent to education, culture, art, and information media. Consumption concepts and patterns featured as simple, moderate, green and low carbon are widely disseminated and spontaneously practiced. The good social tendencies and practices of green, low-carbon, and conservative consumption are internalized by the entire society.

(2) **A substantial increase in the supply of green consumer products and services**: Enterprises' contributions to and capacities for green product design, research and development, and manufacturing continue to grow. Substantial increases in both the types and market shares of various green products and services such as labeled energy-saving and environmentally-friendly products, environmentally labeled products, and certified green and organic products are realized. The circulation channels and sales networks for green products are increasingly improving, with a number of exemplary green product markets and sales platforms in shopping malls and supermarkets having formed. The penetration rate of green consumption on major e-commerce platforms has risen significantly, and the provision of infrastructure that underpins green consumption has noticeably been enhanced.

(3) **The preliminary formation of green and low-carbon consumption patterns**: Such irrational practices as extravagant consumption, excessive consumption, and food waste are largely corrected. Green consumption is fully implemented in all areas of household consumption, including clothing, food, homes, mobility, and daily appliances. Green transport modes like walking, cycling and public transportation are further developed and practiced. The ability of public institutions to lead and promote green consumption is further strengthened. The scope of green products for government procurement and the corresponding procurement scale continue to expand. A social atmosphere for the recycling and reuse of idle resources is basically formed. The recycling rate of domestic waste is remarkably increased, and the amount of domestic waste generated is drastically reduced.

(4) **Establishment and improvement of a green consumption policy system with both incentives and constraints**: Further improvements are witnessed in the standard system and green labeling and certification system for green products and services. Laws and regulations pertinent to green consumption are gradually perfected to form basically an economic policy framework for green consumption that can encourage and discourage consumer behaviors.

6.4.1.2 Establish Specific Indicators for Monitoring and Assessing Green Consumption

In 2016, the NDRC and three other line ministries issued the "Green Development Index System" to put forward evaluation indicators for gauging the greenness of life-related areas. Major indicators include: per capita energy consumption reduction rate for public institutions, green product market share (high-efficiency and energy-saving product market share), the growth in the number of new energy vehicles, green mobility (public transport passenger volume per 10,000 urban population), the proportion of green buildings in new buildings in urban areas, the green space rate in urban built-up areas, the penetration rate of tap water in rural areas, and the penetration rate of sanitary toilets in rural areas. These indicators do not thoroughly cover all areas of consumption, and thus have limited capacity to truly and scientifically measure and reflect the progress and status of green consumption at the national or regional levels.

There are also practices of sustainable consumption at the international level. The UN Sustainable Development Goal 12 is designed to ensure sustainable consumption and production patterns, including the following targets: by 2030, halve per capita global food waste and substantially reduce waste generation. The German sustainable consumption indicators mainly include: per capita carbon dioxide emissions, per capita carbon dioxide emission related to food consumption, and food waste generation; the potential of residential heating for carbon dioxide emission reduction; per capita paper consumption and the proportion of recycled paper in office paper; carbon dioxide emissions from the transportation sector, and especially the area of aviation. Swedish sustainable consumption indicators mainly include: carbon emissions of food consumption, food labels; energy labels for residential building materials, renewable electricity, and improvement of public transportation infrastructure. The indicators related to green consumption in the European Green Deal mainly cover the following aspects: greenhouse gases emission reduction targets, improved energy efficiency, energy consumption of buildings, greenhouse gases emissions from transportation, zero- and low-emission vehicle ownership, and public recharging and refueling stations.

Building on domestic progress and with reference to international experience, China's green consumption index and indicator system should be expected to reflect the overall status and level of green consumption, and at the same time implement and advance green consumption-related work. The indicator system should include both result-oriented indicators and process-oriented indicators.

Therefore, the indicator system needs to cover the entire chain and process of green consumption, including: green product supply, green consumption processes and patterns, and the endpoint of consumption (Fig. 6.7). Green product supply affects the production process and subsequent resource and environmental impacts caused by consumption, mainly through the product supply structure, including the proportion of green products. Green consumption refers to the process whereby a consumer consumes a green product, such as in the areas of clothing, food, housing, household appliances, and mobility. Consumption patterns relate to the intensity of resource and environment consumption. The endpoint of green consumption is mainly reflected in the domestic waste generated by consumption and corresponding waste disposal, which represents the impact of the consumption endpoint on resources and the environment.

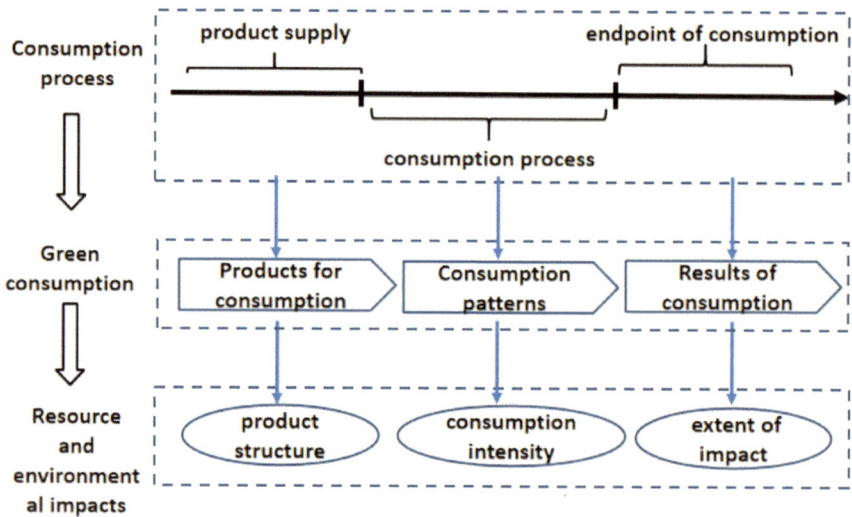

Fig. 6.7 Green consumption process

Based on the above processes, as well as principles of scientificity, comprehensiveness, policy relevance, data availability; with an eye to the future; and taking into consideration the overall strategic goals and key areas for furthering green consumption, the SPS team selected 13 specific indicators. These include an overall index and domain-specific indicators. Combined they establish a framework for China's green consumption indicator system (Table 6.3). Further efforts can be made to improve the selection and establishment of indicators on the basis of this indicator system framework. At the same time, on the basis of this current framework, it is essential to set the corresponding target values for certain periods of time in the future, such as the 14th FYP period. The setting of targets should follow the principle that the green consumption promotion "can only get better, not worse" and ensure that positive indictor trends continue and the negative indicator trends continue to decline.

6.4.2 Key Areas and Major Tasks for Pushing Forward Green Consumption

The concept of green consumption widely recognized internationally includes the following aspects: First, it is important to mobilize consumers to choose green products, meaning they are not polluting and are beneficial to public health when consumed. Second, emphasis should be placed on the disposal of waste generated during the consumption process. Third, efforts are needed in guiding consumers to transform their consumption philosophy and switch to the practice of sustainable consumption which advocates the respect to nature and the pursuit of health, and a balanced emphasis on environmental protection, resource and energy conservation in the process of pursuing a life of comfort. The China Consumers Association outlined three dimensions to define green consumption in 2001. The first one is the content of consumption (i.e. consumers choose green products that are non-polluting and beneficial to public health). The second refers to the consumption process (i.e. emphasis on minimizing environmental pollution and close attention to waste disposal). The third dimension is about the idea, understanding, or philosophy of consumption (i.e. a balanced emphasis on environmental protection and resource and energy conservation in the process of pursuing a life of comfort, so as to achieve sustainable consumption). These basic understandings provide essential rules to follow in determining key areas and main tasks for realizing green consumption.

6.4.2.1 International Experience and National Practice in the Key Areas for Green Consumption

From an international perspective, the focus areas for sustainable consumption or green consumption in economies such as the EU, Germany, and Sweden include food, housing, mobility, daily necessities, and public procurement. The EU's focus is put on sectors of food, mobility, housing, consumer goods, leisure and tourism, as well as cross-sectoral areas such as energy, water resources, and waste. Although the national priorities of EU countries may differ, these main areas often are at the top of the policy agenda and ultimately form the top priorities of sustainable consumption policies. In Sweden, greenhouse gas emissions from consumption have always been an important criterion for evaluating the impact of consumption. Strong efforts in Sweden have been invested in the establishment of consumption indicators and collection of greenhouse gas emission data covering products' life cycles. In terms of household consumption, the areas that contribute the most greenhouse gases to total emissions and thus are of particular importance are food, transport and housing. In Sweden, emissions from private consumption account for 60% of total emissions, while those from public consumption make up 11%. In private consumption, contributions of food, transport, and housing are 15%, 20%, and 10%, respectively. The remaining 29% of total emissions is attributed to investment. Thus, food, transport and housing are the priority fields for sustainable consumption policies. Germany focuses on the six major consumption fields of mobility, food, home, workplace and office, clothing, as well as leisure and tourism. Transport accounts for 26% of total greenhouse gas emissions, food contributes 13%, and consumption at home makes up 36%.

From the perspective of relevant domestic policies in China, as set out in the "Guiding Opinions on Promoting Green Consumption" and the "Opinions on Accelerating the Establishment of Legal and Policy Framework for Green Production and Consumption" issued by NDRC and nine other departments, the key areas for green consumption are essentially comprised of fields such as old clothes recycling, green home, green mobility, green office, green procurement, and green product supply. The NDRC's "Overall Plan for Creating a Green Life Actions" proposes to launch coordinated actions in seven key areas, including energy-saving public institutions, green family, green school, green community, green mobility, green shopping mall, and green building.

From the perspective of China's concrete practices, pilots and efforts are conducted in a number of key areas. For example, in the building and construction field, there are actions to promote the application of green building standards to new buildings and green renovation of existing old communities. In the field of automobile and transportation, vigorous development and application of new energy vehicles has occurred. In the power sector, renewable energy and green power consumption has been promoted. In terms of new business forms, people are piloting the establishment of digital platforms for low-carbon lifestyles. In the sector of logistics, practices include green packaging, green transportation and distribution, and green recycling. In the field of food, efforts are focusing on the establishment of a sustainable food supply chain and consumption system. In financial sector, there are continuous innovations and provisions of financial products for green consumption. All of these practices provide a policy and practical foundation for identifying future key areas for green consumption in China.

6.4.2.2 Identification of Key Areas for Green Consumption

Based on international experience and China's domestic foundation, the SPS team explored three different dimensions to analyze and identify the key areas that China's green consumption should focus on, including expenditures and growth of various consumption sectors, resource and environmental impacts of various consumption sectors, and the pulling effect on economic growth of various consumption sectors.

(1) Expenditure and Growth of Various Consumption Sectors

As found in China's statistical system and practice, household consumption basically falls into eight categories, including food, tobacco, and liquor; clothing; residence; articles for daily use and services; transport and communications; education, culture and recreation; health care; and other articles and services. In 2018, food expenditure in China fell to 28.4%; and spending in residences made up 23.5%. Expenditure on transport and communications increased rapidly before 2010, gradually stabilizing thereafter, and reached 13.5% in 2018. 2018 expenditures on articles for household use and services accounted for 6.2%. The respective share of health care, clothing, articles for daily use and services, as well as other articles and services remained stable in the range of 6–8%.

As shown by the estimation results from scenario models, household consumption in China will speed up its shift from a subsistence-based model to a well-off one, and the consumption structure will change dynamically. The proportion of expenditure on food and clothing will show a downward trend; both the share of housing expenditure and that of transport and communications will decline slightly; and the proportion of spending in fields of health care, household appliances and items, education, culture and recreation, and others will continue to rise. The forecasted expenditure in eight categories of consumption in the future is displayed in the following tables. In general, in the next 15 years, there will be no substantive changes in the structure of household consumption; and food, residences, and mobility will continue to dominate household consumption in China (Tables 6.4 and 6.5).

Table 6.4 Expenditure in eight categories of consumption (Unit: 1,000 million yuan)

Category	2015	2020	2025	2035
Food	79,072	109,175	146,615	269,298
Clothing	21,151	27,309	33,336	67,325
Residences	58,760	87,734	129,462	269,298
Household appliances and items	16,244	34,824	68,615	134,649
Transport and communications	35,999	61,143	102,275	161,579
Education, culture and recreation	29,627	48,229	77,677	188,509
Health care	17,946	30,738	51,785	134,649
Other	7181	17,716	37,544	121,184

Table 6.5 Share of expenditure in eight categories of consumption

	2015	2020	2025	2035
Food (%)	30	26	23	20
Clothing (%)	8	7	5	5
Residences (%)	22	21	20	20
Household appliances and items (%)	6	8	11	10
Transport and communications (%)	14	15	16	12
Education, culture and recreation (%)	11	12	12	14
Health care (%)	7	7	8	10
Other (%)	3	4	6	9

(2) **Resource and Environmental Impacts of Various Consumption Sectors**

Two types of energy and environmental effects of consumption can be identified. There are the direct energy and environmental impacts caused by the energy consumption and environmental pollutants emitted by industrial activities. This direct impact is limited to energy consumption and environmental pollutants directly generated by industries. In addition, there are the estimated total energy consumption and environmental impacts caused by consumption as derived from input–output models. This component of consumption's impacts include not only the energy consumption and environmental pollutants emitted directly by household consumption, but also the energy consumption and environmental pollutants emitted during the production of various products to be consumed. The complete impacts are estimated as described below:

In 2015, the total energy consumption per unit of expenditure in categories of food, tobacco and liquor; clothing; residences; articles for daily use and services; transport and communications; education, culture and recreation; health care; and other articles and services was 122.44 kg/10,000 yuan, 170.99 kg/10,000 yuan, 125.19 kg/10,000 yuan, 166.80 kg/10,000 yuan, 220.61 kg/10,000 yuan, 138.73 kg/10,000 yuan, 201.27 kg/10,000 yuan, and 147.71 kg/10,000 yuan, respectively; per unit expenditures for energy consumption were highest in the categories of transport, communications and health care (Fig. 6.8).

The pollutants emitted per unit of expenditure in the eight major categories of consumption in 2015 are shown in Table 6.6. As revealed by the table, the category food, tobacco and liquor had the largest chemical oxygen demand (COD) per unit of expenditure; the category of transport and communications had the lowest COD per unit of expenditure; and there was little difference in COD among the remaining

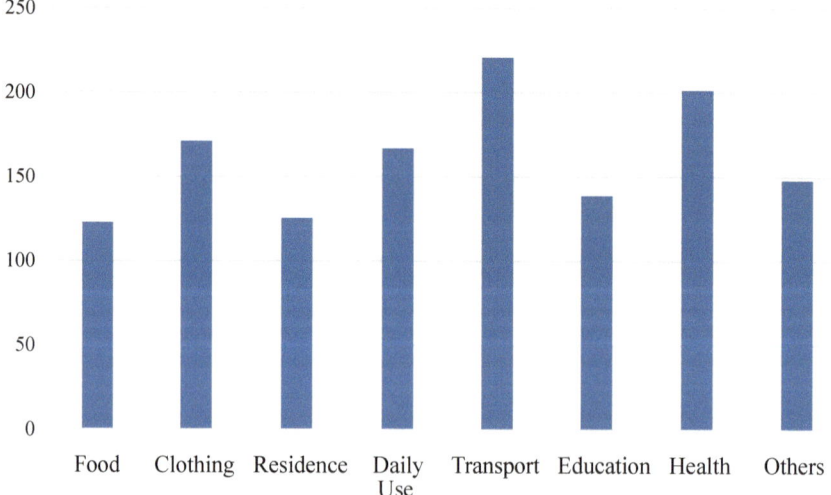

Fig. 6.8 Total energy consumption per unit of expenditure in 8 categories of consumption

Table 6.6 Pollutant emissions per unit of expenditure in eight categories of consumption (kg/10,000 yuan)

	COD	Ammonia nitrogen	SO_2	NO_X
Food, tobacco and liquor	3.15	0.37	0.50	0.26
Clothing	1.24	0.14	0.67	0.36
Residence	1.11	0.10	2.66	2.32
Articles for daily use and services	1.01	0.10	0.79	0.48
Transport and communications	0.59	0.06	0.79	0.58
Education, culture and recreation	1.22	0.11	0.73	0.38
Health care	1.11	0.11	0.87	0.48
Other articles and services	1.31	0.13	0.72	0.40

categories. Similarly, the category food, tobacco and liquor had the largest level of ammonia nitrogen emissions; while the category transport and communications had the lowest level of ammonia nitrogen emissions. Sulfur dioxide and nitrogen oxide emissions per unit of expenditure for residences were the largest, with little difference in the emission amounts of the remaining categories, which is primarily attributed to the residential consumption of coal.

(3) **Pulling Effect on Economic Growth of Various Consumption Sectors**

Based on the 2017 input–output table covering 149 sectors, the SPS team analyzed the economic pull effect of green consumption in key areas. As a first step, based on survey results, the SPS team separated green products and services by industries: agricultural products, processed foods, selected household appliances, household goods, automobile manufacturing, construction and decoration, wholesale and retail trade in the input–output table; it then used this categorization in measuring the economic pull effect of green consumption in these various fields. The production cost structure of green products/services in various categories is different from that of traditional products/services. Of the eight categories of consumption, the products and services included in the table are the categories of food, residence, articles for daily use and services, as well as transport and communications.

The calculation results show that one unit of green consumer goods in the field of food, tobacco and liquor has an economic pulling coefficient of 2.5; one unit of green building in the residence category has an economic pulling coefficient of 3, the pulling coefficient for household items and services is 3.8, ranking top; and electric vehicles in the field of transport and communications enjoy a pulling coefficient of 2.7. Built on the proportion of various types of consumption in total household consumption, and the economic pulling effect of green consumption in different fields, preliminary estimates show that under the current household consumption structure, green consumption in categories of food and residence have the strongest comprehensive pulling effect on the economy, followed by the field of transport and communications, and with the category of articles for daily use and services ranking third (Table 6.7).

Table 6.7 Economy-pulling effect of green consumption growth in 8 categories consumption

	Food, tobacco and liquor	Clothing	Residence	Articles for daily use and services	Transport and communications	Education, culture and recreation	Health care	Others
Ratio in total household consumption (%)	28	6	23	6	13	11	8	2
Economic pulling coefficient for per unit of green consumption	2.5	–	3.0	3.8	2.7	–	–	–
Comprehensive pulling effect	0.7	–	0.7	0.2	0.4	–	–	–

6.4.2.3   Main Tasks for Advancing Green Consumption

According to the above calculation and analytical results, China's green consumption during the 14th FYP period should focus on key areas such as food, residence, mobility, household appliances, clothing, and tourism, Main tasks should be identified to facilitate transformation of consumption in a green, low-carbon and conservation-oriented direction, and speed up the formation of green consumption, with the purpose of effectively boosting ecological environment quality and high-quality development.

(1)   **Food: Promote Green Diet**

The EU aims at a Farm to Fork Strategy to reduce pollution caused by eutrophication. Germany calls for a green diet and lifestyle to reduce food waste and encourage the use of low-packaging or zero-packaging products. Sweden is concerned about carbon emissions and labeling for food consumption. The main tasks for China to promote green diet may include the following aspects. First, resolutely fight against food waste; initiate anti-food waste actions covering the entire chain of storage, transportation, retail, and dining-table; and advocate scientific and enlightened food consumption practices. Second, push canteens and dining service providers for government agencies, state-owned enterprises, and public institutions to reduce the wasting of food and generation of food waste; encourage consumers to order based on needs and take leftover food home; and encourage catering enterprises to set reasonable charging standards for food waste in buffet-style meals. Third, fully implement green take-out plans for the catering industry; and support catering enterprises, food retail enterprises, and food takeout industries to use simplified packaging and recyclable packaging to reduce excessive packaging and the use of plastic boxes. Fourth, unify and strengthen green and organic food certification systems and standards; and expand the effective supply of green food.

(2)   **Residence: Promote Green Buildings**

The EU supports renovation of public and private buildings to improve the energy efficiency of household heating systems. Germany supports consumers in purchasing energy-efficient household appliances, home supplies, water, electricity, and heating services, etc., and endeavors to improve energy labeling schemes. Sweden promotes the energy labeling schemes for housing materials and the use of renewable electricity. China can take the following major steps to promote green building. First, guide capable cities and localities to fully apply green building standards to their new buildings; expand the scope for mandatory use of green building; and promote the full application of green building standards in public construction projects that are sponsored by government funds. Second, push forward the application of green building standards in the renovation of old communities. Third, implement the action plan for production and application of green building materials; and promote the use of green building materials and environmentally-friendly decoration materials such as energy-saving doors and windows, and products from recycled construction waste. Fourth, promote the design, construction and operation of green buildings in a comprehensive way, including high-quality planning and construction of green municipal infrastructure systems, such as water, electricity, and gas utilities, and waste disposal; carry out the construction and renovation of energy-efficient residences; and promote methods and technologies for green rural housing construction. Fifth, strengthen environmental labeling, especially energy efficiency labeling and certification for green household appliances; and enhance the effective supply of energy-efficient green household products.

(3)   **Mobility: Promote Green Mobility**

The EU is accelerating the shift to sustainable and smart mobility by using connected multimodal mobility to improve transportation efficiency, and setting the price of transport to reflect its impact on the environment and health. Germany is using car labels to provide information about vehicle efficiency, and supporting efforts to make public transport more attractive, such as upgrading local public transport networks. Sweden continuously improves its public transport infrastructure. China should initiate the following efforts to promote green mobility. First, increase the proportion of public transportation systems in urban planning and construction, create smart cities, and improve the efficiency of public transportation systems. Encourage the use of low-carbon transport modes such as walking, cycling, and public transportation. Strengthen the construction of urban slow-moving systems such as bicycle lanes and pedestrian walkways to improve conditions for cycling and walking. Second, strengthen efforts in promoting new energy automobiles; accelerate the construction of electric vehicle charging infrastructure; and advocate modes of shared transport such as car sharing and car-pooling. Third, encourage the use of new and clean energy vehicles in expanding and upgrading fleets for public transportation, sanitation, taxis, commuting, urban express mail service, and urban logistics. Fourth, reinforce the promotion and use of new energy vehicles in such areas as national Ecological Civilization pilot zones and key areas for air pollution prevention and control.

(4)   **Household items and Service: Promote Green Household Appliances**

The EU encourages consumers to choose reusable, durable and repairable products; while Japan boosts the recycling and reuse of household goods. China's major efforts in the area should consist of the following steps. First, encourage consumers to choose green products such as energy-efficient household appliances, high-efficiency lighting products, water-saving utensils, and green building materials. Promote the use of new energy vehicles. Call for the return of cloth bags and shopping baskets; promote the repeated use of environmentally friendly shopping bags; and reduce the use of disposable daily necessities. Second, encourage businesses to provide reusable, durable, and maintainable products and enable consumers to choose such products. Encourage textile, construction, electronics and other industries to carry out design of recyclable products that can reduce material use or promote reuse. Advocate the long-term use of furniture, electronics, and electrical appliances. Third, support the development of sharing economy and encourage the effective recycling of personal idle resources. Orderly develop on-line reservation for car-pooling, private vehicle rental, bed and breakfast rental, and exchange of old objects, etc. Fourth, improve the recycling system of social renewable resources. Encourage the provision of development, upgrade and maintenance services for information electronic device and products. Fifth, boost the greening, reducing and recycling practice in express packaging. Sixth, improve the efficiency in the use of office equipment, assets and supplies; strictly implement the government's priority procurement and mandatory procurement system for energy-saving and environmentally-friendly products; and expand the scope and scale of government green procurement; and improve the evaluation standards of energy and material-saving public institutions.

(5)   Clothing: Promote Green Clothing

Germany supports and facilitates resource-efficient development of the apparel and textile industry to reduce the potential risks and impacts of textiles on health and the environment. China should launch the following major tasks to promote green clothing. First, carry out "zero-discard" activities and "clothes reborn" activities for old and used clothes. Facilitate the development and improvement of waste and used textile recycling systems for communities; regulate the waste and used textile recycling, sorting, and hierarchical utilization mechanisms, so as to orderly promote the reuse of second-hand clothes. Second, boycott fur and leather products made of rare animals to conserve biodiversity. Support and facilitate textile and apparel companies to build green supply chains. Use renewable raw materials and reduce the potential health and environmental risks of new functional textiles. Third, improve the recycling and reuse of waste and old textiles in earth structures, building materials, automobiles, and home decoration sectors. Fourth, enhance the efforts in environmental labeling and certification of textiles and apparels. Substantially increase the effective supply of green textiles and apparels.

Tourism: Promote green tourism

Green tourism is on the rise worldwide. Main steps for China to take to promote green tourism could include: First, develop and release green tourism and consumption conventions and guidelines. Second, encourage tourist hotels, restaurants, and management agencies of scenic areas to introduce incentive measures for green tourism. Third, formulate and/or revise appraisal rules and standards for green services including *inter alia* green markets, green hotels, green restaurants, and green tourism. Fourth, star-rated hotels and chain hotels should gradually reduce the free provision of disposable toiletries and supplies and pilot demand-oriented provision. Fifth, publish such green tourism information on relevant tourism promotion websites and platforms and encourage consumers to bring their own toiletries. Six, endeavor to integrate biodiversity conservation into tourism-related standards and certification programs.

Employ Building a Green Life actions as a channel to push forward the implementation of tasks for advancing green consumption

Link green consumerism to the implementation of the "Overall Plan for Creating Green Life Actions" issued by NDRC. Effectively incorporate tasks for advancing green consumption into actions for developing energy-saving public institutions, green families, green schools, green communities, green mobility, green shopping malls, and green buildings. Expect social systems, such as the media, to truly and effectively facilitate information provision related to the formation of a simple, moderate, green and low-carbon, civilized and healthy life philosophy and lifestyle.

6.4.2.4 Overall Policy Framework for Promoting Green Consumption

At present, there are a number of green consumption related policies in China, but they are quite fragmented and have not been integrated into a systematic and effective policy framework. Specific observations can be made: (1) There is a lack of systematic planning and top-level design. Most green consumption policies are conceptual, guiding and voluntary in nature, with incomplete categories, limited policy impacts and enforcement efficacy, and insufficient operability. (2) In relation to green consumption policies, the most emphasis has been placed on resource and energy conservation; less attention has been given to eco-environmental protection. There are insufficient economic incentives in these policies, leading to limited regulatory effectiveness. (3) Government functions and responsibilities related to green consumption are scattered in different agencies. The role of environmental authorities needs to be strengthened. The fragmentation of policies and management is quite prominent. If no systematic design and integration of related policies occurs, the environmental and economic effects of green consumption will be greatly weakened. The expectation of future high-quality development and green economic transformation have put forward demands and higher requirements for the development of green consumption policies. It is necessary to constantly improve, strengthen and innovate policies in this area during the 14th FYP period to form a more holistic and integrated policy framework.

Consumption is an economic behavior that must respect economic laws. Consumption is a social behavior that involves each and every member of society. Consumption is also a cultural behavior, influenced by factors such as values and customs. The design of policies needs overall consideration of government interventions, the economic, social, and cultural attributes of consumption, as well as factors including incentive mechanisms, supervision and management, publicity and education, and so on. The overall policy framework to promote green consumption should consider the internal linkage and transmission mechanism between production and consumption, cover measures for both the supply and demand side, and mobilize multiple stakeholders such as government, enterprises, consumers, and social organizations to take joint actions (Fig. 6.9).

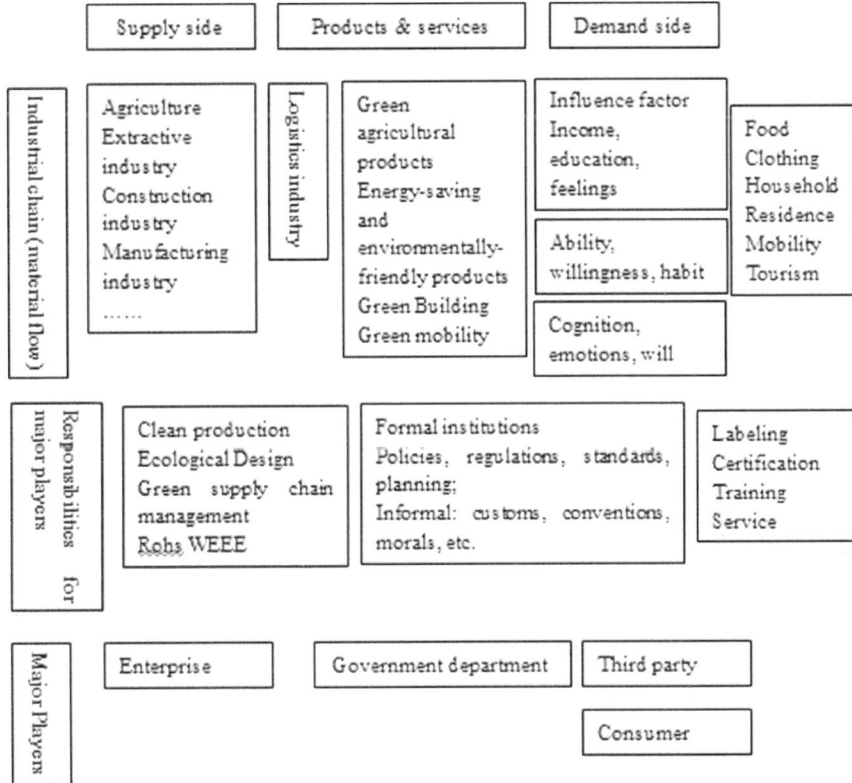

Fig. 6.9 Schematic diagram of green production and consumption governance structure

Design of policy measures for the supply side

Green production provides green consumers with quality consumer goods and promotes healthy consumption patterns. Policy objectives on the supply side should enhance the diversity of supply of green products and services, so as to address the problem that there are no green options for consumers to choose from; ensure the quality of products and services; and guarantee the regulated operation of the market. On the supply side of products and services, both government (as policy maker, promoter, and supervisor) and enterprises (large enterprises, medium- and small-sized enterprises, individually owned enterprise, small and micro enterprises, and farmers in production, logistics, and services industries) can intervene in the market. The way these actors can influence the supply of green consumption in the market are discussed below.

Governmental interventions mainly include: formulation of regulations and standards to form relevant institutional arrangements to promote green consumption; use industrial policies, fiscal and taxation policies, price policies and other policy measures to encourage or mobilize consumers' willingness and support for green consumption; develop and implement standards for technologies, products, and quality, in particular the Leader/Top Runner standards system can spur the continuous improvement of products and services; ensure an open, fair and just market, and regulate market operations through inspection, supervision and management.

The responsibilities and functions of enterprises mainly include: contribute to the reduction in prices of green products through technological innovation in order to expand the scale of green consumption; assume eco-environmental protection responsibility and corporate social responsibility (CSR), and carry out product and service life cycle assessments (LCA), green supply chain management, and clean production to reduce the negative environmental impacts in the life cycle of consumer products; emphasize dematerialization in the production of energy-saving, environmentally friendly and low-carbon products; develop smart logistics, and reduce the logistics costs of green consumer goods through system optimization and management of green consumer products (quantity and quality), brands, storage, transportation routes, and transport modes, in this way contribution to the upgrading of consumption.

The policy measures for the supply side of green consumption can be summarized as found in the following diagram (Fig. 6.10).

Design of policy measures for the demand side

On the demand side of green consumption, governments and enterprises are the main consumers. They can promote the consumption of bulk green products through green procurement, and can thus serve as a model. Consumers can be a third party, a mass organization, a household, or an individual. The policy goals on the demand side of green consumption are to use green products and services to the maximum extent; reduce the waste of consumer products and improve utilization efficiency during the consumption process; at the endpoint of consumption, encourage consumers to participate in the construction and action of recycling systems for idle products and waste products, and reduce littering to relieve environmental pressures.

Measures on the demand side of green consumption should be centered on consumers; follow the problem-oriented principle, highlight key points, pursue systematic coordination, take into consideration applicability and feasibility, and follow a step-by-step approach; target efficient resource utilization, environmental quality improvements, and climate friendliness; establish and improve relevant laws, standards and policies to promote resource use reduction, clean production, resource recycling, and end-of-life governance; and form a green consumption pattern in the whole of society. Such measures can be summarized as seen in the following diagram (Fig. 6.11).

6.4.2.5   Key Policy Measures to Promote Green Consumption

Under the overall policy framework for advancing green consumption, China may consider further strengthening some key policy measures during the 14th FYP period.

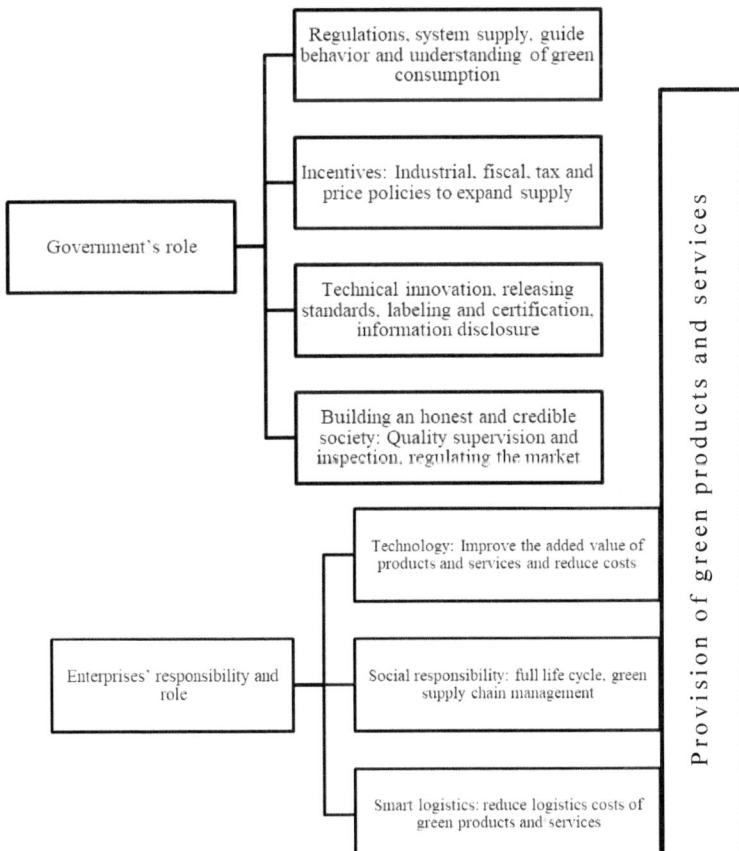

Fig. 6.10 Framework of policy measures for the supply side of green consumption

(1)   Improve Long-Term Incentive Policies for Promoting Green Consumption

Amend relevant laws and regulations. First, consider how to lead the public to appreciate the concept and follow the behavior that "waste no resources no matter how rich you are" through the formulation and implementation of laws and regulations. Second, revise the *Government Procurement Law* to promote mandatory green product procurement. Enterprises and public institutions that use public finances should purchase and use office supplies in accordance with government green procurement regulations. Encourage other social organizations to practice green procurement.

Improve market cultivation and economic incentive policies for green consumption. Establish economic incentives and market-driven systems through prices, taxation, credits, supervision, and market credits to expand the supply of green and ecological products and enrich the green options available for household consumption. Such measures as levying taxes on products with high energy consumption

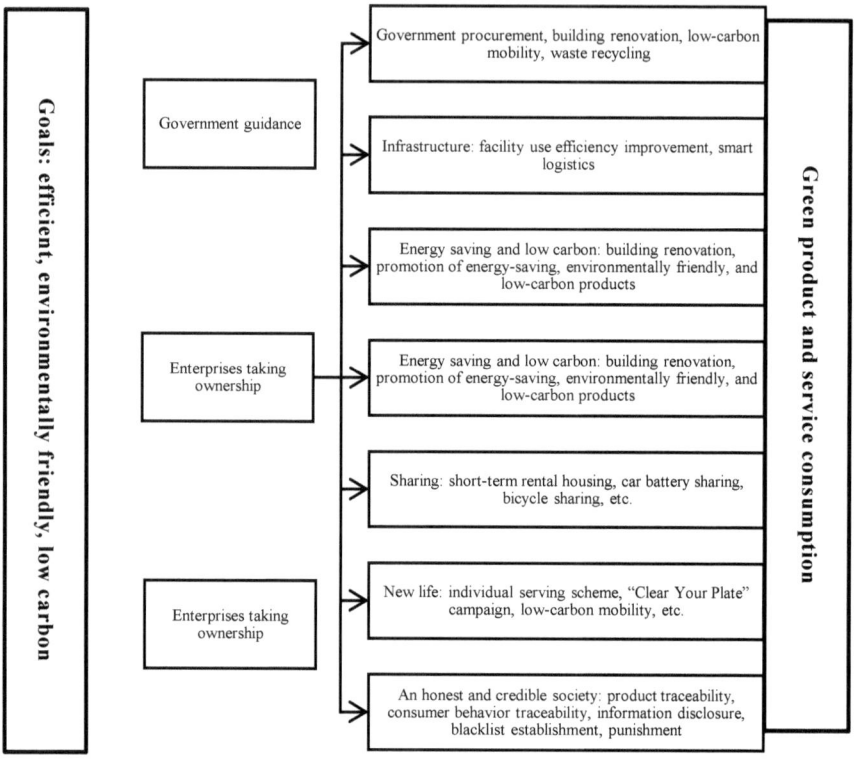

Fig. 6.11 Framework of policy measures for the demand side of green consumption

or that are highly polluting and resource intensive can be used to limit developments in corresponding markets. Encourage consumers to purchase alternative products rather than conventional fuel vehicles –by increasing the cost of using conventional fuel vehicles or by giving financial subsidies to electric vehicles, in this way easing the pressure on China's energy supply, reducing environmental pollution, and producing a direct "green effect". Consider the establishment of a waste tax to change the public's littering behavior and avoid corresponding environmental pollution. Research the point redemption scheme for green consumption, and carry out a step-by-step green consumption point exchange pilot in the financial field, so as to lay a foundation for the future implementation of a point redemption scheme and point exchange at a larger scale.

(2) **Reinforce efforts to Further Promote the Development of a Circular Economy**

Push forward mandatory implementation of the extended producer responsibility system in the electronics, home appliances, and the express delivery and logistics industries, to lead the reduction in the generation of the huge volume of waste associated with consumption in these areas. At the same time, extend the resource and environmental responsibility of producers to their products from the production process through the entire life cycle to cover product design, production process control, smart logistics, recycling, and waste disposal. Adopt green consumer goods and services measures including ecological design, clean production, and green supply chain management to promote aggregated resource consumption and minimized emissions of pollutants and greenhouse gases; make green consumer goods and services affordable for household consumption, and in this way foster green consumption habits.

(3)   **Speed up The Establishment of a Finance System for Green Consumption**

Develop financial standards for green consumption and improve financial incentive mechanisms to advance green consumption. Establish financial standards and statistical systems for green consumption, in which financial management and regulatory agencies incorporate personal green consumption credits into the scope of green credit and green finance, so as to guide commercial banks to innovate and promote green consumer crediting and expand market scale. Set up a system and mechanism for developing green consumer crediting and continue to create innovative green consumer credit products for the targeted market, with the intent of boosting the penetration of green consumer credit into all areas of society, to make green consumer credit more attractive and practical, and to improve people's capacity to engage in green consumption. Encourage and guide financial institutions to provide low-interest or interest-free green consumer loans for the purchase of new energy vehicles, energy-saving and environmentally friendly household appliances, green building materials, and other products that have been accredited by energy-efficiency and environmental certification systems.

(4)   **Improve the Governance System for Green Consumption That Is Built, Governed and Shared by All**

Clarify the role of relevant government departments in advancing green consumption; develop a corresponding list of green consumption responsibilities for government departments; and establish a cross-sectoral mechanism for coordinated action, so as to form a green consumption governance structure that is built, governed and shared by all. For example, the eco-environment authorities are responsible for developing green standards and supervision; the development and reform departments perform the role of planning and macro-level regulating; the industry and information departments takes charge of facilitating the production of energy-saving, environmentally friendly and low-carbon products and corresponding services; the commerce departments are responsible for logistics and market construction; and commodity quality supervision departments are to regulate the market. Strengthen the role of consumer associations in promoting green consumption, encourage enterprises to assume more environmental and social responsibilities, and set up a carrot and stick approach to encourage green consumption by the public. All of these efforts will help form a governance structure for green consumption and achieve the modernization of governance capacities.

(5) **Propose the Launch of a Nationwide Green Consumption and New Lifestyle Campaign**

The government should give effective play to the leading role of positive stars and social celebrities in demonstrating green lifestyles, and enable green consumption to become a social fashion. Integrate the concept of green consumption into related education and training programs for various institutions at various levels, including families, schools, governments, and businesses. Incorporate green consumption initiatives into thematic publicity and education events such as the national energy-saving week, science promotion week, national low-carbon day, and environment day. Establish a green consumption incentive and disciplinary system for the general public; strengthen green consumption information disclosure and public participation; advocate a simple, moderate, green and low-carbon production system and lifestyle; oppose extravagant consumption, excessive consumption (waste), and irrational consumption; and raise green consumption awareness in the whole of society.

(6) **Strengthen the Infrastructure and Capacity Building for Green Consumption**

Liberalize market access for green and eco-labeled products and services, encourage all types of capital to invest in green industries, and use the "Internet plus" initiative to promote green consumption. Strengthen the standard system and green certification systems for green products and services, step up implementation of the Leader/Top Runner system for energy efficiency, environmental friendliness, and water efficiency, and push forward environmental product labeling schemes. Strengthen the certification of green and ecological products, and improve the standard system and green certification systems for green and ecological products and services.

 Establish a sound statistical indicator system for green consumption and enhance monitoring, data collection, statistics and evaluation reporting on green consumption Set up a nationally unified information platform for green consumption; leverage big data resources, publish information on green products and services, improve the transparency in green product production and consumption, and encourage stakeholders to recognize the credibility of green product and service certification / evaluation results. Strengthen capacity building and training on green consumption for governments, social organizations, enterprises and the public; build multi-stakeholder partnership networks; and push forward the participation of multiple stakeholders. Carry out environmental impact and social risk assessment and integrate green consumption into global purchasing chains and value chains in international infrastructure development projects such as the Belt and Road Initiative and other South-South cooperative programs to further the greening of infrastructure development processes.

6.5 Conclusions and Policy Recommendations

After two years of research, the SPS team has put forward "8 general + 8 specific" policy recommendations and outlined the key research findings that underpin these recommendations.

I. **Fully considering the plans to advance high-quality development, Ecological Civilization, and post-COVID-19 pandemic green recovery, the Chinese government should give green consumption and lifestyle a prominent strategic position on its agenda and comprehensively promote relevant practices through the implementation of the 14thFYP.** There are at least six reasons to support this recommendation.

First, acknowledging the scale of the country's consumption, China plans to comprehensively transform consumption through quality upgrading. This will be a window of opportunity for fostering a new green consumption and lifestyle model. Lessons from other industrialized countries suggest that should this window be missed; it will be difficult to reverse a situation where a pattern of mass consumption and mass waste disposal forms.

Second, quantitative assessments have found that, since 2012, the decline in the resource and environmental performance of the consumption sector in China has partially offset improvements made in the resource and environmental performance of the production sector, thereby slowing down the speed of the country's green economic transition. With expanding consumption, growing pressure on natural resources and the environment can be expected to intensify. Precaution is needed to counter this tendency.

Third, in recent years, final consumption has maintained its place as the top driving force fuelling China's economic growth. As CGE models show, green consumption can be expected to have long-term positive effects on economic growth and employment. Achieving green consumption in the clothing, food, residential, and mobility sectors would act as a new driving force, providing momentum for speedy growth in related industries. Greening of food manufacturing, a transition to electric automobiles, and the greening of wholesale and retail practices will have the most significant pulling effects on green development in corresponding industries. This conclusion suggests crucial elements for a green recovery from the COVID-19 pandemic.

Fourth, a good social foundation is ready for China to fully boost the formation of green consumption. As shown by the *Survey Report on the Status of the General Public's Green Consumption in China* (2019), the concept of green consumption is becoming increasingly popular in the public's understanding of daily consumption. Around 83.34% of respondents expressed support for green consumerism, and 46.75% of these were "highly supportive." According to the *Report on Green Consumption Trends and Developments 2019*, the year-on-year increase in the sales volume of green consumption related products on JD.com was 18% higher than the overall sales increase of all products on the platform in 2019. In the face of the COVID-19 pandemic, an unprecedentedly large share of the public is reflecting in a profound way on the human-nature relationship; this will further enhance people's willingness to consume in a greener way.

Fifth, consumption is an activity all citizens and groups engage in. Green consumption is a concrete action that can be taken by everyone in order to contribute toward building an Ecological Civilization. Promoting the formation of green consumption and lifestyles is doubtless an effective approach towards a governance system to be built, governed, and shared by all.

Sixth, internationally the European Union (EU) and countries such as Germany and Sweden have incorporated sustainable consumption into their overall development strategies as a new engine for economic growth and improvement of people's well-being.

Therefore, China should seize the window of opportunity to upgrade and transform consumption patterns, incorporating concrete green consumption and lifestyle actions and green development practices into the development of 14th FYP and thereby translating the central government's strong political will to accelerate the formation of Ecological Civilization over the next five years.

II.   **Establish goals and indicators for China to promote green consumption during the 14th FYP**

At present, China has not yet established an explicit system of medium- and long-term goals and definite monitoring and evaluation indicators that are specific to green consumption. Taking into account existing progress in the development of green consumption policies and practices, as well as the requirements for achieving high-quality development and the building of an Ecological Civilization, it is recommended that the overall goals for China to accelerate green consumption during the 14th FYP period should adhere to the philosophy of Ecological Civilization, substantially improve the level of green consumption, and speed up the formation of green production patterns to foster a novel internal impetus for eco-environmental quality improvement and high-quality development. More specific targets may include a significant increase in the awareness of green consumption throughout society, a substantial increase in the supply of green consumer products, the preliminary formation of a green and low-carbon consumption pattern and lifestyle, and a fundamental green consumption policy system that makes use of both incentives and constraints.

Drawing on the UN 2030 Sustainable Development Goals on Consumption, and with reference to the experiences and learnings in such countries as Germany and Sweden, China should establish a green consumption indicator system, making use of both qualitative and quantitative methods, in order to monitor and evaluate the overall status and level of green consumption. Such an indicator system can also be used to follow up corresponding target values during the 14th FYP period. Green consumption indicators can be divided into an overall index and domain-specific indicators. The overall index could be composed of indicators for changes in per capita CO_2 emissions from daily life, daily domestic water consumption per capita, the output value of major green products, the government's green procurement ratio, etc. Domain-specific indicators could be established for clothing, food, residential, mobility, household appliances, and tourism sectors to reflect their resource demands and environmental performance, making use of the best available data.

III. **Promoting green consumption in the clothing, food, residential, mobility, household items and services, and tourism sectors during China's 14th Five-Year Period**

The clothing, food, residential, mobility (including communications), and daily utility products (household items and services) sectors account for 76% of individual consumption in China. As shown by the analysis from the CGE model built by the SPS team, this structure will not change significantly in the next 15 years. The above-mentioned five areas are the ones with the largest resource and environmental impacts among all domains of individuals' consumption. For every one unit of green product consumed, the economic output coefficient is 2.5 for the food sector, 3.0 for the residential sector, 3.8 for household items and services, and 2.7 for mobility and communication. These figures show the strong positive contributions these areas can have for the economy and environmental performance. Germany, Sweden and like-minded countries have identified food, housing, and mobility (including tourism) as key areas for sustainable consumption, based on their respective Greenhouse Gas emission contributions.

To this end, China should designate clothing, food, housing, mobility, household appliances, and tourism as key areas for promoting green consumption during the 14th FYP period and beyond. The main tasks will be to give priority to increasing the effective supply of green products and services in related fields while boosting the practice of "reduce, reuse and recycle." Steps to do this are to:

(I)   **Promote a green diet**. Initiate anti-food waste actions covering the entire food chain from storage to transportation, retail to the dining table; implement comprehensive plans for green take-out for the catering industry; unify and strengthen green and organic food certification systems and standards; and expand the supply of green food.

(II) **Promote green buildings**. Steer cities and localities with adequate capacities towards fully applying green building standards for new buildings and expand the scope where green building is mandated. Push forward the application of green building standards in the renovation of old communities. Implement an action plan for the production and use of green building materials; promote the comprehensive design, construction and operation of green buildings; strengthen environmental labelling, especially energy-efficiency labelling and certification for green household appliances; and enhance the effective supply of energy-efficient green household products.

(III) **Promote green mobility**. Encourage the use of low-carbon or net-zero transport modes such as walking, cycling, and public transportation; strengthen efforts to promote new energy-efficient automobiles; encourage the use of new and clean energy vehicles in expanding and upgrading fleets for public transportation, sanitation, taxis, commuting, urban express mail services, and urban logistics; reinforce the promotion and use of new energy vehicles in such areas as national Ecological Civilization pilot zones and key areas for air pollution prevention and control.

(IV)   **Promote green household appliances.** Encourage consumers to choose green products such as energy-efficient household appliances, high-efficiency lighting products, water-saving utensils, and green building materials. Encourage businesses to provide reusable, durable, and maintainable products and enable consumers to choose such products. Support the development of a sharing economy. Encourage the effective recycling of personal resources not being used; improve the recycling system of social renewable resources; boost greening, reducing, and recycling practices in express packaging; strictly implement the government's priority procurement and mandatory procurement system for energy-saving and environmentally friendly products; and expand the scope and scale of green government procurement.

(V)   **Promote green clothing.** Carry out "zero-discard" activities and "clothes reborn" activities for old and used clothes. Boycott fur and leather products made of rare animals to conserve biodiversity. Support and facilitate textile and apparel companies so they can build green supply chains. Improve the recycling and reuse of waste and old textiles in earth structures, building materials, automobiles, and home decoration sectors. Enhance the efforts in environmental labelling and certification of textiles and apparel. Substantially increase the effective supply of green textiles and apparel.

(VI)   **Promote green tourism.** Develop and publish green tourism and consumption conventions and guidelines. Encourage tourist hotels, restaurants, and management agencies in scenic areas to introduce incentive measures for green tourism. Formulate and/or revise appraisal rules and standards for green services, including *inter alia* green markets, green hotels, green restaurants, and green tourism. Star-rated hotels and chain hotels should gradually reduce the free provision of disposable toiletries and supplies and pilot demand-oriented provision. Publish green tourism information on relevant tourism promotion websites and platforms and encourage consumers to bring their own toiletries. Endeavour to integrate biodiversity conservation into tourism-related standards and certification programs.

IV.   **Build a green consumption policy system with equal emphasis on both the supply side and sides, balanced use of incentives and constraints, and the principle that the system should be built, governed, and shared by governments, businesses, and consumers.**

Consumption is an economic behaviour involving both supply and demand. The design of green consumption policy should respect economic principles. Consumption is a social behaviour that involves each and every member of society. Consumption is also a cultural behaviour, influenced by factors such as values and customs. The design of consumption policies needs to incorporate the clearly defined responsibilities and obligations of each actor and take into account incentive mechanisms, supervision and management, publicity and education, and so on.

The government should establish institutional arrangements for promoting green consumption by formulating legislation and standards. Encourage or mobilize consumers' willingness and behaviour to practice green consumption through industrial policies, fiscal and taxation policies, and price policies. Lead the constant increase in product and service quality by formulating and implementing standard systems of technology, products, quality, etc., and in particular, the "top runner" standards system. Guarantee the openness, fairness, and justice of the market, and regulate market operations via inspection, monitoring, and management.

Enterprises should facilitate the reduction of product prices through technological innovation to expand the supply of green products; practice eco-environmental responsibilities and corporate social responsibility (CSR); carry out such measures as product and service life-cycle assessment (LCA); green supply chain management, clean production, innovative business and consumer models and circular economy to reduce the negative environmental impacts in the life cycle of consumer products; emphasize a material reduction in the production of energy-saving, environmentally friendly, and low-carbon products; develop smart logistics and reduce the logistics costs of green consumer goods by system optimization and management of green consumer products (quantity and quality), brands, storage, transportation routes, and transport modes, so as to meet the needs coming from consumption upgrading.

In a context where sound incentives and constraints are in place, and with an enabling social atmosphere and favourable market environment, consumers will either consciously or unconsciously fulfill their responsibilities and obligations to protect the ecological environment, practice green consumer behaviour, and form a green lifestyle.

V. **Establish green consumption promotion systems, mechanisms, and technical support institutions with clearly defined rights and responsibilities, and give full play to the unique role of women, youth, and social organizations in promoting green lifestyles.**

The Chinese government should further define the role of government agencies, such as integrated economic management authorities, industrial/sectoral management authorities, and administrative authorities for eco-environmental protection, in advancing green consumption. It should also develop a corresponding list of green consumption responsibilities for government departments and agencies and establish a cross-sectoral collaboration mechanism to create synergies. At the same time, China should set up a technical support organization that is dedicated to promoting green consumption and in charge of specific operations, including green consumption research, information disclosure, monitoring and evaluation, communication and education, capacity building, etc. Meanwhile, the role of social organizations such as the China Consumer Association in promoting green consumption should be highlighted and given full play.

China should give full play to the unique role of women and youth in promoting a green lifestyle. According to relevant surveys, 80% of household consumption decisions are made by women, and female consumers have become the pioneers and main force in green consumption. Young people are strongly positive toward ecological environment protection and green consumption, and thus are indispensable to furthering green lifestyle practices.

Gender-focused and youth-driven practices are quite common in countries like Germany and Sweden.

VI.   **Build on public reflection as a result of COVID-19 to launch a nationwide green consumption and new lifestyle campaign.**

The government should leverage the positive roles that stars and social celebrities can have in demonstrating a green lifestyle to help green consumption become socially fashionable. Integrate the concept of green consumption into related education and training programs for various institutions at various levels, including families, schools, governments, and businesses. Strengthen awareness-raising efforts and incorporate green consumption initiatives into thematic publicity and education events, such as the National Energy-Saving Week, Science Promotion Week, National Low-Carbon Day, and Environment Day. Establish a green consumption incentive and penalty system for the general public; strengthen green consumption information disclosure and public participation; advocate simple, moderate, green and low-carbon production and lifestyles; oppose extravagant consumption, excessive consumption (waste), and irrational consumption; and raise awareness throughout society.

VII.   **Strengthen infrastructure and capacity building for green consumption.**

Build a green consumption statistics database and carry out monitoring, data collection, and statistical and evaluation reporting on green consumption. Establish a national unified green consumption information platform to publish information on green products and services, improve transparency in green product production and consumption, and encourage stakeholders to recognize the credibility of green product and service certification/evaluation results. Strengthen capacity building and training on green consumption in government, social organizations, enterprises, and the general public; build multistakeholder partnership networks; and push forward the participation of multiple stakeholders.

VIII.   **Further develop a national green consumption action plan.**

According to the experience of Germany, Sweden and other countries, in addition to the use of 14th FYP as an overarching instrument to lead related tasks, it is necessary to further develop a corresponding national action plan specific to green consumption. This should be a medium- and long-term action program to promote the formation of green consumption and lifestyle in a more comprehensive, in-depth, and systematic manner.

IX.   **Specific policy recommendations on green production and consumption that deserve close attention.**

(I)   **Establish sound green building standards, and incorporate energy-saving and environmental protection requirements into the ongoing renovation of**

old communities in China to guarantee green renovation. The renovation of old communities should be integrated into the creation of smart cities and waste-free cities. New buildings must thoroughly follow green building standards.

Statistics show that housing and residential expenditures account for 23.5% of residents' consumption in China, and building energy consumption makes up over one third of the total societal energy consumption. China's existing buildings have nearly 60 billion square metres of floor area, more than 95% of which belongs to high-energy-consumption buildings, with per unit energy consumption 2–3 times higher than that in developed countries with the same climate conditions. The recovery rate of construction waste in China hardly reaches 5%, far below the level of 90% in developed countries. In addition, some predictions indicate that the growth rate of the cumulative residential area of old communities in China will accelerate significantly in the next decade. An SPS CGE model analysis shows that a moderate increase in green construction investments for the green renovation of old communities and new green buildings will have positive effects on economic growth, employment, and resource use and environmental conditions in the short term. Related research also reveals that green buildings can save about 30% in energy demand compared to conventional buildings.

Therefore, the Chinese government should attach great importance to the development of green buildings and seize the particular opportunity brought by the ongoing large-scale renovation of old communities to fully push forward green renovation. The goal of green innovation can be achieved by establishing and improving governance mechanisms for the renovation of old communities, improving the green standard system and supervision system, and employing such means as smart technologies to greatly improve the quality of the green renovation.

(II)   **Comprehensively study and formulate a policy system on green production and consumption for the automobile industry.**

The automobile industry has become the pillar industry of China's economy. Since 2009, the volume of car sales in China ranked top in the world for 10 consecutive years. At present, the number of employees in automobile-related industries has exceeded one sixth of the total employed population in China. However, the resource and environmental problems caused by the use of automobiles are becoming increasingly prominent. In 2017, the transportation sector in China accounted for 46% and 66% of the total national consumption of gasoline and diesel, respectively. In 2018, automobile-emitted NO_X amounted to 43.6% of the total NO_X emissions in the country; yet the contribution of the automobile sector to China's total NO_X emission reductions was less than 20%. Therefore, the automobile industry is an important area to promote green consumption and production.

From the perspective of fuel efficiency and pollution emission standards, the Chinese government has made significant progress in boosting the green transformation of the automotive industry in the areas of automobile consumption and production, transportation, and energy policies. In the field of new energy vehicles in particular, remarkable achievements have been made. According to the analysis from the SPS team, China's policies for the new energy vehicle industry are generally effective. Of all measures, purchase subsidies have made the highest contribution to the development of the new energy vehicle industry, with a ratio close to 50%; the purchase subsidy policies also have had the most significant effects on promoting technology progress, cost reduction, and market growth.

However, a green consumption and production policy system for China's auto industry has not yet taken shape. A number of issues have seriously hindered green consumption and production practices in the auto industry, such as the imbalance in tax collection, loose links with energy conservation and emission reductions, an overemphasis on subsidies for purchase, and so on. Corresponding reform should be directed to the establishment and improvement of a sound policy system covering the entire automotive industry chain to support green consumption and production. Specific actions should focus on the production process, the development and application of non-HFC (Hydrofluorocarbons) alternatives, and encouraging alternative technologies. In the procurement process, tax reform should be promoted to enhance the effect taxes can play in leading energy conservation and emission reductions, reducing the cost of purchasing green car products, and encouraging green consumption. With regard to actual use, efforts are needed to make green car products easier to use and reduce corresponding costs. In terms of vehicle scrappage and recycling, it is important to advance the improvement of battery recycling policies and standards, improve policies related to the remanufacturing industry, and enable the integrated development of the remanufacturing and insurance industries so as to facilitate the development of the remanufacturing industry.

In terms of tax reform, the following steps could be considered. From 2021 to 2025, gradually phase out the current purchase tax exemption policy for new energy vehicles. From 2026 onward, implement a preferential tax policy based on fuel consumption, and set up a dynamic policy adjustment mechanism to respond to changes in fuel consumption regulations. From 2031 to 2035, raise the threshold for preferential policies and introduce a punitive tax system.

(III)   **Strengthen the reform of the green power consumption market.**

By the end of 2019, the installed capacity of renewable energy-based power generation in China reached 794 million kWh, making up 39.5% of total installed power; renewable energy power generation amounted to 2.04 trillion kWh, accounting for 27.9% of the total power generation. It is anticipated that, by the end of the 14th FYP period (2021–2025), renewable energy-based power generation will approach 40% of China's total power generation.

Therefore, it is of great significance to create a green power consumption market and unlock the demand for green power from enterprises and other users. The following specific measures can be taken. First, promote the use of power purchase agreements (PPA) and virtual power purchase agreements (VPPA) to further articulate the specific rules and regulations for various power sources, including renewable energy, to be engaged in market-based transactions. Second, cut down improper administrative interventions from local governments, liberalize power generation plans and use plans, and respect users' choices. Third, facilitate market-oriented transactions between power users and clean energy-based power generation companies such as hydropower, wind power, and solar power. Fourth, improve the policy and market environment for various users to jointly develop and use distributed renewable energy-generated power. Fifth, gradually expand the pilot of direct trade of renewable energy-based power. Sixth, recognize the environmental attributes of renewable energy certificates to enhance enterprises' confidence in the trading market. Seventh, establish a communication platform that includes various stakeholders to strengthen communication and cooperation.

(IV) **Formulate a national action plan for the development of a green logistics industry.**

As of the end of 2018, China's express delivery volume had reached 50.71 billion pieces, more than the developed countries and economies of the United States, Japan, and Europe combined. In 2018, the express delivery industry consumed 50 billion pieces of express waybills, 24.5 billion plastic bags, and 5.7 billion envelopes, 14.3 billion packaging boxes, 5.3 billion woven bags, and 43 billion metres of tape, resulting in a cost of nearly 1.4 billion yuan for waste disposal, including landfill and incineration. At the same time, transportation in China's logistics industry is still dominated by conventional fuel vehicles. Nearly 20 million vehicles are running for the logistics industry, consuming large amounts of gasoline and diesel while emitting pollutants.

Over the last few years, China has seen a number of good practices develop in green logistics, with promising experiences gained. However, in general, the lack of systematic policy support has been a chief factor restricting the development of a green logistics sector. Weaknesses include: corresponding legislation is lagging behind; management responsibilities are scattered in many government departments; responsibilities of relevant market players are unclear; existing legislative efforts focus more on macro-level guidance rather than concrete instruments; relevant standards, evaluation systems, and practical guidelines are missing; and inputs and the influence of pilot projects are not sufficiently strong. Therefore, policies to promote the development of a green logistics industry in China should be directed to the formulation of a special action plan at the national level as a package solution to the above policy issues, to comprehensively push forward the green development of the sector and systematically address the resource and environmental problems brought about by the booming growth of the sector.

(V) **Fully employ digital technologies to support green and low-carbon lifestyles.**

In recent years, there have been lots of projects (platforms) to pilot and boost digital low-carbon lifestyles. Schemes (and apps), including Ant Forest, Carbon Generalized System of Preferences (CGSP), the Zero Carbon Group app, and the Bean Sprouts App, have achieved positive results in innovating tools and mechanisms to lead a low-carbon life. Among them, the enterprise-led Ant Forest and the government-led CGSP are characteristic.

Based on these kinds of experiences, and with the support of the government, China is capable of building a digital platform for a green and low-carbon lifestyle with national influence and unified applicable standards to support individual consumers and groups to follow green and low-carbon behaviour. Such a unified platform would tackle a series of difficulties faced by the existing independent, decentralized, spontaneously organized, small platforms. For example, due to the lack of special policy support, platforms solely operated by enterprises are often unsustainable. Considerations of privacy and data security prevent current platforms from obtaining large volumes of information on effective emission reductions. Due to discrepancies in accounting standards for green and low-carbon activities and a lack of unified supervision, carbon emission reductions arising from users' low-carbon activities may be double-counted. A unified national digital platform can also provide technical support for large-scale green consumption actions by governments and groups, such as the carbon neutrality plan for conferences and events.

(VI) **Speed up the development of standards for green products and services, strengthen the certification and recognition efforts, and enhance the effective supply of green products and services.**

Green products and services are the basis of green consumption. A priority should be placed on accelerating the development of standards for green products and services, including environmental labelling, energy conservation, water conservation, and green buildings, and reinforcing third-party, independent certification and recognition efforts. Green products and service standards and corresponding certification and recognition not only link the consumer to the producer but also leverage green consumption and green production. Thus, they deserve to be highlighted and given close attention.

(VII) **Public institutions such as government and state-owned enterprises should take the lead in green procurement and carbon neutrality actions to play a stronger demonstrating role.**

Revise the Government Procurement Law to include major actors such as government departments, public institutions, and state-owned enterprises at all levels in the scope for green procurement. Expand the range of products and services for green procurement, and pilot the implementation of a mandatory green procurement system. Establish incentive policies to encourage other social organizations and businesses to practise green procurement. Pilot the establishment of carbon neutrality schemes for large-scale events (including conferences and sports competitions) by governments, public institutions, and state-owned enterprises at all levels, and encourage other social actors to take carbon neutrality actions. Support various carbon neutrality actions through the national digital low-carbon platform and establishment of a carbon neutrality fund.

(VIII) **Issue green consumption vouchers/coupons to stimulate and lead green consumption.**

In recent days, to spur consumption in the face of the COVID-19 pandemic, local governments in Nanjing, Hefei, Hangzhou, Zhengzhou, and other cities in China have issued a variety of vouchers/coupons for food and beverages, supermarkets, rural tourism, and car subsidies, yielding positive results. For instance, as of April 9, 2020, the written-off vouchers in Hangzhou have amounted to 220 million yuan, resulting in consumer spending of 2.37 billion yuan, with a multiplier effect of 10.7 times. Recently, similar approaches have been used in the United States.

Based on these practices, it is necessary for China to study the feasibility of issuing green consumption vouchers/coupons, not only as a stimulus for green recovery in the face of the COVID-19 pandemic but also as a normal practice that includes various forms of vouchers. The vouchers can be given by governments, product manufactures and sellers, and even groups that are interested. The scope of preferential treatment will be limited to the consumption of green products and services to attract consumers with green targets so as to leverage green consumption. Groups that are willing to push forward green consumption should be encouraged to carry out pilot projects.

References

1. CCICED Special Policy Study on Green Transition and Sustainable Social Governance, June 2019. http://en.cciced.net/POLICY/rr/prr/2019/201908/P020190830114076694525.pdf
2. "Avoiding Consumer Scapegoatism: Towards a Political Economy of Sustainable Living": https://helda.helsinki.fi/bitstream/handle/10138/303978/Avoiding.pdf.
3. Behavioural Study on Consumers' Engagement in the Circular Economy: https://ec.europa.eu/info/sites/info/files/ec_circular_economy_executive_summary_0.pdf.
4. Information on EU Circular Economy Action Plan: https://ec.europa.eu/environment/circular-economy/.
5. The Environmental Objectives System. http://www.swedishepa.se/Environmental-objectives-and-cooperation/Swedens-environmental-objectives/The-environmental-objectives-system/.
6. "A National Food Strategy for Sweden – More Jobs and Sustainable Growth throughout the Country. Short Version of Government Bill 2016/17:104." Regeringskansliet, 20 Apr. 2017, www.government.se/information-material/2017/04/a-national-food-strategy-for-sweden--more-jobs-and-sustainable-growth-throughout-the-country.-short-version-of-government-bill-201617104/.
7. "Science Council of Japan Releases Policy Recommendation:" JFS Japan for Sustainability, www.japanfs.org/en/news/archives/news_id035986.html.
8. EN-Consilium. register.consilium.europa.eu/doc/srv?l=EN&f=ST%2010917%202006%20INIT.
9. "Legislation." Official Journal of the European Union, vol. 52, 31 Oct. 2009, pp. 1–48.
10. "Communication from the Commission to the European Parliament, The Council, The European Economic AND Social Committee and the Committee of the Regions." Commission of the European Communities, 16 July 2008, pp. 1–12.
11. "A New Circular Economy Action Plan For a Cleaner and More Competitive Europe." Communication from the Commission to the European Parliament, The Council, The European Economic AND Social Committee and the Committee of the Regions, 11 Mar. 2020, pp. 1–19.
12. "European Circular Economy Stakeholder Platform." European Circular Economy Stakeholder Platform I A Joint Initiative by the European Commission and the European Economic and Social Committee, 7 June 2021, circulareconomy.europa.eu/platform/.
13. National Program for Sustainable Consumption: https://www.bmu.de/fileadmin/Daten_BMU/Download_PDF/Produkte_und_Umwelt/nat_programm_konsum_bf.pdf.

14. Consumer Scapegoatism and Limits to Green Consumerism: https://doi.org/10.1016/j.jclepro. 2013.05.022.
15. Swedish national Strategy for Sustainable Consumption: https://www.government.se/4a9932/ globalassets/government/dokument/finansdepartementet/pdf/publikationer-infomtrl-rappor ter/en-strategy-for-sustainable-consumption--tillganglighetsanpassadx.pdf.
16. Akenji, L, et al. "Sustainable Consumption and Production in Asia — Aligning Human Development and Environmental Protection in International Development Cooperation." World Scientific Publishing Company and Distributed under the Terms of the Creative Commons Attribution Non-Commercial, 2017, pp. 17–43.
17. EU Seventh Framework Programme for Research and Technological Development (FP7): https://wayback.archive-it.org/12090/20191127213419/https:/ec.europa.eu/research/fp7/ index_en.cfm.
18. EU Horizon 2020: https://ec.europa.eu/programmes/horizon2020/en/what-horizon-2020..
19. More detailed information on the implementation of 10YFP (target 12.1) is accessible through the One Planet Network platform: https://www.oneplanetnetwork.org/.
20. "Sök." Naturvårdsverket, 8 Apr. 2015, www.naturvardsverket.se/Global-meny/Sok/?query= styrmedel%2Bf%C3%B6r%2Bh%C3%A5llbar%2Bkonsumtion.
21. Miljömålsrådet, http://www.sverigesmiljomal.se/miljomalsradet (Environmental Objective Council).
22. "Oj, Vi Missade Målet!" Sveriges Miljömål. https://www.sverigesmiljomal.se/contentassets/ f2f66cba53f745398381eb7346a215a6/miljomalsradets-atgardslista-2020.pdf.
23. "Forum För Miljösmart Konsumtion: Konsumentverket." Forum För Miljösmart Konsumtion | Konsumentverket, www.forummiljosmart.se/.
24. New data build on new methodology developed within the PRINCE project: https://www.pri nce-project.se/publications/environmental-impacts-from-swedish-consumption-new-indica tors-for-follow-up-prince-final-report/.
25. "Forskningsprogrammet PRINCE." YouTube, YouTube, www.youtube.com/playlist?list=PLg GFtRVUTORQspUzwN7xGX1pKkMz4okum.
26. "Generationsmålet." Generationsmålet - Sveriges Miljömål, sverigesmiljomal.se/miljomalen/ generationsmalet/.
27. "Regeringsuppdrag Mätmetoder Och Indikatorer För Att Följa Upp Konsumtionens Klimat- påverkan." Naturvårdsverket, www.naturvardsverket.se/Miljoarbete-i-samhallet/Miljoarbete- i-Sverige/Regeringsuppdrag/Redovisade-2019/Matmetoder-for-konsumtionens-klimatpav erkan/.
28. Annual evaluation of the Environmental Quality Objectives 2019 (report in progress.)
29. "Konsumenterna Och Miljön 2018." 2018. https://www.konsumentverket.se/globalassets/pub likationer/var-verksamhet/konsumenterna-och-miljon-2018-17-konsumentverket.pdf.
30. Dagmar Vinz, Gender and Sustainable Development: A German Environmental Perspective, European Journal of Women's Studies, 1 May 2019, https://doi.org/10.1177/135050680810 1764.
31. OECD (2008) "Promoting Sustainable Consumption: Good Practices in OECD Countries"; OECD (2018) "Policy Coherence for Sustainable Development and Gender Equality Fostering an Integrated Policy Agenda".
32. " Konsumentbeteende Och Genus – En Forskningsöversikt." Vol. 59, 2017, pp. 1–58. https:// www.konsumentverket.se/globalassets/publikationer/var-verksamhet/konsumentbeteende- och-genus-en-forskningsoversikt-konsumentverket.pdf
33. "New Measures of Wellbeing." Regeringskansliet, www.government.se/articles/2017/08/new- measures-of-wellbeing/.
34. "Hungary's Ombudsman for Future Generations." Environmental Rights Database, environmentalrightsdatabase.org/hungarys-ombudsman-for-future-generations/.
35. "A Parliamentary Ombudsman for Future Generations?" Intergenerational Foundation, 16 Aug. 2011, www.if.org.uk/2011/08/16/a-parliamentry-ombudsman-for-future-generations/.

Chapter 7
Major Green Technology Innovation and Implementation Mechanisms

7.1 Introduction

Ecological civilization. The Chinese government has established ecological civilization as a basic strategy guiding China's future development. The principles of "green development for a beautiful China" and "people-centered development" constitute the core values and visions for China in the new era of civilized development with a sound ecosystem.

The fundamental purpose of ecological civilization is to ensure ecological safety through lower and more efficient use of natural resources, exerting the service function of ecological capital. With reduced emission of carbon and other pollutants, extensive and unsustainable growth mode should be changed to an ecologically friendly development path. Through green development, green prosperity can be realized and people can share benefits of green development, reaching the harmonious coexistence of humanity and nature.

Respond to climate change. By 2015, three international legal instruments addressing climate change—namely the United Nations Framework Convention on Climate Change (UNFCCC), Kyoto Protocol, and the Paris Agreement—had set the framework for global climate governance and low-carbon green development. The Paris Agreement determined "to keep the increase in global average temperature to well below 2 °C above pre-industrial levels," and stipulated the "common but differentiated responsibilities" among countries. As a signatory, China has made a commitment to reach a peak of carbon emissions by 2030 and to reduce carbon emission intensity by 60–65% compared with 2005 so as to undertake its responsibilities as a major country in dealing with climate change.

© The Author(s) 2022
China Council for International Cooperation on Environment
and Development (CCICED) Secretariat,
Green Consensus and High Quality Development,
https://doi.org/10.1007/978-981-16-4799-4_7

Green development. Western countries, especially the EU countries, pay close attention to green development. According to the European Green Deal issued in 2019, by 2050, Europe will become the world's first "carbon–neutral" area. The newly elected EU institutions came up with the concept of "Green New Deal." For the first time, China's 13th FYP clearly proposed the green development goal of "overall improvement of ecological environment" by promoting the implementation of resource conservation and intensive utilization, increasing comprehensive environmental governance, strengthening ecological protection and restoration, actively responding to global climate change, and developing green and environmentally friendly industries—it also identified 27 green development projects in four categories. Green development has become one of the five concepts and a strategic and programmatic idea guiding the development of China in the new era.

"Green city". Many European and North American cities have put forward goals and action plans of building green cities and carbon–neutral communities, and published important documents to promote urban green development, such as the *Aalborg Charter, Aalborg Commitment, Leipzig Charter* and *Freiburg Charter*. In its Vancouver: 2020 action plan of the greenest city, Vancouver has proposed the goal of "zero-carbon, zero-waste, and sound ecosystem" and a series of target areas, including energy, water and transportation, to build a green city of prosperity and development.

Chinese cities are areas with concentrated population and economy and intensive resource consumption and emission; green development becomes the only choice for Chinese cities. In recent years, the Chinese government and relevant departments have promoted development action plans such as "sponge city," "ecological city," "green new area," "low-carbon city," "low-carbon community," "ecological restoration," "urban repair," and "waste-free city," carried out many pilot and demonstration projects, and explored the ways of urban green development.

Green technology. The term "green technology" refers to technologies that can reduce consumption, decrease emission, and improve ecological environment, including technologies applied in exploitation, production, manufacturing, construction, planning, design, utilization, maintenance, and management. A wide application of green technology is necessary for cities' development. China's international commitments to addressing climate change and its emission reduction policies, as well as its increasingly stringent environmental governance policy requirements are driving the application of green technology in urban areas on a larger scale.

The 14th FYP period is a critical one for China's urbanization and urban development transformation. Boosting new urbanization with the development of green and low-carbon technology and promoting green development and application of green technology during new urbanization are core tasks for China to cope with climate change, as well as to stimulate the green economy after COVID-19.

Popularization and application of green technology. In 2018, the Chinese government proposed that we should promote green development, accelerate the building of legal institutions and policy guidance for green production and consumption, establish and improve the economic system of green, low-carbon, and circular development. It also proposed that we build a market-oriented green technology innovation system to develop green finance. According to the *Guidance on Building a Market-Oriented Green Technology Innovation System* issued in 2019, the market-oriented green technology innovation system should be basically completed by 2022.

Green technology has good positive externalities, good social and environmental benefits; green technology has a certain public welfare nature, and has huge potential market demand, and should be supported by society and government. Under the traditional resource utilization model and market pricing mechanism, green technology often lacks price competitiveness. We should adopt a resource pricing mechanism and ask producers or users to cover all the costs (including external costs) so as to realize fair competition between green and non-green technologies and products. Or we can give support to green technologies and products through proper fiscal subsidies. Green technology should continue to innovate so as to lower costs and increase market competitiveness.

Green technology is an emerging high-tech and innovative sector attracting a lot of attention. We should actively encourage and support its development and build a credible information distribution system and full life-cycle assessment procedure to avoid falling into the trap of blind promotion.

Popularization of green technology should involve wide participation from the government, higher learning institutions, professional organizations, enterprises, social entities, and the public. The government should establish a complete system of laws, regulations and policies to encourage universities and professional institutions to participate in the research and development of green technology; encourage enterprises to actively research and develop, produce, and apply green technology, while paying more attention to the differentiated demands and consumption abilities of different gender, age, and capability groups; advocate green lifestyle and green consumption concepts to the society and the public; and form a good environment for the whole society to participate in green development and apply green technology.

The research emphasis of green technology SPS. The SPS on *Major green technology innovation and its implementation mechanism*, issued by China Council for International Cooperation on Environment and Development (CCICED) is committed to the research of green development, promotion and application of green technology, evaluation methods, laws, and policies in urban areas. Through research, it recommends 10–20 scalable green innovation technologies, conducts a comprehensive assessment from the perspective of the full life cycle to provide technical support for the green development policy of the 14th FYP, and at the same time, puts forward suggestions on the policy guarantee system of green technology promotion.

7.1.1   Characteristics, Problems and Opportunities of Urbanization and Urban Development

Urbanization and urban population growth have created an increasing amount of pressure and opportunities for urban transformation. With the urban population accounting for 60% of the total population in 2019, China has transformed from a traditional agricultural society to an urban society, and it will see a further increase in its urban population by about 150 million over the next 10 years. Among the floating population who are already working and living in urban areas but not officially registered as urban residents, about 100–150 million will settle down in cities and counties to enjoy family reunion. The population increase and the need to settle down will generate huge new demand and resource consumption, and also provide the cities with opportunity to change the development mode and optimize resource allocation.

Cities are where most resources are consumed and most carbon is emitted, but also the core places for green and low-carbon development. Industry, construction, and transportation are the three major resource consumers and are mainly concentrated in cities. Cities are where most emissions are generated and also the core places for achieving green and low-carbon development throughout society. Under the influence of GDP-oriented policies and existing financing and taxation policies, a model featured with extensive use of resources and blind expansion of land was formed in Chinese urban areas. Fragmented urban layout, inefficient use of land, dislocation of infrastructure and public services, and inefficient transportation operations have led to a perpetuating high level of carbon footprint and created a "lock-in effect."

Increases in expansion of consumption are putting greater pressure on supply, and also driving a transformation of lifestyle and consumption. Rising per capita income has witnessed a continuous expansion of China's middle-income group which now accounts for about 30% of the total population. The driving force for social development has shifted from meeting the needs for food and clothing to pursuing a better life. The values and lifestyles of the middle class are taking shape. On the one hand, increased consumption power of urban residents is leading to more resource consumption for material and cultural needs and more emissions. On the other hand, the emerging middle class and well-educated young people are more open to green and low-carbon value and rational lifestyles. It is a good opportunity to advocate low-carbon lifestyles and consumption, and to promote green development and green technology in cities.

Policy change from focusing on seeking additional resources to focusing on utilizing existing resources has created new space for applying green technologies. Chinese cities have accumulated a large amount of existing assets, including inefficiently used or idle land, infrastructure, industrial, and civil buildings. China is asking its cities to shift the focus of their development from increasing additional assets to utilizing existing assets efficiently. It proposes new ways and scenarios of promoting green technologies, including but not limited to multifunctional and hybrid land development, low-carbon reconstruction and multiple use of buildings, organic renewal of cities, development of low-carbon communities, green slow-moving transportation, distributed energy supply, and the principles of the circular economy.

Huge differences between cities require different green development strategies and green technology supplies. China's natural resource and population are extremely unevenly distributed, there is a gap of natural condition and development level among the eastern coastal, central and western regions, and the northern and southern regions. There are also big differences in development levels, needs, and capabilities between large cities, mid-size cities, and small towns. Different areas respond differently to natural disasters caused by extreme weather, rising sea levels caused by climate change, etc. Urban green development should have a "common but different" strategy and path; green technology research and promotion should be highly targeted and adaptable. It also provides various driving forces for the development and innovation of green technology.

7.1.2 Vision and Criteria for Urban Green Development

The vision of urban green development is to make green industrial production and everyday life the mainstream choice of society through green technology innovation, promotion, and application, and to **build beautiful cities that are green and prosperous, low-carbon and intensive, recycling and reusing, fair and inclusive, safe and healthy to set a "Chinese example" for global sustainable development.**

Green and prosperous cities: We should build a high-quality green economy, promote the development of low-carbon industries, circular economy, and green consumption, and make green development the core competency of cities and the core value of society.

Low-carbon intensive cities: We should optimize urban layout, advocate mixed land use, encourage green transportation, promote green buildings, green infrastructure and low-carbon communities, and completely abandon the urban development model of high consumption and high emission.

Recyclable and reusing cities: Cities should aim to use all resources to their maximum efficiency, including underused city and building space, transportation that shares resources such as bikes and cars. Waste can be recycled or reused. Resource recycling replaces the expanding and incremental development mode.

Fair and inclusive cities: We should provide residents with fair access to quality life and ecological capital services regardless of gender, status, age and identity (*hukou*) status; provide residents with equal access to and participation in green development.

Safe and healthy cities: With a continued focus on ecological conservation, we should mitigate conflicts between cities and nature, protect valuable natural resources and ecological capital for contemporary and future generations of residents, so as to create a healthy and livable environment, and improve the health and well-being of urban residents.

In order to achieve the vision, urbanization and urban development should follow:

Criterion 1: Urban development serves to meet people's need for a good life and to achieve harmony between humanity and nature, not just for economic growth;

Criterion 2: We must pursue prosperity, but just as importantly we must make direct contributions to the reduction of global resource consumption and carbon emissions while adopting climate-resilient development strategies to cope with changes;

Criterion 3: Urban green development not only requires the government to make great efforts but also should give full play to market entities and encourage broad participation by enterprises and the public;

Criterion 4: Urban green development should be oriented toward fairness and justice, and should not harm the welfare of women, the elderly, children, and mobile residents or deprive vulnerable groups of their development rights;

Criterion 5: Urban green development should establish differentiated strategies and paths to adapt to all natural conditions, different stages of development, and development needs and capabilities of cities and regions at different levels;

Criterion 6: Urban green development should be oriented toward effectiveness and applicability, and requires multi-dimension life-cycle assessments of green technologies to avoid following suit and steer away from "new technology pitfalls."

7.1.3   Goals and Approach for Urban Green Development

General goal: Based on the concept of "ecological civilization" and new development outlook, we aim to realize modernization in a way that human and nature live in harmony, with such spatial layout and economic activity distribution pattern that are conducive to conserving resources and protecting the environment. "Green and low-carbon" will become the theme of both work and life, and the idea of green development will be implemented in all aspects and processes of urban economy, as well as social development and environmental evolution. Ultimately, green and low-carbon development will become the norm, and the vision of "beautiful China" will be materialized.

2020–2025 (The 14th FYP in China): Priority is given to the green development and green technology application in key sectors including water, energy, architecture, urban planning, transport, food, etc. to overcome the lock-in effect. Starting with pilot programs, policy measures and framework of standards should be improved;

meanwhile, a green finance market that harmonizes with its international counterparts should be put in place. Promotion of green solutions and products among industry and consumers will be boosted through the constant improvement in measures such as emission permitting, supervision, reporting, monitoring, auditing, and standards formulation. During this stage, the total emissions of greenhouse gases will be significantly cut and the eco-environment quality notably improved.

2025–2030: Put in place a world-leading technical system for green development; popularize the concept of urban green development and the application of green technology. Sophisticated legal, institutional, and policy frameworks should be established so that the way of production and life is directed to shift toward a greener style. China should aim for the early-attainment and over-achievement of its national commitments under "2030 UN Sustainable Development Goals" and the Paris Agreement.

Post-2030: With reference to the national development objectives and the goals of 2 and 1.5 °C in the Paris Agreement, drawing lessons from the EU's zero-emission goals and roadmap for 2050, green technology should be applied in cities across China. An economic system featuring green development will be established. Resource conservation, circular use of materials, and in-depth emission cutting should become the universal themes. Relevant legal instruments and policy measures should be institutionalized to guide and support green production and consumption. By 2050, green development will be the paradigm of choice for the whole of China, where according to the plan, the vision of "beautiful China" will be realized. With these achievements, China will make its outstanding contributions to the global 1.5 °C goal.

In accordance with the general goals and stage goals stated above, the objectives for urban green development in six key sectors are listed in Table 7.1.

To achieve the goals above, a green technical system should take into account the economic, social, and cultural features of cities. In order to cover all relevant processes in life, work, and leisure of urban citizens, the system should account for each link of the value chain: "resource conservation and utilization; production and building; consumption and use; decomposition and recycling." The green technical system would facilitate the realization of green development goals by addressing the salient issues of "not green enough" and high emissions in urban development in China. For a green technical system to be established, technologies on the following fronts should be developed and promoted:

Conservation and reduction in the use of shared natural resources. Ecological red lines must be strictly observed, and the natural landscape should be left undisturbed as much as possible. The energy-intensive, high-emission urban development approach will be abolished; and shared natural resources should be used in a rational and frugal manner and be fairly distributed.

"Circular" production and building. In the process of production and building, resources should be used in such a way that maximizes efficiency and minimizes

Table 7.1 Goals for each stage in six key sectors of developing green technology

	2025	2030	Post-2030
Water management	Build a well functioning network for sewage collection and treatment	All urban sewage is collectively treated; more water is recycled, to a higher technical standard	Water environment quality is significantly improved; sustainable use of water is on a par with the standards in major developed countries
Energy	Put in place a green, intelligent and networked energy system. Incremental demand is primarily met by clean energy. Realize the equal price of renewable energy and coal	Renewable energy and nuclear energy enjoy an increased proportion in the energy mix, while fossil fuels are decreasing significantly; clean energy becomes the mainstream	Energy efficiency reaches a level that represents world-leading standards; energy consumption peaks; fossil fuel is completely replaced by new and renewable energy
Transport and mobility	Formulate frameworks for green solutions, standards, and application support measures; constantly optimize the energy structure and make tangible achievements in emission cutting in the transport sector. Encourage electric car and motor vehicle to improve ridership	Reach a leading international standard for smart and low-carbon technology; a sophisticated system for green solutions, standards, and application support measures is established. All public transport and delivery vehicles are electrified; significant emission cutting is achieved	A complete system of green, ultra-low-carbon, intelligent transport is put in place
Architecture	Green building standards are applied on all newly built civil architectures; buildings sport cleaner and greener energy systems	All newly built civil architectures feature ultra-low energy consumption level or even better	All newly built civil architectures are zero-energy consumption. All existing buildings are upgraded into energy-efficient buildings
Land use and planning	Green development concept is implemented in all processes of planning and construction management; the pilot programs of green cities and green communities are scaled up	Prioritize the "green transformation" of settlement environment in key urbanization areas. A number of world-class green cities and green communities take shape	The overall living environment in urban areas and neighbourhoods features the quality of green, low-carbon, livable, and prosperous. "Ecological civilization" is achieved

(continued)

Table 7.1 (continued)

	2025	2030	Post-2030
Food	Address the issue of food safety; cut emissions from food production	Improve the health of the food chain; advocate greener diet structure	Build a green, low-carbon, nutritional, and fair system of food supply and demand

consumption and emissions, while producing high-quality, high value-added green products.

Responsible consumption and use. Cultivate the awareness of responsible consumption and the good behaviour of careful use and protection of materials, so as to reduce the life-cycle consumption of resources, carbon emission, and environmental pollution, while improving the standard of living and net well-being benefits for urban citizens.

Safe decomposition of waste and resource regeneration. Taking zero-carbon emission as the ultimate goal, high-standard treatment and recycling of solid waste, waste water, and exhaust gases will be rolled out. Consequently, the vision of "zero-waste city" and "zero-pollution environment" will be realized.

7.2 Existing Policies and Problems

7.2.1 Existing Policies for Urban Green Development

After the 1992 United Nations Conference on Environment and Development, as a party to the Convention, China issued the "China Agenda 21—China's twenty-first Century White Paper on Population, Environment and Development." Since then, it has continued to advance through laws, policies, planning, and standards, actively building and improving laws and policy systems related to sustainable development, energy conservation and emission reduction, green development and low-carbon development.

At the macro level, in recent years, the state has issued a number of important policy documents on ecological civilization construction, sustainable development, new urbanization construction, urban planning and construction management, etc., to promote the sustainable development of ecological civilization in the field of urbanization and urban development. The policy document emphasizes that urban construction must be "people-oriented," taking into account the multi-dimensional development goals of economy, society, resources, environment, and culture, and attaching importance to the role of reform and innovation and technological progress. These policy documents show that the Chinese government attaches great importance to urban green development and also proposes goals and requirements for urban green development (Tables 7.2).

In response to climate change, promoting energy conservation, emission reduction, and low-carbon development, the Energy Conservation Law was promulgated

Table 7.2 Selected macro-level policy documents for urban green development

Time	Title of document	Key points
2014.3	"National New Urbanization Planning (2014–2020)"	Enhance urban sustainability; build livable, dynamic, and modern cities with unique characteristics
2015.4	"Opinions on Accelerating the Development of Ecological Civilization by the Central Committee of CPC and the State Council"	Building a society featuring resource conservation and environmental friendliness; mainstream the values of "ecological civilization"
2015.12	"Gazette for the 2015 City Work Conference of CPC Central Committee and the State Council"	Sustainable urban development and livability
2016.2	"Opinions on Driving New Urbanization to Greater Depths by the State Council"	New urbanization, sustainable and healthy economic growth
2016.2	"Opinions on Improving the Management of Urban Planning and Construction"	Manage orderly construction, moderate land development, and efficient operation of cities; build livable and dynamic urban environment with unique characteristics
2016.3	"Outline for the Thirteenth Five Year Plan"	"Urbanize the people"; optimize urban spatial layout and morphology; build harmonious, livable cities; achieve coordinated development of urban and rural areas
2016.9	"China's National Plan on Implementation of the 2030 Agenda for Sustainable Development"	Build inclusive, safe, disaster-resilient, and sustainable cities and human settlements
2016.12	"China's National Plan on Implementation of the 2030 Agenda for Sustainable Development"	Build innovation demonstration zones for National Sustainable Development Agenda
2017.4	"Chapter on Urbanization and Technology Innovation for Urban Development" of the Thirteenth FYP	Technological innovation system in urban development
2017.10	Report of the CPC 19th National Congress	New type industrialization, informatisation, urbanization, and modernization of agriculture; shared responsibility and benefits of growth; harmonious coexistence of humans and nature
2018.8	"Opinions on Developing Safe Cities by the Central Committee of CPC and the State Council"	Urban security, disaster prevention, and public security

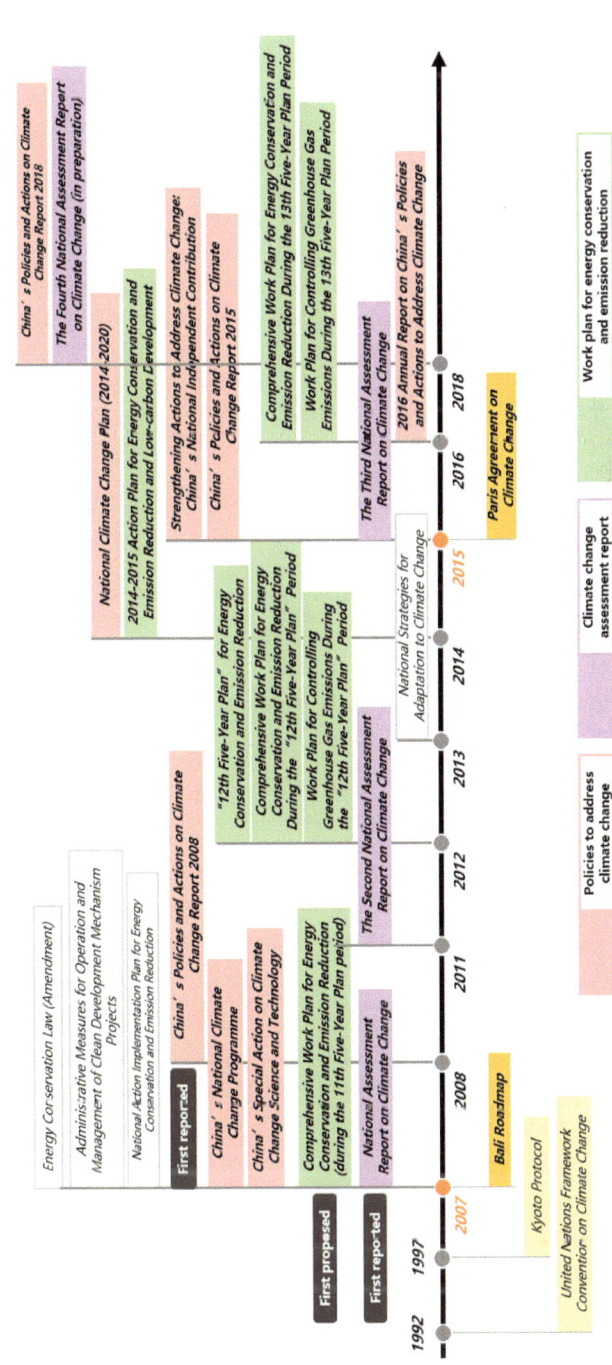

Fig. 7.1 The development course of low-carbon development policy in China

in 1998, and the Renewable Energy Law was promulgated in 2006, gradually forming three types of systemic policy tools: policies to address climate change, climate change assessment report, energy-saving and emission reduction work plan. The country has successively issued the "China National Plan to Address Climate Change," "The 13th Five-Year Comprehensive Work Plan on Energy Conservation and Emission Reduction," "The Medium- and Long-Term Development Program for Energy," "Administrative Measures for the Operation of the Clean Development Mechanism Project," and "Green Travel Action Plan" along with many other policy documents. In terms of standard specifications and technical regulations, national or local technical regulations have been formulated for cities, communities, buildings and other fields, including "Green Building Evaluation Standards," "Guidelines for Corporate Greenhouse Gas Accounting and Reporting," and "Low-Carbon Community Pilot" Construction Guide, etc. However, compared with the EU's green and low-carbon laws, policy financing, and market standards that complement each other, China's urban green development and green technology promotion and protection systems are still not perfect (Figs. 7.1 and 7.2).

7.2.2   Analysis of Existing Problems of Urban Green Development

Despite the consensus on green development among cities, some problems remain prominent in its implementation.

Legislation lags behind practice. When it comes to green development, China lacks a systematic legislation framework. Existing laws and regulations are primarily single-purpose for issues such as environmental protection, pollution prevention and treatment, or energy use, etc., with stress on the limits. There are no laws to direct comprehensive green development and carbon reduction. Green development is more driven by government policy rather than continuously bolstered by laws and regulations.

Goals and indicators for emission reductions are absent. The currently binding targets for emissions reductions are too conservative and fail to match the commitments China made at the Paris Climate Conference and the global trend of going green. The lack of systematic objectives has made it difficult to allocate tasks and responsibilities to local governments.

Coordination between departments is weak. Green development is a systematic cause that requires the coordination of multiple departments. However, because the division of responsibilities is not clearly defined, and the coordination mechanism is weak, no one embraces the responsibility. Take the shared bike service as an example: it involves over 10 departments and various layers of the government. Some departments welcome the service while some others prefer putting restraints.

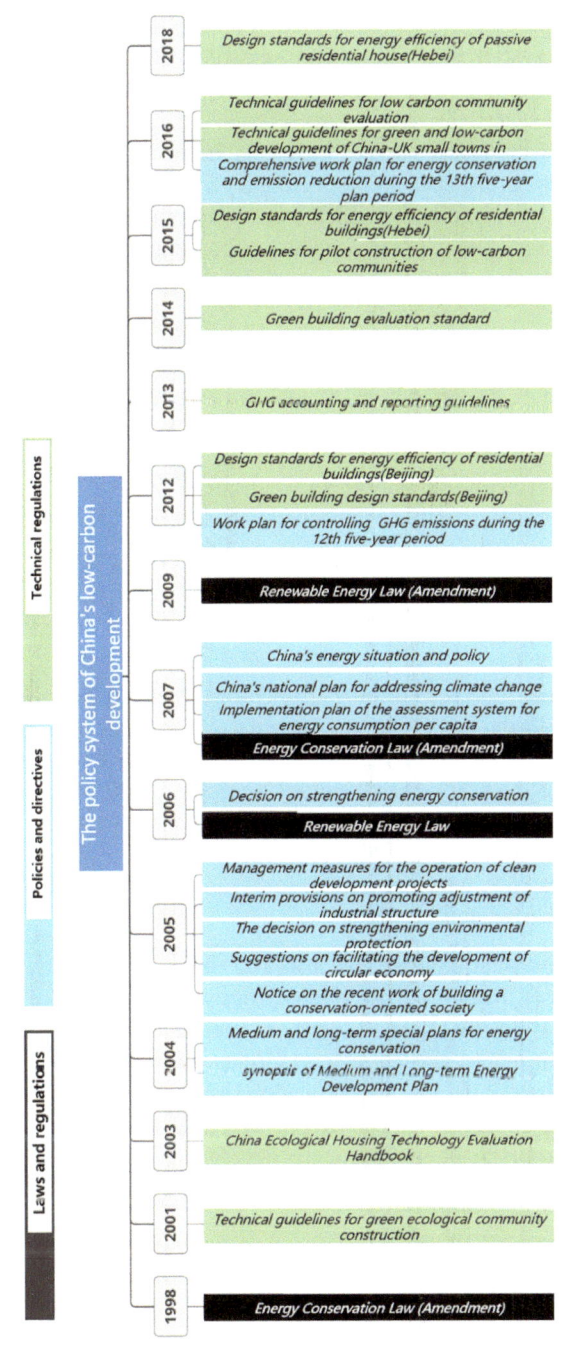

Fig. 7.2 The major law and policy of China's energy-saving and low-carbon development

Such inconsistency among different decision-makers finally gave rise to a large-scale "quick flood and ebb" phenomenon.

The green financing system is still next to non-existent. China has seen a significant increase in green and environmentally themed investment year-on-year over the last couple of years. However, the source of green funding is rather limited: companies relied almost solely on themselves to raise the capital needed since there are few other options available. Green finance is still in its infancy in China. Problems such as the inconsistency in green accreditation and the confusion in claiming green finance collateral have kept financial institutions away from funding green projects. Besides, there are no expressly stated principles and guidance for green investment in the financial sector.

Businesses are not keen on the idea, and thus the green value chain has not taken shape. At present, the popular way of encouraging businesses to participate in green technology innovation and application is through fiscal subsidy. But candidate technologies were often not evaluated in a scientific manner; consequently, many projects that are truly "green" are not properly subsidized. Also, China doesn't have a mature carbon trading market that can generate enough profit as reward. The poor liquidity and low pricing of carbon in the current market have held companies back from taking part. Besides, China doesn't have the policy tools to encourage and facilitate knowledge transfer, commercialization, consumption, and application of green technology.

The support system for green technology innovation is not well developed. Technological innovation is the key underpinning pillar for green development. Unfortunately, on that front, the current situation in China is that while universities are enthusiastic about R&D, businesses take quite the opposite attitude when it comes to green technology and products. International collaboration is another weak point. The status quo makes the innovation community slow in picking up market demands and incapable of effectively translating R&D results into solutions and products that address business and household needs.

Public awareness and participation are low. The values of green and low-carbon have not yet taken root in society. The reasons for this include: the general public don't know where and how to take part in urban green development; there is a lack of a trustworthy platform to disclose green performance and emissions data; there is not yet a professional and inviting social education system; etc. On the other hand, as people's incomes rise, consumerism and hedonism are becoming widespread, which has aggravated the problems of environmental degradation, resource depletion, and carbon emission.

7.2.3 Challenges in the Innovation and Application of Green Technology

Urban green development has great bearing on the fulfilment of national emissions commitments. Over the past two decades, urban development in China has been moved toward heavy investment, large consumption, and high emissions. Tremendous resources have been invested in the water, energy, and transport sectors. This approach has addressed the supply shortage problem but did not give much consideration to the green transformation of supply and services. In urban planning and land use, the expansion of built-up areas has consumed huge amounts of land and construction materials. As a result, while work and life conditions have improved, the problems of misallocation of land supply, overcrowding of residents and buildings, and an excessive number of high-rises have become more serious. Studies reveal that material production, construction, and operation of buildings in China account for as much as 40% of its total carbon emissions—and this percentage is still rising sharply as urban development continues. It is imperative that China adopt more stringent carbon emission standards and renew its commitment to shifting development mode toward urban green development and wider application of green technologies. That is the path China must take in order to fulfil its "intended national independent contribution" to the Paris Agreement.

Urban development mode is built upon the old approach and short of motives and measures for adopting green development. Under current circumstances, including the fiscal and tax regime and the GDP-based performance assessment method, city governments cannot help keeping economic growth high on their agenda. Cities continue to source capital for infrastructure construction from the proceeds of land supply, instead of actively seeking to change modes and pursue green development. Though green development slogans were occasionally invented, many efforts were aborted once a city faced setbacks. On one inspection tour made by a central government task force to detect environmental breaches, a total of RMB 1.43 billion fines were ordered and over 18,000 cadres were punished. Those figures suggest that local governments haven't done a perfect job on pollution control and the implementation of green development concept.

The lack of scientific evaluation and performance assessment mechanisms has led to the uncoordinated application of green technologies. When green technologies and green products were scaled up, more credit is given to their economic performance over green features. More often than not, they were pitched in "promotion campaigns" that ignore the systematic nature of a city and the life-cycle value flow of green products. Some so-called "green" technologies had not been evaluated on their life-cycle performance and potential externalities before they were commercialized, thus causing waste of resources. Examples to this point include shared bikes and integrated underground corridors. In the "Opinions on Nurturing a Market-Oriented Innovation System for Green Technology" issued by the central government in 2019, it was stressed that a good evaluation system should be established to step up evaluation of green technology, demonstrating the importance of proper evaluation and the urgency to catch up on that front.

Priorities should be set for the whole range of green technologies. Green city and technology have countless aspects, including "technology clusters" in various areas, e.g., material technology, biotechnology, pollutant treatment, resource recycling, clean production, etc. which are relevant to carbon emissions, environment, energy, resources, manufacturing, transport, infrastructure, building, food, land, and other areas. On the other hand, cities are suffering increasingly from problems such as carbon emission, energy consumption and resource depletion, environmental degradation and pollution, heat island effect, big city malaise, etc. Therefore, while a comprehensive approach is needed in promoting green technologies, priorities should be set to address pressing issues and the near-term demands of citizens. Technical breakthroughs can thus be made by selecting key issues and concentrating resources on solving them.

7.2.4   Green Technology Key Area Identification

The realization of urban green development needs to focus on material green technologies and methods in all aspects of urban planning, construction, operation, and maintenance. There is a need to identify and select green technologies that can solve current outstanding problems while being forward-looking, comprehensive, innovative, and practical.

In order to promote the city's "full process and full chain" green development and ultimately achieve the goal of zero-carbon emissions, we should comprehensively sort out and analyze the green technology system from the four dimensions of "resource protection and utilization, production and construction, consumption and use, decomposition and recycling." Under the dimension, many technical fields are further identified (Table 7.3).

Table 7.3 Main fields of green technology

Resource protection	Production processes	Consumption processes	Decomposition processes
Water resource conservation	Cyclic economy	Energy structure	Solid Waste Treatment Technology
Economical use of land	Energy Technology Manufacturing Technology	Energy Supply Technology	Water Treatment Technology
Sustainable energy	Construct technology	Transportation Technology	Waste Gas Purification
Eco capital service	Green Building	Operation and maintenance of buildings	Renewable resource recycling technology
Use of meteorological resource	Material Technology Agricultural Technology	Urban infrastructure	Kitchen waste collection and recycling technology
urban public space climate adaptation model	Food Production	Food supply chain	
...	...	...	...

The European landmark urban green development program documents: *Aalborg Charter* (1994), *Aalborg Commitment* (2004), *Leipzig Charter: Sustainable European Cities* (2007) and the *Freiburg Charter* (2010) demonstrate that for more than 20 years, energy, transportation, construction, water environment and food (agriculture) have been important areas of low-carbon green development in Europe. In the "European Green Agreement" issued by the European Commission in December 2019, special attention was also paid to energy, construction, transportation, agriculture, pollution prevention, biodiversity, and sustainable industry (Fig. 7.3).

Preliminary research has been focused on analyzing the outstanding problems in various fields and the development needs of green technology, i.e. the innovation and maturity of green technology, the wide range of applications, and the sense of urban residents' gains. As agreed by the Chinese and foreign team leaders, the Chief Counselor of the China Council for Foreign Cooperation agreed that before May 2020, Green Technology SPS focused its research on the six fields of **water, energy, transportation, building, land use and planning, and food**. In these six fields, single or comprehensive technology screening and evaluation will be carried out, and a list of green technology application promotion lists will be proposed to provide advice and support for the promotion and application of green technologies in Chinese cities during the 14th FYP period (Table 7.4).

Through a comparative analysis of China's and Europe's key policy documents on green development, we believe that it is necessary to systematically analyze and identify green development areas and that the six key areas selected at the first stage are appropriate.

7.3 Assessment Methods, Development Directions and Implementation in Key Sectors for Green Technology

7.3.1 Major Green Technology Assessment Method

The actual effect of the application of green technology has always been a matter of great concern and continuous debate. The lack of prior and necessary objective assessment in practice combined with blindly promoted green technology has caused a great waste of resources and negative effects. Constructing a comprehensive green technology evaluation system and broad, multi-dimensional valuation of technologies is crucial and urgent for the promotion of green technologies.

	Aalborg Charter	Aalborg Promise	Leipzig Charter	Freiburg Charter
Core issues and concerns	**Sustainable development of urban economy:** investment to protect existing natural resources **Sustainable land use structure:** promoting the development of efficient public transportation and efficient energy supply, maintaining a humane scale increasing the density, striving to achieve a mix of functions **Sustainable urban transport structures:** giving priority to eco-friendly modes of transport **Avoid pollution of ecosystems:** air, **water,** soil and **food**	**Public natural resources:** renewable energy; efficient use of water; ecological agriculture and sustainable forestry economy **Consumption and lifestyle with a sense of responsibility:** waste reduction and recycling; Improve energy efficiency at the end of consumption **Urban planning and urban development:** transformation and renewal of declining or disadvantaged areas; reasonable urban density to avoid urban expansion; Mix the functions of buildings and development areas priority to residential development in the city centre **Improve the way of travel, reduce the traffic:** reduce the need for private motor vehicle travel; Increase the proportion of public short-distance transport routes to pedestrian or bicycle routes; Promoting the transformation of motor vehicles into vehicles free of hazardous emissions; Formulate comprehensive and sustainable local transport development plans **From local to global:** integrating climate protection policies into relevant strategies and regulations in the field of energy, transportation, procurement of waste, agriculture and forestry	**Create and secure high-quality public spaces:** **Enhance the modernization of infrastructure networks and improve energy efficiency:** urban transportation systems should be coordinated with regional transportation systems; urban transportation must be coordinated with the functional requirements of housing, employment, environment and public space; Improve energy efficiency; Create compact residential structure : Stock building in vulnerable urban areas should be improved; **Promote high-performance, low-cost urban transport:**	**Mixed, safe and inclusive cities:** build different living and working spaces for all residents/promote the development of innovative living forms **Short-distance cities:** Facilities and institutions that have good accessibility to cities and regional centers should be further developed. Follow the development concept of "compact, distributed city"; **Urban development/ density model along public short-distance transportation:** increase building density in areas along public short-distance transportation

Focuses of this SPS: **Land use and planning, energy, transportation, construction, water, food**

Fig. 7.3 Four programmatic documents of the European sustainable towns movement

Table 7.4 The six key research areas

Areas	Explanation
Water treatment and water resources	Protecting water quality, managing water pollution, and promoting recycling, safe and efficient use of water resources
Clean and sustainable city energy	Reduce ore energy use and greenhouse gas emissions, promoting the use of renewable energy and clean energy
Improving city transportation	Promoting green travel, promoting energy conversion and emission reduction of motorized transportation, and optimizing transportation
Developing green building	Promoting the development of high-quality, low-consumption, low-emission buildings and construction technologies
Optimizing land use and planning	Improving land use patterns, optimizing urban layout, building low-carbon communities, and promoting mixed-function planning, and keeping development at a human scale
Urban food production and supply	Exploring city food production systems, protecting biodiversity and urban green production space

7.3.1.1 Comparative Study of Evaluation Approaches

At present, there are two mature mainstream approaches in use globally for evaluating green technologies. One is systematic scoring based on a detailed indicator framework, with weight being given to each indicator. Evaluation conclusions are based on the final score. This approach applies mostly to clearly defined evaluation objects. Examples include the U.S.'s Leadership in Energy and Environmental Design (LEED) standards in green architecture evaluation [1], the DGNB system of Germany [2], the Green Building (GB) Tool in Canada [3], etc. The other approach is framework evaluation, in which the priorities of evaluation are preset with an evaluation framework. This approach is mostly applied to the evaluation of green technologies in the broad sense. For example, the S&P Global Ratings Green Evaluation takes transparency, governance, and mitigation/adaptability as the primary dimensions; under each, there are secondary evaluation items to be scored or rated against [4]. The evaluation is then summarized around the three topics of low-carbon, finance, and governance/management. This approach can be used for seven sectors, including green architecture, green energy, water resource, green mobility, energy efficiency, nuclear energy, and fossil fuel-fired power generation.

The promotion of green technologies in China has mostly taken the form of promotion catalogues, and no evaluation method or system has been established for green technologies. The *Evaluation Standards for Green Architecture* published by the Ministry of Housing and Urban and Rural Development have not been widely applied nor substantiated by the capabilities of professional technical institutions. In the absence of a comprehensive life-cycle all-in-cost evaluation, many so-called green technologies in China have actually created negative impacts such as waste of resource and safety hazards once scaled up. Examples include shared bikes, EVs, and integrated infrastructure corridors.

7.3.1.2 Recommended Framework for Evaluation

Evaluation of green technologies should be done based on measures of life cycle and all-in cost to investigate their effectiveness, economic feasibility, scalability of emissions, and consumption reductions as well as user acceptability.

Life-cycle evaluation: Before, during, and after the application. The focus of green technologies should not be just on their application per se with no regard given to pre- and post-application green performance. We should use the life-cycle evaluation of green technologies from their conception to scrapping. That is, conception and design, resource pooling, operation, maintenance, decommissioning, and decomposition should be investigated to demonstrate the technology's performance in terms of emissions reductions.

Multi-dimensional, all-in cost evaluation: Incremental cost versus incremental benefits. When defining the dimensions for green technologies, consideration must be given to economic returns and financial feasibility because that is the foundation for scaling up; in the meantime, the acceptance and recognition from the general public must be considered in addition to the carbon-reduction benefits and economic returns. Benefits should thus be understood as a comprehensive package of economic, environmental, and social returns. In view of that, we propose a multi-dimensional evaluation framework that covers the aspects of economics, environment, and social, in which the breakeven is defined based on overall cost.

An evaluation framework involving three stages and three core dimensions. The life-cycle evaluation of a green technology covers the three stages of **before, during, and after its application** and the three main aspects—**economic, environmental, and social.** The evaluation makes overall cost and benefits analysis for green technologies through formulating a comprehensive evaluation framework.

The economic dimension looks into the incremental cost and incremental benefits of a green technology in its three stages of before, during, and after the application. The environmental dimension examines the economic returns in terms of emissions cutting and energy conservation in the three stages. The social dimension involves government support, feedback from users/citizens, and willingness of businesses to embrace the technology.

Therefore, the life-cycle evaluation framework for green technologies adopts **quantitative technical assessment in the economic dimension and environmental dimension combined with qualitative assessment in the social dimension.** Recommendations about future promotion will then be made based on the evaluation results.

7.3.1.3 The Formulation of Evaluation Procedures

Technical evaluation. In view of the various characteristics of a green technology, technical feasibility should be demonstrated through a quantitative evaluation from the perspectives of economics and environment using a life-cycle approach. First, taking recommendations from industry experts and companies as the main source, absorbing international experience, screen key areas, summarize all known green technologies, and include them in the candidate green technology library to form a long list of green technologies in the field. Secondly, from the three aspects of technology readiness, problem targeting, and promotion feasibility, select the green technologies that are likely to be adopted by the national 14th FYP, to form a list of green technologies. Third, establish quantitative evaluation indicators from the economic and environmental dimensions, forming a four-quadrant matrix evaluation based on carbon reduction and financial evaluation. From this matrix evaluation, a shortlist of green technologies was further selected (Figs. 7.4 and 7.5; Table 7.5).

Technologies recommended. For candidates on the shortlist, a qualitative assessment will be made on the social dimension. Three groups, namely businesses, citizens/users, and industry experts will be consulted for their comments on the candidate technologies. Expert validations will then be synthesized into evaluation conclusions and recommendations: candidates with good prospects for scaling up will be nominated.

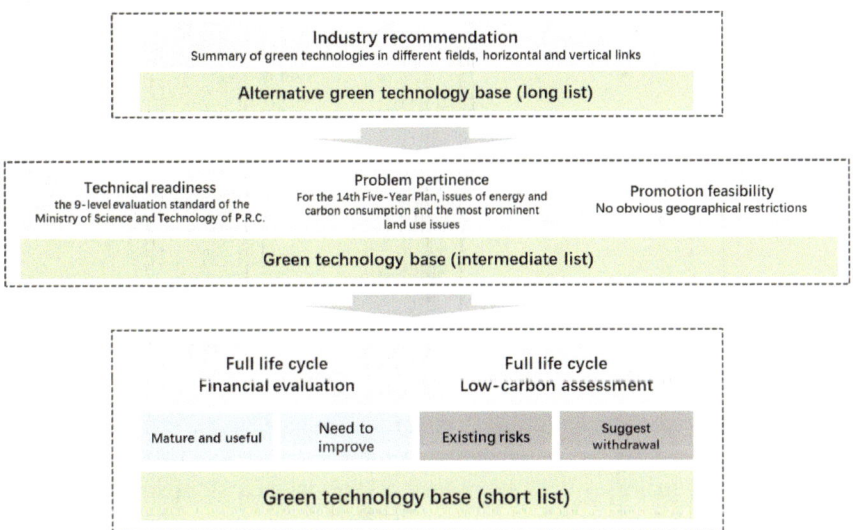

Fig. 7.4 Process of the green technology "long list—intermediate list—short list" 3- steps technology assessment

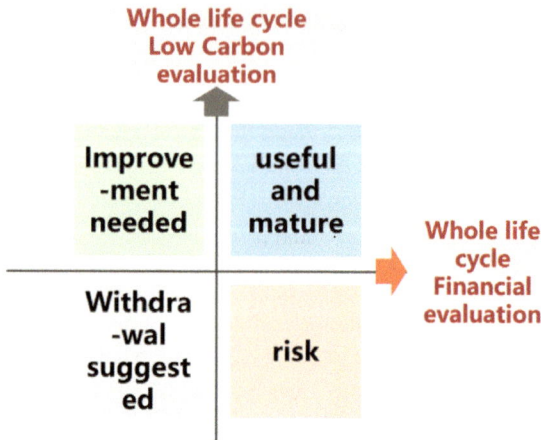

Fig. 7.5 Green technology full life-cycle finance—low-carbon 4-quadrant evaluation system

Table 7.5 Green technology finance—low-carbon evaluation system

	Index dimension	Major index
Economic dimension-financial evaluation	Incremental cost	Research and design, production, processing and construction, operation and maintenance management, disposal costs, etc
	Incremental benefit	Resource savings (water savings, energy savings, material savings, land savings, etc.), economic benefits
Environment dimension-low-carbon evaluation	Carbon emission	Amount of carbon emissions
	Environment quality	Emissions of other air pollutants, water and soil environmental impact

Businesses responsiveness: Understand the advantages and hurdles in product manufacturing and/or application of a technology; comprehensively evaluate their acceptance and willingness to adopt a green technology.

Diverse citizens/users acceptability: Collect data from end users on indicators such as comfort, safety, fairness, and convenience; conduct surveys to get feedback on various green technologies in terms of their acceptance and "sense of gaining" by the public.

Government policy: Review existing policies for the degree of support, comprehensively evaluate related resource investments and governance capabilities.

Industry experts validation: Validate the data and process of the evaluation; review the environmental benefits and application prospect of a green technology; propose a final list of recommended technologies.

7.3.1.4 Assessment of the Evaluation Methods

Green technology involves many fields, and there are many types of technology with significant differences. In researching this project, the aforementioned evaluation methods have been applied to the green technology recommendation process in the six pillars of water, energy, transportation, construction, land, and food.

In the evaluation stage, it is important to first fully identify the green connotation in various fields, classify it according to its outstanding green utility, and define the technical level that can be evaluated. Second, it is recommended to adjust the specific evaluation indicators of the economic and environmental dimensions according to the specificity of the field. For example, the technical focus in the field of land use and planning should be to improve environmental and life benefits. In addition, highly comprehensive technologies are split into multiple individual technologies, which are evaluated separately and combined afterwards. Third, the focus differs according to each field. For example, assessment of technologies in the field of water and energy focuses on enterprise responsiveness; land use and planning mainly consider government policies and expert evaluations.

7.3.2 Chinese and Foreign Experiences and Emerging Best Practices

7.3.2.1 Water Sector

In terms of the water sector, China has always been faced with problems such as insufficient water resources, uneven distribution of water resources, pollution-induced water shortages and rapid growth in water demand. Meanwhile, China is also confronted with many challenges, such as the insufficient treatment efficiency of sewage and low utilization efficiency of recycled water. Therefore, the Chinese government launched the "Major Science and Technology Project of Water Pollution Control and Governance"[1] in 2016, started the work "Pilot of Construction of Sponge City" in 2015 and proposed the "Three-year Action Plan for Quality and Efficiency Improvement of Urban Sewage Treatment" in 2019 which focuses on comprehensively improving the efficiency of the sewage treatment system [5].

---

[1] Known as the "Special Project of Water" for short, it is listed in the major technical breakthrough project in the outline of national mid- and long-term scientific and technological development planning.

Sewage Treatment and Water Recycling Economy. More than 80% of the world's fresh water eventually becomes sewage, and the discharge of sewage that is not managed properly can cause health hazards and a shortage of water resources. Sewage treatment technology consists of a combination of physical, chemical and biological processes to remove solids, organic matter, nutrients and pathogens from sewage. It can solve the problem of water shortages. The byproducts of sewage treatment can generate energy [6], and it can not only enhance a city's water recycling but also decreases the cost of downstream water use [7]. Furthermore, it can recycle up to 90% of the methane gas, which can be used for clean power generation and heating. China has launched an urban sewage treatment program and black and odorous water regulation with carrying out the pilot in 60 cities.

Utilization of Reclaimed Water. Generally, developed countries attach great importance to the utilization of high-quality reclaimed water. The core of the utilization of reclaimed water is the safety of reclaimed water quality. In 2009, Japan published its *White Paper on Sewers*, which stipulates that the government shall strictly control the quality of reclaimed water and regularly disclose the monitoring results to the public, so as to promote the safe utilization of reclaimed water. In Australia, Sydney has adopted a high-tech remote control system to automatically and continuously monitor the operation of recycled reclaimed water to ensure the safety of reclaimed water.

Non-Revenue Water Management (NRW). Non-revenue water is one of the most serious problems of the urban water supply system. It reduces the amount of water that is lost before it reaches the customers by adopting comprehensive methods, including leak detection, pipeline assessment, pressure management and hydraulic modelling. It reduces energy consumption and emissions, and positively affects the financial performance of water utilities by reducing the amount of water lost during conveying, treating, and distributing [8, 9]. The International Water Association summarized a series of methods, procedures and evaluation indicators that have drawn broad attention from many water supply enterprises in China to guide the control of the NRW leakage and loss from pipe network.

7.3.2.2 Energy Sector

For China to achieve its goal of reaching a carbon dioxide emissions peak in 2030, it must control total energy consumption, improve energy utilization efficiency, and vigorously develop renewable energy. Currently, the utilization of new energy technology in China is seeing the rapid development of innovation breakthroughs, industrialization demonstrations, application promotion and demonstration. However, there are still key bottlenecks and obstacles for the safe transmission, distribution, storage, and high-efficiency utilization of new energy.

Integrated Green Energy Grid [10]. This type of grid means that variable supply can be realized through technologies such as solar photovoltaic (PV), wind, battery, and pumped hydro energy storage, smart grids and electricity meters, using abundant wind energy in the winter and solar in the summer to cover heating and cooling demand peaks respectively, using short-term battery storage to account for variable supply and longer-term storage such as hydrogen to solve the problem of the balance of electricity generation and demand during peak and off-peak hours [11]. At present, the application of an integrated green energy grid based mainly on the microgrid in communities has become an important method for solving power problems in countries such as the United States and Japan: demonstration projects in the United States account for about 50% of all global projects [12]. Moreover, the United States, the European Union and the United Kingdom have formulated policies to encourage the scaling of the green grid. Investment in energy storage technology will be key to unlocking the potential of the green grid.

Internet of Energy (IoE). It means that Internet of Things (IoT) technology and processes will be applied to the energy industry to improve energy efficiency and reduce waste. These include smart metering, remote control and automation systems, and smart sensor and demand response systems. According to the International Energy Agency, the IoE can reduce the operating costs of power systems by 2–11% and reduce the demand for electricity generation from fossil fuels by 30% [13], which is helpful for improving grid reliability and safety as well as asset utilization.

Near-Zero-Emission Cooling and Heating. This is a technical method that fully considers emission reductions, energy conservation, and pollution reductions in the cooling and heating systems of buildings and industries, including solar power generation for buildings, combined heat and power generation, and waste heat utilization in urban areas, ground source heat pumps in communities, and heat insulation of buildings etc. [14]. The geothermal heating in Xiongxian, Hebei Province and Xi'an City, Shaanxi Province are successful practices in China. Near-zero emissions equipment often has higher upfront costs, but are cheaper over its lifetime; for this reason, consumers need to be incentivized by policies to make upfront investments.

7.3.2.3 Transportation Sector

Transportation in urban areas in China faces a prominent mismatch between supply and demand, with problems such as insufficient development of green and intensive transportation, the rapid growth of transportation energy consumption and emissions, and increased air pollution through vehicle emissions [15]. Therefore, China has proposed coping strategies, including promoting the new energy-oriented means of transportation, intelligence and information-oriented development of transportation in urban areas, priority development of public transportation, and sustainable development of transportation.

Intelligent Transportation System. It is the main development direction for the transportation sector in urban areas in China to promote the construction of connected vehicles, IoT and public cloud services, and the in-depth integration of new technologies such as big data, Internet, and artificial intelligence with the transportation sector [16]. In recent years, Europe and the United States have implemented many successful pilots in the seamless integration of the travel system.

New Energy Vehicles and Supporting Facilities. China is promoting the development of electric and new energy-oriented public transport vehicles and vigorously developing blade electric vehicles and supporting charging facilities. Moreover, electric vehicles with hydrogen cells are also a focus of attention. An electric vehicle's emissions are up to 43% lower than a fuel-driven vehicle's [17]. The United States and Japan have applied hydrogen energy to large-capacity passenger service vehicles and heavy freight vehicles and have carried out a large number of technological developments. In Europe, the promotion of electric vehicles through various policy measures in Norway, for example, is a practical case of great value.

Transportation Demand Management and Cycling Trip. This can effectively reduce the number of vehicle trips, alleviating traffic congestion and carbon emissions through the transportation demand management to promote changes in traffic behaviour. Strategies include road pricing, parking pricing, as well as providing better alternatives to driving. For example, China has clearly proposed the strategy of developing the rail transit in a metropolis, giving priority to public transportation, and encouraging shared transportation [18]. The European Union promotes bicycling, walking, public transportation and shared transportation [19]. The European Union also attaches more importance to the demonstration and guidance of the trips by bicycle, and the Bicycle Expressway Project in Copenhagen is a renowned best practice in the world.

7.3.2.4 Building Sector

Energy consumption for building operation (20%) and construction (20%) account for about 40% of society's total energy consumption, and this is projected to grow. Therefore, its reduction is a necessity if China is to achieve its low-carbon goal. Meanwhile, with the development of a social economy, the requirements for buildings to be safe, comfortable, and healthy are increasingly higher.

Green Building. This refers to providing a healthy, suitable, and highly efficient building space with saving resources and protecting the environment throughout the entire life cycle. In 2019, China expanded the definition of green building performance from conservation and environmental protection to five aspects: safety and durability, health and comfort, convenience of life, resource conservation, and livable environment, which brought new development prospects.

Near-Zero Energy Building. This refers to adopt passive building design that reduces the need for heating, air conditioning, and lighting by taking advantages of natural conditions and natural forces, or takes active technical measures to maximize the utilization efficiency of energy equipment and systems while creating and fully utilizing the renewable energy and realizing low energy consumption. Developed countries have implemented a large number of practices of near-zero energy buildings and zero-energy buildings, such as the SDE4 building with net-zero energy consumption by the School of Design and Environment, National University of Singapore. In 2019, China issued the *Technical Standard for Near-Zero Energy Buildings*.

Healthy Building. This is a recently emerging technology field that focuses on the characteristics of the building environment that affect human health and welfare, such as air, water, nutrition, light, health, and degree of comfort. In 2012, the United States proposed certification standards for WELL (healthy) buildings. In 2016, China clearly proposed the initiative "creating a green and safe healthy environment and reducing the incidence of diseases"[2] and released the *Assessment Standard for Healthy Building* and *Evaluating Standard for Healthy Housing*. The Chuangye Building in Shenzhen Bay, as one of the optimal practical cases in China, has obtained the certification by WELL.

Smart Operation and Maintenance of Building. A system which carries out the operation, optimization, and control for the building and equipment based on the increased use of AI technology, monitoring and analysis of information such as building environmental quality, energy and water resources consumption, and user needs; it can achieve the energy conservation for equipment and behaviour mode through the digital management and providing people with good training to ensure the building is green, low-carbon, and efficient. The "Building Management System" falls into this category. The emerging optimal practices include the Edge Building in Amsterdam and Tencent Binhai Building in Shenzhen.

7.3.2.5   Land Use and Planning

In the past 30 years, land in urban areas in China has experienced a series of problems such as excessive expansion, disorderly layout, large-scale demolition and construction, and imbalanced supporting facilities, all of which have affected the ecological environment and the rapid increase in carbon emissions. The planning and control for the scale, form, intensity, network, and node of land in urban areas serve as the source technology to make urban areas green and low-carbon.

Green Urban Form. Advanced cities around the world reduce carbon emission and respond to climate change to achieve green development through the urban planning such as increasing urban compactness, reasonably planning population density and accessibility, a green space layout model, along with mixed land use and transit-oriented development. The optimal practices internationally include the overall planning and urban form control in Copenhagen, Denmark [20].

Green and Livable Block. The practice of green community in international advanced cities mainly revolves around the "3D Principles" (Density, Diversity,

---

[2] Quoted from *Plan of Health China 2030*.

Design) [21], which include a series of technical measures such as providing a slow-moving environment, improving the utilization of mixed land, attaching importance to street interconnection and protecting public open spaces, to achieve the building of a livable environment and the reduction of comprehensive energy consumption. The international and domestic emerging optimal practices include the Hammarby Sjöstad community in Stockholm, Sweden, the Gauteng Community in Utrecht, Netherlands, and the China-Singapore Tianjin Eco-City etc.

7.3.2.6 Food Sector

China's food supply is not only facing the challenges of increasing food demand, arable land, water resources, climate change and other factors; it is also facing the problems of food security, long-distance transportation, imported animal feed, high food waste rate, malnutrition caused by overnutrition and other urbanization factors. Therefore, the food sector needs to be more closely integrated with urbanization to guarantee green, low-carbon, nutritional, and safe urban food.

Food Traceability. The traceability of foods by technical means can comprehensively track the environmental, economic, health, and social impacts at all stages including food production, sales, and consumption, so as to better identify and respond to food safety issues, ensure consumers' rights and make enterprises earnestly fulfill the promise of sustainability, support supply chain optimization, reduce food loss, and safety of food supply. This can be achieved, for instance, via the IoT, which enables comprehensive data collection and blockchain tracking, aggregates and shares data from supply chains, and allows food sensing technologies to identify safety and authenticity.

Smart Agriculture. Control and optimize agricultural production, services, and sales through IoT, Internet and AI technology, through such methods as analyzing data, ambient temperature, rainfall, and soil salinity etc. Smart agriculture can realize more thorough agricultural information perception, deeper agricultural intelligent control, and more direct and transparent public services. China has proposed a modern agriculture strategy focusing on "Internet + ," and Tencent is actively engaging in the "AI + Agriculture" and has made preliminary progress.

Urban Agriculture. This refers to activities that plant and cultivate agricultural products using the land, water, and building space scattered in various corners of urban areas or suburbs and processes and sells agricultural products with the purpose of satisfying the demand of customers in urban areas. Urban agriculture can enable consumers in urban areas to obtain healthy and nutritious foods while improving the land utilization rate, reducing carbon emissions from food transportation and improving the ecological environment. Urban agriculture is well developed in countries such as Singapore and Japan. For example, Singapore has proposed the "Vision 30.30" action which aims to achieve the goal of self-sufficiency for 30% foods by 2030 through developing urban agriculture.

7.3.3 Technology Recommendation in the Six Pillars During the 14th FYP

Based on primary technology development directions, considering the outstanding problems in China's water sector, combined with Chinese and foreign practices, Chinese and foreign experts jointly recommend the following green technologies in each sector, during the 14th FYP.

7.3.3.1 Water Sector

Technology to improve quality and to increase the efficiency of sewage treatment and the integrated system of plants, networks, and rivers: This technology aims to construct, coordinate, and manage sewage treatment plants, drainage pipe networks, water treatment, sludge resource utilization, bank waste recycling system construction, and reclaimed water utilization, etc. It should solve the problem of low-influent concentration of pollutants in sewage treatment plant while improving the total amount of pollutant emission reduction of sewage treatment facilities, while ensuring their safe and efficient operation.

This is the intersection of sewage treatment and the sponge city. It not only pays attention to the problems of traditional sewage fields such as sewage treatment plants and water treatment; it is also concerned with the problems of the water recycling economy such as sludge resource utilization and recycled water utilization. Besides, the unified scheduling and management of drainage facilities (as well as the integration with sponge city technologies such as the initial stormwater pollution treatment) are coordinated to solve the problem of sewage collection, so as to improve sewage treatment efficiency.

Smart operation technology of reclaimed water system: Combined with urban topography, reclaimed water utilization and urban land layout, this technology optimizes the service scope, site selection, and water intake mode of reclaimed water plants. Using optimization technology to improve water resource allocation and the pipe network system should help save water resources and reduce energy consumption and operation costs; establish a dynamic simulation system of reclaimed water transmission and distribution based on the digital simulation model; and reduce the leakage of pipe networks while improving operational efficiency.

This is the intersection of reclaimed water and non-revenue water management technology. Through big data analysis and smart simulation methods, it can provide optimized solutions to major problems in the field of reclaimed water, such as service scope, site selection, and water intake methods of reclaimed water plants. Establishing a dynamic simulation system and optimization technology for reclaimed water transmission and distribution can improve the allocation of water resources and the pipeline network system, reducing leakages in the pipeline network, and increasing the benefits of non-revenue water management.

Water quality assurance technology of the reclaimed water system. This establishes an intelligent management and control platform to realize the intelligent management and control of reclaimed water system quality and ensure the reliability of reclaimed water quality, quantity, and pressure. The system is based on model simulation, automation, and big data technology.

This is the most important technology in the field of reclaimed water. It can intelligently and efficiently control the quality of reclaimed water and realize high-quality reclaimed water to supplement urban water supplies. The technology can also establish a water quality traceability and emergency response mechanism for any new coronavirus epidemic situation and other special situations, to ensure the level of disinfectant in the reclaimed water system, in order to respond to major public safety risks.

7.3.3.2 Energy Sector

Mid-deep geothermal utilization technology: Through exploration and drilling, geothermal water extraction or underground heat exchange is used to extract underground heat energy, combined with heat exchange and heat pump technology to heat and cool surrounding buildings as necessary.

This is a major technology in the field of near-zero-emission refrigeration and heating. The northern and middle regions of China are rich in geothermal resources, and more than 7 billion square metres of buildings (about 50%) have not yet used clean energy for heating [22]. As a highly efficient and stable clean energy heating method, this technology has broad prospects. In addition, although the construction and installation costs of this technology are high, operation and maintenance costs are low, and the overall economic benefits are superior to heating in large boiler rooms.

Energy Internet-integrated management platform technology: based on blockchain, AI, smart grid, big data, etc., this technology coordinates the relationship between energy supply and demand; a comprehensive service platform is established from the combination of energy planning and design, engineering construction, transmission, and distribution and control, data interaction, energy efficiency monitoring, smart operation and maintenance, and market transactions.

This is the first step in the practice of "energy Internet," and it is also the intersection of integrated green energy grid and energy Internet. On the one hand, the integrated management platform can break the original information island, realize the intelligent energy consumption method of full-service data sharing, and promote the construction of an integrated green energy grid; on the other hand, through Internet technology, it can provide services such as measurement certification, market transactions, energy finance, smart dispatch, and operation optimization, thereby maximizing energy efficiency [23]. In order to enable this technology, there is a need for more dynamic pricing of energy, along with attention to the management of cybersecurity risks.

Microgrid technology: This technology consists of distributed power sources, energy storage devices, energy conversion devices, power distribution facilities, electrical loads, monitoring and protection devices, etc., and is an intelligent DC or AC power generation and distribution system in a certain area.

This is the key technology in the field of integrated green energy power grids [24]. The microgrid features investment savings, green and efficient production, flexible operation, good resilience, integration of power generation and storage, and convenient access and application of renewable energy. It can make up for the shortcomings of traditional large power grids, such as difficulty in peak shaving and poor stability; it is most suitable for new urban areas and remote areas. China has launched microgrid demonstration applications since 2017.

Industrial waste heat central heating technology: Large-scale industrial enterprises such as power plants will emit a large amount of waste containing more low-grade heat energy during the production process. By extracting the heat energy from the waste, centralized heating can be achieved for the surrounding areas.

This is another major technology in the field of near-zero-emission refrigeration and heating. This technology can make full use of industrial waste heat in China's energy-intensive industries, while effectively controlling environmental and atmospheric pollution and reducing carbon emissions. Using this technology for urban heating does not need to consume a lot of high-quality energy such as natural gas and electricity, and can reduce heating costs and emission reduction pressures. It is a good alternative low-carbon method compared to China's "coal to power" and "coal to gas" [25].

7.3.3.3 Transportation

Mobility as a service (MaaS) technology: Through a single platform and one-stop service, it connects government management departments, links users and transportation companies, and effectively integrates many different transportation resources such as cars, buses, bicycles, sidewalks, and shared transportation. By auto-pricing and other means, it provides travellers with a variety of travel packages.

This is a major emerging technology in the field of smart transportation systems. It is cleaner and more efficient than traditional transportation systems. It can reduce costs and save time while increasing capacity. Studies have shown that with the full implementation of this technology, the cost of a single trip can be reduced by 25–35%, capacity increased by 30%, and travel time reduced by 10%. At present, China is advancing the research and development of this technology, and cities such as Shenzhen and Beijing have already carried out pilot projects [26].

Hydrogen energy vehicle technology: A new energy vehicle driven by fuel cells with hydrogen as the main energy source. This technology is a major technology emerging in the field of new energy vehicles and supporting facilities in recent years. At this stage, there are two types: hydrogen energy vehicles and hydrogen energy rail transit vehicles.

Hydrogen energy vehicles rely entirely on hydrogen energy, with zero carbon emissions and no pollution. In addition, it has the advantages of high energy density and conversion rate of hydrogen fuel cells, long life, easy recovery of raw materials, and small space occupied by hydrogenation facilities. Hydrogen infrastructure can also be used for energy storage and to heat homes.

Hydrogen energy rail transit vehicles consist of hydrogen and fuel cells to form a power system. This technology can get rid of the traditional line traction power supply system, reduce investment, and has the characteristics of low noise, low pollution and long service life [27]. Pilot projects for hydrogen energy rail technology have been deployed in the United States, Japan, Spain, and Germany.

Intelligent charging system technology: Based on the IoT, Internet of Vehicles, artificial intelligence, and energy demand management, this technology coordinates the power supply side, charging side, and transmission and distribution network to realize the digitalized, scenario-based and intelligent operation of new energy vehicle charging services. It is composed of intelligent charging pile, vehicle pile network, intelligent charging information management platform and so on.

This is another important technology in the field of new energy vehicles and supporting facilities. The lack of charging supporting facilities is one of the main obstacles for the current large-scale promotion of pure electric vehicles in China. This technology can alleviate the contradiction between supply and demand for vehicle charging, and at the same time better balance the load of the power grid. In the future, through vehicle-to-grid (V2G) technology, electric vehicles will be integrated into the energy storage infrastructure and can give power to the grid during peak times.

Bicycle special road technology: This is a special road facility technology for bicycle traffic. It can be flexibly adopted in the form of elevated or ground layout to ensure the independent special road rights of bicycles.

This is also one of the specific technical practices and applications in the Travel Demand Management (TDM) series of technologies, and has a significant role in improving the bicycle using rate and reducing carbon emissions. According to relevant research, the rate of bicycle trips along the route can be increased by 10–20%, and carbon emissions can be reduced by 60–70 tons/km per year [28]. In addition, the technology has good social recognition, high comprehensive social and economic benefits, and is urgently needed in China [29]. Beijing, Xiamen, and other cities have already conducted preliminary trials.

7.3.3.4 Building Sector

"Steel structure + modular internal space" technology: "Modular internal space" refers to the design of the building as a large column network, high-rise high-standard modules. It can realize the free division of internal space, the separation of building structure and building equipment pipelines, as well as the adaptation of equipment and facilities to changes in functional space, thereby making the building more durable. The steel structure can make a large space that is safe and economical, while also more conducive to the realization of "modular internal space."

This is one of the key technologies in the field of green building. China newly builds about 2 billion square metres every year. The traditional construction method consumes large amounts of building materials, the construction site environment is poor, the construction time is long, and the construction quality is uneven. As a result, the average lifespan of Chinese buildings is only 30 years. This technology has the advantages of recycling materials, saving building materials, reducing construction waste, reducing on-site operations and workers' needs, improving the construction environment, shortening the construction period, etc., while further solving the problem of space supply and demand matching, and space safety supply, thereby improving building life.

Building three-dimensional greening technology: This refers to the use of building roofs, overhead floors, balconies, window sills, walls and other building parts for greening technology, including plant selection and matching, building structure, maintenance management system, etc.

This is a key technology in the cross fields of green building and healthy building. The density and intensity of urban construction in China are relatively high, and the green space per capita is limited. The development of three-dimensional greening is a realistic choice. This technology has multiple benefits, such as improving building thermal performance, improving building microclimate, beautifying human settlements, and rebuilding biodiversity. Studies have shown that with this technology, the duration of natural ventilation that can be applied to buildings increases by 35.3%, and the cumulative amount of cooling load that can be treated by ventilation increases by 8.81%; the external surface temperature, internal surface temperature, and indoor air temperature of the summer wall of the building are respectively lowered by 21.6, 5.7 and 5.2 °C; the air conditioning power rate can reach 39.97%.

Photovoltaic, building-integrated photovoltaic (BIPV), distributed energy storage, and DC power supply technology: This technology integrates photovoltaic power generation with distributed energy storage and DC power supply in buildings. It is a necessary technology in the field of near-zero energy consumption.

It combines the advantages of renewable energy utilization and distributed energy storage to improve energy security and DC power supply to improve energy supply and demand matching.

Photovoltaic distributed energy storage can effectively solve the contradiction between the uncertainty on the power source side and the peak–valley changes on the load side, and realize low-carbon energy. With the improvement of photovoltaic efficiency and cost reduction, as well as the development of battery technology, photovoltaic and distributed energy storage has significant potential for application in buildings. DC power supply has more advantages in terms of safety, efficiency, reliability, and distributed power supply coordination and constant power supply. At the same time, because of the use of a safe voltage power supply, it is more friendly to children and the elderly. The combination of the two has broad technology and market prospects.

Group intelligent building system technology: This is a new-generation building intelligent platform technology, which realizes intelligent control of the building environment and electromechanical equipment, improves user comfort, improves the operating efficiency of building systems, and reduces building energy consumption. The swarm intelligence system takes the space and source equipment as standardized units and connects them into a computing network covering buildings and cities according to their spatial location. It adopts a decentralized architecture and designing a parallel computer system based on the changing process of the physical field, so that the computing process and the physical field achieve deep integration. In addition, through a self-organizing intelligent community, it optimizes overall functioning.

This is an emerging technology in the field of smart building operations. Existing intelligent building systems generally have problems such as inaccurate correspondence between data and physical systems, a long engineering period, and monitoring without control, which makes it difficult to adapt flexibly to changes in urban systems. Group intelligent building system technology can control electromechanical equipment through monitoring and intelligent algorithms to achieve accurate and efficient matching of building environment supply and demand. Research from other countries shows that relying on swarm intelligence algorithms (including typical particle swarm intelligence algorithms and simplified swarm intelligence algorithms) can reduce building energy consumption by more than 25%, while the construction project cycle is only three to four weeks, with an initial investment cost equivalent to the general building management systems.

7.3.3.5 Land Use and Planning

Green city form technology package: This mainly includes two technical tools: urban development boundary delineation and city main function layout along the bus corridor. The delineation of urban development boundaries is mainly through defining the external boundaries of urban growth, strictly controlling the large-scale extensive expansion of cities, and promoting intensive urban development; the main functions of cities are developed along public transport corridors, mainly through the coordinated development of urban functions and public transportation, which increases the overall public transportation trip ratio and reduces the carbon emissions of trips through a rational function layout [30].

The delineation of urban development boundaries can help control the size of the city and reduce transportation energy consumption. At the same time, it forces more resources to be transferred to the city's internal development, guiding the government and enterprises to phase out high-pollution and high-emission industries, further reducing industrial energy consumption, and improving land use efficiency. From 1990 to 2013, the city of Portland in the United States delineated its development boundary: population growth was cut nearly in half, and carbon emissions fell by 14% (during the same period, carbon emissions in the United States increased by 6%) [31].

Guiding the development of the city's main functions along the bus corridors, locating residents' commuting, shopping, and use of urban public services around public transportation stations, increases residents' willingness to use public transportation and reduces the energy consumption of private cars [32]. Stockholm, Sweden, guides the development of the city's main functions along the bus corridor, allowing more than 60% of residents to use public transportation for travel, much higher than other European cities [33].

Green livable carbon–neutral block technology package: This mainly includes three technical tools: multi-layer high-coverage building layout; street space design of a dense road network in small blocks; and comprehensive implementation of bus-oriented development in combination with public transportation stations. A multi-layer high-coverage building layout limits the height of residential buildings and matches the corresponding floor area ratio and building density indicators, comprehensively using neighbourhood scale optimization, building regression control, and open space optimization methods to achieve a relative balance between residential density and human settlement environment quality. The street space design of the dense road network in small blocks focuses on improving the interconnectivity of streets, increasing road density, creating a comfortable and convenient walking environment, and reducing the use of private cars. Combined with public transportation stations, we will fully implement bus-oriented development to improve the convenience of connecting public transportation in the neighbourhood, the quality of the slow-moving environment on the street, and overall public services in the community.

Compared with high-rise and super high-rise buildings, the multi-layer high-coverage building layout has multiple advantages, such as low life-cycle cost, high quality of human settlement environment, and reduced fire safety hazards. It also helps reduce industrial energy consumption in the construction, operation, maintenance, and demolition processes.

The street space design of the dense road network in small blocks helps create a more comfortable and convenient pedestrian environment, guide residents to choose green transportation, and reduce traffic energy consumption. According to research, for a block of about 1 square kilometre, using this technology, the walking distance of residents from the door of the house to the centre of the block can be reduced by more than 30% [34].

Combined with public transportation stations, comprehensively implement bus-oriented development, guide community service facilities around rails or bus stations, and provide diversified green connection methods to help reduce car use and transportation carbon emissions.

7.3.3.6   Food Sector

Food safety information monitoring and tracking technology: Through the integration of a series of technologies such as RFID or electronic QR code information collection, WSN IoT, EPC global product electronic code system, logistics tracking and positioning, etc., comprehensively track and share information of all aspects of food production, sales, and consumption to clarify responsibilities and protect rights and interests.

This is a key technology for the concern of food traceability and is of great significance to food safety and self-discipline in the food industry. This technology can achieve a full record of food information at the production site and circulation links so that it can be documented; at the same time, each link is interlocked to avoid data loss or human intervention in the circulation process, to ensure food safety and reliability. It allows consumers and managers to easily and quickly understand food sources and transportation processes, as well as enhance food safety monitoring.

Vertical agricultural technology: This is a combination of environmental control technology and architectural agriculture integration, that is, in urban buildings, making full use of renewable energy and greenhouse technology, with the help of hydroponic cultivation, modern LED lighting and seed selection and other innovative technologies, to increase agricultural production and land utilization.

This is a major technology for urban agriculture and a popular investment field in developed countries. Vertical farms have the advantages of having a small footprint and high yield per unit area; intensive use of water and fertilizers, no heavy metals and pesticide residues; localized production and distribution to make food fresher, etc., which can achieve efficient planting under resource shortage conditions and meet green food demand. It can also improve environmental efficiency, with multiple benefits. The United States and Japan have already carried out pilot projects, and Singapore has adopted it as one of the main technologies for realizing food self-sufficiency with a target of producing 30% of food within the city by 2030.

Digital food platform technology: This platform will connect the various supply chain links from production to consumption, allowing consumers to directly connect with producers to ensure the fresh and convenient supply of agricultural products and the efficient allocation of resources.

This technology is a major one in the field of transportation and sales in the field of smart agriculture. Research by the International Food Policy Research Institute shows that from production to consumption, the global food waste rate is as high as about 30% due to the influence of middlemen, but the freshness is difficult to guarantee. This technology can eliminate intermediate links and directly target agricultural products and food procurement. Studies have shown that if implemented properly, the loss rate can be reduced by 50%. In addition, the technology played an important role during the COVID-19.

7.4 Cross-Cutting Issues and Enablers on Green Technology Promotion and Gender Equity

7.4.1 Cross-Cutting Issues and Enablers on Green Technology Promotion

7.4.1.1 Fourth Industrial Revolution

The Importance of the Fourth Industrial Revolution for Green Urban Development: The Fourth Industrial Revolution (4IR) is a confluence of new technologies, including artificial intelligence, robotics, IoT, autonomous vehicles, 3-D printing, nanotechnology, biotechnology, materials science, energy storage, and quantum computing [35], which is changing the way people live, work, interact, and access urban services. 4IR-enabled cities have the potential to deliver a sustainable future for all. However, positive change is not inevitable, and city leadership must be forward-looking and agile enough to steward this change for the benefit of all society.

In cities, the 4IR can be thought of as the layering of a series of technologies that brings physical infrastructure into the digital realm. Ubiquitous connectivity is provided by the Internet and the high bandwidth of the 5G network; this enables the IoT of a city, which consists of sensors that can detect and digitize everything from the temperature of buildings to leaks in pipes and the prevalence of viruses in wastewater. The digitizing of the physical space creates a digital twin of the city in the form of data streams that can be manipulated using AI algorithms or logged and traded on blockchains. This manipulation of digital data, in turn, controls the physical infrastructure of the city, opening huge possibilities.

The 4IR Supports Green Technology Implementation Across the Six Pillars: The 4IR can positively affect urban areas in a number of ways: Firstly, the development of mixed land use can significantly benefit from 4IR technologies. For example, AI coupled with online platforms could monitor and adapt spaces based on local citizens' habits and consumer demands.

Second, intelligent urban assets can unlock the **circular economy** potential, reducing waste and improving resource efficiency for societies [36]. The IoT can provide data on the location, condition and availability of an asset (from the location and availability of a shared bike or the condition of a water pipe), this data enables extending the life of an asset (through predictive maintenance), greater utilization (by sharing) or more use cycles (through product reuse) [37].

Third, as outlined in the **energy pillar**, 4IR technology can enable the transition to a smart grid or the IoE with decentralized renewable power generation systems, including from BIPV.

Blockchain can contribute to resolving **water scarcity**. A blockchain-based smart water market could effectively allocate water resources by providing accounting, auditing and trading platform replacing intermediaries. Based on a study modelled on Los Angeles County, blockchain-enabled trading could reduce water inequality by facilitating water trades between systems with a surplus and those with a deficit. It could also create incentives for wastewater recycling [8].

Fourth, smarter risk forecasting and regenerative materials can anticipate and reduce the hazards of **climate shocks and natural disasters** [38]. Predictive AI analytics, IoT and sensors can help early identification of cities' tremors or sea-level changes [39]. Advanced materials, such as self-healing concrete, can absorb energy and thus help buildings resist earthquakes [40].

Lastly, in the not-too-distant future, a new generation of quantum sensors will also vastly increase what can be sensed and digitized in a city. Quantum sensors are able to measure minute changes in gravitational and magnetic fields by manipulating and sensing atoms. Early uses of these could be the ability to probe deep underground, creating an underground map of where existing pipes and cables are located allowing for better urban **construction** and maintenance; and building better lidar (a base technology for autonomous vehicles) based on photons rather than lasers [41].

Benefiting from the investment capacity and clear long-term vision of the government, China has emerged as a global power in the transition toward 4IR and smart urban development. China is specializing in "breakthrough" innovations and is home to many companies which are experts in AI and 5G technology, such as Huawei, the world's leading telecom hardware provider [42].

So far, 31 Chinese provinces have invested more than USD 7 trillion in 22,000 projects for new smart infrastructure construction. In March 2020, China launched the "New Infrastructure" initiative, including the 5G network, big data, ultra-high voltage transmission, inter-city transportation, artificial intelligence, industrial IoT, and new energy vehicle charging stations.

Governance Approaches to Successfully Implement the 4IR: The complex, transformative and dynamic nature of the 4IR requires new governance approaches to address the interlinked dynamics of emerging technologies and to accelerate the positive societal implications of digital transformation while minimizing the potential drawbacks [43]. Over the next few years, Chinese governance will be inevitably called to face two major challenges:

(i)   the development of a long-term human-centred vision around technology integration;
(ii)   the development of an agile approach to embrace, rather than hinder, innovation.

The rapid technological change of the 4IR calls for a new model of more purposeful technology integration that puts citizens at the centre [44]. For example, in the case of urban sprawl, the advent of the autonomous vehicle might push it to the extreme level by providing the option for people to live much further away and use the commute time to work or sleep. Planners need to anticipate this and plan for more dense cities that serve the population as a whole.

Second, 4IR technologies mature at a rapid pace and therefore require an agile approach to governance, which can involve prototyping new governance structure and mechanism as well as working closely with other stakeholders such as the private sector and academia. China can adopt an agile and proactive approach to harnessing 4IR technologies. Some tools useful for this purpose can be:

Pilot cities: Working with the private sector and academia, policy-makers can use the data gathered in pilot cities to replicate the innovations elsewhere as well as to support policy-making. Pilot cities could be an innovative way of testing the technologies outlined in this report.

Policy labs: Initiatives aimed at designing new policies and public services to steer emerging innovations toward sustainability and inclusion [45].

Regulatory sandboxes: Safe spaces for companies to test innovative products, services, and business models without needing to face the normal regulatory and financial hurdles (i.e., licensing) of engaging in their experimental activities [45]. Examples of jurisdictions and their regulatory sandboxes include Sweden for autonomous vehicles (Drive Sweden), Bahrain for financial technologies, and Singapore for energy innovation [46–48].

7.4.1.2   Circular Economy

The Importance of the Circular Economy in Green Urban Development. The circular economy is a system of production and consumption that aims to keep all products and materials at their highest value at all times and in which the output of one process becomes an input to another, eliminating waste from the system and minimizing the need for the extraction of virgin materials while eliminating the use of toxic materials [49]. China was among the first countries in the world to legislate for a circular economy, implementing the Circular Economy Promotion Law in 2009 and has continued to be a leader in the field. A recent study by the Ellen MacArthur Foundation found that the further application of circular economy principles in Chinese cities could save Chinese consumers RMB 70 trillion (around USD 9.9 trillion) and reduce greenhouse gases by 23% through 2040 [50].

Circular Economy Support Green Technologies Across the Six Pillars. The circular economy touches all pillars outlined in the report, as it deals with the flows of resources around a city. These flows are complex systems that intersect across different aspects of the city. Therefore, the implementation of the circular economy is intimately tied to systems analysis and thinking. Illustrations of this approach in food, building, and consumer waste show how various changes at different points of the cycle can shift the whole system.

For example, the urban food system touches on the food, water, mobility, and energy pillar and is a good example of where the circular economy can be applied. Key raw material inputs into food production are nitrogen, phosphorous, and potash, of which China consumed 54.16 million tons in 2015 [51]. China is self-sufficient in nitrogen and phosphorus but dependent on potash imports (43.8% dependence in 2017) mostly from Russia, Canada, and Belarus [52, 53]. Food then enters the city by truck, affecting mobility by adding to traffic and pollution. Before it is eaten, globally, one third of all calories are lost in the form of food waste either at the transport, processing, or retail stage or in individual households [54].

Under a circular system enabled by green technology, land and water would be used as efficiently as possible with some degree of vertical farming in cities that could be built in under-utilized space in the urban environment and save water by using hydroponics in a closed water system while decreasing traffic from transport. Food waste would be minimized along the value chain and in the home, in part by using digital platforms. Any unavoidable waste would be converted into usable products or processed industrially using anaerobic digestion. This would produce an output of methane gas for renewable energy production and compost to be returned to food production. Sewage sludge can also be harvested and used as an input into agriculture. For example, in the Netherlands, the Amersfoort sewage treatment plant produces 900 tonnes of high-grade fertilizer per annum as well as purifying wastewater for reuse in cities. According to government figures, wastewater in China contains nearly 120,000 tonnes of phosphorus, which could be better captured with the implementation of the right technology [5].

In construction, which globally consumes 42.4 billion tonnes of material per year, there are many interventions and technologies which could better use construction waste, including the 3D printing of new buildings using waste materials. However, the most important lever comes at the design stage. Buildings such as Circl, a mixed-use restaurant and office building in Amsterdam, are designed and built with eventual disassembly in mind—a concept known as "buildings as material banks." To accompany these buildings and record the valuable materials inside, developers create materials passports; these log a blueprint of the building, including the value and location of valuable components, to allow for easy disassembly and assessment of embedded material value, leading to higher recycling rates. Using recycled steel uses only 16–20% of the energy of virgin steel, while in the case of aluminum, the number goes down to 5% [55].

At a global level, only 20% of Waste of Electrical Electronic Equipment (WEEE) is formally collected and recycled [56]. These products can be a valuable source of scarce and valuable materials. The global value of WEEE is estimated to be more than USD 60 billion [57]. A study by the World Economic Forum and Tsinghua University found that in China only 10% of aluminum, 6% of tin, 0.6% of cobalt and 0% of rare earths are captured from scrap electronics products. If 100% of these metals were recycled, the material value alone would be worth USD 3.3 billion by 2030 [58].

China has also outlined ambitious policies for the circular economy in industry, including recycling of 50% of key products by 2025 and the inclusion of 20% recycled materials in all new products. Many companies with production facilities in China have also made commitments to the circular economy and the use of recycled material in products. There is a significant opportunity for public and private stakeholders to come together around this goal [58].

Incorporating the Circular Economy Into Urban Development Planning. There are a number of ways that the circular economy can be implemented into urban planning and the application of green technology.

Firstly, it is important to have a circular economy plan based on a system analysis of the city. This should outline the opportunities for all stakeholders including startups, research institutes, government departments, urban planners, and the private sector, as well as aiming to engage citizens in the circular economy.

For example, the London Circular Economy Roadmap provides guidance for the acceleration of the British capital's transition to become a circular city. The Roadmap could bring London net benefits worth GBP 7 billion every year by 2036, mainly in the sectors of the built environment, food, textiles, electricals, and plastics [59].

Secondly, policy and taxation should be aligned with the goals of the circular economy. These could include: charges that take into account the negative externalities of single-use items such as plastic bags; tax breaks for the use of recycled materials in products; and incentive mechanisms such as extended producer responsibility. Consider also unproductive policies, such as those which inhibit or tax waste movements in and out of manufacturing zones or subsidize virgin feedstocks.

Thirdly, facilitate investment with government funds or blended finance models into innovation for the circular economy. Government funding can help de-risk investment in circular business models [60]. Banks can also be encouraged to set up innovation funds (or companies innovation challenges) to spur entrepreneurship.

Governance That Facilitates the Circular Economy. Governance for the circular economy can be complex: as outlined above, material often flows in cities across many different areas and therefore different departments. Changing an entire system to implement the circular economy in cities requires collaboration with many stakeholders including the private sector, responsible for much of the innovation and implementation; citizens, who need to change how they use resources; and academia, who have specialized knowledge. Due to the cross-cutting nature of the circular economy, many cities, regions, or countries have made use of a platform approach which brings together all of the key players in a structured way to collaborate on implementing the circular economy in a city or region or around a particular challenge, examples include:

- The European Sustainable Phosphorus Platform (ESPP): The ESPP works with a diverse range of stakeholders and ensures knowledge sharing; creates network opportunities in the phosphorus management field; and addresses regulatory obstacles [61].
- The Platform BAMB—Buildings as Material Banks—connecting 15 partners from seven European countries to create circular solutions in the building sector. Through design and circular value chains, BAMB aims at increasing the value of building materials. This way, at the end of their life, buildings become banks of valuable materials instead of being wasted [62].
- The Platform for Accelerating the Circular Economy (PACE) is a global convening platform and project accelerator to help speed the transition to a circular economy. It was launched in Davos in 2017 and is now hosted in the Netherlands. The platform provides a leadership platform for CEOs, ministers and heads of international organizations to come together to collaborate; it runs and supports high-impact projects around the world and works with partners on sharing knowledge and learnings [63].
- The Ellen MacArthur Food Initiative brings together key actors to stimulate a global shift toward a regenerative food system based on the principles of a circular economy.

7.4.1.3 Data Governance

The Importance of Data Governance for Green Urban Development. Data has been called the "new oil" and is the engine of the 4IR. Each day more than 2.5 quintillion bytes of data are generated, with much of it coming from IoT devices in urban areas. This data is a potentially rich source of information that could be used to improve the delivery of urban services, the management of urban systems, and the quality of life of citizens. Unfortunately, only a very small amount (less than 1%) is used to drive decisions and create value. Data is normally held by many different players, stored on different systems, and lacks interoperability, meaning it is unable to release its full potential to generate immense social and economic value [64]. Building better data governance by creating well-designed, regulated, and trusted frameworks that enable data opening, connection, and sharing has the potential to unlock this enormous social and economic value for cities.

Data Governance Supports Green Technology Across the Six Pillars. There are essentially two types of data generated in cities. **Public sector data** refers to data "generated, collected and stored by international, national, regional and local governments and other public institutions, as well as data created by external agencies for the government or related to government programs and services." **Private sector data** refers to information "generated, collected and owned by private companies or individuals, such as customer activity data, personal data, business operational data and industrial data." [65]

In the case of public sector data, governments can implement open platforms to share data for free, covering critical sectors such as geographical data, climate data, water resources data, road structure, traffic maps, buildings, energy, air pollution, etc. [66].

The city of Berlin, as part of its Open Data Initiative, has created an open data platform that has 935 datasets freely available. Within this, datasets on **mobility** cover everything from real-time public transportation data to the location of bicycle accidents. This data platform allows companies to build applications that act as an interface to help citizens navigate the city [67].

Uber Movement, a software app, is giving city planners and members free access to anonymous data gathered from millions of Uber trips in more than 700 cities across the world. The data shared by the software enables urban planners to better address city **mobility** challenges [68].

Creating an IoE will mean that energy utility companies will have to manage not only the energy grid but also a data grid that must be interoperable with IoT devices from numerous different manufacturers. Smart appliances; smart meters; EVs and renewable energy generated at a building or household level all create data that needs to be understood and analyzed by the utility. Strong data governance and standards will be needed to facilitate sharing between this network of devices, the utilities, and third parties [69].

Governments Should Take Actions to Effectively Implement Data Governance. Effective data governance implementation is possible only if stakeholders trust the data-sharing platforms that are being used. Thus, the proper management of data requires that **necessary restrictions and regulations** are in place [65]. In fact, opening public sector data without restrictions and allowing private sector data-trading activities without regulation could decrease the general level of confidence in data-sharing platforms. Some of the essential elements that characterize a robust data regulatory framework are:

- **Data privacy**, which should be guaranteed throughout the whole process of data collection, sharing, and use;
- **Data security** to avoid cyber threats such as unauthorized access to data and data impairment;
- **Data interoperability**, which allows data sharing and uses across systems, platforms, locations and jurisdictions;
- **Data accountability**, which should be addressed by "validating and declaring the data provider, evaluating for potential bias, securing transparency and traceability of the data source and data flow";
- **Eligibility of operators**, which ensures that whoever operates on the platform has the legal right to do so and is constantly monitored by legislation;
- **Promote further unlocking of value** by creating platforms for the sharing of private sector data to improve green development in cities. While also encouraging companies and industry associations to facilitate the sharing of data [65].

7.4.2 Gender Perspectives in Green Technology Promotion and Implementation

7.4.2.1 The Importance of Gender Perspectives in Urban Green Development

In the UN's "Rio Declaration on Environment and Development," special highlight is placed on women's significant role in environmental management and sustainable development. Most often, women take charge of the housework, making themselves important users and participants in the sustainable consumption of natural resources. Women also value green, safety, and health more than men—studies show that the penetration of green consumption among women is higher than men. In education, women exert a vital influence on forging and raising green awareness with children. Women also play a part in improving community adhesiveness, thus strengthening the capacity of communities in natural disaster prevention and management.

That said, in the course of achieving urban green development, the fact is that gender perspectives are not always accounted for. Many women feel unsafe around cities, due to challenges such as: public transport systems that are designed primarily to facilitate commuting rather than meeting gender-responsive needs; lack of proper lighting in public places; challenges in sanitary infrastructure etc. These all come down to the absence of consideration of gender perspectives.

7.4.2.2 Experience on Adopting Gender Perspective in Urban Development from the International Experience

Transport: Studies by the World Bank suggest that the public transport system is usually designed to meet the needs of working men while ignoring women's need to travel in non-commuting hours. Similarly, most automobiles are designed with male users in mind. For instance, the dummy used in a crash test is modelled after adult men: that is at least part of the reason that the likelihood for women to get injured in a car crash is 73% higher than men [70].

Land use and planning: Mixed land use can reduce travel distance, therefore benefiting citizens who use public transport less frequently (e.g., women). Mixed use can also help women to better balance home caring tasks and their jobs. In addition, better design and management of public spaces will give women a greater sense of safety. To that end, a new tool for urban environment assessment called "women safety audit" has already emerged [71].

Architecture: Indoor temperature should be adjusted to make women feel comfortable. Stairs are often designed too wide or too high that they don't fit the gait of women. A UN report suggests that including gender perspectives in architecture design will only increase the building cost by less than 1% [72].

Other topics that attract attention from the international community include: **mobility**—to facilitate safe, convenient, and affordable mobility in and around cities; **safety and protection from violence**—protect women from practical or perceivable dangers in both public and private domains; **health and sanitary considerations**—lead a positive life in environments free of health risks [73].

7.4.2.3 Gender Perspective in Urban Green Development

Creating a gender-responsive environment and realizing the full potential of women is one of the paramount goals for urban green development in China. The key principles in this respect are: know and account for women's role in green development; understand that women and men have equal rights in getting their different needs met; take a gender-sensitive approach in planning activities involving women, etc. Therefore, we define the following three dimensions as the priorities in this regard:

Formulation and governance of green policy: Make sure that women are duly represented in the formulation and decision making of green policy. More women should be encouraged to take part in urban green development and governance. The performance evaluation of green policy should be more gender responsive.

Education and employment on green technology: Women should be provided with more opportunities for professional training and research on green technology. More jobs are to be created for women in the sector of green technology. In the production of green solutions and products, a gender-sensitive policy must be applied.

Consumption and use of green products: Surveys on the demand data of women should be included in green product R&D. In addition, women are encouraged to participate in the promotion and scaling up of green solutions and green products.

7.4.2.4 Gender Perspective in the Six Key Sectors

Water: Technologies are developed to ensure good quality and smart management of recycled water and to reduce the technical instability in water treatment. Budget for gender-responsive measures will be embedded in local water resource management. These will help address the issue that women may not be able to use and manage water resources in the same way men do due to their physiological differences.

Energy: Mainstream the gender perspective in energy policy-making. Women do not only have a significant role to play in managing household primary energy use, but also function as facilitators for the technical revolution toward sustainable energy.

Women should be better trained in the acquisition, installation, operation, and maintenance of sustainable energy solutions. Moreover, taking advantage of their gender strength, women can promote green technology and clean energy to other women, and educate others in the community on how to use them.

Transport: The different mobility demands of women and men should be regarded as the baseline for transport policy research and the preconditions for transport planning. Women must be better represented in transport decision making and management. In all key processes in the transport sector, including standard formulation, research, decision making and management, transport service operation, etc., women should be ensured an important part to play.

Architecture: Understand women's differentiated demands for building space; develop technical codes and standards for building design that meet the needs of women.

Land use and planning: Women have dual roles both as working labour and as home carers, so they need more urban and community functions integrated within a limited time frame. In addition, women's opinions should be sought in public engagement for urban planning and community governance decision making so that their needs can be duly addressed.

Food: Women usually have more say in deciding what food to buy and how it is prepared. In the promotion of food safety technology, then, consideration must be given to the different roles, needs, and opinions of women and men, and a women-friendly approach will be adopted in technology design.

7.5   Policy Recommendations

7.5.1   List of Recommended Major Green Technology During the 14th FYP Period

Based on the prominent problems, visions, and goals of green development in Chinese cities, and combined with Chinese and foreign practices and current technical progress, Chinese and foreign experts jointly proposed the technical development direction of six major sectors, and recommended the following 20 technologies during the 14th FYP which shall refer to Table 7.6 for details.

Table 7.6 Recommendations for green technology in six major sectors during the 14th FYP

Major sectors	Technical development director	Recommended technology
Water	Sewage Treatment and Water Recycling Economy	Sewage Treatment and Plant, Network and River Integrated Quality and Efficiency Improvement Technology
	Utilization of Reclaimed Water	Water Quality Guarantee Technology for Recycled Water System
	Utilization of Reclaimed Water and Non-Revenue Water Management	Smart Operation Technology for Recycled Water
Energy	Integrated Green Energy Grid	Microgrid Technology
	Near-Zero-Emission Cooling and Heating	Industrial Waste Heat Central Heating Technology
		Middle-deep Geothermal Heating Utilization Technology
	Energy Internet	Integrated Energy Internet Management Platform Technology
Transportation	Intelligent Transportation System	MaaS Travel Service Technology
	New Energy Vehicles and Supporting Facilities	Hydrogen-powered Vehicle Technology
		Intelligent Charging System Technology
	Transportation Demand Management and Cycling Trip	Bicycle Special Road Technology
Building	Healthy Building	Building Three-dimensional Greening Technology
	Green Building	"Steel Structure + Modular Internal Space" Technology
	Near-Zero Energy Building	Photovoltaic, BIPV, Distributed Energy Storage and DC Power Supply Technology
	Smart Operation and Maintenance of Building	Intelligent Building Cluster System Technology
Land Utilization and Planning	Green Urban Form	Technical Package for Green Urban Form
	Green, Livable, and Carbon–Neutral Block	Technical Package for Green, Livable, and Carbon–Neutral Block

(continued)

Table 7.6 (continued)

Major sectors	Technical development director	Recommended technology
Food	Food Traceability	Food Safety Information Monitoring and Tracking Technology
	Urban Agriculture	Vertical Agricultural Technology
	Smart Agriculture	Digital Food Platform Technology

7.5.2   Policy Recommendations for Green Development and Technological Innovation During the 14th FYP Period

Lessons were drawn from developed countries (e.g., EU members, Japan, the United States) on how they successfully promoted green development at the state level to municipal and then to community levels. Recommendations are therefore made on how China should improve its green development and technological innovation, not least by developing favourable legal, policy, and institutional measures. The recommendations are hereby elaborated from the following four perspectives: legislation, government policy and management, market player contributions, and public engagement.

The National Strategy and Legislative Security: Green development has become a statement of the country's development strategy. We should also clearly propose the overall national green and low-carbon strategy and the overall goal of low-carbon development. We should also accelerate the construction of related legal systems.

 Firstly, a national plan for green development and low-carbon development should be created, which proposes to fulfill the commitments of the Paris Agreement, to achieve the 2 °C goal and to move toward a carbon–neutral systemic plan.

 Secondly, it is necessary to clarify the total amount of carbon emissions in stages before 2050, to control the target, timetable, and road map for achieving the target, and to break the total amount down according to provinces and cities. Economically developed regions and cities should be encouraged to assume more responsibility for reducing emissions.

 Thirdly, we have to accelerate the development of a legal system for green and low-carbon development and resource consumption and emission control. This should encourage cities to formulate and implement carbon-reduction targets, achieve green development, and promote local laws and regulations on green technologies.

The mechanism construction from government perspective

A complete framework of administrative, fiscal, and tax policy should be built to fully engage the government in promoting, incentivizing, and disciplining green development and green technology.

Put in place a quota system for carbon emissions. In such a system, emissions quotas should be determined based on the existing carbon intensity control program and assigned to various levels from state to provinces to cities and counties. The carbon emissions quota will be included in the emissions cutting program and annual work plan of governments at all levels. The quota system can be piloted in more-developed provinces and cities before being rolled out more widely.

Stress the role of strategic planning as the guidance. Establish an international alliance of green technologies and innovations to build a communication platform for domestic and foreign companies, decision-makers, and expert groups, and promote continuous communication and joint resolution of green development issues in Chinese cities.

The mechanism construction from market perspective

A healthy market should be nurtured in order to allow businesses to play their due role in promoting green development and green technology.

Encourage the market to lead the trend. The role of businesses as the main market players and the determinant in resource allocation should be respected. A market system that mobilizes businesses to work dynamically on developing green technology and manufacturing green products should be formed.

Enhance financial support. The private sector should be encouraged to partake in the green development course. Fundraising by green technology companies should be made easier. The finance market should uphold green principles and devise a complete green financial system.

Promote research and innovation. A market-oriented green technology innovation system will be developed in order to attract businesses and professional institutions far and wide to join the cause.

The mechanism construction from society perspective

Citizens and communities will be educated to follow a green lifestyle; the green well-being of cities and communities will be safeguarded and upheld. A green governance system that has extensive engagement with the public will be championed.

Implement the emissions data publication system; expand sources for environment data acquisition; and improve transparency. Catalogues will be published to disclose the environmental performance of businesses and cities. Incentives and punishment will be given according to transparency performance. Statistics from various departments will be integrated into one platform to put all entities under the monitoring eyes of the general public.

Establish institutional mechanisms that encourage public participation. We need to shift low-carbon green development from administrative management to social governance, to clarify citizens' responsibilities, powers, and interests in protecting the environment, to use social media to explain to the public the importance of the promotion and application of green technologies, and how behaviour changes can achieve better green development, thereby raising the awareness of green and low-carbon development in the whole society. Issue laws and regulations to ensure that the public and social organizations participate in decisions related to low-carbon emissions reduction, and introduce relevant policies to encourage the public and society to participate in green and low-carbon development.

7.5.3   Policy Recommendations for Green Technology Implementation in the Six Pillars During the 14th FYP Period

On the basis of the overall policy recommendations, combined with the two working methods of technical standards and norms and pilot projects, further specific policy recommendations in six key areas are presented, see Table 7.7.

7.5.4   Policy Recommendations Based on International Best Practices

The policy recommendations based on international best practices aim to support greater efforts to strengthen the people-centric urban green development goals that promote eco-sustainability, resilience, equity, and quality of life—while also helping steer the Chinese economy toward high-quality growth, particularly through investment in new infrastructure and green technology.

Create guidance for city- and building-level design. Develop a guide for green and low-carbon development and transit-oriented development in Chinese cities; set and promote a standard for green buildings based on best international practices.

Invest in new urban infrastructure. Continue with the New Infrastructure plan to help stimulate the economy post COVID-19. Add three key urban green technologies to the list of new infrastructure investments: building-integrated photovoltaic (BIPV), water treatment technology and energy storage, with ambitious targets for each. In order to build an integrated green energy grid (IGEG), double annual investment in wind, solar, and energy storage.

Table 7.7 Policy recommendations for green technology promotion and application in six major sectors during the 14th FYP

Sector	Policy Recommendations	
Water	Laws and regulations	Establish a multi-level legal system for reclaimed water and improve management methods for the utilization of reclaimed water
	Department policy	Prepare the planning of reclaimed water; propose the plant, network, and river integrated water governance and quality and efficiency improvement mechanism, and establish the management system for reclaimed water; strengthen the inspection of the implementation of urban sewage treatment and drainage regulations, and establish the assessment mechanism for an index of sponge city, sewage treatment, and reclaimed water
	Technical standards and specifications	Improve the classification of sewage treatment and drainage standards and technical standards of facility; revise the standards for the utilization of reclaimed water; prepare the treatment and drainage standards for water pollutants in specific areas
	Financial revenue	Provide financial support and VAT relief for energy-saving technology, establish, guide and standardize multiple funding channels, and encourage franchise system and financing methods such as BOT and TOT; endogenous financing methods that obtain funds through various fee collections and tax refunds
	Pilot Project	Establish the pilot demonstration city for black and odorous water treatment and quality and efficiency improvement of sewage and water-saving city
Energy	Laws and regulations	Draw up the Priority Law of Renewable Energy, the Promotion Law of Heating by Renewable Energy, and the Promotion Law of Compulsory Utilization of Renewable Materials for Plastic Packaging Waste
	Department policy	Establish an organization of low-carbon energy management and technology promotion; promote the high-proportion development of low-carbon energy and form the multi-energy and complementary supply system; vigorously promote the garbage classification, reform of energy product price and carbon tax policy. Establish and implement accountability measures and indicators

(continued)

Table 7.7 (continued)

Sector	Policy Recommendations	
	Technical standards and specifications	Establish national standards and service systems for low-carbon energy technology; formulate the carbon emission assessment specifications throughout the life cycle; prepare the comprehensive energy planning; promote the WELL certification standard
	Financial revenue	Establish a subsidy mechanism for R&D and pilot applications, expand diversified financing channels, promote the reform of green taxation mechanism, and provide loan guarantees for renewable energy. Introduce a more flexible energy pricing system and implement the feed-in tariff subsidy policy. Innovate the hybrid financing mechanism to mobilize the investment of social capital. Reduce initial costs through fiscal and taxation policies and innovative financial products, and encourage the application of photovoltaic integrated technology
	Participation of the public	Guide the public to gradually shift to green energy consumption through publicity and education
	Laws and regulations	Establish the regulation system of hydrogen energy management and safety and the battery recycling management system, and supplement the proposed Promotion Law of Effective Utilization of Resources
	Department policy	Strengthen the management of the production, sales, and use of electric bicycles; establish the assessment mechanism for recycling management of lithium battery; promote cross-border alliances of industries such as green transportation, mobile payment and financial industry
Transportation	Technical standards and specifications	Formulate technical standards for MaaS travel services, technical specifications for sharing of traffic data resources, technical specifications for safety of electric bicycles, planning specifications and standards for charging facilities, guidelines on planning and layout of battery recycling stations, specifications and standards for planning, construction and management of supporting facilities of hydrogen energy

(continued)

Table 7.7 (continued)

Sector	Policy Recommendations	
	Financial revenue	Formulate the MaaS freight price system and subsidy policy, tax preferential policy for R&D of new battery technology, tax preferential policy for R&D of preparation technology, storage technology and transportation technology of hydrogen energy; set up special subsidy for the construction of infrastructures of hydrogen energy and special funds for the construction of special roads for bicycles
	Pilot project	Promote the demonstration project of MaaS travel service and the demonstration project of special roads for bicycles. Promote the use of zero-emission traffic areas and congestion charging pilot projects
	Participation of the public	Strengthen the guidance for the public and establish the sharing and open mechanism of travel service data
Building	Department policy	Implement the target of dual control for total energy consumption and energy consumption intensity; implement the planning of performance goal system and multi-objective optimization for green building project, and establish the smart operation and maintenance system; include the air quality indicators in the completion acceptance process (or combine with the inspection procedures of fire protection); establish the management methods for cyclic utilization of building materials and clarify the main responsibilities of all parties involved in the construction and supervision of green buildings
	Technical standards and specifications	With reference to LEED and WELL, include the performance target, total energy consumption and intensity indicators of green building into the urban planning standards, and formulate the technical standards for pre-planning and post-assessment of green building and guidelines of adaptive design of building; formulate the action plan of green building and prefabricated building during the 14th FYP
	Financial revenue	Include the individual purchase of and residence in (including the lease of) green building into the special deduction for personal income tax, and formulate the tax preferential policy for enterprises in using green buildings

(continued)

Table 7.7 (continued)

Sector	Policy Recommendations	
	Pilot project	Build the pilot demonstration city/urban area with buildings of near-zero energy consumption, demonstration project of near-zero carbon emission zone and pilots of energy-saving transactions of public building
Land utilization and planning	Laws and regulations	Establish the Development and Protection Law for Space of National Land and the Natural Reserve Law, and grant the legal status to urban design
	Department policy	Establish a quick-response mechanism for rules and regulations and price adjustment, and formulate the stipulations for intensive and intensive utilization of land, the disposal methods for idle land and the management methods for land reserve; implement supervision, inspection and full-process management for high-rise residential projects; establish the planning assessment and supervision system for the planning of a low-carbon pilot city and pilot community, and the performance assessment and dynamic assessment system for industrial land
	Technical standards and specifications	Formulate relevant standards and norms for mixed utilization of land, guidelines and relevant norms for construction of residential areas, and planning standards for low-carbon and emission-reducing city/community/public space; formulate relevant guidelines to ensure the connectivity of blocks in terms of community services such as planning, design, construction and management
	Financial revenue	Study and formulate environmental tax policy to encourage environmentally friendly development for land
	Pilot project	Build the pilot demonstration city/urban area with buildings of near-zero energy consumption, the demonstration project of near-zero carbon emission zone and the pilot of energy-saving transactions of public building

(continued)

Table 7.7 (continued)

Sector	Policy Recommendations	
Food	Laws and regulations	Establish laws to guarantee the basic quality standards of farmland soil and water environment; improve the legal guarantee of the food safety and tracking system
	Department policy	Promote the whole process management of food from the place of production to consumers, establish a traceability system for the development of the entire industrial chain of agricultural products; strengthen the construction of rural broadband networks, establish agricultural product information platforms, improve the links between farmers and the market; develop e-commerce for farmers Skills training on the platform; formulate policies to promote the conversion of abandoned land and brown land in suburbs into smart agricultural land
	Technical standards and specifications	Establish green food standards for the entire life cycle; develop technical guidelines for vertical agriculture
	Financial revenue	Establish a fiscal transfer mechanism for regional fair price payment; formulate a tax incentive policies for the production and consumption of green food
	Pilot project	Promote the township/village pilot of green food safety information and tracking technology; establish a pilot community and evaluation system for vertical agriculture

Continue to drive the adoption of electric vehicles. Strengthen the construction of electric vehicle infrastructure; focus on promoting the electrification of high-mileage commercial vehicles, promote shared vehicles, and implement transport demand management.

Promote digital innovation along the food value chain. Enable robust innovation ecosystems, apply digital innovation to the whole food value chain, improve traceability in food supply chains, produce and adopt healthier, more nutritious and sustainable diets, and promote urban indoor farming.

Build carbon–neutral communities. Make clear targets and a shared roadmap for carbon–neutral, circular communities, mobilize government departments, the private sector and all stakeholders to participate in the construction of carbon–neutral and circular communities.

Delivery mechanisms. Three strategies that will help to achieve a green transition in Chinese cities include: Pilot cities and policy sandboxes; a cross-border alliance on green technology and engagement with the public.

References

1. "Leadership in Energy and Environmental Design (LEED®)." *Leadership in Energy and Environmental Design*, www.cement.org/sustainability/leadership-in-energy-design-(leed).
2. "DGNB – German Sustainable Building Council." *Deutsch*, www.dgnb.de/en/.
3. *GB Tools - Home Page*, www.gbtoolsltd.com/.
4. Standard & Poor's Financial Services LLC,Global Ratings Green Evaluation Report, 2017.
5. China Statistical Yearbook (2018). Available at: http://www.stats.gov.cn/tjsj/ndsj/2018/ind exeh.htm.
6. M. Swilling, M. Hajer, T. Baynes, J. Bergesen, F. Labbé, J.K. Musango, A. Ramaswami, B. Robinson, S. Salat, S. Suh, P. Currie, A. Fang, A. Hanson, K. Kruit, M. Reiner, S. Smit, S. Tabory (2018), "The Weight of Cities: Resource Requirements of Future Urbanization", A Report by the International Resource Panel. United Nations Environment Programme, Nairobi, Kenya. Available at: https://www.resourcepanel.org/reports/weight-cities.
7. Nutrient Platform, "Phosphorus From Wastewater In Amersfoort". Available at: https://www.nutrientplatform.org/en/success-stories/phosphorus-from-wastewater-in-amersfoort/.
8. New America, "The Development of Smart Water Markets Using Blockchain Technology". Available at: https://www.newamerica.org/fellows/reports/anthology-working-papers-new-americas-us-india-fellows/the-development-of-smart-water-markets-using-blockchain-techno logy-aditya-k-kaushik/.
9. Great Lakes Echo (2020), "Water sensors, data collaboration make Great Lakes smarter". Available at: https://greatlakesecho.org/2020/02/21/water-sensors-data-collaboration-make-great-lakes-smarter/.
10. International Energy Agency. *2010 Energy Technology Outlook: Scenarios and Strategies for 2050*[M]. Beijing:Tsinghua University Press, 2011.
11. International Energy Agency. *Tracking Clean Energy Progress 2017*. Beijing: Science Press, 2018:17–61.
12. *Summary of Microgrid Demonstration Project* by Wang Chengshan, Zhou Yue, Distribution & Utilization, January 2015.
13. International Energy Agency (2019), *China Power System Transformation*, https://www.iea.org/reports/china-power-system-transformation.

14. *13th Five-Year Plan for Building Energy Conservation and Green Building Development,* Ministry of Housing and Urban-Rural Development of the People's Republic of China. No. Building Science [2017]53 March 1, 2017.

15. *China Vehicle Environmental Management Annual Report,* issued by Ministry of Ecology and Environment of People's Republic of China; http://www.gov.cn/guoqing/2019-04/09/con tent_5380744.htm Retrieved on March 5, 2020.

16. *Green Travel Action Plan (2019–2022)* issued by Ministry of Transport and other ministries; http://www.gov.cn/xinwen/2019-06/03/content_5397034.htm Retrieved on October 3, 2019.

17. Global EV Outlook 2019:Scaling-up the transition to electric mobility,May 2019 https://www. iea.org/reports/global-ev-outlook-2019 Retrieved on November 25, 2019.

18. *Outline of Transportation Power Building,* issued by the Central Committee of the Communist Party of China and the State Council; http://www.gov.cn/zhengce/2019-09/19/content_5431 432.htm Retrieved on November 8, 2019.

19. European MaaS Roadmap 2025. MAASiFiE project funded by CEDR; https://www.resear chgate.net/publication/317416483_Deliverable_2_European_MaaS_Roadmap_2025_MAA SiFiE_project_funded_by_CEDR/link/5939f82baca272bcd1e29417/download Retrieved on August 11, 2019.

20. *Urban Form and Low Carbon City: Research Progress and Planning Strategy* by Liu Zhilin, Qin Bo, *Urban Planning International,* Issue 02, 2013.

21. Robert Cervero,*TOD and Sustainable Development, Urban Transport of China,* Issue 01, 2011.

22. Annual Report on China's Energy Development 2018 by Lin Boqiang, Beijing: Peking University Press, 2019.

23. National Development and Reform Commission, National Energy Administration. *Energy Technology Revolution and Innovation Action Plan (2016–2030),* No. Development, Reform and Energy [2016]513. April 7, 2016.

24. National Development and Reform Commission. *National Key Energy Conservation and Low Carbon Technology Promotion Catalogue (2017 Edition, Energy Conservation Section).* No. 3 of 2018 Announcement of National Development and Reform Commission. February 28, 2018.

25. National Development and Reform Commission, National Energy Administration. *13th Five-Year Plan for Energy development* No. Development, Reform and Energy [2016] 2744. December 26, 2016.

26. Interoperable Transit Data: Enabling a Shift to Mobility as a Service. Rocky Mountain Institute,October 2015. http://www.rmi.org/mobility_ITD Retrieved on September 21, 2019.

27. The Future of Hydrogen:Seizing today' s opportunities,Report prepared by the IEA for the G20, Japan, June 2019. https://webstore.iea.org/download/direct/2803 Retrieved on December 21, 2019.

28. CO2 EMISSIONS FROM FUEL COMBUSTION Highlights(2019 edition),International Energy Agency;www.iea.org Retrieved on September 1, 2019.

29. *Evaluation of Walking-Friendliness in Chinese Cities,* jointly issued by Natural Resources Defense Council and School of Architecture, Tsinghua Universityhttp://nrdc.cn/Public/upl oads/2017-12-15/5a336e65f0aba.pdf ;Retrieved on September 23, 2019.

30. *Study on the Impact of Household Travel Energy Consumption in Built Environment of Residential Areas of Jinan City* by Zhang Jie, Yang Yang, Chen Xiao, Mao Qizhi, *Urban Studies,* Issue 07, 2013.

31. *Portland: We can we have both environmental protection and economy, Economic Information Daily,* July 25, 2016, Page A04

32. *Study on the Impact of Urban Block Form on Residents' Travel Energy Consumption* by Jiang Yang, He Dongquan, ZEGRAS Christopher, *Urban Transport of China,* Issue 09, 2011.

33. *Low Carbon Oriented Urban Spatial Structure- A New Mode of Urban Transportation and Land Use* by Pan Haixiao, *Urban Studies,* Issue 01, 2010.

34. *Road Network Planning and Road Design Based on the Mode of "Dense Road Network, Small Block-Taking the Core Area Planning of Chenggong New District in Kunming as an Example* by Shen Feng, Li Liang, Zhai Hui, *City Planning Review, Issue 05, 2016.*

35. The World Economic Forum (2016), "The Fourth Industrial Revolution: what it means, how to respond". Available at: https://www.weforum.org/agenda/2016/01/the-fourth-industrial-revolution-what-it-means-and-how-to-respond/.
36. Ibid.
37. The World Economic Forum (2016), "Intelligent Assets Unlocking the Circular Economy Potential". Available at: http://www3.weforum.org/docs/WEF_Intelligent_Assets_Unlocking_the_Cricular_Economy.pdf.
38. The World Economic Forum (2018), "Harnessing the Fourth Industrial Revolution for sustainable emerging cities". Available at: http://www3.weforum.org/docs/WEF_Harnessing_the_4IR_for_Sustainable_Emerging_Cities.pdf.
39. Ibid.
40. Flextregrity. Available at: http://www.flextegrity.com/.
41. Battersby (2019), "Core Concept: Quantum sensors probe uncharted territories, from Earth's crust to the human brain" PNAS August 20, 2019 116 (34) 16663–16665; https://doi.org/10.1073/pnas.1912326116.
42. The Economist (2018), "Chinese Tech vs American Tech: which of the world's two superpowers has the most powerful technology industry?". Available at: https://www.economist.com/business/2018/02/15/how-does-chinese-tech-stack-up-against-american-tech.
43. The World Economic Forum (2018), "Agile Governance: Reimagining Policy-making in the Fourth Industrial Revolution". Available at: http://www3.weforum.org/docs/WEF_Agile_Governance_Reimagining_Policy-making_4IR_report.pdf.
44. The World Economic Forum (2018), "Rethinking Technological Development in the Fourth Industrial Revolution". Available at: http://www3.weforum.org/docs/WEF_WP_Values_Ethics_Innovation_2018.pdf.
45. Chari, Vasant, et al. "Blog Policy Lab." Policy Lab, 22 May 2020, openpolicy.blog.gov.uk/category/policy-lab/.
46. "Hittar Inte Sidan." Drive Sweden, www.drivesweden.net/en%20and%20www.testsitesweden.com/en/projects-1/driveme.
47. "Welcome to the Central Bank of Bahrain." Central Bank of Bahrain, www.cbb.gov.bh/assets/Regulatory%20Sandbox/Regulatory%20Sandbox%20FrameworkAmended28Aug2017.pdf.
48. "Regulatory Sandbox." EMA; Regulatory; Sandbox; Regulatory Sandbox, 2019, www.ema.gov.sg/Sandbox.aspx.
49. World Economic Forum and Ellen MacArthur Foundation (2014), "Towards the Circular Economy" http://www3.weforum.org/docs/WEF_ENV_TowardsCircularEconomy_Report_2014.pdf.
50. Ellen MacArthur Foundation (2018), "The Circular Economy Opportunity for Urban and Industrial Innovation in China". Available at: https://www.ellenmacarthurfoundation.org/publications/chinareport.
51. J, Cai, X. Xia, H. Chen, T. Wang, H. Zhang (2018), "Decomposition of Fertilizer Use Intensity and Its Environmental Risk in China's Grain Production Process". Available at: https://www.researchgate.net/publication/323152227.
52. S. Dong (2019), "Reduce Potash Import Dependence in China". Available at: https://iad.ucdavis.edu/sites/g/files/dgvnsk4906/files/inline-files/Sisi%20Dong_capstone%202019_1.pdf.
53. GPCA (2018), "China Fertilizer Industry Outlook". Available at: https://gpca.org.ae/wp-content/uploads/2018/07/China-Fertilizer-Industry-Outlook.pdf.
54. FAO (2020), "Food Loss and Food Waste". Available at: http://www.fao.org/food-loss-and-food-waste/en/.
55. BAMB (2019), "Metals Value Chain Report". Available at: https://www.bamb2020.eu/wp-content/uploads/2019/02/Metals-Value-Chain.pdf.
56. ITU (2017), Global E-waste Monitor, https://www.itu.int/en/ITU-D/Climate-Change/Pages/Global-E-waste-Monitor-2017.aspx.
57. World Economic Forum (2019), "A New Circular Vision for Electronics". Available at: http://www3.weforum.org/docs/WEF_A_New_Circular_Vision_for_Electronics.pdf.

58. World Economic Forum (2019), "Recovery of Key Metals in the Electronics Industry in the People's Republic of China". Available at: https://www.weforum.org/reports/recovery-of-key-metals-in-the-electronics-industry-in-the-people-s-republic-of-china.
59. LWARB (2017), "London's circular economy route map". Available at: https://www.lwarb.gov.uk/wp-content/uploads/2015/04/LWARB-London%E2%80%99s-CE-route-map_16.6.17a_singlepages_sml.pdf.
60. World Economic Forum (2019), Harnessing the Fourth Industrial Revolution for the Circular Economy Consumer Electronics and Plastics Packaging . http://www3.weforum.org/docs/WEF_Harnessing_4IR_Circular_Economy_report_2018.pdf.
61. European Sustainable Phosphorus Platform, "About the European Sustainable Phosphorus Platform (ESPP)". Available at: https://www.phosphorusplatform.eu/platform/about-espp.
62. BAMB (2020), "About BAMB". Available at: https://www.bamb2020.eu/about-bamb/.
63. About PACE (2020) accessed at: https://pacecircular.org/.
64. SmartImpact (2018), "Data Governance & Integration for Smart Cities". Available at: https://smartimpact-project.eu/app/uploads/2018/02/SmartImpact_Data-Gov-and-Intergration_A4_AW.pdf.
65. World Economic Forum (Forthcoming), "Protocol - Unlocking the shared value of dynamic IoT data in smart city with trusted platforms".
66. Public sector data refers to data "generated, collected and stored by international, national, regional and local governments and other public institutions, as well as data created by external agencies for the government or related to government programs and services". Source: World Economic Forum (Forthcoming), "Protocol - Unlocking the shared value of dynamic IoT data in smart city with trusted platforms".
67. European Data Portal (2017) "Analytical Report number 6" Available at: https://www.europeandataportal.eu/sites/default/files/edp_analytical_report_n6_-_open_data_in_cities_2_-_final-clean.pdf.
68. Uber Movement. Available at: https://movement.uber.com/?lang=en-US.
69. Kotagiri, Sunil (2019) "Data Quality and Governance Critical for Utilities" Available at: https://www.tdworld.com/smart-utility/data-analytics/article/20972300/data-quality-and-governance-critical-for-utilities.
70. Including Gender in the World Bank Transport Strategy, World Bank, 2006, http://documents.worldbank.org/curated/en/968841468147567926/pdf/841800WP0Trans0Box0382094B00PUBLIC0.pdf.
71. Women's safety audits, what Works and Where? UN-Habita, 2009, Available at: https://unhabitat.org/womens-safety-audit-what-works-and-where.
72. Disabilities, Office of the High Commissioner for Human Rights, United Nations, 2007, https://www.ohchr.org/Documents/Publications/training14en.pdf.
73. 'Sorry, we didn't take women' needs into consideration during product design', March 9th, 2020. Avaiable at: https://xw.qq.com/cmsid/20200310A000I900.

Chapter 8
Green BRI and 2030 Agenda for Sustainable Development

Aligning with Sustainable Development Goal 15 to Promote Global Biodiversity Conservation

The Belt and Road Initiative (BRI) promises to create new opportunities for shared growth among countries through policy coordination, connectivity, unimpeded trade, financial integration, and people-to-people connections. It takes on new and deeper relevance amidst the global pandemic that has stricken the world. The fight against COVID-19 pandemic has made it abundantly clear that the global community is inescapably interconnected and needs stronger international collaboration through shared institutions and economic growth paths that are resilient, inclusive, and sustainable. The BRI has the potential to make major contributions to these needs.

The BRI has significant potential to boost the incomes of BRI countries and the world at large. According to the World Bank, the BRI could increase trade in BRI countries by 9.7% and foreign direct investment (FDI) by 7.6%, which would lead to an increase in real income for Belt and Road economies by up to 3.4%. Increases of standards of living in the BRI countries also benefit the rest of the world, which according to the World Bank would grow by up to an additional 2.9% due to the BRI. These estimates stand in sharp contrast with similar estimates for the Trans-Pacific Partnership, which would have boosted the growth of its membership by just 1.1% and the rest of the world by 0.4% [1].

Alongside the significant benefits associated with major infrastructure financing, large infrastructure finance is also endemic to a set of sustainability-related risks, including biodiversity risk, and the BRI is no exception. Some studies show that the BRI may become associated with losses in wildlife movement and mortality through habitat loss, the spread of invasive species, increases in illegal logging, poaching, and fires; and cause deforestation through the construction of roads, power lines and power plants, and subsequent mining activity. For these reasons, it is important to incorporate eco-environmental risk mitigation and management into the "green BRI" framework to align it with the 2030 Agenda for Sustainable Development.

© The Author(s) 2022
China Council for International Cooperation on Environment
and Development (CCICED) Secretariat,
Green Consensus and High Quality Development,
https://doi.org/10.1007/978-981-16-4799-4_8

With the aim of fulfilling these commitments, this book examines how both China and the international community have learned over time to prevent and mitigate such risks. China's Ecological Red Line standards and analogous international practices offer a number of models that can be adapted to green the BRI with respect to biological diversity. The book includes further strategic principles for aligning the BRI with the Sustainable Development Goals (SDGs) and the Paris Agreement in general and establishing the green BRI Roadmap, which links three frameworks from the strategic aspect: the green BRI, the 2030 Agenda for Sustainable Development, and development goals of BRI participating countries. Specifically, this roadmap includes 4 major approaches. First, enhance policy communication. It is important to take the green BRI as an important practice of realizing SDGs and facilitating global environment governance reform with green development as the shared principle. Give full play to the role of BRI International Green Development Coalition (BRIGC) and other cooperation platforms. Second, enhance strategic alignment. It is suggested establishing the mechanism for linking Green BRI with the 2030 Agenda for Sustainable Development, actively promoting the alignment of environmental policies, planning, standards and technologies, and strengthening information sharing with the help of the BRI Environmental Big Data Platform. Third, improve project management. Establish and improve the mechanism for project management on green Belt and Road to further reinforce environmental management in BRI projects and prevent ecological and environmental risks from the development of BRI projects. Fourth, improve capacity building. It is recommended to jointly conduct green capacity building programs, such as the Green Silk Road Envoys Program, to create people-to-people bond in building green BRI.

Under the framework outlined by the above Roadmap for building a green BRI, with a special focus on Sustainable Development Goal 15 (SDG 15) and biodiversity conservation, more specific policy recommendations have been proposed to better align BRI, SDG 15, and CBD. This SPS recommends that China:

First, apply international norms and standards to facilitate the use of stricter environmental standards in BRI projects. It is recommended to actively align BRI efforts with the fulfillment of international and national commitments to international conventions, including the CBD and UNFCCC. **Second**, focus on environmental impacts and carry out assessment and classification-oriented management of BRI projects. It is recommended to boost the development of the guidance on assessment and classification of BRI projects, based on the on-going Joint Research on Green Development Guidance for BRI Projects undertaken by BRIGC, which could provide green solutions to BRI participating countries and projects. **Third**, improve policy instruments to prevent and control the eco-environmental risks related to BRI projects. It is recommended to carry out environmental impact assessment for key BRI sectors and projects and establish a regular environmental risk regulatory mechanism that incorporates environmental pollution, biodiversity conservation and climate change as important factors for assessment. It is important to make full use of green finance instruments and environmental risk assessment tools, and take ecological redlining as a key instrument. **Fourth**, improve the coordination mechanism and facilitate effective linkage and alignment among different SDGs using Nation-based Solutions (NBS). It is necessary to create synergies with efforts for SDG 13 of Climate Action.

8.1 Linkages Between the Green Belt and Road and the 2030 Agenda for Sustainable Development

8.1.1 Background and Progress of Building the Green Belt and Road Initiative

8.1.1.1 The Background, Goal and Achievement of the Belt and Road Initiative

Since the financial crisis in 2008, the world has recognized the need to forge new sources and patterns of economic growth. In this context, the Belt and Road Initiative (BRI) was proposed as China's contribution to a comprehensive solution for sustainable development. Pursuing the principles of extensive consultation, joint contribution and shared benefits, the BRI promises to create new opportunities for shared growth and prosperity among countries through policy coordination, connectivity, unimpeded trade, financial integration, and people-to-people connections. It takes on new and deeper relevance amidst the global pandemic that has stricken the world, as it has become acutely clear that major international efforts like the BRI can help bolster cooperation against pandemics and other international challenges like financial crises, climate change, and global biodiversity loss.

The accomplishments thus far have been impressive. From 2013 through 2019, cumulative commodity trade between China and countries along the Belt and Road, defined in the broadest terms, exceeded USD 7.8 trillion; direct investment to countries along the Belt and Road approximated USD 110 billion; and the value of new project contracts reached nearly USD 800 billion [2]. As estimated by the World Bank [3], implementing BRI projects will reduce the aggregate costs for trade among BRI participating economies by 3.5% and those for the trade between BRI participating economies with the rest of the world by 2.8%. By November 2019, the investment from Chinese enterprises in building economic and trade cooperation zones overseas in BRI countries amounted to USD 34 billion, creating tax revenue of over USD 3 billion and 320,000 local jobs [4]. According to the World Bank [3], the implementation of the Belt and Road Initiative has the potential to raise real income gains raise incomes in BRI countries by 3.4% and increase global real income by up to 2.9% for the rest of the world. The BRI has been recognized by the United Nations as a solution for facilitating the implementation of the 2030 Agenda for Sustainable Development.

However, the BRI has even greater potential, specifically in the area of supporting biodiversity through high-quality infrastructure investment and global coordination. In April 2019, research findings and recommendation reports from the Advisory Council of the Belt and Road Forum (BRF) for International Cooperation (2019) highlighted that the Belt and Road Initiative and UN 2030 Agenda for Sustainable Development shared common ground in terms of facilitating cooperation, implementation instruments and measures, among others, which could achieve greater synergy.

8.1.1.2 Progress of the Development of the Green BRI

Since its inception, building the Belt and Road into a pathway for green development has been the aspiration and expectation of the Chinese government as well as the shared goal of all participating countries. China has accelerated its progress in building an ecological civilization, making unprecedented efforts in recent decades. The concepts of "putting ecological progress in the first place" and "green development" have been widely accepted by Chinese society as a consensus, and economic growth is shifting from a conventional model of "development first and green later" to high-quality development led by ecological civilization. By jointly building a green BRI with participant countries, China is creating a platform for countries to share and learn from one another the experience of green transitions and sustainable development. Over the past six years, China has been working closely with BRI participating countries in areas of environmental governance, biodiversity conservation and climate change mitigation and adaptation via bilateral and regional cooperation. It has witnessed positive and concrete results in building a green BRI and implementing the 2030 Agenda for Sustainable Development.

First, China has improved the BRI's top-level design and enhanced its cooperation mechanisms. In March 2015, the National Development and Reform Commission (NDRC), the Ministry of Foreign Affairs (MFA) and the Ministry of Commerce (MOFCOM) jointly issued their "Vision and Actions on Jointly Building Silk Road Economic Belt and 21st-Century Maritime Silk Road." The document proposes that China should promote ecological progress in conducting investment and trade, increase cooperation in ecological conservation, biodiversity protection, and climate change mitigation and adaptation. In 2017, the Ministry of Ecology and Environment (MEE, then Ministry of Environmental Protection) issued the "Belt and Road Ecological and Environmental Cooperation Plan" and launched the "Guidance on Promoting Green Belt and Road," which identified the roadmap for the development of a green BRI, together with the MFA, NDRC and MOFCOM.

As the BRI gradually unfolds, the green BRI framework is gaining a positive response from the international community. Currently, the MEE has signed nearly 50 bilateral and multilateral environmental cooperation agreements and has launched BRI International Green Development Coalition (BRIGC). The BRIGC was proposed by Chinese President Xi Jinping during the First Belt and Road Forum for International Cooperation (BRF), officially launched on the Thematic Forum of Green Silk Road of the Second BRF, and listed as one of the sectoral multilateral cooperation initiatives and platforms in the Joint Communique of the Leaders' Roundtable of the Second BRF. The main goal of BRIGC is to promote international consensus, understanding, cooperation and concerted actions to achieve green development of the BRI. To date, more than 150 Chinese and international organizations from over 40 countries have confirmed their partnership, including more than 70 overseas institutions such as government departments of BRI participating countries, international organizations, think tanks and businesses. Currently, BRIGC is actively promoting policy dialogues, thematic partnerships, and champion projects. The flagship research on BRI Green Development Report, the Joint Research on the "Green Development Guidance on BRI Projects" (the "Green Light" System), and the joint study on BRI Green Development Case Studies have been launched.

Second, platforms and modes for cooperation have been enriched to be more pragmatic. China has expanded platforms for collaboration, including the China-Cambodia Environmental Cooperation Center and China-Laos Environmental Cooperation Office, which actively promote capacity building programs and champion projects. The Belt and Road Environmental Technology Exchange and Transfer Center (Shenzhen) was established to take advantage of the industrial resources of the area to promote innovative development and international transfer of environmental technologies. These platforms will facilitate environmental cooperation along the Belt and Road on regional and national levels. The BRI Environmental Big Data Platform (referred to as "the Big Data Platform") was officially launched. It has developed its own application (APP) for information updates, which helps to improve the "One-Map" system for integrated data services. With the help of information technologies, such as "Internet+" and big data, the Big Data Platform is designed to be an open platform for the exchange of ecological and environmental information through sharing and collaboration. It will provide environmental data support to BRI participating countries, including ecological environmental protection concepts, laws, regulations and standards, environmental policies and management measures, etc.

Third, China has promoted in-depth policy communication to build consensus on green development. China has made full use of existing international and regional cooperation mechanisms to share its vision, experience, and achievements in ecological civilization and green development, through the UN Environment Assembly, CEEC Ministers' Conference on Environmental Cooperation, and other international events. Meanwhile, the MEE is also engaged in opening up new channels for dialogue and communication. It held the Thematic Forum of Green Silk Road of the Second BRF for International Cooperation, organized sideline events on Green BRI during World Environment Day Celebrations, UN Climate Action Summit, and China-ASEAN Environmental Cooperation Forum, and sponsored more than 20 thematic forums each year on biodiversity conservation, climate change mitigation and adaptation, and eco-friendly cities with the attendance of more than 800 people from BRI participating countries and regions.

Fourth, these cooperation projects have borne fruit. For example, the Chinese government has established the Green Silk Road Envoys Program to promote capacity building in environmental governance in China and BRI participating countries. This program has trained more than 2000 government officials, technological staff, youth, and scholars from 120 BRI participating countries. According to the List of Deliverables of the Second BRF, the Chinese government will continue to implement the Green Silk Road Envoys Program, which expects to train 1500 environmental officials from the BRI participating countries in the next three years. The Chinese government has also worked with relevant countries to jointly implement the Belt and Road South-South Cooperation Initiative on Climate Change to improve the capacity of BRI participating countries in addressing climate change and promote the implementation of the Paris Agreement. Moreover, China is also engaged in helping BRI participating countries in climate change mitigation and adaptation and energy transition, and promoting Chinese environmental technologies, standards, and low-carbon and energy-saving products in the international market through building low-carbon demonstration zones and organizing capacity building activities based on the reality and demands of BRI participating countries.

8.1.2   The Focus on SDG 15

In May 2019, the Intergovernmental Science-Policy Platform on Biodiversity and Ecosystem Services (IPBES) released the Global Assessment Report on Biodiversity and Ecosystem Services. The report evaluated the influence of biodiversity and ecosystem services on the economy, well-being, food security and life qualities. The report revealed that, over the past 50 years, the speed of biodiversity loss is unprecedented across human history. The top direct drivers for the most drastic biodiversity loss include changes in the use of land and sea, direct exploitation, climate change, and invasive alien species; while values and behaviors such as demographic and sociocultural changes, economic and technological factors, as well as institutions and governance are considered as critical indirect drivers for biodiversity loss. Overall, 75% of the terrestrial environment has been severely changed by human behavior and activities. The pressures brought by the above drivers made it difficult to attain the related goals set by the Convention on Biological Diversity (CBD) and the UN Framework Convention on Climate Change (UNFCCC), unless more revolutionary actions are taken. Similarly, to realize relevant goals and targets in the 2030 Agenda, revolutionary changes from the status quo protection speed and measures have to be implemented.

The year 2021 will mark an important turning point. The 15th meeting of the Conference of the Parties (COP 15) to the CBD will take place in Kunming, China in 2021, with the theme of "Ecological Civilization: Building a Shared Future for All Life on Earth." COP 15 will review the Post-2020 Global Biodiversity Framework, set up 2030 objectives and targets for the conservation of global biodiversity, formulate the strategy for the conservation of global biodiversity in a new decade (2021–2030), and launch the new course of post-2020 global biodiversity conservation.

The 2030 Agenda has highlighted the significance of biodiversity, with SDG 14 "Conserve and sustainably use the oceans, seas and marine resources for sustainable development" to deal with marine biodiversity and SDG 15 "Protect, restore and promote sustainable use of terrestrial ecosystems, sustainably manage forests, combat desertification, and halt and reverse land degradation and halt biodiversity loss" to address issues with terrestrial biodiversity. In this sense, CBD COP 15 could be considered as a key window of opportunity to speed up the attainment of biodiversity-related SDGs.

Built on the results of the first phase of the Special Policy Study (SPS) on Green Belt and Road and 2030 Agenda for Sustainable Development, this SPS, as the second phase of the series, will take a goal-by-goal and step-by-step approach to the alignment of BRI and biodiversity-related SDGs. Given the severity of terrestrial ecosystem degradation and biodiversity loss, this SPS will primarily focus on SDG 15 as the entry point and propose policy recommendations for COP 15 on how to encourage BRI participating countries to better implement SDGs with the help of BRI.

8.1.3 Progress of Countries Along the Belt and Road in Implementing SDG 15

Progress is still lacking in achieving SDG 15 across BRI participating countries. The Sustainable Development Report 2019, published by the UN Sustainable Development Solutions Network (SDSN) and Bertelsmann Stiftung, evaluated progress among 193 countries in realizing SDG 13 (climate action), SDG 14 (life below water), and SDG 15 (life on land). It concludes that "trends on greenhouse gas emissions and, even more so, on threatened species are moving in the wrong direction."

The SDSN assesses the progress of 139 + 1 countries along the Belt and Road towards realizing SDGs. The report selects five indicators to evaluate the implementation of SDG 15, including the mean area that is protected in terrestrial sites important to biodiversity (%), the mean area that is protected in freshwater sites important to biodiversity (%), the Red List Index of species survival, permanent deforestation (5 years average annual %), and imported biodiversity threats (per million population).

SDSN finds particularly strong challenges in the geographic regions most closely associated with Belt and Road corridors: the Association of Southeast Asian Nations (ASEAN) as well as West and South Asian countries. These results are discussed below. The detailed evaluation results are shown in Fig. 8.1.

From the perspective of implementing SDG 15, SDSN finds that only four Central and Eastern European countries out of 140 countries have realized "Goal Achievement" of SDG 15: Poland, Hungary, Romania, and Bulgaria. The implementation of

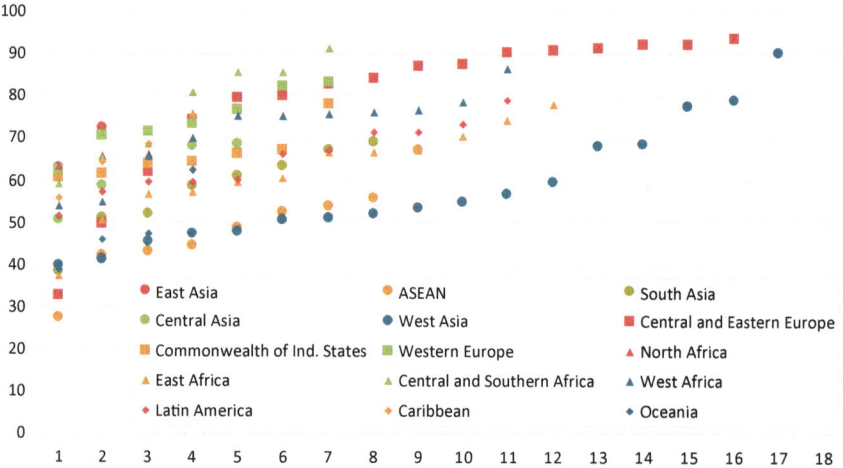

Fig. 8.1   Score of countries along the belt and road on SDG 15. *Note* Circles indicate Asia, squares indicate Europe, triangles indicate Africa, and diamonds indicate other regions

SDG 15 in Central and Eastern Europe is generally better than in other regions. For countries in other regions, there are various degrees of risks in the implementation of SDG 15. "Major Challenges" exist for three ASEAN Member States (Malaysia, Indonesia and Viet Nam), four in South and West Asia (Afghanistan, Iraq, Turkey, and Syria), four in East Africa (Djibouti, Madagascar, Seychelles, and Somalia) and four in Oceania (Fiji, Micronesia, the Solomon Islands, and Vanuatu).

Regarding the time sequence of implementing SDG 15, SDSN finds once again that Central Eastern European countries exhibit better performance than other regions. Ten out of 16 Central and Eastern European countries are on track or maintaining achievement, four countries show a moderately improving trend and two countries showed stagnation in their work. ASEAN Member States and countries in South Asia are the main areas facing challenges. The scores of SDG 15 in half of the 10 ASEAN Member States are decreasing, while two countries are in stagnation. Four countries out of eight in South Asia demonstrated a declining trend in implementation performance. Most countries in Central Asia and Commonwealth of Independence States (CIS) revealed stagnation in implementation, including five Central Asian countries and five out of seven CIS countries.

From the perspective of specific indicators, the most impactful indicator for ASEAN and South Asian countries in implementing SDG 15 is the Red List Index. Following a time sequence, the performance on this indicator in ASEAN and South Asian countries exhibited a decreasing trend. In addition, for ASEAN Member States, permanent deforestation also brings tremendous risks in implementing SDG 15.

8.1.4 Benefits and Biodiversity-Related Risks of BRI

The BRI has the potential to close major infrastructure gaps, accelerate regional integration, and increase economic growth in a manner that advances progress towards the SDGs. Indeed, there is certain evidence that after just a few years the BRI is contributing to the achievement of some of these goals. Any large-scale development effort also has potential risks, and the key to the BRI's success will be to maximize the potential benefits while minimizing the potential risks. One such risk is the biodiversity decline that is often associated with major infrastructure investments in ecologically fragile areas with insufficient risk assessment in advance or risk management in operation. When accentuated, biodiversity loss can even jeopardize the economic returns of infrastructure investments.

Chinese President Xi Jinping exhibited China's commitment to biodiversity when he unveiled the Beijing Call for Biodiversity and Climate Change alongside French President Emmanuel Jean-Michel Frédéric Macron in late 2019. In the call, China and France pledge to lead by example to

> Mobilize additional resources from all sources, both public and private, at the domestic and at the international level, towards both climate adaptation and mitigation; make finance flows consistent with pathways towards low greenhouse emissions and climate-resilient development, as well as for the conservation and sustainable use of biodiversity, the conservation of oceans, land degradation amongst others; ***ensure that international financing, particularly in the infrastructure field, is compatible with the Sustainable Development Goals (SDGs) and the Paris climate agreement***. (China Daily, 2019, emphasis added)

This SPS is intended to conduct evidence-based research in order to formulate a framework of policies that will help the BRI be compatible with SDG 15. This section outlines the potential and realized benefits of the BRI and the potential biodiversity loss risks associated with the BRI.

8.1.4.1 Benefits of the BRI

The world community faces a financing gap of 2.1% of global GDP annually to 2030 in order to provide the infrastructure that is needed to meet the SDGs [5]. The China-led BRI has the potential to take a leading role in closing those gaps in a manner that is aligned with the SDGs. According to estimates from the World Bank [3], the transport corridors of the BRI will significantly increase economic growth in BRI countries. New transport corridors can increase the speed and efficiency of trade routes, connect isolated human settlements, and create better access to markets by facilitating the transportation of goods, services, and people across the world. When infrastructure is completed, there are boundless possibilities for "spillover effects" where new forms of economic activity arise that would not have without the infrastructure investment [6].

The BRI has significant potential to boost the incomes of BRI countries and the world at large. According to the World Bank, the BRI could increase trade in BRI countries by 9.7% and foreign direct investment (FDI) by 7.6%, which would lead to an increase in real income for Belt and Road economies by up to 3.4% and by up to an additional 2.9% for other countries. In contrast, estimates for the Trans-Pacific Partnership (TPP) show that TPP would have boosted the growth of its membership by just 1.1% and the rest of the world by 0.4% [1]. The BRI then may have the largest potential to boost economic prosperity among participant countries and across the world.

These benefits are already being realized. Dreher et al. [7] looked at the impact of China's overseas projects financed by the China Development Bank, Export–Import Bank of China, and other Chinese financial institutions on economic growth in 138 countries. The authors found that on average a Chinese-financed project yields a 0.7% increase in economic growth two years after the project is committed.

8.1.4.2 Biodiversity Risks and the BRI

Alongside the significant benefits associated with major infrastructure financing, large infrastructure finance is endemic to a set of sustainability-related risks, including biodiversity risk, and the BRI is no exception. A handful of studies have already identified some of the potential biodiversity risks of the BRI. In a recent article in *Conservation Biology*, Hughes [8] spatially located proposed road and rail projects of the BRI (defined as those along the BRI corridors) and examined the extent to which such projects will be proximate to Key Biological Diversity Areas (KBAs) across the world. The author estimates that 16% of the world's KBAs are within 50 km of BRI proposed road projects and 60.6% of the world's KBAs lie within the BRI's proposed rail routes. The author also found that, 0.2% and 14.9% of KBAs are just 1 km from BRI road and rail routes respectively. In all, the author predicts that the BRI could endanger 4138 animal and 7371 plant species along the BRI [8]. A study published in *Current Biology* led by Xuan Liu [9] from the Chinese Academy of Sciences looks at the potential for the BRI to accentuate invasive species. They find that BRI countries fall in 27 of 35 recognized global biodiversity hotspots across the world and that the proportion of areas of high invasive species potential in BRI covered regions is 1.6 times larger than in non-BRI regions.

The earliest study was conducted by the World Wildlife Fund for Nature (WWF). According to WWF's analysis, BRI corridors in Eurasia overlap with the range of 265 threatened species including 39 critically endangered species and 81 endangered species, with 1739 Important Bird Areas or KBAs and 46 biodiversity hotspots or Global 200 Ecoregions. WWF finds the potentially most impacted areas to be the China-Indochina Peninsula Economic Corridor, the Bangladesh-China-India-Myanmar Corridor, and the China-Mongolia-Russia Economic Corridor. A background study for the World Bank analysis discussed above came to similar conclusions. The China-Indochina Peninsula Economic Corridor and China-Mongolia-Russia Economic Corridor are facing the highest risks of biodiversity loss due to deforestation [10].

To appropriately address these risks, China's development finance institutions, which provide the bulk of the lending necessary for BRI projects to move forward, can institute safeguards that work with BRI signatory countries to screen, assess, and oversee the operation to ensure best practices. A 2020 study in *Nature Sustainability* evaluated policies in financiers associated with the BRI: 35 Chinese and 30 international institutions. The authors find that only 17 of these lenders require biodiversity impact mitigation, and only one of those is Chinese: the China-ASEAN Investment Cooperation Fund [11]. As a result, China faces potentially severe challenges in establishing cooperative mechanisms to oversee and mitigate biodiversity risks associated with specific BRI projects. This SPS explores lender safeguards and biodiversity risk mitigation in more detail below, in order to explore the potential for advancement in these areas.

Biodiversity loss also reduces economic well-being. A study published in the journal *Global Environmental Change* found that between 1997 and 2011, the world economy lost between USD 4 trillion and USD 20 trillion per year in ecosystem services from land cover changes [12]. A 2019 World Bank study examining the economic impact of conservation efforts in Kenya shows that biodiversity management can make the difference between infrastructure projects having positive or negative economic impacts, because of impacts on ecosystem services for surrounding communities [13].

Risks to biodiversity clearly carry potential impacts for human communities, but those impacts can manifest differently across gender lines, which can severely curtail the effectiveness of conservation planning if it is not taken into account. In many rural, poor settings, biodiversity loss impacts women to a greater extent than men, especially in communities where women are tasked with collecting water, firewood, and wild foods, which is common in developing countries globally [14, 15]. If forests and riverine ecosystems are damaged, their tasks become more onerous, requiring farther travel in often insecure areas.

If the BRI does not develop and institutionalize a strategic set of appropriate policies and standards to mitigate the biodiversity risks, it could encounter financial, social, environmental, and political risks as well that may further erode the maximum potential of the BRI. Fragile ecosystems can jeopardize the integrity of infrastructure projects, reduce financial rates of return, and accentuate debt-driven macroeconomic stress in host governments and on the balance sheets of Chinese financiers. Furthermore, increased degradation of biodiversity can lead to social conflict and reputational risks that can also threaten the geo-political relationships that are so important to the BRI as well. For these reasons and more it is important to control biodiversity risks associated with the BRI.

8.2 An Analysis of Relevant Policies and Standards on SDG 15

8.2.1 Research and Evaluation of China's Experience

8.2.1.1 Biodiversity Conservation in China

China is among the world's megadiverse countries, yet its biodiversity is seriously threatened. To strengthen biodiversity conservation, China has been conducting biodiversity surveys, assessments of endangered categories of ecosystems and species and in-situ and ex-situ conservation, as well as developing policies and regulations on biodiversity conservation.

In terms of in-situ and ex-situ conservation, China has established a natural protected area system pivoting on national parks and also including nature reserves, scenic areas, forest parks, geographic parks, wetland parks, and cultural and natural heritage sites, among others. To supplement the natural protected areas, China has also established key ecological function zones and priority areas for biodiversity conservation. Currently, China has more than 10,000 protected areas, including national parks, nature reserves, forest parks, scenic areas, geographic parks, wetland parks, drinking water sources, and so on, covering 18% of the national land territory. At the same time, China has proposed an ecological function zoning scheme that consists of large-scale ecological function zones of different levels (including national key ecological function zones, important ecological function zones, bio-sensitive zones and vulnerable zones), which has played a significant role in protecting biodiversity and safeguarding national ecological security. However, even with these measures in place, China has still witnessed severe ecosystem degradation and accelerated biodiversity loss due to a lack of clear identification of natural protected areas' boundaries. The drawing of ecological redlines could identify areas with unique ecological functions, which must be strictly protected in order to realize centralized management of the eco-space.

8.2.1.2 Practices of Ecological Redlining in China

(1) The Drawing and Management of Ecological Redlines
   In October 2011, the State Council of China released the "Opinions on Strengthening Major Environmental Protection Work" to put forward ecological redlining for the first time. The document articulates the drawing of ecological redlines in major ecological function areas, sensitive areas and vulnerable areas for permanent conservation. In February 2017, the General Office of the CPC Central Committee and the General Office of the State Council jointly issued and circulated "Opinions on Drawing and Strictly Following Ecological Redlines," which established the framework, basic principles and overall goal

of delineating and observing the ecological redlines. The release of this document represented a new phase of accelerated development of the ecological redline system in China.

(2) The Development of Scientific Methodology for the Drawing of Ecological Redlines

Scientific assessment is necessary before drawing ecological redlines. The aim of this step is to identify the spatial distribution of areas with critical ecological functions (such as water conservation, biodiversity protection, and water and soil preservation) and areas sensitive or vulnerable to water loss and soil erosion, desertification and salinization. The next step is to conduct a spatial mapping analysis of the two categories of areas and draw a redline for ecological protection that encompasses all development-prohibited areas at national and provincial levels and other protected areas in need of strict protection.

The design of ecological redlines aims to bring almost all rare and endangered species in China and their habitats under protection, with due consideration to China's own reality. Ecological redlining doesn't equal identifying new protected areas, but rather, constructing and optimizing the systems for ecological protection with a more scientific, comprehensive and systematic approach. It could turn existing protected areas into an integrated ecological protection system that is easy to manage. It contains both established protected areas of all kinds and areas that lack protection.

(3) The Establishment of the System for Delineating and Observing Ecological Redlines

In drawing the ecological redlines, the national government develops technical guidelines for provincial governments to decide the areas to be covered autonomously. Based on the "Methods for the Management of Ecological Redlines" issued by the Central Government, provincial governments develop their own methods with reference to local reality with detailed regulations on environmental access, the sustainable utilization of resources, ecological conservation and restoration, compensation for ecological protection and assessment and evaluation. Governments of all levels should take the responsibility of managing and regulating the ecological redlines.

(4) Significant effects have been achieved

In January 2018, the State Council approved the redline drawing plans from 15 provinces (autonomous regions and municipalities), including Beijing, Tianjin, Hebei, provinces and municipalities in the Yangtze River Economic Belt, and Ningxia. All these plans have been promulgated and implemented. In October 2018, the MEE and the Ministry of Natural Resources of China organized review meetings, principally approving the plans of drawing ecological redlines in 16 other provinces (autonomous regions and municipalities). The areas and sites covered by ecological redlines should be specified and demarcated after surveys. Still, based on the drawing plan, the ecological redline areas nationwide account for one-third of the national territory. Major ecological land within the redline boundaries, including forests, grasslands, and wetlands, accounts for 55% of the major ecological land nationwide. The natural protected area system pivoting on national parks has covered more than 18% of China's national land territory, surpassing the ratio of 17% set out by the 2020 Aichi Biodiversity Targets. The wild population of certain rare and endangered species such as the giant panda, crested ibis, and Tibetan antelope, has steadily increased. The major ecological land protected by the redlines covers the catchment areas of the Yangtze River, the Yellow River, and the Pearl River, among other major rivers at and above Category III in China, as well as all biodiversity-rich areas identified at the national level and the vast majority of biodiversity-rich areas defined at provincial levels. Redlining has also protected most river and lake water sources as well as some underground water sources, all the distribution areas of species on the List of Wildlife under Special State Protection, as well as the areas where protected fauna and flora are mostly distributed.

8.2.1.3   The Experience of China in Biodiversity Conservation Through the Ecological Redline Policy (ERP)

Ecological redlines help with biodiversity conservation through bringing areas with rich biodiversity and of importance under protection. In this way, habitats within the ecological redline can be preserved and restored, and in-situ and ex-situ biodiversity conservation can be realized.

(1)   The drawing of ecological redlines should be scientific and rational
An integrated and systematic approach should be taken to drawing ecological redlines. Scientific assessment is needed for the identification of different areas based on the importance of ecological functions and the sensitivity and vulnerability of eco-environment. Areas within the ecological redline include all development prohibited areas on the national and provincial levels and other protected areas where strict protection is necessary.

(2)   Human activities should be strictly controlled in areas protected by ecological redlines
In terms of functional positioning, ecological redlines are of great significance to maintaining ecological equilibrium and supporting sustainable economic and social development. Areas within ecological redlines are land with critical ecological functions, the use of which must be strictly controlled. In terms of conservation, ecological redlines represent the critical point and baseline for safeguarding ecological security. Areas within ecological redlines should never be allowed to see degradation in their function, shrinking in their size or change in their nature. In principle, ecological redlines should be managed the same way as "development prohibited" areas, with all development activities not in line with the function positioning of the areas being strictly prohibited.

1)   The management of protected areas, including national parks, nature reserves, scenic areas, forest parks, geographic parks, world natural

heritage sites, wetland parks, and drinking water sources, should follow related laws and regulations.

2) For other areas within ecological redlines, the following human activities are prohibited: mining activities; land reclamation, sand quarrying and other activities that may destroy coastlines; large scale agricultural activities, including wasteland reclamation, animal husbandry of scale and fishing; textile, printing and dyeing, leather manufacturing, paper-making and other manufacturing activities; real estate development; the construction of passenger and freight stations, ports and airports; coal-fired power and nuclear power generation and hazardous articles warehousing; the production of products with heavy pollution and high environmental risks listed in the Comprehensive Directory of Environmental Protection (2017), and production and operation activities with high environmental risks identified by Measures for the Administration of Compulsory Liability Insurance for Environmental Pollution.

(3) Ecological restoration and ecological compensation should be conducted in areas protected by ecological redlines

1) Conducting ecological restoration
China will soon develop plans for ecological conservation and restoration within ecological redlines that prioritize the protection of sound ecosystems and major habitats, restore damaged ecosystems, establish ecological corridors and sites, and improve the integrity and connectivity of ecosystems. Ecological restoration in areas protected by ecological redlines is identified as an important component in the protection and restoration of ecosystems, including mountains, waters, forests, lakes, and grasslands. The government is determined to effectively provide financial resources for ecological conservation and restoration by coordinating funding channels for various conversation and restoration projects within ecological redlines, such as programs on water and soil conservation, natural forest conservation, and comprehensive improvement of land and resources. Ecological restoration within marine ecological redlines will be conducted, based on the principle of integrated governance of the land and the sea, with special emphasis on the comprehensive management of estuaries, littoral zones, islands and polluted waters.

2) Introducing an ecological compensation mechanism integrating government funding and funding from other sources
Governments of all levels should increase their funding for areas protected by ecological redlines. Local governments are encouraged to launch fiscal, credit, financial, and tax policies to facilitate the implementation of ecological redline policies and establish ecological compensation mechanisms.

Local governments should develop diversified investment and financing mechanisms guided by the government with extensive public engagement to pool in resources from all sides. Governments are also encouraged to launch pilot programs on payment for ecosystem services and develop market-based mechanisms to realize the value of ecological products.

(4) Integrated monitoring should be developed and continuously improved for ecological redlines

It is important to access real-time statistics, improve the capability of the integrated analysis and application of monitoring data, be informed of the composition, distribution and dynamic change of ecosystems protected by ecological redlines, and keep track of human interference. Administrative decisions should be made in a scientific way with illegal acts being checked and handled in time.

(5) An accountability system should be established for safeguarding ecological redlines

1) Strengthening supervision for law enforcement

With the establishment of enforcement mechanisms for ecological redlining, regular supervision and inspection of law enforcement should be conducted to identify and punish illegal acts damaging ecological redlines. Should any violation appear, it should be investigated.

2) Establishing an assessment mechanism

Assessment of the performance of local governments in implementing ecological redlines should be conducted. The results of the assessment shall be a reference in determining the political achievements of local governments.

3) Strengthening the accountability system

Government officials whose decisions/actions cause severe damage to the ecological environment and resources shall be held accountable in all cases, regardless of their current positions.

4) Launching an incentive mechanism

Rewards should be given to organizations and individuals that have outstanding performance in protecting ecological redlines. It is recommended that personnel be assigned for the promotion of ecological redlines to improve the engagement of local residents.

5) Improving information transparency and public engagement

Governments should release information concerning ecological redlines, including their distribution and adjustment, to safeguard people's right to know, participate and supervise, and give full play to the role of the media, non-governmental organizations (NGOs) and volunteers in promoting ecological redlines.

8.2.2 Research and Evaluation from International Experience

Adopting a set of harmonized standards for SDG 15 across the BRI can help minimize the risk and maximize the benefits and legitimacy of all actors involved, through bolstering environmental and social risk management (ESRM). This section reviews the international standards related to SDG 15 as practiced by the major multilateral financiers of infrastructure, integration, and development finance across the world, in two sections. First is a short note about the benefits of putting standards in place. Second is a comparative analysis of some of the major policies practiced by international actors.

Over the last few decades, environmental assessment and oversight systems have proliferated in the realm of international finance and investment. This section identifies the international actors that serve as peers for the Chinese financial institutions most active in BRI project finance and provides a survey of common practices among them. BRI projects predominantly receive financing through Chinese official entities such as the Silk Road Fund, the China Development Bank, and the Export–Import Bank of China, though not exclusively so [16]. Thus, the international equivalent for the sake of environmental governance of cross-border infrastructure development is the cohort of multilateral development finance institutions (DFIs) that have been traditional sources of support for BRI signatory countries.

8.2.2.1 Benefits of Developing Green Standards and Safeguards Across the BRI

Developing green standards can ensure that the BRI is calibrated to the SDGs while bringing benefits to virtually all of the stakeholders in the BRI. High-level or best-in-class environmental standards should thus take into account the preferences of Chinese and the other multitude of stakeholders engaged in the BRI to ensure that the BRI can provide public goods to the global economy as a whole (Table 8.1).

Standards can also increase project performance and profitability of projects. For example, in 2018 the International Finance Corporation (IFC) found that establishing standards across each of the common norms noted above were correlated with strong financial performance (measured by return on assets and return on equity) and financial risk ratings in 656 IFC projects representing USD 37 billion [19]. Risk instruments based on debt sustainability analysis (DSA) can help ensure that Chinese actors do not have to bear the risk of default on projects. While full assessments of the costs and benefits of ESRM are hard to quantify, the Independent Evaluation Group (IEG) of the World Bank (an independent monitoring group) conducted an assessment of the costs and benefits of ESRM in 2010 and concluded that benefits from the "environmental safeguards far outweigh the incremental costs" [17]. Weighing risks and benefits from a sample of bank projects, the World Bank found that most sensitive projects yielded "low cost—low benefits or high cost—high benefits for recipient

Table 8.1 Benefits of standardizing the SDGs in the BRI

Benefits of standards across the BRI	
Chinese actors	Expansion of markets
	Greater project effectiveness
	Prevention from default risk
	Prevention and mitigation of environmental and social risk
	Prevention and mitigation of reputational risk
Host countries	Improved management of fiscal resources
	Better management of natural resources
	Strengthening of institutional capacities
	Prevention and mitigation of environmental and social risk
	Prevention and management of reputational risk
Local communities	Reduced likelihood of social conflict
	Enhanced voice and ownership
	Reduced vulnerability
	Improved livelihoods
Global	Equitable use of resources
	Enhancement of global public goods
	Interconnectivity and global growth
	Leadership and legitimacy

Source Authors' adaptation base on World Bank [17], China Development Bank-UNDP [18]

countries." In the same IEG survey mentioned above, the World Bank also found that over half of the "task team leaders surveyed reported that the Bank's safeguards increased acceptability of the project among beneficiaries, and the safeguard policies also increased acceptability among nearly 30% of co-financiers" [17].

Box 8.1. Case Study: Incorporating ESRM into Chinese Mining Enterprises in Peru

Chinese financiers, firms, and the government can benefit substantially from establishing a set of harmonized standards around these common norms. First, these tools can help Chinese banks and firms expand and maintain market share overseas. China's experience in Peru is a case in point. Because of a lack of ESRM on the part of Chinese investors and the Peruvian government, China's first foray into Peru was a costly one. Chinese firms struggled to work with workers and local communities over worker health and safety, emergency preparedness, and biodiversity concerns. Though some of the issues were actually due to a lack of enforcement of host country systems rather than the Chinese firm, Chinese firms in general suffered reputational damage. Indeed, it became more difficult for Chinese firms to win contracts for mining and exploration in that country because of the perception that Chinese firms and financiers did not have proper risk management strategies. Later, Chinese copper firms devised significant ESRMs and participated in stakeholder consultations during the design stage. Such activity helped get market access and enhance China's reputation rather than worsen it. Indeed, when an accident did occur, ESRM plans allowed the company and host country to respond in such a way as to mitigate the worst damage [20, 21].

Standards can also benefit local communities close to projects. Engaging with local laborers and communities about a project beforehand can help identify concerns before they turn into conflicts. In Bolivia, Chinese tin companies took part in a prior informed consent engagement with local communities that rejected the location of the tin company. Bolivia found another community more suited and equipped for the project, likely deferring social conflicts that would have hurt the companies' business prospects and damaging China's reputation in general [21].

Box 8.2. Beyond DFIs: Environmental Governance Systems in the United Nations
Through United Nations mechanisms, nations have developed parallel systems to the systems of governance established by the DFIs profiled here. In this context, the CBD has long been a global platform for efforts to raise and harmonize national standards. CBD guidance is highly compatible with the "green BRI" framework, in that it encourages countries to collaborate in information sharing and capacity building to develop their own standards and practices (CBD 1992, Article 14).

In 2006, the CBD established voluntary guidelines for biodiversity-inclusive environmental impact assessment, including substantial upstream attention to identifying potential areas of concern. The guidelines encourage parties to focus upstream effort—before projects are proposed—in developing biodiversity mapping resources, such as the ones developed in China's recent history of demarcation of conservation priority areas. Individual project proposals can then be screened to ensure that all likely risks will be adequately addressed in the assessment stage. Impact assessments should be conducted with full participation by all stakeholders, to the extent possible. After individual projects' impact assessment, accountability mechanisms should be established to monitor and manage those projects' risks, and oversee any necessary mitigation [22]. CBD has also called for harmonization of standards among biodiversity financing mechanisms. Includes standards to apply in all cases, including but not limited to: highlighting and prioritizing the intrinsic value of biodiversity and its role in local livelihoods, effective public participation by project stakeholders, the establishment of institutional frameworks to oversee safeguard implementation.

The Global Environment Facility (GEF) has been another important source of guidance on environmental standards. The GEF does not finance projects independently but rather works through co-financing. As such, its standards can "crowd in" other lenders and enable a broader reach. The GEF has nine minimum standards for projects, including assessment, accountability mechanisms, conservation practices, and restrictions on land use and the involuntary resettlement of existing communities. The first minimum standard, on environmental and social assessment, management, and monitoring, echoes CBD guidance in its requirement for project screening as early as possible to establish which risks—among those covered by this standard as well as the remaining eight—may apply to each project. The second standard requires the establishment of institutional mechanisms such as those described below, to address problems that may arise in an accountable and transparent fashion. While the scope of these safeguards represents a crucial element in the environmental management of international development finance, its scale is modest. GEF's current 4-year work cycle draws on $4.1 billion in pledged funding [23]. That represents a tiny fraction of the development finance issued through major development finance institutions. For comparison, the World Bank has approved over $120 billion in projects over the last four years [24]. For this reason, the international section of this paper focuses on the largest DFIs, which are the traditional sources of infrastructure finance in developing countries, as a comparison point for BRI projects.

Box 8.3. Beyond DFIs: Environmental Governance Systems in the Private Sector
In addition to the multilateral approaches profiled in this section, systems for private investment and finance have also made significant advances in recent years. Perhaps best-known are the Equator Principles, for use by private financial institutions in evaluating proposals for support. These begin with an emphasis on early review and categorization of projects, to ensure that project-level assessments adequately address all of the salient environmental and social risks, in a way that ensures the broadest possible public participation. They also include the importance of well-designed institutional accountability mechanisms, which work in conjunction with national judiciary remedies to ensure appropriate project management in practice [25]. Complementary to the Equator Principles are the International Organization of Standards' environmental management tools, collected under the title ISO-14000. These systems do not specify specific safeguards but cover the extent to which institutions have established their own standards, with a commitment to employee training and auditing to ensure compliance.

While these frameworks can be important tools for private lenders and investors to better select and manage projects, they are not strictly analogous to projects financed under the BRI, which involve cooperation among national governments. Thus, this section focuses on common practices among development finance institutions, which have traditionally represented the bulk of infrastructure finance for developing countries.

8.2.2.2 Comparative Analysis of Biodiversity Policy for International Financial Institutions

This section of the report surveys the practices of eleven major international institutions financing infrastructure across the world with respect to biodiversity. What immediately emerges from such an analysis is a remarkable convergence with respect to the objectives and guiding principles across these institutions. Virtually all institutions seek to minimize the risks to biodiversity and aim to have "no net loss" or even a "net gain" in biodiversity. Moreover, most institutions also require biodiversity assessments tied to mitigation measures, and entail stakeholder engagement and consultation in the assessment and management of biodiversity. A detailed analysis of specific operations and policies also shows that there are major similarities across institutions as well.

The majority of the international financial institutions have established the goal of biodiversity as a core of their activities. The Asian Infrastructure Investment Bank (AIIB), the Development Bank of Latin America (CAF), as well as the World Bank (WB) and International Finance Corporation (IFC) all recognize the need to "integrate conservation needs and development priorities; through sustainable use of the multiple economic, social and cultural values of biodiversity and natural resources in an optimized manner." To measure and calibrate such goals, institutions range from a policy of "no net loss" of biodiversity (such as the AIIB) or alternatively "no net loss or a net gain in biodiversity" (such as the European Investment Bank, EIB, Asian Development Bank (ADB), German Development Bank, and the CAF).

The majority of the international financial institutions also converge significantly with respect to overarching principles and policy operations for biodiversity protection. Virtually all of the institutions require these five traits:

- Alignment with international commitments and national legal requirements;
- Exclusionary lists of categorically ineligible projects due to biodiversity.
- Requirements for biodiversity screening and impact assessments;
- Application of a subsequent "mitigation hierarchy" for no net loss or a net gain to biodiversity; and
- Meaningful stakeholder engagement and consultation in the assessment and management of biodiversity.

Table 8.2 Operational requirements for biodiversity safeguards applied by DFIs

International best practice for biodiversity conservation

	Alignment with international and national commitments	Exclusionary list of categorically ineligible projects	Biodiversity impact assessments	Adopt mitigation hierarchy	Stakeholder engagement and consultation
ADB	X	X	X		X
AfDB	X	X	X	X	X
AIIB	X	X	X	X	X
BNDES	X		X	X	
CAF	X	X	X		X
EBRD	X	X	X	X	X
EIB	X	X	X	X	X
IADB	X	X	X	X	X
IFC	X	X	X	X	X
KFW	X	X	X		X
WB	X	X	X	X	X

Source Authors' analysis of official documents and interviews. *Note ADB* Asian Development Bank; *AfDB* African Development Bank; *AIIB* Asian Infrastructure Investment Bank; *BNDES* Brazilian Development Bank; *CAF* Development Bank of Latin America; *EBRD* European Bank for Reconstruction and Development; *EIB* European Investment Bank; *IADB* Inter-American Development Bank; *IFC* International Finance Corporation; *KfW* German Development Bank; *WB* World Bank

These policies are exhibited in Table 8.3. In Table 8.3, international institutions are listed vertically and specific biodiversity measures are listed horizontally across the table. It should be noted however that while these institutions have these policies, they are not always executed, which can thus lead to negative outcomes for projects, biodiversity, and communities alike [26] (Table 8.2).

Table 8.3 shows that there is a great deal of practice with respect to biodiversity across the largest international development finance institutions in the world economy. For the purposes of clarity, we identified five core areas of commonality listed above in the following: alignment with international commitments and national legal requirements; requirements for screening and assessments with specific biodiversity measures (and their related social impacts) that are fully disclosed; application of a subsequent "mitigation hierarchy" for no net loss or a net gain to biodiversity; entail stakeholder engagement and consultation in the assessment and management of biodiversity; and have an exclusionary list of categorically ineligible projects. This section of the paper highlights some of those programs.

(1)   Alignment with international commitments and national legal requirements
      A common trait across all of the international institutions is to align the practices of the institution with specific global or national commitments and legal

Table 8.3 Best practices in incorporating gender into biodiversity finance

Project stage	Best practices
Upstream: planning	In planning for expected local biodiversity losses and changes to community access to local ecosystems, disaggregate the expected impact on local livelihoods by gender. Ensure that women are not disproportionately hurt by greater difficulty in carrying out traditional gathering roles. This practice is particularly effective in contexts where women and men have different traditional work roles
	In arranging stakeholder engagement processes, ensure that women can participate fully. This practice helps planners understand the potentially different ways that a project may impact men and women differently. In contexts where women do not traditionally participate in mixed-gender public discussions, consider designing women-only engagement spaces
Midstream: implementation	In projects where communities receive monetary compensation for a loss of access to local ecosystems, ensure that the financial compensation is distributed in such a way that it does not worsen women's well being. This practice is particularly relevant in contexts where women traditionally control resources they gather from local ecosystems but men control financial resources
Downstream: monitoring and accountability	Account for changes in men's and women's use of time as well as financial resources. In contexts where women serve as local stewards of crop biodiversity through the cultivation of heirloom crop varieties in household or village gardens, this practice can ensure that biodiversity does not suffer. Garden crop biodiversity can be key to the resilience of local food systems during extreme weather events or economic turmoil
	Ensure that accountability and grievance mechanisms are fully accessible to women. This practice is particularly important in contexts where women lack equal property rights, have limited access to local judicial systems, or do not traditionally participate in mixed-gender public discussions. Women's participation in accountability mechanisms can allow project overseers and sponsors to monitor impacts on women's traditional role of crop biodiversity caretaker
	As part of the post-project evaluation, develop a "tip sheet" for incorporating gender into future project planning in this particular context. This running collection of wisdom will help ensure that future development projects in this cultural context will be able to fully incorporate lessons learned through this project

requirements. Most of the institutions surveyed have language such as the following from the AIIB: "The Bank will not knowingly finance Projects involving the following….The production of, or trade in, any product or activity deemed illegal under national laws or regulations of the country in which the Project is located, or international conventions and agreements, or subject to international phase out or bans" [27]. The AIIB and others then provide an illustrative list of the kinds of international and national commitments they mean to adhere to (discussed below in "Exclusionary lists").

(2)   Exclusionary lists of categorically ineligible projects due to biodiversity
      Often linked to the alignment language, the AIIB and others then provide an illustrative list of the kinds of international and national commitments they mean to adhere to. In the case of the AIIB they list the following [27]:

- "Trade in wildlife or production of, or trade in, wildlife products regulated under the Convention on International Trade in Endangered Species of Wild Fauna and Flora (CITES),"
- "Activities prohibited by legislation of the country in which the Project is located or by international conventions relating to the protection of biodiversity resources or cultural resources, such as, Bonn Convention, Ramsar Convention, World Heritage Convention and Convention on Biological Diversity."
- Commercial logging operations or the purchase of logging equipment for use in primary tropical moist forests or old-growth forests.
- "Production or trade in wood or other forestry products other than from sustainably managed forests."
- "Marine and coastal fishing practices, such as large-scale pelagic drift net fishing and fine mesh net fishing, harmful to vulnerable and protected species in large numbers and damaging to marine biodiversity and habitats."

Most of the international institutions in the survey extend the possibility of excluding a project beyond these international and national commitments to cases where screening and environmental impact assessments may warrant it. Most have similar language on this matter. The African Development Bank's policy reads that "If the Bank finds that the environmental or social impacts of any of its investments are not likely to be adequately addressed, the Bank may choose not to proceed with the investment…. When the habitat/biodiversity implications of a project would appear to be particularly severe, the Bank may decide not to finance the project" [28].

(3)   Requirements for biodiversity screening and impact assessments
      All of the major international institutions surveyed also perform analyses of biodiversity impacts as part of broader environmental impact assessments. With respect to biodiversity, these policies charge the institution to consider the direct, indirect and cumulative project-related impacts on the habitats and the biodiversity they support. The World Bank considers threats to biodiversity, for example habitat loss, degradation and fragmentation, invasive alien species,

overexploitation, hydrological changes, nutrient loading, pollution and incidental take, as well as projected climate change impacts. The World Bank also determines the significance of biodiversity or habitats based on their vulnerability and irreplaceability at the global, regional or national levels and will also take into account the differing values attached to biodiversity and habitats by project-affected parties and other interested parties [29]. Similar or identical language is found in the policies of most of the institutions studied here (see Table 8.3).

The CAF is one institution with a slightly different language and scope. Its policy states that it will examine "Relevant physical, biological, and socioeconomic conditions within the study area, In particular, environment-related aspects likely to be significantly affected by the proposed development, including, in particular, population, fauna, flora, soil, water, air, climatic factors, material assets, including the architectural and archaeological heritage, landscape and the interrelationship between factors above. Current and proposed development activities within the project's area of influence, including those not directly connected to the project" [30].

The Inter-American Development Bank (IADB) operates in such biodiverse places such as the Amazon basin which is home to many nations. The IADB's policy also addresses transboundary biodiversity issues associated with a project. The environmental assessment process for the IADB seeks to identify, early in the project cycle, transboundary issues associated with the operation. The environmental assessment process for operations with potentially significant transboundary environmental and associated social impacts, such as operations affecting another country's use of waterways, watersheds, coastal marine resources, biological corridors, regional air sheds, and aquifers, will address the following issues: (i) notification to the affected country or countries of the critical transboundary impacts; (ii) implementation of an appropriate framework for consultation of affected parties; and (iii) appropriate environmental mitigation and/or monitoring measures, to the bank's satisfaction.

In addition to estimating biodiversity impacts, international bodies recommend that economic impacts be differentiated on a gender basis, in order to estimate the indirect impact on women's work as stewards of crop biodiversity. The CBD's 2015–2020 Gender Action Plan calls for calculating project costs and benefits should be estimated differently for women and men, rather than collectively, as do the Green Climate Fund and Climate Investment Funds [31–33].

(4) Adopting a mitigation hierarchy to address identified concerns

To the extent that the compulsory biodiversity impact assessments identify issues that may impact biodiversity, Table 8.3 shows that most of the major international financial institutions (eight of the 11 surveyed) require a "mitigation hierarchy" to meet the overall objective of "no net loss" or a "net gain" in biodiversity. The mitigation hierarchy has the following four pillars:

- **Avoidance**: measures taken to avoid creating impacts from the outset, such as careful spatial or temporal placement of elements of infrastructure, in order to completely avoid impacts on certain components of biodiversity.
- **Minimization**: measures taken to reduce the duration, intensity and/or extent of impacts (including direct, indirect and cumulative impacts, as appropriate) that cannot be completely avoided, as far as is practically feasible.
- **Rehabilitation/restoration**: measures taken to rehabilitate degraded ecosystems or restore cleared ecosystems following exposure to impacts that cannot be completely avoided and/ or minimized.
- **Compensation:** measures, such as offsets, taken to compensate for any residual significant, adverse impacts that cannot be avoided, minimized and/or rehabilitated or restored, in order to achieve no net loss or a net gain of biodiversity. Offsets can take the form of positive management interventions such as restoration of degraded habitat, arrested degradation or averted risk, protecting areas where there is imminent or projected loss of biodiversity.

(5)   Stakeholder engagement, consultation and disclosure:
All of the institutions surveyed for this SPS require stakeholder engagement and consultation in the assessment and management of biodiversity. Each of the institutions makes some commitment to carry out consultations with affected peoples and communities and seek their informed participation throughout the project cycle.

As noted earlier, the CAF is perhaps the most engaged in major infrastructure projects in areas where there are significant concerns about biodiversity in areas that inhabit large and often vulnerable populations. The CAF requires that consultations with project-affected groups be held early in the environmental impact assessment process and maintained throughout the project cycle. Throughout the project cycle important information is supposed to be disclosed in a timely manner to affected groups, civil society organizations, and other key stakeholders. The CAF also requires that "The potential impact of projects over forests and natural habitats, and the rights of access to and use of resources for the welfare of the communities shall be evaluated as a part to the Environmental and Social Assessment" [30], 64.

The IFC requires that borrowers go so far as implementing a *Stakeholder Engagement Plan*. Where applicable, the Stakeholder Engagement Plan will include differentiated measures to allow the effective participation of those identified as disadvantaged or vulnerable. When the stakeholder engagement process depends substantially on community representatives, the client is required to make every reasonable effort to verify that such persons do in fact represent the views of affected communities and that they can be relied upon to faithfully communicate the results of consultations to their constituents. When affected communities are subject to identified risks and adverse impacts from a project, the client will undertake a process of consultation in a manner that provides the affected communities with opportunities to express their views on project risks, impacts and mitigation measures, and allows the client to consider and respond to them [34].

Development finance institutions have learned the importance of ensuring that their stakeholder engagement plans incorporate the voices of women, particularly in cases where communities may be facing displacement. As mentioned in Sect. 8.1.4.2 above, in many rural, poor settings around the world women do not customarily take part in public discussions but do bear the brunt of biodiversity losses, which can curtail their ability to serve as stewards of crop biodiversity, further potential biodiversity losses, and limit the benefits of conservation projects. For example, an inter-bank working group with representatives from the AIIB, ADB, AfDB, EBRD, EIB, IADB, NDB and the World Bank recently published joint recommendations on meaningful stakeholder engagement, which encourage project planners to ensure that these processes are designed specifically to prioritize the participation of women and other disadvantaged groups, and if necessary, disaggregate stakeholder engagement processes by gender [35].

8.3 Analysis of SDG 15 Related Investments Possibilities

Working towards SDG 15 is no small task. Biodiversity is fragile and necessary for the lives and livelihoods of global communities, and if damaged, difficult or impossible to regenerate. To prioritize it among the ever-accelerating world of international finance and investment, the field of biodiversity finance has emerged.

The need is certainly present and pressing. At a 2015 workshop in Beijing, the Intergovernmental Science-Policy Platform on Biodiversity and Ecosystem Services (IPBES) concluded that "urgent and concerted action" was needed to avert ecosystem degradation globally, for the sake of 3.2 billion people currently impacted by degraded lands [36]. Economically, they estimate that the losses caused by this biodiversity degradation amount to 10% of global GDP. These same authors track successful ecosystem restoration across every region and continent of the globe.

Biodiversity conservation is, by definition, an act that prioritizes long-term well-being over short-term booms. It requires investing in the natural capital necessary to support future economic production and human health. It also requires investing in activities that will pay off in positive externalities distributed throughout a wide array of communities, which the investor will not be able to completely reap themselves. Thus, it needs external encouragement in order to flourish, in the form of an enabling policy environment, preferential financial arrangements, and impact investors motivated to fuel positive change not only for their own portfolios but for the communities where they operate.

8.3.1 Survey and Assessment of the Chinese Experience

The SDG 15 aims to protect, restore and promote the sustainable use of terrestrial ecosystems. In recent years, China has continuously increased its financial input in ecological compensation mechanisms, transfer payments to ecological function areas, grassland compensation, subsidies for returning farmlands to forests, subsidies for wetland protection and restoration, and other programs. In the meantime, China has continued improving the property right system of natural resources, exploring new ways of cooperation among governments, businesses and environmental organizations, promoting sustainable forest management, combating desertification, halting and reversing land degradation, and halting biodiversity loss. In 2018, China scored 62.7 on SDG 15, up by 7% compared with 2017, indicating that certain progress has been achieved in terms of terrestrial ecosystem protection.

The ecological compensation mechanism continues to improve [37]. The Chinese government attaches great importance to the development of the ecological compensation mechanism and launched policy documents such as *Suggestions on Improving the Ecological Compensation Mechanism, Guidelines for Accelerating the Development of a Horizontal Ecological Compensation Mechanism for Upper and Lower Reaches of Rivers, Action Plan for the Establishment of a Market-Oriented, Diversified Ecological Compensation Mechanism, Guidelines for Establishing and Improving the Long-term Mechanism for Ecological Compensation and Conservation in the Yangtze River Economic Belt,* and *Plan for A Pilot Program of Establishing a Comprehensive Ecological Compensation Mechanism.* These documents establish the framework of an ecological compensation mechanism with Chinese characteristics. China had a fiscal input of nearly RMB 200 billion yuan in ecological compensation in 2019. Meanwhile, both the central and local governments have been taking market-oriented approaches to expand the source of funding for improving the ecological compensation mechanism. For example, the water source areas of the middle route of the South-to-North Water Diversion Project established ecological compensation through pairing cooperation; Jinhua City and Pan'an City in Zhejiang Province took the lead to adopt off-site development as a means of compensation; the drainage areas of Xin'an River engaged the private sector in ecological compensation programs; Moutai Group plans to invest a total of RMB 500 million yuan in 10 years starting from 2014 in water environment compensation in the drainage areas of Chishui River; and China Three Gorges Corporation has been playing an active role in the protection of the Yangtze River while exploring for market-oriented approaches to improve the compensation mechanism.

Transfer payments to ecological function areas have been increasing. To guide local governments to intensify the efforts to protect the ecological environment, and improve the capacity of local governments in places with national key ecological function areas to provide basic public services, the central government established the transfer payment system for key national ecological function areas in 2018 to support the protection of these areas. By the end of 2019, the central government has made transfers amounting to RMB 524.2 billion yuan to key national ecological function areas, of which RMB 81.1 billion yuan was made in 2019, RMB 9 billion yuan more than in the previous year, registering an increase of 12.5%. Meanwhile, China has kept expanding the coverage of key national ecological function areas to 819 counties. Once included in national ecological function areas, local governments will receive financial and policy support as long as they strictly implement the negative

RMB 100 million yuan

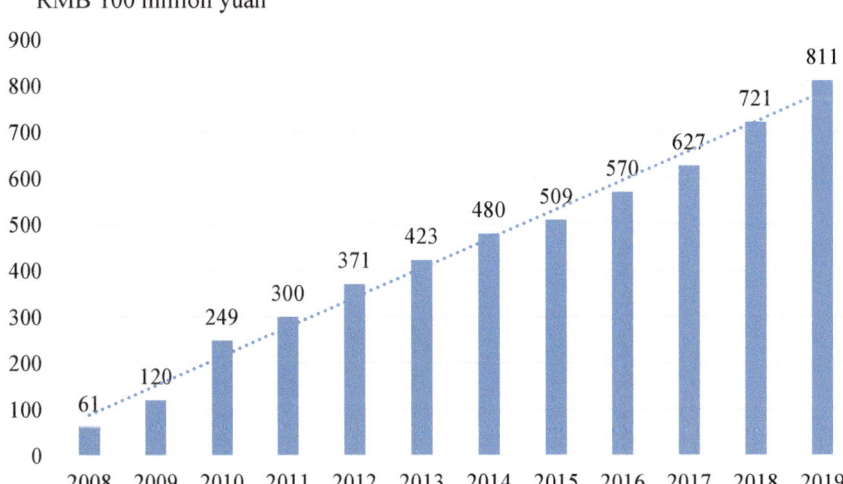

Fig. 8.2 Transfer payment to key national ecological function areas (2008–2018) [38] *Source* Dong et al, 39; Ministry of Finance, [40]

list system for industrial access. According to relevant regulations, a region counted as a key national ecological function area needs to strengthen ecological protection and restoration, regulate the boundaries of industrialization and urbanization, and enhance the supply capacity of eco-products (Fig. 8.2).

The standards for the compensation for ecological services of forests have been rising. In recent years, the central government has been increasing its input in compensation for ecological services of forests and raising the standards for compensation year by year. Starting from 2010, the standards of compensation for state-level non-commercial forests have varied according to their ownership. The compensation rate for state-owned state-level non-commercial forests was RMB 5 yuan/year mu (1 mu = 0.067 ha) in 2010, while that of privately-owned and community-owned state-level non-commercial forests has increased from RMB 5 yuan/year mu to RMB 10 yuan/year mu. In 2013, the compensation rate for privately-owned and community-owned state-level non-commercial forests was raised to RMB 15 yuan/year mu. In 2015, 2016, and 2017, the rate for state-owned state-level non-commercial forests increased step by step, reaching RMB 6 yuan/year mu, RMB 8 yuan/year mu and RMB 10 yuan/year mu respectively. As the central government increases its fiscal input and raises the standards for compensation, local governments are expected to positively improve compensation system for ecological services of forests in local areas.

Policies for fiscal support to ecological protection and restoration of wetlands continue to improve. China attaches great importance to the protection of wetlands with increasing fiscal input in accelerating the development and optimization of policies concerning fiscal support to ecological protection and restoration. From 2013 to

2016, the central government allocated RMB 5 billion yuan to protect wetlands in China and continued to provide support through the Funds for Reform and Development of Forestry afterwards. In 2014, the Ministry of Finance and the State Forestry Administration launched the pilot program of wetland ecological benefit compensation. For important wetlands on the route of migratory birds managed by the forestry system, their loss due to the protection of birds and other wild animals will be properly compensated. Currently, the central government allocates fiscal input to local governments, who will then decide the scope of wetland ecological benefit compensation and the areas to be protected.

Box 8.4. Measures Taken by the Funds for Reform and Development of Forestry to Support Wetland Protection and Restoration
The first measure entails supporting the protection and restoration of wetlands. For wetlands of international/national importance, national wetland parks at important ecological locations, and national wetland nature reserves at or above the provincial level managed by the forestry system, efforts will be made to protect and restore the wetlands, improve the current ecological status, and maintain the health of the local ecosystem.

The second measure entails supporting the restoration of farmland to wetland. It is encouraged to return farmlands to wetlands within the wetlands of international importance, national wetland nature reserves, and provincial nature reserves within wetlands of national importance managed by the forestry system, so as to expand the area of wetlands and improve the surrounding ecological status. The third measure entails supporting the wetland ecological benefit compensation. For important wetlands on the route of migratory birds managed by the forestry system, their loss due to the protection of birds and other wild animals will be properly compensated. In so doing, all parties are motivated to protect wetlands and maintain the wetlands' ecosystem service functions.

The grassland ecological protection subsidy incentive policy has been continuously promoted. To protect grassland ecosystem, guarantee the supply of meat and dairy products, and increase the income of herders, the Chinese government implemented the grassland ecological protection subsidy incentive policy. Currently, it covers eight major pastoral provinces (autonomous regions), including Inner Mongolia, Xinjiang, Tibet, Qinghai, Sichuan, Gansu, Ningxia and Yunnan, and five non-major pastoral provinces, such as Heilongjiang. RMB 152.033 billion yuan has been given as subsidies to 268 pastoral and farming-pastoral counties in the above provinces. In 2019, a new round of grassland ecological protection subsidy incentive of RMB 18.76 billion yuan was included in the central budget to support the banned grazing area of 1.206 billion mu and the grass-animal balance area of 2.605 billion mu (Fig. 8.3).

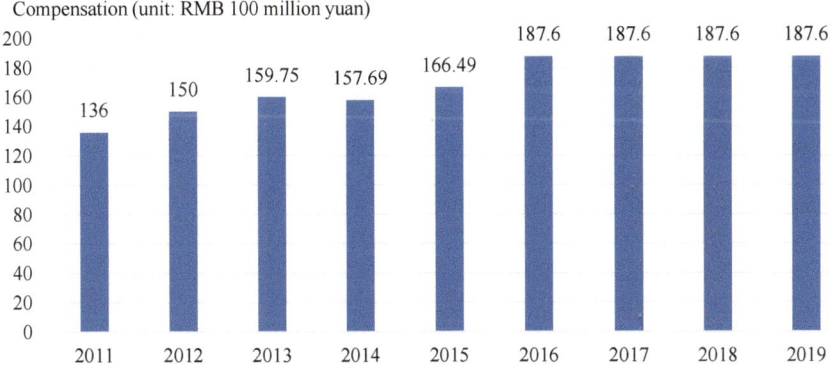

Compensation (unit: RMB 100 million yuan)

Fig. 8.3 Grassland ecological protection subsidy incentive (2011–2019). *Source* Chinese Academy of Environmental Planning [41]; Ministry of Finance website

The unified confirmation and registration of natural resources has been rolled out. The confirmation and registration of natural resources is important in promoting the reform of the property rights mechanism of natural resource assets, which is a key part of China's ecological civilization framework. By the end of October 2018, 1191 natural resource registration units have been established in 12 provinces (autonomous regions) and 32 pilot areas, and the total registered area has reached 186,727 km^2. The state also focused on exploring the confirmation and registration of national parks, wetlands, water flows, and proven reserves of mineral resources. Starting from the end of 2018, the confirmation and registration of natural resources in key areas has been implemented nationwide step by step. It is planned that within 5 years, the unified confirmation and registration of natural resources in nature reserves will be completed, such as national and provincial key parks, natural reserves and various natural parks (scenic areas, wetland parks, natural heritage, geo-parks, etc.). At the same time, the unified confirmation and registration of individual natural resources with complete ecological functions owned by the public will be conducted, such as major rivers and lakes, key wetlands, key national forests, important grasslands, and other areas.

Innovative approaches have been taken to promote cooperation between governments, environmental organizations, and large enterprises. "Debt-for-nature swaps" can be traced back to the 1980s. In a debt for nature swap, a nation agrees to swap the preservation of the natural environment for some of its debt. This benefits the nation because it brings its overall debt level down, and it benefits the environment by creating more protected habitat for animals and plants. Debt-for-nature swaps may be organized by conservation organizations or by governmental organizations concerned with environmental preservation. Currently, China doesn't have any recorded cases of debt-for-nature swaps. However, there are many cases in which international organizations or large enterprises cooperate with governments in environmental protection for win–win benefits, mostly in the form of Payments for Ecosystem Services (PES). The project for the protection of water source jointly operated by The Nature Conservancy (TNC) and Longwu, Zhejiang Province is a case of such a practice. The project, with TNC as the consultant, protects water source areas with funding from trust agencies. The forest land is operated and managed in an integrated manner. The income from such operations is used to cover the compensation to farmers and the cost of the protection and management of the water source area. In this way, the water source could be properly protected and the trust fund company could get a fair share of economic return.

Box 8.5. Water Source Protection Program Jointly Operated by TNC and Longwu, Zhejiang Province in the Form of Trust [42, 43]
On January 15, 2015, TNC and the government of Huanghu County, Zhejiang Province, signed an agreement on the protection of Longwu Reservoir, a county-level water source. According to the agreement, the goal of the program is to reduce the factors that may cause water quality deterioration and improve the water quality of the reservoir from Class II to Class I. It is a good attempt to adopt a win–win model towards ecological conservation that could benefit both the environment and the community. Funded by Alibaba Foundation, it is the first water source protection program of TNC in China in the form of trust. In September 2015, Wanxiang Trust-TNC Charitable Trust decided to invest RMB 330,000 yuan to support the program [42] (Liu Liwen 2015).

In November 2015, Wanxiang Trust launched the first water fund trust in China—Wanxiang Trust-Shanshui Fund No. 1. Shanshui Fund Trust invited TNC to be the consultant. In the same month, the first water source protection and management project supported by the Trust—Longwu Small Water Source Protection Program was officially launched.

The program is an innovation that integrates social resources and engages multiple actors, including farmers, financial institutions, charity organizations, local organizations, businesses in the lower stream of the industrial chain related to agriculture in its daily operations. In this way, it could generate positive outcomes through interaction, collaboration and sharing. The program effectively addressed the issue of pollution caused by human activities that could benefit the whole community; it also established a sustainable funding mechanism that could bring environmental benefits and economic returns for investors.

Mode: Trust

Target of compensation: Residents in the nearby community

Main provider of compensation: Alibaba Foundation

Means of compensation: Farmers entrust the forest land to the trust and get steady income as compensation

Operation mechanism: The trust agency operates and manages the forest land while promoting the growing of bamboo shoot and ecological tourism. The income from such operations is used to cover the compensation to farmers and the cost of the protection and management of water source area.

8.3.2 Survey and Assessment of International Practices

Globally, biodiversity finance has taken a variety of approaches, often in the form of PES or investments in soil and water health to ensure their ability to support future agriculture. The former category has been explored among developed as well as developing countries, under a variety of terms including "eco-compensation" in China and "agri-environmental programs" in the European Union [44]. A recent global survey in the journal *Nature Sustainability* estimates that over $36 billion is invested in PES annually, with approximately one-third of that amount taking place within China [45]. Of the conservative USD 36 billion estimate, the great majority is estimated to be concentrated in watershed subsidies, which attract approximately USD 23.7 billion per year. One of the better-known examples of watershed PES programs among developing countries is found in Ecuador, where the capital city of Quito established the world's first municipal water fund, with the cooperation of TNC, in 2000 [46]. Quito's groundbreaking PES program (named FONAG for the Spanish acronym for Water Protection Fund) adds a surcharge on water users and included a bottled water plant in order to fund the conservation of the watershed that provides water for the municipality of Quito.

A second common approach—and one of particular importance for the context of a green BRI—is the use of biodiversity offsets. These financial arrangements seek to mitigate the *net* biodiversity impact (or create a positive net biodiversity impact if possible) due to new project construction by financing separate conservation efforts. Enabling policies for this type of biodiversity financing exist throughout Asia, Europe, and the Americas, but range widely in definition [47–49]. A recent study in *Nature Sustainability* which considered only those projects implemented under a "no net loss" policy, found nearly 13,000 such projects globally, cover an area of roughly 153,679 km^2. Some of the largest existing such programs have occurred in very disparate countries, including: Mongolia, Brazil, and Uzbekistan [50]. However, as Gardner et al. [51] demonstrate in a *Conservation Biology* article, the "no net loss" standard is highly ambitious in practice, requiring comparable gains of new biodiversity (not simply conservation) relative to the biodiversity losses that are to be offset, and requiring that those gains be maintained over the long term. Accomplishing these goals requires strong institutional support and the involvement of a wide array of geographic locations, in order to effectively "hedge" the risks of partial project failure.

More broadly, biodiversity offsets take place within the mitigation hierarchy described in Sect. 8.2. As part of the "compensation" stage, offsets serve as a last option if the earlier options of avoiding, minimizing, and rehabilitating/restoring ecosystems and the communities that rely on them are not feasible or insufficient. For example, Villarroyo et al. [52] review the national policies regarding biodiversity offsets in Latin America and find that three countries' national regulations (Chile, Colombia, and Mexico) specifically make mention of both the mitigation hierarchy and offsets in relation to their EIA processes. However, many scholars note that significant institutional capacity building work remains in governments that seek to support offset schemes, particularly in establishing the scientific bases for "ecological equivalence" between geographic areas, in order to be assured of the net biodiversity impact of offset schemes [51–54].

Nonetheless, Luck et al. [55] find that global biodiversity finance has suffered from a severe geographical mismatch: flows have been directed mostly to low-priority ecosystems, while the most important ecosystems have been overlooked. Over half of all flows are focused on the United States, Canada, and Europe, despite the fact that these ecosystems are of "mutual low priority" for the two goals of preserving ecosystem services and biodiversity. Meanwhile, areas of high priority for both of those goals are concentrated in Southeast Asia and South America, which have attracted less than 15% of global biodiversity collectively. This mismatch is assuredly related to the fact that approximately half of all biodiversity finance stems from domestic government funding, and so funding is staying within wealthy countries. Thus, if the world is to make progress toward SDG 15, international biodiversity cooperation—through investment and especially aid—will be key.

Incorporating Commercial Investors

Traditionally, biodiversity finance has been limited to aid: official development aid (ODA) and philanthropy. However, opportunities for commercial investors have also been developing in recent years. Many types of biodiversity-maintaining or biodiversity-enhancing activities will pay for themselves in the medium-to-long term, though they require new sources of upfront financing to launch. By preserving or enhancing existing natural capital, these investment possibilities allow for reduced costs in economic production in the long term. For example, Burian et al. [56] advocate for agricultural investments aimed at building soil health and resilience, which will bring economic benefits in increased crop yields and decreased expenses on agrichemical inputs. IPBES estimates that the economic benefits of soil preservation are an average of 10 times greater than the cost of those efforts [36]. Finally, these benefits are multiplied as they impact downstream ecosystems through less-polluted waterways that better support both urban and rural life.

However, to succeed, biodiversity finance efforts must be well-matched with local needs, well-designed with local input, and well-managed by local governments. Clark et al. [57] find that the sector is beset by potential "greenwashing," in which commercial investors' activities are not actually biodiversity-enhancing or biodiversity-protecting, but market themselves as such in order to access advantageous financing and public reputational enhancements. While a few such investments may do no harm, allowing this type of activity to flourish under the banner of "biodiversity finance" brings risks to the entire sector, threatening the legitimacy of its claims and with it, its access to the favorable financing that will ensure its continued viability.

Bearing in mind the potential benefits as well as the potential risks, the United Nations Development Programme (UNDP)'s Biodiversity Finance Initiative (BIOFIN) has developed five areas of focus for developing frameworks for commercial biodiversity finance:

- Policy and Institutional Review, examining the ways in which national institutions are sufficiency robust and well-designed to encourage biodiversity finance, as well as what areas can benefit from reforms (with the added optional aspect of identifying economic drivers of biodiversity loss);
- Expenditure Review, calculating current expenditures to support biodiversity;
- Needs Assessment, estimating the total amount necessary in biodiversity-supporting expenditures and the gap in actual spending;
- Finance Plan, setting targets and finding potential sources for this funding;
- Finance Solutions, establishing and carrying out a plan to address the institutional and financial gaps discovered in previous steps.

Bilateral BIOFIN Cooperation Between Partner Governments

As the sector of biodiversity finance continues to expand, and particularly as it opens to commercial activities, China has the opportunity to establish itself as a global leader in the field. The globally networked nature of the BRI highlights the importance of working to preserve biodiversity in the "hotspots" along the network, to ensure that the entire enterprise brings net benefits to the communities and the ecosystems that support them.

Two such "hotspots" stand out among potential partners for Chinese conservation finance, one on each side of the Pacific Ocean: Indonesia and Ecuador—two countries with BRI MOUs with China. Both are among the 17 mega-biodiverse countries, who collectively boast 17% of the world's species. They represent the peak of global terrestrial and marine biodiversity, respectively. Ecuador is home to the most biodiverse section of the Amazon rainforest, often called the "lungs of the earth" [58]. The Ecuadorian Amazon rainforest sits at the headwaters of the Amazon River, and ecosystem preservation there has the potential to benefit the downstream Amazon biome. Indonesia sits in the center of the world's marine biodiversity, sometimes called the "Amazon of the Seas" or the Coral Triangle, for the tremendous density and diversity of coral species there [59]. Both countries have made significant progress in their UNDP BIOFIN process, preparing to host and manage biodiversity finance successfully.

Furthermore, both Indonesia and Ecuador have strong economic ties with China. According to FDI Markets, China has created more new investment in Indonesia than in any other country in the last decade: over USD 52 billion. Ecuador, while a much smaller country, has also built up an important friendship with China. For the last decade, China has been Ecuador's most important creditor, creating deep goodwill between the two governments. In 2019, Ecuador became the first Latin American or Caribbean nation to become a full member of the AIIB, signaling its interest and institutional readiness to continue to strengthen its financial ties with Asia and in particular, with China.

One major obstacle to biodiversity preservation is a simple matter of geography: biodiversity hotspots are disproportionately located in developing countries, with limited fiscal space to design and carry out long-term projects that will not yield financial benefits for many years. One way to circumvent this problem is for these countries to collaborate in a bilateral or plurilateral fashion with their most important strategic partners, either creditor nations or major sources of investment, to ensure that the growth in economic activity between their nations does not bring environmental degradation. Three main models are common in the area of bilateral biodiversity finance: debt-for-nature swaps, National Environmental Funds (NEFs), and bilateral sustainable development banks.

In debt-for-nature swaps, creditors cancel a share of the debt in exchange for the debt service payments being redirected to maintain biodiversity. Alternatively, impact investors or international non-profit organizations play a pivotal role, negotiating a deal in which they buy a country's debt at a discount, work with the debtor nation to build the institutional infrastructure to oversee the biodiversity plan, and help establish a fund to support these activities. These deals can effectively cut off a vicious cycle of fiscal constraints leading to insufficient environmental management, hurting economic outcomes, reinforcing fiscal constraints.

When implemented well, debt-for-nature swaps can allow chronically indebted countries an alternative to environmentally-damaging activities to pay down debt. They can also create an institutional structure to oversee the establishment of definitions of sustainable economic activities appropriate for the newly protected areas, and the fiscal space to ensure that the new protections are well-managed, with adequate participation from local communities to ensure enforcement. However, debt-for-nature swaps are not quick fixes for serious debt problems, nor can they bring a sudden stop to ongoing ecological disasters. As the case of Seychelles demonstrates, establishing the conservation areas is a process of multiple years. Thus, rather than being used as a last resort or rescue option for disaster scenarios, it is best considered as a long-term, proactive approach to conservation.

NEFs share many of the same characteristics of debt-for-nature swaps, but with less intervention on the part of outside partners. NEFs are locally-managed funds set up in collaboration with external partners, that support conservation efforts domestically. The "trust fund" nature of NEFs can make them particularly appropriate funding instruments for projects that need medium- or long-term investments, such as the delineation, establishment, and maintenance of national parks. For example, Brazil's Amazon Fund supports non-deforestation livelihood projects for forest-dwelling communities [60]. Other Amazon-basin countries including Bolivia, Colombia and Peru have NEFs to support their national systems of protected areas. In Asia, Bhutan and the Philippines both have similar funds [61].

While NEFs are managed by national governments, they can be established in conjunction with strategic partners overseas. For example, the Foundation for the Philippine Environment has been supported through debt swaps from the United States and Japan. In these cases, NEFs are similar to the debt-for-nature swaps profiled above, without the same level of detailed conditionality. Instead of debtor nations agreeing to set aside particular tracts of land, they establish general support for the nationally-defined conservation strategies. The fact that the local governments oversee the funds and their management makes them suitable for bilateral cooperation with partners that prefer to allow as much local direction as possible.

Finally, bilateral conservation financing can take the form of special-purpose development banks. For example, the North American Development Bank is a project of the governments of the United States and Mexico, and was brought into being as part of the negotiations for the North American Free Trade Agreement (NAFTA), with the objective of ensuring that the U.S.-Mexico border would not be degraded due to the higher economic activity expected under NAFTA. It finances sustainable development projects on both sides of the border [62]. It has financed USD 1.2 billion in projects as of year-end 2018 [63]. This model may be particularly attractive in the establishment of cross-border transit corridors such as those in the BRI, or in partnerships between countries that expect to see significant increases in investment and trade.

Box 8.6. Debt-for-Nature Swap in the Seychelles

NatureVest, the biodiversity finance platform of TNC was founded in 2014 to mobilize private capital for conservation. In 2016, in conjunction with other private funders, NatureVest signed a deal with the Seychelles' Paris Club creditors, to buy a portion of Seychelles' debt at a discounted rate, spending approximately $22 million for approximately $25 million in debt.

In cooperation with the government of Seychelles, this debt relief will allow the establishment and maintenance of approximately 400,000 km^2 of ocean. As of this writing, roughly half of that area has already been set aside, in the form of two new protected areas. The remainder is expected to be added within a year of this writing.

Two factors have contributed to the success of this debt-for-nature swap: the leadership of the government of Seychelles and the unhurried nature of the planning process. Through both of these strategies, this Seychelles project has been able to earn the local support necessary for management and enforcement over the years to come.

This process represents a culmination of existing national government goals announced in 2012, when it announced a plan to increase protected areas to include 30% of its marine EEZ ("The Initiative," 2019). Seychelles adopted a mapping methodology using international best practices, adapting UNESCO [64] recommendations. To determine which marine areas would be protected and what sustainable activities would be permitted, the MSP incorporated the input of 10 ministries and 100 public stakeholders who participated through 9 public workshops and 60 consultations.

To ensure the program's viability, the government of the Seychelles led a mapping process, the Seychelles Marine Spatial Planning (MSP) initiative, beginning in 2014 ("Seychelles Marine," n.d.). The MSP has been a deliberately time-consuming process in order to ensure an evidence-based outcome with sufficient public input. In fact, while Phase I was completed in 2018, protecting 15% of the Exclusive Economic Zone (EEZ), Phase II is not expected to finish until the end of 2020 ("The Initiative," 2019).

8.4   Analysis of SDG15-Related Governance Structure

8.4.1   Survey and Assessment of Practices in China

8.4.1.1   Governance Structure

China has integrated biodiversity into the target system for building an Ecological Civilization, and has been constantly improving the system and institutional settings for biodiversity conservation. It employs a system characterized by unified national regulation and division of responsibilities and cooperation among different sectors to protect biodiversity. In particular, right after it approved the CBD in 1993, China established a Coordinating Group for the Implementation of CBD, with the former State Environmental Protection Administration (SEPA) as the leading agency and the participation of 20 departments/line ministries under the Chinese State Council. It founded a CBD implementation office in the then SEPA and identified the national focal points for CBD implementation, biodiversity clearing-house mechanism, and biosafety respectively. An inter-ministerial Joint Meeting for Protection of Biological Resources was set up at the same time. The Coordinating Group meets every year to develop an annual work plan for CBD implementation and launches a variety of activities. So far, an initial national working mechanism has been formed for biodiversity conservation and CBD implementation. China started the development of a China Biodiversity Conservation Action Plan in 1992 and released the finalized document in 1994. This Action Plan has identified both the location of ecosystems and the list of species for priority conservation, and set out the goals for seven areas of biodiversity conservation in China.

In 2010, the State Council of China founded the "China National Committee for the 2020 International Year of Biodiversity". During its meeting, the State Council reviewed and approved the China Action Plan for the International Year of Biodiversity and the China National Biodiversity Conservation Strategy and Action Plan (2011–2030). In June 2011, the State Council decided to rename the "China National Committee for the International Year of Biodiversity 2010" as the "China National Committee for Biodiversity Conservation", and designated the Vice Premier of the State Council in charge of environment as the Director of the Committee. At present, this Committee has 23 member departments/ministries. It is mandated to coordinate all biodiversity conservation efforts in China, and direct the implementation of China actions for UN Decade on Biodiversity. The establishment of the China National Committee for Biodiversity Conservation shows China's determination to strengthen environmental protection and promote sustainable development, and its commitment to the international community. Since 2015, China has promulgated or revised 56 policies, laws and regulations related to biodiversity conservation, with the policy and legislative system for biodiversity conservation in China gradually improving.

In addition to the governance structure at central level, the environmental protection agencies of the governments at provincial level have also been carrying out reforms to better protect biodiversity. In 2008, SEPA was upgraded to the Ministry of Environmental Protection (MEP) and became a department of the State Council. All the provinces, autonomous regions, and municipalities have upgraded their environmental protection agencies to departments, building a unified environmental protection system. Referring to the responsibility and function orientation related to biodiversity conservation at the national level, some provincial governments have established relevant coordination mechanisms, specifying the leading role of environmental protection departments in biodiversity conservation and the corresponding responsibilities of multiple internal agencies within those departments. To go with the actual local conditions, some provinces have set up administration agencies in line with the needs of local biodiversity conservation. For example, Yunnan Province has set up a Lake Protection and Administration Division, showing the local features of its institutional reform and biodiversity conservation. In 2018, in accordance with the "Decision of the Central Committee of the Communist Party of China on Deepening the Reform of the Party and Government Institutions", the Chinese State Council established the new MEE to practice the holistic thinking of integrated management of mountains, waters, forests, farmlands, lakes, and grasslands. All provinces, autonomous regions and municipalities have formed their new Department of Ecological Environment to comprehensively guide, coordinate and supervise the work of eco-environmental protection.

Other major institutions mainly include the China Biodiversity Conservation and Green Development Foundation (CBCGDF), the China National Committee for Biodiversity Conservation, and the Biodiversity Committee of the China Academy of Sciences.

8.4.1.2 Green Belt and Road and Biodiversity Conservation

In the light of China's experience in biodiversity conservation and need for building a green BRI, early efforts to align green BRI and biodiversity conservation have seen growth in the areas of governance mechanisms, governance system, information, technology development and scientific research, as well as green investment and finance, so as to jointly promote biodiversity conservation and implementation of SDG 15 in BRI participating countries.

First, there was a focus on establishing a mechanism and platform for cooperative governance to enable the improvement of a governance system for biodiversity conservation in BRI participating countries. Important progress began to show in integrating the existing bilateral and multilateral international cooperation mechanisms with green Belt and Road, building a network for biodiversity conservation, innovating cooperative models, as well as formulating a cooperation platform with inclusive participation of multiple stakeholders, including national governments, think tanks, business, civil societies and the wide participation of the public. Meanwhile, it is necessary to give full play to the mechanisms established for China-ASEAN cooperation, the Shanghai Cooperation Organization (SCO), the Lancang-Mekong cooperation, the Conference on Interaction and Confidence Building Measures in

Asia, Euro-Asia Economic Forum, the Forum on China-Africa Cooperation, and the China-Arab States Cooperation Forum among other cooperative platforms. Efforts are still needed to facilitate the establishment of environmental cooperation platforms for the six major Economic Corridors, and expand cooperation with relevant international organizations and agencies, so as to promote the effective implementation of SDG 15.

Second, efforts have been made to enhance cooperation on green technologies and research and development. There is a growing demand for the transfer of green, advanced and applicable technologies in developing countries along the Belt and Road, as well as need for joint research and development, promotion and application of cutting-edge technologies on the conservation of biodiversity. Specifically, further actions are to be taken for a platform on scientific research and technology development across scientific and research institutions and think tanks. Joint research with relevant countries and regions on biodiversity is a favorable opportunity for the conservation of global biodiversity. With the scientific study over the biodiversity of countries and regions along the Belt and Road, it would contribute to the analysis on the biodiversity evolution mechanisms and its characteristics and patterns on geographical distribution in these regions, expedite the scientific research on global diversity and help to provide training and capacity building for young officials and scientists in countries along the Belt and Road.

Third, steps are emerging to promote information exchanges, including biodiversity-related information sharing and disclosure, as well as provision of comprehensive information as decision-making support and safeguard. Growing needs are observed for enhancing the construction of biodiversity information base on the BRI Environmental Big Data Platform; for the full inclusion of national spatial and information infrastructure; for the exchange and sharing on environmental laws and regulations, policy standards and practices and experience; for enhanced comprehensive cooperation among different national departments and the sharing and disclosure on the ecological and environmental information; and for the improved capacity on risk evaluation and prevention targeting at BRI projects overseas. It is necessary to facilitate cooperation on the ecological and environmental information products, technologies and service to provide comprehensive information support and safeguard for building a green Belt and Road.

Fourth, promising progress has been observed in the development of systems on green investment, green trade and green finance. Green finance systems help to build up the foundation for the long-term run of BRI projects. A good example is the Social Responsibility and Environmental Protection Guidelines for Investments in the ASEAN Region released by China-ASEAN Investment Cooperation Fund (CAF). The document prescribes that when CAF provides consulting services for businesses on overseas investment based on its Environmental and Social Management System Arrangements (ESMS), it could refer to the Performance Standards to identify and manage the impacts of environmental and social risks, clarify the evaluation metrics during the investment process, and continuously monitor the later-stage investment, as a way to facilitate invested enterprises to avoid, ease and manage risks via a sustainable operation way. This Performance Standards covers eight areas including biodiversity conservation and the sustainable management of biological and natural resources, which jointly composed the standards that clients should meet on sustainable management of biodiversity through the overseas investment process. Specifically, it includes: (1) checking whether the company understands and deals with the impacts of the project on biodiversity; (2) checking whether the company carries out activities in regions under legal protection; (3) checking whether alien species are introduced in the process of project execution, and checking whether the company has the approval or permission from competent authorities if there are plans on introducing alien species; and (4) checking whether the natural resources, forest and vegetation, fresh water and marine resources utilized by the project can be regenerated and whether the company is dedicated to managing them in a sustainable way.

Fifth, such documents as the Green Investment Principle (GIP) for BRI have been released to enhance green guidance on business activities and encourage businesses to adopt voluntary measures for environmental protection and sustainable development. It is important to motivate environmental business to explore the national markets in BRI participating countries, and guide competitive environmental companies to "go global" in clusters with reference to China's experience and standard in building demonstrative ecological industrial parks, so as to enhance biodiversity conservation, prioritize the in-situ conservation and protection in proximity, and take actions for ecological restoration. Meanwhile, efforts have been taken to guide business to augment the research, development and application of major technologies in addressing climate change.

Sixth, there is an increasing necessity to promote gender equality in the BRI cooperation and strengthen female leadership in biodiversity conservation. Biodiversity and gender are hot topics at the international level. Promoting gender mainstreaming in biodiversity conservation has gained widespread attention from the international community in recent years. Biodiversity and gender have been included in the CBD as a key issue. However, such problems as imperfect mechanisms and weak awareness related to gender exist in biodiversity research in China. In view of such problems, the following steps are thus recommended: set up gender focal points in all departments and establish a cross-sectoral communication and cooperation mechanism for gender mainstreaming to comprehensively enhance institutional capacity building; conduct gender mainstreaming training in biodiversity management departments and institutions to raise basic awareness of staff; as well as consider gender in the policies related to eco-environmental protection and green Belt and Road development, and set up gender indicator in the evaluation system for specific projects. Such practice will also help BRI projects to meet the gender-related international standards and requirements of the host countries, promote people-to-people bond, and enable the development of BRI to move forward steadily.

8.4.2 Survey and Assessment of International Practices

As mentioned in Sect. 8.2, the environmental management in the international systems has evolved rapidly over the last few decades. The same can be said for the enforcement and accountability of those systems. Just as Sect. 8.2.2 profiled the screening and assessment systems of international DFIs, this section profiles the accountability mechanisms of those same bodies.

Across the world, DFIs have mobilized to address SDG 15 and ensure that their activities protect project-affected biodiversity. While Sect. 8.2.2 explained standards and guidelines, this section explains the DFI governance structures that have been adapted to ensure that conservation is sufficiently considered. It compares governance structures as adopted by Chinese policy banks' peers: major DFIs, both multilateral as well as national in nature. It includes: AfDB, ADB, AIIB, EBRD, EIB, IADB, IFC, KfW, and the World Bank.

8.4.2.1 Governance for Biodiversity: Incorporating SDG15 into DFI Decision-Making

As explained in Sect. 8.2.2, most major DFIs incorporate biodiversity considerations into their operations through the use of set standards, mitigation hierarchy deployment, and consultations with affected stakeholders, who are likely to depend upon the local ecosystem for their livelihoods and therefore be particularly attuned to any biodiversity threats. In addition to these processes, several DFIs also incorporate other steps to mainstream SDG 15. These approaches are varied across DFIs. However, commonalities do arise in the requirements that DFIs set for themselves in this aspect, including:

- Incorporating expertise into assessments: the AfDB and AIIB require input from qualified experts to identify potentially-impacted ecosystems and ecosystem services.
- Empowering project implementers to adapt to changing conditions: AfDB, AIIB, EIB, and the World Bank all require the use of adaptive management in their projects. In this approach, borrowers and clients must allow for the possibility that as they develop their projects, conditions will not be what they initially expected. Newly-discovered species or other biodiversity-related project impacts may emerge. Project plans should specify what types of challenges may arise, and how the project implementers will adapt to these changing circumstances. With this planning done, implementers are empowered to change plans during the course of the project. In the case of the AIIB, major changes require additional environmental assessments to ensure that they are adapting their plans adequately.

8.4.2.2 Policy Implementation: Monitoring and Reporting

Borrowers and clients may commit to responsible environmental management, and DFIs strive to consider the implications for biodiversity, but actual performance will determine final outcomes. To this end, DFIs often institute monitoring and reporting requirements for their borrowers and clients. In doing so, DFIs often emphasize their respect for the national sovereignty of borrowing nations, devising methods that prioritize collaboration between lender and borrower for the best possible outcomes. Several different approaches emerge among DFIs, ranging from those that give borrowers the most responsibility in monitoring to those that utilize outside auditors.

8.4.2.3 Policy Implementation: Grievance Mechanisms

Many DFIs—multilateral as well as national—have instituted policies for stakeholders, including independent NGOs and project beneficiaries, to file grievances and request an investigation if they suspect that biodiversity has been harmed in the pursuit of DFI-supported projects. By developing institutional mechanisms for hearing, investigating, and ruling on these claims, DFIs can ensure that their borrowers and grantees are living up to the terms of the agreement, prevent small harms from ballooning into larger harms, protect their own reputation globally, learn from their experiences, and incorporate these lessons into future activities.

These grievance mechanisms can be at the DFI level, the project level, or both. *Project-level grievance mechanisms* allow greater flexibility, by promoting the resolution of problems in a way that is often faster and more accessible for stakeholders than relying on one centralized system for claims from projects all around the world. However, they can be more cumbersome for DFIs to manage, requiring oversight of processes in many different countries. The common elements in the design of project-level grievance mechanisms include: design architecture, institutional location, processes, and treatment of claimants.

DFI-level grievance mechanisms allow for stakeholders to bring a claim to the central DFI body, or its designated complaint mechanism, for consideration. These mechanisms can be simpler to manage for DFIs, as they only entail the creation and management of one body. However, they can be less accessible for project-affected stakeholders, and may mean that some rulings take more time than they would in project-level mechanisms.

Table 4.1, in Annex 4, shows the various policy elements that DFIs incorporate into their project-level grievance mechanisms. A wide variety of arrangements exist, enabling DFIs to learn from these examples in designing their own mechanisms.

All of the DFIs listed in Table 4.1 also have DFI-level grievance mechanisms, though their design is too varied to display in table form. In addition to these DFIs, several other major multilateral and national development banks have these mechanisms, including the IADB, CAF, and Brazil's National Bank for Economic and Social Development (BNDES).

8.4.2.4 Incorporating Gender

Regardless of the venue used, international DFIs have learned the importance of ensuring that accountability mechanisms are accessible for women. In many rural, poor settings, women's property rights are limited, such that ownership is recorded through their fathers, husbands, or sons. In these contexts, national justice systems may not recognize their standing to bring a complaint through local courts, as they may not be able to demonstrate a loss to the value of their property. However, if their concerns are not heard, gender-based biodiversity risks may be unheeded and worsen. Both the ADB and the World Bank have recommended that their projects ensure accessibility for women to their accountability mechanisms, regardless of property [65]. This stage completes the upstream-to-downstream inclusion of gender considerations in biodiversity finance, to ensure that women are not disproportionately impacted in ways that can limit their ability to act as biodiversity stewards at the local level. Table 8.3 collects best practices from international DFIs on incorporating gender throughout the entire project cycle. It is not intended to be a comprehensive list but rather a collection of common best practices as recorded by research and evaluation staff at DFIs worldwide.

8.5 Policy Recommendation: Construction Roadmap of Green Belt and Road

The research team has described in the previous content the progress that has been made within China and among China's peers in balancing the benefits of investment with the risks to communities and the ecosystems that support them. Given the speed and scope of this institutional progress, it is crucial to harness all of the lessons possible for additional growth, in order to ensure that the BRI fulfills its potential to support sustainable development globally. Built on the results of the first phase of SPS on Green Belt and Road and 2030 Agenda for Sustainable Development, this section further improves the roadmap for building a green BRI and proposes policy recommendations for aligning BRI and SDG 15.

8.5.1 Roadmap for Building a Green BRI

8.5.1.1 Enhancing Policy Communication, and Taking the Green Belt and Road Initiative as an Important Practice of Realizing Sustainable Development Goals and Facilitating Global Environment Governance Reform

It is important to set green development as the fundamental principles of building the Belt and Road. China has the opportunity to integrate green development and ecological civilization through the "five connectivities" in building the Belt and Road, facilitate the construction of green infrastructure, green investment and green finance, and build the Belt and Road into one route towards green and sustainable development so as to establish a community with shared future for mankind on the basis of green development.

China should augment cooperation in the domain of environmental protection on international multilateral platforms for BRI. Specifically, it is necessary to incorporate the Thematic Forum on Green Silk Road as a fixed thematic forum within the schedule of the Belt and Road Forum for International Cooperation (BRF). It is necessary to bring into full play the role of the BRI International Green Development Coalition and the Belt and Road Sustainable Cities Alliance in serving as the international platforms for jointly developing the Green Silk Road, facilitating the realization of SDGs, and improving global environmental governance system. Efforts are recommended to disseminate the concepts and practice of green development in BRI participating countries via champion countries, pilot cities, and demo projects. In addition, China should make good use of BRI's strengths in the five connectivities to jointly facilitate the implementation of policies on ecological and environmental protection, biodiversity conservation, climate change mitigation and adaptation etc. to bolster support for existing international conventions such as CBD, the Convention on International Trade in Endangered Species of Wild Fauna and Flora (CITES), UNFCCC, etc.

8.5.1.2 Enhancing Strategic Alignment and Establishing the Mechanism for Linking Green BRI with the 2030 Agenda for Sustainable Development

Given that a green BRI is a crucial tool in realizing the 2030 Agenda for Sustainable Development and in particular promoting international biodiversity conservation, this SPS recommends the following steps to strategically align planning with biodiversity goals:

Enhance policy design. This report recommends that China incorporate the implementation of the 2030 Agenda for Sustainable Development (SDGs) as an important task into building a green Belt and Road. When signing MOUs on jointly building the

Belt and Road with relevant countries and international organizations, China needs to include jointly building a green Belt and Road and expediting the alignment of BRI and the 2030 Agenda as an important part of these MOUs.

Establish a mechanism for implementation. This report recommends setting up working groups/expert teams with partners based on the situations in different countries and jointly draft strategies for building a Green Silk Road together, identifying the priority areas for cooperation in both the short and long terms and fostering sound linkages between different national plans based on the practical needs of implementing SDGs in BRI participating countries.

Develop mechanisms for participation and feedback. A network should be built with government guidance, business support and public participation, prioritizing the perfection of mechanisms on the involvement of international organizations. Initiate the mechanism for whole-process participation, covering negotiation, decision making and dynamic feedback, in order to make sure the successful alignment of building green Belt and Road and implementing the 2030 Agenda for Sustainable Development under open and transparent circumstances.

Establish professional mechanisms for cooperation for cities and localities along the Belt and Road. Cities along the Belt and Road should be encouraged to consider their own industrial structure, advantages, and development goals, and create a policy framework favorable for addressing issues of common concern to explore opportunities for cooperation and guide private sector in participating BRI cooperation.

8.5.1.3   Improving Project Management, Establishing and Improving the Mechanism for Project Management on Green Belt and Road

To incorporate the above-mentioned strategies into BRI project management, this SPS recommends the following steps:

Establish a mechanism for risk evaluation and management of BRI projects. It is important to strengthen communication and coordination between China and BRI participating countries and that among different Chinese government agencies. China should establish science-based risk evaluation and management mechanisms for their projects to respond to various risks, strictly follow the host countries' norms and standards in such procedures as project design, construction, operation, procurement, and bidding. An encouraging environment should be created for BRI projects to apply the principles, standards, and customary practices for environmental protection that are used by international organizations and multilateral financial institutions, and strive to realize goals that are made with high standards, beneficial for people's livelihoods and sustainable. China needs to support its financial institutions to incorporate the ecological and environmental impacts of projects as a key factor in their project rating and risk rating systems, and put forward evaluation methods and instruments on the environmental and social risks for BRI projects as an important metric for granting governmental support, development financial support, and policy financial support. Practice in commercial finance is encouraged to adopt similar standards.

Call for wide application of green finance instruments under the Belt and Road framework. First, establishment of the Belt and Road Green Development Fund needs to be explored, with priority given to projects in support of the development of ecological and environmental infrastructure, capacity building and green industries in countries along the Belt and Road. Second, it is necessary to establish guarantee agencies on green investment and financing under the BRI with the wide participation of different countries, in order to share risks and mobilize social capital in green domains. Third, there is a need to establish the mechanism for environmental information disclosure, and enhance the transparency of information based on the BRI Environmental Big Data Platform.

Speed up facilitation of trade in environmental products and services. Improve the opening level of environmental products and services market, encourage enhanced import and export of environmental products and services such as pollution prevention and treatment technologies and services, and help foster green industrial development in BRI participating countries.

8.5.1.4 Improving Capacity Building, Jointly Conducting Green Capacity Building Programs with BRI Participating Countries

With regard to public engagement, this SPS recommends that BRI project planners take the following steps:

Enhance people-to-people bonds among BRI participating countries. The Green Silk Road Envoys Program should be expanded into a flagship program on capacity building under the Belt and Road framework, which aims to enhance ecological and environmental cooperation and communications and share the ideas and practices of building an Ecological Civilization and green development in China via such activities as capacity building workshops for environmental officials, managers, and practitioners, consultation for policy development, etc.

Support and facilitate the exchange and cooperation of environmental organizations from China and BRI participating countries. The first step is to clarify the leading and responsible government department, and then guide/support environmental organizations to build up their own cooperation networks. In addition, efforts are needed to perfect the mechanism for the involvement of environmental social organizations and come up with a list of items on international communication with the participation of environmental organizations.

Facilitate gender mainstreaming and augment women's leadership roles. There is a necessity to improve gender consciousness among policy makers and women communities and facilitate the mainstreaming of gender consciousness in the process of policy formulation and project implementation for building a green Belt and Road. This report recommends enhancing institutional capacity building on gender mainstreaming in environmental protection related agencies, and explore the possibility of establishing a cross-sectoral communication mechanism to facilitate gender mainstreaming. With the help of the Green Silk Road Envoys Program, China could organize thematic capacity building programs and seminars on improving women's leadership roles in green development upon inviting the participation of female officials, experts, scholars and youth in the domain of environmental protection from BRI participating countries, and share methods and experience in gender mainstreaming with BRI partners.

8.5.2   Policy Instruments for Aligning the BRI with SDG 15

Under the framework and in the spirit of the general roadmap for building a green BRI outlined above, this session recommends the policy directions for aligning BRI, SDG 15, and CBD. The proposed policy recommendations are built with full consideration of the main objective and approaches internationally used for biodiversity conservation. The main objective is focusing on the establishment and mainstreaming of global standards, which are primarily practiced through the establishment of operational risk management strategies to protect institutional reputations and the cooperative relationships. The corresponding major approaches for operationalizing standards include: (1) aligning institutional practices with international or national commitments, (2) using exclusionary lists of categorically ineligible projects, (3) requiring projects to undergo biodiversity impact assessments, (4) adopting a mitigation hierarchy to do no harm and if possible benefit local ecosystems, and (5) incorporate local stakeholders.

The following four policy directions are thus proposed:

First, apply international norms and standards to facilitate the use of stricter environmental standards in BRI projects. The BRI projects should comply with the environmental laws, regulations and standards of the host country. These projects are encouraged to adopt environmental protection principles, standards and practices implemented by international organizations and multi-lateral financial institutions. It is recommended to actively align BRI efforts with the fulfillment of international and national commitments to international conventions, including the Convention on Biological Convention and the United Nations Framework Convention on Climate Change. It is also necessary to align BRI with other biodiversity related international conventions that China is a signatory to such as the International Convention for the Protection of New Varieties of Plants, Convention Concerning the Protection of the World Cultural and Natural Heritage, Convention on International Trade in Endangered Species of Wild Fauna and Flora, Convention on Wetlands of International Importance Especially as Waterfowl Habitat, and achieve synergies with climate related conventions like UNFCCC.

Second, focus on environmental impacts and carry out assessment and classification-oriented management of BRI projects. It is recommended to boost the development of the guidance on assessment and classification of BRI projects, which should include clearly defined positive and negative lists, in order to give adequate attention to the projects' potential impacts related to environmental pollution, biodiversity conservation, and climate change based on the ongoing Joint Research on Green Development Guidance for BRI Projects undertaken by BRIGC, which could provide green solutions to BRI participating countries and projects and provide green credit guidelines to financial institutions. It is recommended to pilot concept and practice of green development via champion countries, pilot cities and demo projects and enhance BRI green development case studies and experience sharing. The research on Green Development Guidance for BRI Projects shows that the assessment- and classification-oriented management should consider the various international and national commitments of the host countries, and meet the host countries' needs for economic growth and environmental protection. Such management should guide and assist businesses to incorporate environmental impact assessment (EIA), as well as biodiversity conservation and impact mitigation measures at the stage of project design.

Besides, it is important to adopt a mitigation hierarchy for those projects identified as having significant biodiversity risks as a result of strategic environmental assessment (SEA). Given such international practice, drawing on its own experience with the ecological redlining, biodiversity offsets, ecological restoration, and ecological compensation schemes, China is recommended to develop a standardized biodiversity conservation hierarchy that should include four components of "avoidance", "mitigation", "restoration", and "compensation".

Third, improve policy instruments to prevent and control the eco-environmental risks related to BRI projects. These instruments include the following three aspects.

- It is recommended to carry out environmental impact assessment for key BRI sectors and projects and establish a regular environmental risk regulatory mechanism. The biodiversity analysis should fully examine the ecological and socioeconomic conditions of the project locality, gauge the direct, indirect and cumulative impacts of the project on wildlife habitats and biodiversity, and consider how the project-affected stakeholders value biodiversity and wildlife habitats.
- It is recommended to make full use of green finance instruments and environmental risk assessment tools, establish biodiversity conservation governance and financing framework, and give full play to the proactive role of financial institutions in guiding green investment. Given that biodiversity conservation requires an enabling policy environment, the MEE should be charged to work with the NDRC and other administrative bodies to design biodiversity impacts mitigation strategies and to jointly design financing mechanisms for mitigation, compensation, and restoration schemes in consultation with various stakeholders in the Chinese government, host countries, and other affected parties and partners.
- It is recommended to take ecological redlining as a key instrument to align the BRI and SDG 15 and share best practices in ecological redlining with participating

countries. It is important to support the BRI participating countries to build on the experience of ecological redlining in developing similar land use strategic planning.

Fourth, improve the coordination mechanism and facilitate effective linkage and alignment among different SDGs using Nation-based Solutions (NBS). It is important to create synergies with efforts for SDG 13 of Climate Action, consider a step-by-step reduction of investments in carbon-intensive industries such as coal-fired power plants to prevent carbon lock-in, and further strengthen investments in green projects on environmental protection and renewable energy to encourage environmentally-sustainable, green and low-carbon projects. The concept of green development should be incorporated into the selection, implementation and management of infrastructure projects with strengthened efforts being made in researching and developing the guidelines on sustainable infrastructure operation.

Annex 1: Supporting Evidence for Chapter 1

Table A1-1 Geographic distribution of countries that have signed BRI memorandums of understanding

Region	BRI Countries
East Asia	Republic of Korea, Mongolia
ASEAN countries (10 countries)	Singapore, Malaysia, Indonesia, Myanmar, Thailand, Laos, Cambodia, Viet Nam, Brunei, Philippines
West Asia (17 countries)	Iran, Iraq, Turkey, Syria, Jordan, Lebanon, Israel, Saudi Arabia, Yemen, Oman, UAE, Qatar, Kuwait, Bahrain, Greece, Cyprus, and Sinai Peninsula of Egypt
South Asia (8 countries)	India, Pakistan, Bangladesh, Afghanistan, Sri Lanka, Maldives, Nepal, Bhutan
Central Asia (5 countries)	Kazakhstan, Uzbekistan, Turkmenistan, Tajikistan, Kyrgyzstan
Commonwealth of Independent States (7 countries)	Russia, Ukraine, Belarus, Georgia, Azerbaijan, Armenia, Moldova
Central and Eastern Europe (16 countries)	Poland, Lithuania, Estonia, Latvia, Czech Republic, Slovakia, Hungary, Slovenia, Croatia, Bosnia and Herzegovina, Montenegro, Serbia, Albania, Romania, Bulgaria, Macedonia
Western Europe (7 countries)	Austria, Finland, France, Italy, Luxembourg, Malta, Portugal
North Africa (5 countries)	Algeria, Libya, Mauritania, Morocco, Tunisia

(continued)

(continued)

Region	BRI Countries
West Africa (11 countries)	Cabo Verde, Cote d'Ivoire, The Gambia, Ghana, Guinea, Liberia, Mali, Nigeria, Senegal, Sierra Leone, Togo
Central and Southern Africa (8 countries)	Angola, Cameroon, Chad, Republic of Congo, Equatorial Guinea, Gabon, Namibia, South Africa
East Africa (15 countries)	Burundi, Djibouti, Ethiopia, Kenya, Madagascar, Mozambique, Rwanda, Seychelles, Somalia, South Sudan, Sudan, Tanzania, Uganda, Zambia, Zimbabwe
Latin America (11 countries)	Bolivia, Chile, Costa Rica, Ecuador, El Salvador, Guyana, Panama, Peru, Suriname, Uruguay, Venezuela
Caribbean (8 countries)	Antigua and Barbuda, Barbados, Cuba, Dominica, Dominican Republic, Grenada, Jamaica, Trinidad and Tobago
Oceania (9 countries)	Fiji, Kiribati, Micronesia, New Zealand, Papua New Guinea, Samoa, Solomon Islands, Tonga, Vanuatu

Note Timor-Leste is currently in the process of ASEAN accession

Table A1-2. The state of BRI countries in implementing SDG 15

Region	Country	Goal 15 Implementation	Goal 15 Trend	Region	Country	Goal 15 Implementation	Goal 15 Trend
East Asia	Mongolia		↗	West Asia	Bahrain		.
					Lebanon		⇉
	Republic of Korea		⇉		Qatar		.
ASEAN	Singapore		.		Iran		↓
	Indonesia		↓		Sinai Peninsula, Egypt		⇉
	Malaysia		⇉		Cyprus		.
	Cambodia		↓		Greece		↗
	Vietnam		↗		Jordan		.
	Myanmar		↓	Central and Eastern Europe	Montenegro		↓
	Lao P. D. R.		↓		Serbia		↗
	Philippines		↓		Bosnia and Herz.		⇉
	Thailand		⇉		N. Macedonia		↗
	Brunei		.		Croatia		↗
South Asia	Maldives		.		Albania		↗
	India		↓		Slovenia		↕
	Afghanistan		↓		Romania		↕
	Bhutan		⇉		Slovakia		↕
	Bangladesh		↓		Hungary		↕
	Sri Lanka		↗		Lithuania		↕
	Pakistan		↓		Estonia		↕
	Nepal		⇉		Czech Republic		↕
Central Asia	Turkmenistan		⇉		Poland		↕
	Kazakhstan		⇉		Latvia		↕
	Uzbekistan		⇉		Bulgaria		↕
	Kyrgyzstan		⇉	Comm. of Indep. States	Georgia		⇉
	Tajikistan		⇉		Armenia		↓
West Asia	Iraq		⇉		Ukraine		⇉
	Kuwait		.		Moldova		⇉
	United Arab Emir.		.		Russian.		⇉
	Saudi Arabia		⇉		Azerbaijan		⇉
	Syria		⇉		Belarus		↕
	Israel		↓	Western Europe	Luxembourg		↗
	Yemen		↓		Malta		.
	Oman		.		Austria		↗
	Turkey		⇉		Portugal		↗

(continued)

(continued)

Table A1-2, continued: The state of BRI countries in implementing SDG 15

Region	Country	Goal 15 Implemen-tation	Goal 15 Trend	Region	Country	Goal 15 Implemen-tation	Goal 15 Trend
Western Europe	France		↗	West Africa	Senegal		·
	Finland		↑		The Gambia		·
	Italy		↑		Ghana		·
North Africa	Algeria		⇒		Cote d'Ivoire		↗
	Tunisia		↗		Nigeria		↗
	Mauritania		·		Guinea		↗
	Morocco		⇒		Togo		·
	Libya		·	Latin America	Uruguay		⊥
East Africa	Djibouti		⊥		Panama		·
	Madagascar		⇒		Chile		⊥
	Ethiopia		⇒		Guyana		·
	Tanzania		⇒		Ecuador		⇒
	Kenya		⊥		El Salvador		⇒
	Sudan		↗		Costa Rica		·
	Rwanda		⇒		Peru		·
	Mozambique		⇒		Suriname		↗
	Uganda		↗		Bolivia		↗
	Zambia		⇒		Venezuela		↗
	Burundi		↑	Caribbean	Jamaica		·
	Zimbabwe		↗		Trin. & Tobago		·
	Seychelles		·		Cuba		·
	Somalia		⇒		Dominican Rep.		↗
	South Sudan		↗		Antigua & Barb.		·
Central and Southern Africa	South Africa		↗		Barbados		·
	Angola		⇒		Dominica		·
	Cameroon		⇒		Grenada		·
	Chad		↑	Oceania	Fiji		⊥
	Gabon		↑		Vanuatu		·
	Namibia		↑		New Zealand		⊥
	Congo, Republic		↑		P.N.G.		·
	Equatorial Guinea		·		Kiribati		·
West Africa	Liberia		⇒		Micronesia		·
	Cabo Verde		·		Samoa		·
	Sierra Leone		↑		Solomon Isl.		·
	Mali		⇒		Tonga		·

(continued)

(continued)

Table A1-2, continued: Legend

Colors		Trend Arrows	
Green	Goal Achievement	↑	On track or Maintaining Achievement
Yellow	Challenges Remain	↗	Moderately Increasing
Orange	Significant Challenges	→	Stagnating
Red	Major Challenges	↓	Decreasing
		.	Data not available

Annex 2: Supporting Evidence for Chapter 2

Table A2-1. Assessment of China's progress in implementing SDG 15

SDG15	Main work undertaken by China to achieve SDGs	Indicators	Overall assessments and trends
15.1 By 2020, ensure the conservation, restoration and sustainable use of terrestrial and inland freshwater ecosystems and their services, in particular forests, wetlands, mountains and drylands, in line with obligations under international agreements	Safeguarding the ecological water level of important wetlands and estuaries, protecting and restoring wetland and river and lake ecosystems, establishing systems of wetland protection and degraded wetland protection and restoration, and promoting the rational use of wetlands; promoting the development of the legal system of terrestrial nature reserves and improving the level of protection and utilization of natural resources such as forests; and conducting river and lake health assessments to protect aquatic ecosystems.	National-level protected areas for aquatic germplasm resources	🟢
		Number of wetland parks	🟢
		Percentage of surface water bodies with good quality meeting Classes I-III standards	🟢
15.2 By 2020, promote the implementation of sustainable management of all types of forests, halt deforestation, restore degraded forests and substantially increase afforestation and reforestation globally	Carrying out large-scale land greening, strengthening the implementation of key afforestation projects, improving the natural forest protection system, comprehensively stopping commercial forest logging, and protecting and cultivating forest ecosystems; improving the policy of returning farmlands to forests and grasslands, and exploring the establishment of mechanisms for government-sponsored social services to carry out afforestation and forest protection.	Total forest stock	🟢
		Area of natural forests	🟢

(continued)

(continued)

15.3 By 2030, combat desertification, restore degraded land and soil, including land affected by desertification, drought and floods, and strive to achieve a land degradation-neutral world	Participating in demonstration projects aiming at land degradation neutrality goal under the United Nations Convention to Combat Desertification; promoting the comprehensive control of desertification, rocky desertification and soil erosion, preventing land degradation, continuously expanding the scope of desertification land management, and strengthening the ecological protection and construction of desert areas.	Forest stock in key ecological project areas	🟢
		Grassland vegetation cover rate in key ecological project areas	🟢
		Area of desertified land	🟢
15.4 By 2030, ensure the conservation of mountain ecosystems, including their biodiversity, in order to enhance their capacity to provide benefits that are essential for sustainable development	Comprehensively improving the stability of mountain ecosystems and ecological service functions and building an ecological security barrier, constructing national forest germplasm resource banks and establishing a system of standardized germplasm resource conservation; scientifically optimizing the forest park management system and promoting the sharing and utilization of forest diversity resources.	Number and area of forest parks	🟢
		Total timber standing stock	🟢
		Area of natural forests	🟢
		National investments in ecological conservation	🟢
15.5 Take urgent and significant action to reduce the degradation of natural habitats, halt the loss of biodiversity and, by 2020, protect and prevent the extinction of threatened species	Implementing major projects for biodiversity conservation; strengthening the construction and management of nature reserves, and increasing the protection of typical ecosystems, species, genes and landscape diversity; increasing the investment in ecosystem protection and restoration and carrying out large-scale survey of baselines for species resources in the country; and establishing a national biodiversity observation network.	Red List Index	🔴
		Living Planet Index	🔴
15.6 Promote fair and equitable sharing of the benefits arising from the utilization of genetic resources and promote appropriate access to such resources, as internationally agreed	Gradually establishing and improving laws and regulations on the protection and benefit sharing of genetic resources and promoting the proper access to genetic resources and the fair and equitable sharing and utilization; increasing funding for the conservation of biological genetic resources and participating in international cooperation in access to and use of genetic resources.	Indicators related to access to genetic resources and benefit-sharing	∘ ∘ ∘

(continued)

(continued)

15.7 Take urgent action to end poaching and trafficking of protected species of flora and fauna and address both demand and supply of illegal wildlife products	Seriously implementing the Wild Animal Protection Law and speeding up the improvement of the National List of Key Protected Wild Animals; optimizing the national wildlife protection network, strengthening the import and export management of wild animals and plants, and cracking down on illegal trade in wild animal and plant products such as ivory; restoring and expanding the habitats of endangered wildlife and promoting international cooperation in wildlife conservation.	/	/
15.8 By 2020, introduce measures to prevent the introduction and significantly reduce the impact of invasive alien species on land and water ecosystems and control or eradicate the priority species	Actively participating in international conventions related to the prevention and control of invasive alien species; improving the list of IAS and related risk assessments	Number of newly discovered IAS every decade	🔴
		Batches and number of species of harmful pests intercepted at ports	🔴
		Number of IAS risk assessment standards released.	🟢
15.9 By 2020, integrate ecosystem and biodiversity values into national and local planning, development processes, poverty reduction strategies and accounts	Requiring governments of all levels to undertake ecological conservation and biodiversity conservation taking into account their local circumstances, and to incorporate biodiversity into their long-term and medium term development planning.	Number of sectoral policies related to conservation and sustainable use of biodiversity	🟢
15.a Mobilize and significantly increase financial resources from all sources to conserve and sustainably use biodiversity and ecosystems	Strengthening coordination and increasing funds needed for infrastructure and capacity building	National investments in ecological conservation	🟢

(continued)

(continued)

15.b Mobilize significant resources from all sources and at all levels to finance sustainable forest management and provide adequate incentives to developing countries to advance such management, including for conservation and reforestation	Promoting diversified resource mobilization strategies, guiding enterprises and the public to participate more deeply, and forming a long-term financial mechanism for forest management; helping other developing countries to carry out technical training under the framework of South-South cooperation to improve the rate of utilization of forest resources and the level of forest management; and guiding Chinese companies to carry out sustainable forest management and business operation abroad.	Ecological compensation for forest ecological benefits	🟢
15.c Enhance global support for efforts to combat poaching and trafficking of protected species, including by increasing the capacity of local communities to pursue sustainable livelihood opportunities	Strengthening the review of trade in species restricted by the international trade conventions in which China participates, and strictly managing the certification under the Convention on International Trade in Endangered Species of Wild Fauna and Flora; carrying out special actions to curb the criminal momentum of poaching and illegal trade of wild animals; and encouraging and guiding the development of wild plant artificial cultivation industry.	Number of illegally smuggled or trafficked protected species intercepted or detected	ₒ ₒ ₒ

🟢 Status improving   🔴 Status worsening;
ₒ ₒ ₒ no adequate data   /   no indicators available for assessment
Note: China's Sixth National Report on Implementation of the Convention on Biological Diversity. 2018

Table A2-4 Operational Requirements for Biodiversity Safeguards Applied by DFIs to Clients

	ADB	AfDB	AIIB	BNDES	CAF	EBRD	EIB	IADB	IFC	KFW	WB
Screen and categorize projects for level of impact and risk to biodiversity	X	X	X	X	X	X	X	X	X	X	X
Assess baseline conditions	X	X	X		X	X	X	X	X		X
Assess direct, Indirect, cumulative and Induced impacts and risks to biological resources	X	X	X		X	X	X	X	X		X
Consider trans-boundary Impacts	X	X	X			X	X	X	X		X
Socio-economic impacts of modifications to biodiversity	X	X	X		X	X	X	X	X	X	X
Use of strategic environmental assessment	X	X	X				X	X	X		X
Apply the precautionary approach or principle	X		X		X	X	X	X		X	X
Examine alternatives to project design technology and components	X	X	X		X	X	X	X	X		X
Explicitly incorporate costs of environmental mitigation measures into environmental assessment					X			X			
Apply mitigation hierarchy		X	X	X		X	X	X	X		X
Explicit adherence to national law and host country international commitments	X	X	X	X	X	X	X	X	X		X
Option to use country and/or client systems in lieu of DFI safeguards	X		X					X			X
Engage independent experts and advisory panels		X	X			X					X
Carry out stakeholder consultation during environmental assessment and project implementation	X	X	X		X	X	X	X	X	X	X
Require client to disclose environmental assessments and management plans	X	X	X			X	X		X		X
Prepare Biodiversity Management or action plans			X			X	X				
Enhance Biodiversity	X	X	X		X	X	X	X	X		X

(continued)

(continued)

	ADB	AfDB	AIIB	BNDES	CAF	EBRD	EIB	IADB	IFC	KFW	WB
Use adaptive management procedures to address unanticipated impacts			X				X				
Criteria for Projects in/affecting Critical Habitat	X	X	X		X	X	X	X	X		X
Criteria for Projects in/affecting legally protected and internationally recognized areas	X	X	X	X	X	X	X		X		X
Criteria for Projects in/affecting natural habitat	X	X	X		X	X	X		X		X
Criteria for Projects in/affecting modified habitat	X				X		X		X		X
Use of Offsets	X	X				X	X		X		X
Management of Ecosystem Services		X			X	X	X				X
Sustainable management of natural Living and renewable resources	X	X			X	X	X				X
Control of Invasive Alien Species	X	X	X			X	X		X		X
Genetically Engineered Organisms		X				X					
Environmental Flows		X									
Forest Management	X	X	X		X					X	
Marine Environment		X	X			X				X	
Protection of Indigenous Knowledge and commercial activities		X			X				X	X	X
Supply Chain Management		X				X	X		X		X
Impact of Climate Change on Biodiversity		X				X	X				
List of Categorically Ineligible Projects	X	X		X	X	X	X	X	X	X	X

Source Web pages, official policies, and interviews with individuals at listed international institutions

Annex 3: Evidence from Chapter 3

Detailed descriptions of conservation finance initiatives in China

Increasing transfer payments to ecological function areas. Since 2018 when the central government established the transfer payment system for key national ecological function areas, China has been intensifying the efforts to protect those areas. In 2018, the state made a transfer payment of 72.1 billion yuan to key national ecological function areas, 9.4 billion more than it did the previous year, registering an increase of 15%. Meanwhile, China has kept expanding the scope of key national ecological function areas. Once included in the scope, the area will receive financial and policy support as long as it strictly implements the negative list system for industrial access. According to relevant regulations, a region counted as a key national ecological function area needs to strengthen ecological protection and restoration, regulate the boundaries of industrialization and urbanization, and enhance the supply capacity of eco-products.

Strengthening fiscal support from the central government to forestry ecological protection. On July 27, 2018, the Ministry of Finance and the State Administration of Forestry and Grassland jointly issued the *Management Measures for Forestry Ecological Protection and Recovery Funds*, aiming at regulating the management of Forestry Ecological Protection and Recovery Funds, coordinating the integrated use of such funds, improving the efficiency of utilization and facilitating forestry ecological protection and recovery. According to the Measures, Forestry Ecological Protection and Recovery Funds refer to special transfer payment funds in the central budget for the social insurance and social expenditure of Natural Forest Protection Project (hereinafter referred to as "NFPP"), the cessation of commercial clear-cutting of natural forest, improving relevant policies on returning farmland to forestry and initiating a new round of returning farmland to forestry and grassland. In 2018, a total of 41.604 billion yuan was allocated to several provinces, of which Heilongjiang received the most, 8.595 billion. The Measures has clearly stated that the funds are allocated based on the factor method. The standard of cash subsidy for returning farmland to forestry is as follows: for the Yangtze River Basin and southern areas, 125 yuan per mu each year; for the Yellow River Basin and northern areas, 90 yuan per mu each year. Those returned eco-forests will be subsidized for 8 years, and those returned economic forests for 5 years. As for the new round of returning farmland to forestry and grassland, the returned forests will receive a cash subsidy of 1200 yuan per mu, paid at 3 intervals within 5 years, with 500 yuan in the first year, 300 yuan in the second year, and 400 in the third year; the returned grasslands will receive a cash subsidy of 850 yuan, paid at 2 intervals within 3 years, with 450 yuan in the first year and 400 yuan in the second.

Strengthen fiscal support from the central government to ecological protection and restoration of wetlands. From 2013 to 2016, the central government allocated 5 billion yuan to protect wetlands in China, and continued to provide support through the Funds for Reform and Development of Forestry afterwards. The measures taken include: First, supporting the protection and restoration of wetlands. For wetlands of international/national importance, national wetland parks at important ecological locations, and national wetland nature reserves at or above the provincial level managed by the forestry system, efforts will be made to protect and restore the wetlands, improve the current ecological status, and maintain the health of the local eco-system. Second, supporting returning farmland to wetland. It is encouraged to return farmlands to wetlands within the wetlands of international importance, national wetland nature reserves, and provincial nature reserves within wetlands of national importance managed by the forestry system, so as to expand the area of wetlands and improve the surrounding ecological status. Third, supporting the wetland ecological benefit compensation. For important wetlands on the route of migratory birds managed by the forestry system, their loss due to the protection of birds and other wild animals will be properly compensated. In so doing, all parties are motivated to protect wetlands and maintain the wetlands' ecosystem service functions.

Promote the grassland ecological protection subsidy incentive policy continuously. Since 2011 when the state implemented the grassland ecological protection subsidy incentive policy in 8 major pastoral areas in Inner Mongolia, Xinjiang, Tibet, Qinghai, Sichuan, Gansu, Ningxia and Yunnan and Xinjiang Production and Construction Corps, and gave out a total of 13.6 billion yuan as subsidies, 36 pastoral and agricultural pastoral regions in 5 non-major pastoral provinces including Heilongjiang have been added to the scope, altogether covering 268 pastoral and mixed farming-pastoral counties. In recent years, the state together with the General Bureau of Land Reclamation of Heilongjiang has implemented the grassland subsidy incentive in 13 provinces including Shanxi and Hebei and production and construction corps, achieving remarkable results in improving the grassland ecosystem, the production of animal husbandry and the life of herders. In 2018, a new round of grassland ecological protection subsidy incentive of 18.76 billion yuan was included in the central budget to support the banned grazing area of 1.206 billion mu and the grass-animal balance area of 2.605 billion mu, and award those regions with outstanding performance. The funds were utilized by local governments in grassland management and the transformation and upgrading of the production mode. Besides, the subsidies for banning grazing and incentive for grass-animal balance were required to be given out based on the principle of "to clear targets in a reasonable amount accurately", making sure each target could get their share in time. The distribution of the funds is publicized at the village-level, accepting surveillance by the masses. In addition to supporting the implementation of subsidies for banning grazing and incentives for grass-animal balance, the performance appraisal also requires no less than 70% of the funds should be used in protecting the grassland ecosystem and developing grass-based livestock husbandry, that relevant trails should be conducted in accordance with local realities, and that support to new agricultural operators should be enhanced concerning the development of modern grass-based livestock husbandry.

Launching pilot programs on the unified confirmation and registration of natural resources. The confirmation and registration of natural resources is important to promoting the reform of the property right mechanism of natural resource assets, which, is a key part of China's ecological civilization construction. On July 6, 2018, an evaluation and acceptance meeting for the pilot programs concerning the unified confirmation and registration of natural resources was held in Beijing by seven ministries and commissions, including the Ministry of Natural Resources. At the meeting, pilot programs of several provinces, municipalities and autonomous regions passed the acceptance, indicating that much progress has been made in the field of confirmation and registration of natural resources after over one year's hard work. By the end of October in 2018, 1191 natural resource registration units have been established in 12 provinces and 32 pilot areas, and the total registered area has reached 186,727 km2. Besides, the state also focused on exploring the confirmation and registration of national parks, wetlands, water flows, proven reserves of mineral resources. On the basis of real estate registration, with the core mission being making a clear distinction between national ownership and collective ownership, between national ownership and governments at different levels assuming ownership, between different collective owners, and between different types of natural resources, and bearing in mind the goal of adopting a holistic approach to conserving our mountains, rivers, forests, farmlands, lakes and grasslands, local governments completed the work on investigating resource ownership, establishing registration units, confirming and registering, constructing databases, etc., resulting an effective set of workflow, technical methods and specifications. Starting from the end of 2018, the confirmation and registration of natural resources in key areas has been implemented nationwide step by step. It is planned that within 5 years, the unified confirmation and registration of natural resources in nature reserves will completed, such as national and provincial key parks, natural reserves and various natural parks (scenic spots, wetland parks, natural heritage, geoparks, etc.). At the same time, the unified confirmation and registration of individual natural resources with complete ecological functions owned by the public will be conducted, such as major rivers and lakes, key wetlands, key national forests, important grasslands, etc.

Annex 4: Evidence from Chapter 4

China's Policy Implementation for Biodiversity Conservation

In regard of biodiversity conservation policies, we've already had relevant legislation, technological innovation and international mechanisms in China. A preliminary legal framework for biodiversity conservation has been established, and technological innovation and international collaboration are making continuous progress.

The most important elements include:

• The *Constitution*, which establishes the fundamental law of the state

- National laws and regulations, including the *Marine Environment Protection Law, Water Law, Water Pollution Prevention and Control Law, Water and Soil Conservation Law, Fishery Law, Forest Law, Grassland Law, Wild Animal Conservation Law, Regulations on Wild Plants Protection, Regulations on the Protection of Terrestrial Wild Animals, Regulations on the Protection of New Varieties of Plants, Regulations on Nature Reserves, and Regulations on the Administration of Scenic and Historic Area*
- Nine provinces have established regulations, including Regulations on the Protection of Wetlands in Heilongjiang Province, Regulations on the Protection of Wetlands in Gansu Province, and Regulations on the Protection of Wetlands in Poyang Lake in Jiangxi Province
- International conventions, including the *Convention on Biological Diversity, Convention on Wetlands of International Importance Especially as Waterfowl Habitat, Convention on International Trade in Endangered Species of Wild Fauna and Flora, Convention Concerning the Protection of the World Cultural and Natural Heritage, Declaration of the United Nations Conference on the Human Environment, and Rio Declaration on Environment and Development*
- Judicial measures established through the Supreme People's Court of China, which has organized a division for environmental resources and issued guidelines for biodiversity-related cases.
- Government-sponsored survey, research, and monitoring of biodiversity, producing species catalogs such as the *Flora of China, Fauna of China, Cryptogamia of China, China Red Data Book of Endangered Animals*, and others.
- Public awareness campaigns, both domestically and internationally.

Elements of China's Governance Framework for Conservation

In China, we have the *Constitution*, the fundamental law of the state, the *Environmental Protection Law of the People's Republic of China*, the basis of the environmental law system, and a set of separate laws and administrative regulations on biodiversity conservation issued on the spirit of the above laws, such as the *Marine Environment Protection Law, Water Law, Water Pollution Prevention and Control Law, Water and Soil Conservation Law, Fishery Law, Forest Law, Grassland Law, Wild Animal Conservation Law, Regulations on Wild Plants Protection, Regulations on the Protection of Terrestrial Wild Animals, Regulations on the Protection of New Varieties of Plants, Regulations on Nature Reserves, and Regulations on the Administration of Scenic and Historic Areas*. Besides, we also have local regulations on biodiversity conservation, for instance, *the Regulations on the Protection of Wild Aquatic Animals, the Aquatic Resources Breeding Protection Regulations, Law on the Exclusive Economic Zone and the Continental Shelf, and the Regulations on the Protection of Fishery Resources Breeding of Bohai*. Administrative regulations on biodiversity conservation in wetlands include the *Ramsar Convention, Convention on Biological Diversity*, etc.

In terms of local legislation, 9 provinces have established relevant regulations, including the *Regulations on the Protection of Wetlands in Heilongjiang Province, Regulations on the Protection of Wetlands in Gansu Province, and Regulations on the Protection of Wetlands in Poyang Lake in Jiangxi Province.* In addition, a series of administrative laws and regulations have been issued, including regulations on nature reserves, regulations on the protection of wild plants, regulations on the safety management of agricultural GMOs, regulations on the administration of the import and export of endangered wild flora and fauna, regulations on the protection of wild medicine resources, etc. Some provincial governments and relevant authorities in charge have also formulated corresponding rules and regulations.

China has joined several international conventions related to biodiversity conservation, including the *Convention on Biological Diversity, Convention on Wetlands of International Importance Especially as Waterfowl Habitat, Convention on International Trade in Endangered Species of Wild Fauna and Flora, Convention Concerning the Protection of the World Cultural and Natural Heritage, Declaration of the United Nations Conference on the Human Environment, and Rio Declaration on Environment and Development.* Laws concerning the management of introduced species include the *Law on the Entry and Exit Animal and Plant Quarantine, Animal Epidemic Prevention Law, Marine Environment Protection Law, Regulations on the Prevention of Livestock Epidemics,* etc. As for the emerging safety issues concerning GMOs, the State Council has issued the *Regulations on Administration of Agricultural Genetically Modified Organisms Safety* in 2001. The promulgation of those laws and regulations has efficiently supervised and promoted the conservation of biodiversity in China.

The Supreme People's Court of China has set up a division for environmental resources and issued guidelines on conducting specialized investigations and trials of biodiversity conservation-related cases, so as to guide courts at all levels to classify cases based on different basins or eco-function areas, unify judicial criteria, and improve the multiple-channel dispute settlement mechanism, thus laying a solid foundation for enhancing the juridical protection of environmental resources including biodiversity. Chinese courts give full play to the role of environmental public litigation, trying public interest litigation cases concerning wetlands, forestry, endangered plants, migratory birds in accordance with relevant laws. In the ancient Wucheng Town of Yongxiu County near Poyang Lake, the first biodiversity judicial protection base has been established. Adhering to modern judicial concepts such as strict law enforcement, safeguarding rights and interests, focusing on prevention and restoration and encouraging public participation, the base aims to make the best of judicial services in the process of advancing ecological civilization construction through circuit courts and legal publicity.

Basic surveys, scientific researches and monitoring of biodiversity have been conducted, and technological innovation has been applied to promote the sustainable development of biodiversity. Relevant departments have organized a series of national and regional surveys, researches and monitoring on species and established corresponding databases, and have published several species catalogues such as the *Flora of China, Fauna of China, Cryptogamia of China, China Red Data Book of*

Endangered Animals, etc. China has also drawn on international advanced experience and carried out demonstration projects, strengthened researches on the evaluation and management system of biological genetic resources, and tried to build a mechanism to communicate relevant traditional knowledge and share benefits, thus coordinating the relationship among knowledge protection, expansion and utilization.

China has raised public awareness to participate, and strengthened international cooperation and exchanges. Publicity campaigns on biodiversity conservation in various forms have been launched, and education in this regard has also been enhanced in the campus. Public monitoring and reporting systems for biodiversity conservation have been established and improved. Partnerships on biodiversity conservation have been built in order to give full play to the role of non-governmental non-profit organizations and philanthropic organizations, and mobilize stakeholders both in and out of China to promote the sustainable use of biodiversity resources. Moreover, China always sticks to its commitment to those conventions, introduces advanced experience from abroad, and actively participates in formulating relevant international rules.

Additional Major Institutions with Conservation Management Responsibilities in China

The *China Biodiversity Conservation and Green Development Foundation* (CBCGDF) is a leading nationwide non-profit public foundation and a social legal entity dedicated to biodiversity conservation and green development. The mission of CBCGDF is to mobilize the whole society to care about biodiversity conservation and support the cause of green development, protect strategic resources of the state, promote sustainable economic and social development, promote the construction of ecological civilization and achieve harmony between man and nature, thus building a better home for mankind.

In 2010, the General Assembly of the United Nations declared 2011–2020 the United Nations Decade on Biodiversity. The State Council established the *National Committee for 2010 International Year of Biodiversity*, and held a meeting on which they passed the China Action Plan for 2010 International Year of Biodiversity and China National Biodiversity Conservation Strategy and Action Plan (2011–2030). In the June of 2011, the State Council decided to change the name of the Committee to "China National Committee for Biodiversity Conservation", and it will continue to coordinate the efforts to protect biodiversity and guide China's action plan for the UN Decade on Biodiversity.

In 1992, the *Biodiversity Committee of the Chinese Academy of Sciences* (BC-CAS) was established to coordinate researches on biodiversity. Its responsibilities are as follows: to make biodiversity research policies of CAS; to make a long-term guideline and work plan for CAS's biodiversity researches; to review the rules and regulations on observation and experiments, organizational management mechanisms, and fund allocation plans; to inspect the utilization of funds and the performance of work; to review academic exchanges and training programs; to make plans for domestic and international collaborative researches. BC-CAS will strive to implement the sub-project of "Biodiversity Research and Information Management", an environmental technical assistance project with loans from the World Bank. So far, 30+ databases have been established, 25 of which contain over 140,000 records that can be accessed via Internet.

DFIs Governance Structures for Conservation

- European Bank for Reconstruction and Development (EBRD): Borrowers are tasked with overseeing all management, monitoring, and reporting.
- International Finance Corporation (IFC): IFC works with private-sector borrowers, creating a triangular oversight relationship: IFC, client, and client's national government. Clients are generally tasked with monitoring and reporting, except for situations where national governments have domain over a natural resource or oversight responsibilities. In complex or high-risk scenarios, clients will be required to use the services of outside experts.
- Asian Infrastructure Investment Bank (AIIB) and Development Bank of Latin America (CAF): Borrowers are tasked with monitoring and reporting. The DFI may also carry out periodic site visits and works with implementers to mitigate any harm that has been caused.
- KfW agrees to a monitoring and reporting plan with the borrower or client, who is then empowered to manage that plan.
- Asian Development Bank (ADB): Borrowers compile regular reports, while the ADB maintains responsibility for due diligence in reviewing these reports. The ADB also carries out periodic site visits and works with implementers to mitigate any harm that have been caused.
- African Development Bank (AfDB): The AfDB will occasionally carry out independent audits of projects with substantial risks to biodiversity, including the use of third-party auditors. In cases where problems come to light, it designs action plans with measurable outcomes in conjunction with the borrower, with the aim of strengthening local capacity to monitor and manage projects and mitigate harm.
- Inter-American Development Bank (IADB) and World Bank (WB): These DFIs monitor compliance and oversee reporting.

Table A4-1 Commonalities Among DFI Guidelines for Project-Level Grievance Mechanisms

	AfDB	ADB	AIIB	EBRD	EIB	IFC	KfW	WB
Institutional Location								
It should be independent and monitored by a 3rd party	X							
It may be internal or external, as the DFI deems suitable			X					
Resources								
It should be scaled to the risks and impacts of the project		X	X	X	X	X	X	X
It should be adequately budgeted and staffed					X			
Design and establishment								
It should be designed in cooperation with the borrower/client to ensure legitimacy, accessibility, predictability, and equitability	X							
It should be established as early as possible in the project development process				X				
Process								
It should address affected people's concerns promptly	X	X		X	X	X		
It should use a clear and transparent process		X	X	X	X	X		
It should have a predictable process	X				X			
It should be gender responsive or sensitive		X	X					
It should be culturally appropriate		X	X	X		X	X	
It should be free from manipulation, coercion, or interference				X				
It should have a publicly accessible register of cases and outcomes	X		X					
It should report regularly to the public on its implementation				X	X			

(continued)

(continued)

Table A4-1, continued: Commonalities Among DFI Guidelines for Project-Level Grievance Mechanisms

	AfDB	ADB	AIIB	EBRD	EIB	IFC	KfW	WB
Treatment of complainants								
It should protect complainants from intimidation/retaliation			X	X		X		
It should allow complainants to be remain anonymous if requested			X		X			
It should be free of cost to stakeholders	X				X	X		
It should be readily accessible to all segments of affected people		X	X					
The client should inform stakeholders of its availability			X	X		X		

Note: AfDB: African Development Bank; ADB: Asian Development Bank; AIIB: Asian Infrastructure Investment Bank; EBRD: European Bank for Reconstruction and Development; EIB: European Investment Bank; IFC: International Finance Corporation; KfW: German development bank, originally Kreditanstalt für Wiederaufbau; WB: World Bank.

References

1. PETRI, P.A., PLUMMER, M.G.: The Economic Effects of the Trans-Pacific Partnership: New Estimates [R]. Washington, DC: Peterson Institute for International Economics, (2016).
2. Belt and Road Portal·YIDAIYILU.GOV.CN: Illustration: BRI six years report card [EB/OL]. https://www.yidaiyilu.gov.cn/xwzx/gnxw/102792.htm, (2019-09-09).
3. World Bank: Belt and Road Economics: Opportunities and Risks of Transport Corridors [R]. Washington, D.C: World Bank, (2019).
4. Ministry of Commerce of the people's Republic of China: "Belt and Road" economic and trade cooperation has achieved new development and new breakthrough [EB/OL]. http://www.mofcom.gov.cn/article/ae/ai/202001/20200102928961.shtml, (2020-01-09).
5. BHATTACHARYA, A., GALLAGHER, K.P., MUÑOZ CABRÉ, M., JEONG, M., MA, X.: Aligning G20 Infrastructure Investment with Climate Goals and the 2030 Agenda [R]. Foundations 20 Platform, a report to the G20, (2019).
6. YOSHINO, N., ABIDHADJAEV, U.: Impact of Infrastructure Investment on Tax: Estimating Spillover Effects of the Kyushu High-Speed Rail Line in Japan on Regional Tax Revenue [R]. ADBI Working Papers 574, Asian Development Bank Institute, (2016).
7. DREHER, A., FUCHS, A., PARKS, B., STRANGE, A., TIERNEY, M.J.: Aid, China, and Growth: Evidence from a New Global Development Finance Dataset [R]. Williamsburg, VA: AidData at William & Mary, (2017).
8. HUGHES, A.: Understanding and minimizing environmental impacts of the Belt and Road Initiative [J]. Conserv. Biol. 33(4), 883–894 (2019).
9. Liu et al.: Risks of Biological Invasion on the Belt and Road, Current Biology 29, 499–505, (2019).
10. LOSOS, E.C., PFAFF, A., OLANDER, L.P., MASON, S., MORGAN, S.: Reducing Environmental Risks from Belt and Road Initiative Investments in Transportation Infrastructure [R]. Washington, DC: World Bank Group, (2019).

11. NARAIN, D., MARON, M., TEO, H.C., HUSSEY, K., LECHNER, A.M.: Best-Practice Biodiversity Safeguards for Belt and Road Initiative's Financiers [J]. Nat. Sustain. 3(8), 1–8 (2020).
12. COSTANZA, R., de GROOT, R., SUTTON, P., et al.: Changes in the global value of ecosystem services [J]. Global Environ. Chang. 26, 152–158 (2014).
13. DAMANIA, R., Desbureaux, S., Scandizzo, P.L., Mikou, M., Gohil, D., Said, M.: When Good Conservation Becomes Good Economics [R]. Washington, DC: World Bank, (2019).
14. Global Environment Facility: Mainstreaming Gender at the GEF [R]. Washington, DC: GEF, (2013).
15. ROCHELEAU, D.E.: Gender and Biodiversity: A Feminist Political Ecology Perspective [J]. IDS Bull-I. Dev. Stud. 26(1), 9–16 (1995).
16. XI J.: "Work Together to Build the Silk Road Economic Belt and The 21st Century Maritime Silk Road." Opening Ceremony Speech of the Belt and Road Forum for International Cooperation [J]. Quishi Journal 9(3), 32 (2017).
17. World Bank: Safeguards and Sustainability Policies in a Changing World [R]. Washington, DC: World Bank, Independent Evaluation Group, (2010).
18. China Development Bank and United Nations Development Program.: Harmonizing Investment and Financing Standards towards Sustainable Development along the Belt and Road, Beijing, UNDP. https://www.cn.undp.org/content/china/en/home/library/south-south-cooperation/harmonizing-investment-and-financing-standards-.html, (2019).
19. International Finance Corporation.: Good Practice Handbook: Land Acquisition and Resettlement. Washington, D.C.: IFC. https://www.ifc.org/wps/wcm/connect/74f457f6-ddf7-44ec-87bb-fed991b978fc/Draft_Resettlement+Handbook_Disclosure_March132019_Final.pdf, (2019).
20. IRWIN, A., GALLAGHER, K.P.: Chinese Mining in Latin America: a Comparative Perspective [J]. Journal of Environment and Development 22(2), 207–234 (2013).
21. RAY, R., GALLAGHER, K.P., LOPEZ, A., SANBORN, C.: China in Latin America: Lessons for South South Cooperation for Sustainable Development [R]. Global Development Policy Center, Boston University, (2015).
22. Convention on Biological Diversity.: "Report of the Eighth Meeting of the Parties to the Convention on Biological Diversity." Curitiba, Brazil: Conference of the arties to the CBD. https://www.cbd.int/doc/meetings/cop/cop-08/official/cop-08-31-en.pdf, (2006).
23. Global Environment Facility.: "GEF-7 Replenishment Programming Directions." Stockholm, Sweden: Fourth Meeting of the Seventh Replenishment of the GEF Trust Fund. https://www.thegef.org/sites/default/files/publications/GEF-7%20Programming%20Directions%20-%20GEF_R.7_19.pdf, April 25, 2018.
24. World Bank: Annual Report 2020: Ending Poverty, Investing in Opportunity [R]. Washington, DC: World Bank, (2020).
25. Equator Principles.: "The Equator Principles." https://equator-principles.com/wp-content/uploads/2020/05/The-Equator-Principles-July-2020-v2.pdf, (2020).
26. RAY, R., GALLAGHER, K.P., SANBORN, C.: Development Banks and Sustainability in the Andean Amazon [M]. London: Routledge, (2019).
27. Asian Infrastructure Investment Bank (AIIB): Environmental and Social Framework [EB/OL]. https://www.aiib.org/en/policies-strategies/framework-agreements/environmental-social-framework.html, (2019).
28. African Development Bank (AfDB) Integrated Safeguard System: Policy Statement and Operational Safeguards [R]. Tunis: African Development Bank Group, (2013).
29. World Bank: ESS6: Biodiversity Conservation and Sustainable Management of Living Natural Resources [R]. Washington DC: World Bank, (2018).
30. Development Bank of Latin America (CAF): Environmental and Social Safeguards for CAF/GEF Projects Manual [EB/OL]. https://www.caf.com/media/6742/d0-7_s_e_safeguards_manual_to_caf-gef_projects_may_2015_28.pdf, (2015).
31. Convention on Biological Diversity: 2015–2020 Gender Action Plan [EB/OL]. https://www.cbd.int/gender/action-plan/, (2017-10-02).

32. Climate Investment Funds: CIF Gender Action Plan – Phase 2 [EB/OL]. https://www.climat einvestmentfunds.org/sites/default/files/ctf_scf_decision_by_mail_cif_gender_action_plan_p hase_2_final_revised.pdf, (2016-11-22).
33. Green Climate Fund: Mainstreaming Gender in Green Climate Fund Projects [R]. Yeonsu-gu, South Korea: GCF, (2017).
34. International Finance Corporation: IFC Performance Standards [R]. Washington, DC: IFC, (2012).
35. KVAM, R.: Meaningful Stakeholder Engagement: A Joint Publication of the MFI Working Group on Environmental and Social Standards [R]. Washington, DC: Inter-American Development Bank, (2019).
36. MONTANARELLA, L., SCHOLES, R., BRAINICH, A.: The IPBES assessment report on land degradation and restoration [R]. Bonn, Germany: Intergovernmental Science-Policy Platform on Biodiversity and Ecosystem Services, (2018).
37. LIU, G.: Exploring the ecological compensation system with Chinese characteristics [N]. China Environment News. https://www.gmw.cn/xueshu/2019-12/17/content_33406914.htm, (2019–12–17).
38. DONG, Z., LI, H., GE, C., WANG, J., HAO, C., CHENG, C., LONG, F., LI, X.: Annual report on environmental and economic policy 2017 [J]. Environmental Economy 4, 12–35 (2018).
39. Dong Zhanfeng, Li Hongxiang, Ge Chazhong, Wang Jinnan, Hao Chunxu, Cheng Cuiyun, Long Feng, Li Xiaoliang.: "Environmental Economic Policy Annual Report 2017." Environmental Economics 4: 12–35 (2018).
40. Ministry of Finance.: "Rules of Transfer Payments from the Central Government to Local Key Ecological Function Areas." 86, 2018.6, (2018).
41. Chinese Academy of Environmental Planning.: "China Environmental Economic Policy Progress Annual report: 2017." 2018.1, (2018).
42. The Nature Conservancy (TNC). China TNC cooperates with Zhejiang Longwu in water source protection project [EB/OL]. http://www.tnc.org.cn/#News#schedule#iframe99dc279553caa33 1d70c9f0840779587b1f0c4fddb7a32175cd9319c7a817b5db938ef981a6ed605397fb1, (2015-01-15).
43. Wanxiang Turst. Innovate new business model, Wanxiang Trust launched the first water fund trust in China [EB/OL]. http://biz.zjol.com.cn/system/2015/11/18/020917870.shtml, (2015).
44. SCHOMERS, S., Matzdorf, B.: Payments for Ecosystem Services: A Review and Comparison of Developing and Industrialized Countries [J]. Ecosyst. Serv. 6, 16–30 (2013).
45. SALZMAN, JAMES., BENNETT, G., CARROLL, N., GOLDSTEIN, A., JENKINS, M.: The Global Status and Trends of Payments for Ecosystem Services [J]. Nat. Sustain. 1, 136–144 (2018).
46. Echavarría, Marta.: "Financing Watershed Conservation: The FONAG Water Fund in Quito Ecuador." In Selling Forest Ecosystem Services: Market-Based Mechanisms for Conservation and Development, Stefano Pagiola, Joshua Bishop, and Natasha Landell-Mills, Eds. London: Earthscan, (2002).
47. BULL, J.W., SUTTLE, K.B., GORDON, A., SINGH, N.J., MILNER-GULLAND, E.J.: Biodiversity Offsets in Theory and Practice [J]. Oryx 47(3), 369–380 (2013).
48. GELCICH, S., VARGAS, C., CARRERAS, M.J., CASTILLA, J.C., DONLAN, C.J.: Achieving Biodiversity Benefits with Offsets: Research Gaps, Challenges, and Needs [J]. Ambio 46, 184–189 (2017).
49. MCKENNEY, B.A., KIESECKER, J.M.: Policy Development for Biodiversity Offsets: A Review of Offset Frameworks [J]. Environ. Manage. 45, 165–176 (2010).
50. BULL, J.W., STRANGE, N.: The Global Extent of Biodiversity Offset Implementation under no Net Loss Policies [J]. Nat. Sustain. 1, 790–798 (2018).
51. GARDNER, T.A., VON HASE, A., BROWNLIE, S., et al.: Biodiversity Offsets and the Challenge of Achieving No Net Loss [J]. Conserv. Biol. 27(6), 1254–1264 (2013).
52. Villarroya, Ana, Ana Cristina Barros, and Joseph Kiesecker.: "Policy Development for Environmental Licensing and Biodiversity Offsets in Latin America." PLOS ONE 9(9): e107144. https://doi.org/10.1371/journal.pone.0107144, (2014).

53. BEZOMBES, L., GAUCHERAND, S., KERBIRIOU, C., REINART, M.E., SPIEGEL-BERGER, T.: Ecological Equivalence Assessment Methods: What Trade-Offs between Operationality, Scientific Basis and Comprehensiveness? [J]. Environ. Manage. 60, 216–230 (2017).

54. QUÉTIER, F., LAVOREL, S.: Assessing Ecological Equivalence in Biodiversity Offset Schemes: Key Issues and Solutions [J]. Biol. Conserv. 144(12), 2991–2999 (2011).

55. LUCK, G.W., CHAN, K.M.A., FAY, J.P.: Protecting ecosystem services and biodiversity in the world's watersheds [J]. Conserv. Lett. 2(4), 178–188 (2009).

56. BURIAN, G., SEALE, J., WARNKEN, M., et al.: Business Case for Investing in Soil Health [R]. Geneva: World Business Council for Sustainable Development, (2018).

57. CLARK, R., REED, J., SUNDERLAND, T.: Bridging funding gaps for climate and sustainable development: Pitfalls, progress and potential of private finance [J]. Land Use Policy 71, 335–346 (2018).

58. BASS, M.S., FINER, M., JENKINS, C.N., et al.: Global Conservation Significance of Ecuador's Yasuní National Park [J]. PLOS One 5(1), e8767 (2010).

59. Hoeksema, Bert W. "Delineation of the Indo-Malayan Centre of Maximum Marine Biodiversity: The Coral Triangle." In Renema W. (eds) Biogeography, Time, and Place: Distributions,Barriers, and Islands. Topics In Geobiology, vol 29. Dordrecht: Springer. https://doi.org/10.1007/978-1-4020-6374-9_5, (2007).

60. KLINGER, J.M.: In Their Own Time, on Their Own Terms: Improving development bank project outcomes through community-centered sustainable development partnerships in the Brazilian Amazon [R]. Boston: University Global Development Policy Center, (2019).

61. DILLENBECK, M.: National Environmental Funds: A New Mechanism for Conservation Finance [J]. Parks 4(2), 39–46 (1994).

62. KNOX, J.H.: The Neglected Lessons of the NAFTA Environmental Regime [J]. Wake Forest Law Review 45, 391–424 (2010).

63. North American Development Bank.: North American Development Bank Annual Report [R]. San Antonio, TX: NADBank, (2019).

64. EHLER, C., DOUVERE, F.: Marine spatial planning: a step-by-step approach toward ecosystem-based management [R]. Paris: UNESCO, (2009).

65. Asian Development Bank (ADB).: Building Gender into Climate Finance: ADB Experience with Climate Investment Funds [R]. Manila: ADB, (2016).

Chapter 9
Global Green Value Chains: Greening China's "Soft Commodity" Value Chains

9.1 Introduction

China's economic rise over the past 40 years is one of the major transformational events in modern world history. China has lifted more than 850 million people out of poverty since 1978 [1]. The country's gross domestic product (GDP) has grown from about roughly $200 billion in 1980 to more than $14 trillion in 2019, a 69-fold increase, while the population only grew by 40% (to 1.4 billion) during that time. Merchandise imports in that same period grew from about $20 billion to over $2 trillion, a 100-fold increase, while exports increased 135-fold, from $18 billion to around $2.5 trillion [2].

China's transformation has occurred in tandem with the unprecedented globalization of value chains. While global GDP grew nearly eight-fold from 1980 ($11 trillion) to 2018 ($85 trillion), the global export of goods by value during that period grew nearly tenfold, from $2 trillion to $19 trillion.

As its economy has globalized and matured, China has developed an increasingly holistic vision for the future. China's 13th Five Year Plan (FYP) (2016–2020) has created a blueprint for the nation's future development around five themes: innovation, coordinated development, green growth, openness, and inclusive growth. In 2018, China integrated the concepts of "Ecological Civilization" and "a community of shared future for mankind" into its Constitution. These steps by China accord with the global trend towards green growth and sustainable development, as reflected by the adoption in 2015 of the UN Sustainable Development Goals (SDGs).

Beginning in 1980, global value chains—in which production processes are broken up across countries and among specialized tasks performed by different firms—have become an increasingly important feature of the global economy; they currently account for around half of all global production [3]. While global value chains have many benefits, they also have considerable environmental impacts. The

© The Author(s) 2022
China Council for International Cooperation on Environment
and Development (CCICED) Secretariat,
Green Consensus and High Quality Development,
https://doi.org/10.1007/978-981-16-4799-4_9

global value chains of four soft commodities—soy, beef, palm oil and wood products—are responsible for at least 40% of global deforestation and could lead to biodiversity loss, climate change, and other environmental challenges [4, 5]. The need for greening production, trade, and consumption is increasingly recognized by participants at all stages of these value chains.

As the world's largest exporter and second-largest importer, China is at the centre of global value chains, including for the four soft commodities noted above. China accounted for almost 60% of global soy imports and was the world's second-largest palm oil importer (following India) in 2019 [6]. China's beef imports have grown rapidly and surpassed the United States in 2018, making China the world's largest importer in quantity and second-largest in value (following the United States) [6]. China has also become the world's largest importer of timber, accounting for one third of the value of global timber imports (logs and sawnwood) in 2018 [6].

It is increasingly in China's self-interest to green its value chains, especially those for key soft commodities. As these value chains have become more complex, they are increasingly subject to a variety of risks:

- The COVID-19 pandemic fundamentally disrupted the global economy in just a few months and exposed the vulnerabilities of global value chains to rapid and unexpected change from factors in the natural environment.
- International trade policy shifts and disputes (such as the U.S.–China "trade war") can cause short-term disruptions and increase longer-term uncertainty and instability in global value chains—although the imperatives of mutual economic benefit and stable political relationships are likely to reduce tensions in the longer term.
- Political and economic events in producer countries can also affect the supply and price of export commodities.
- Over-exploitation of a commodity can lead to decreasing availability and/or increasing prices (e.g., some fisheries and timber species).
- Diseases, pests, and invasive species (notably COVID-19 but also African swine fever, avian flu, fire ants, African snails, and locusts) can fundamentally disrupt global value chains.
- Regulatory requirements are becoming more stringent in both producer and end-market countries (e.g., food safety and labour standards, phytosanitary and environmental protections), and this trend is likely to accelerate in light of COVID-19.

The globalization of value chains also presents China with some positive opportunities:

- With such a huge share of the global market, policy reforms by China are likely to trigger comparable changes in other countries. If China becomes an "early mover" in greening its soft commodity value chains, it can turn policy innovations into economic advantages.
- Greening soft commodity value chains is also an opportunity for China to meet its climate change, biodiversity, and sustainable development commitments under environmental treaties and the SDGs.

While economic growth remains a key priority for China, the past decade has seen a gradual shift in policy towards the quality, stability, and sustainability of growth. This is exemplified in China's aspiration to achieve an Ecological Civilization.

The vision of Ecological Civilization extends beyond China's borders. China's economic rise has been accompanied by a significant expansion in international engagement, exemplified by the Belt and Road Initiative (BRI) and China's leading role, with the United States, in catalyzing a successful outcome at the 2015 Paris Climate Summit.

This report seeks to provide a convincing rationale and concrete policy options for Chinese leadership to green its global value chains for soft commodities—particularly those linked to tropical deforestation. This study focuses on soy, beef, palm oil, and forest products (timber, pulp, and paper) and builds on the findings and recommendations of a previous CCICED Special Policy Study, *China's Role in Greening Global Value Chains*, published in 2016 (see Box 9.1).

The exploration of policy options is timely in light of several key upcoming events in 2021 that China will host, including the 15th Conference of the Parties (COP-15) to the Convention on Biological Diversity (CBD), the annual China International Import Expo in Shanghai. The 26th Conference of the Parties to the United Nations Framework Convention on Climate Change (UNFCCC) in the United Kingdom presents an additional opportunity. The study is also timely in light of the process underway during 2020 to finalize China's next FYP (2021–2025).

The focus on these soft commodity value chains is also relevant to China's growing role as a global infrastructure investor under the BRI. The BRI is a major catalyst for infrastructure expansion in many countries, and where and how roads, ports, power grids, and mills are built is a major enabling condition for the expansion of commercial logging and agricultural production into new frontiers. The study has therefore been carried out in close coordination with the work of the BRI International Green Development Coalition, as well as CCICED's previous and ongoing work on greening the BRI.

This study answers the following questions to identify the opportunities and barriers that greening the commodity value chain could bring to China:

- What is the significance of soft commodity value chains for China?
- Why should China pursue green soft commodity value chains?
- How can China "green" its soft commodity value chains?

Three assumptions have guided the formulation of the study's recommendations:

- Recommendations should not interfere with the internal affairs of sovereign nations
- Proposed solutions should be practical and low cost.
- Recommendations should embody a Chinese approach to solving global problems, aligned with the vision of an Ecological Civilization and a community of shared future for humankind.

Box 9.1. Summary Findings and Recommendations of the 2016 CCICED Study

Conclusions

- Global value chains need a green reboot, and China can lead the way.
- Greening global value chains for commodities, in particular, is central to sustainable development.
- It is in China's interest to lead the greening of global value chains for commodities.

Recommendations

- Play a leadership role in promoting the sustainability of global value chains in international governance and policy-making.
- Send a clear policy signal to encourage Chinese companies and multinational companies trading in China to green their global value chains.
- Create an action plan for greening global value chains as a core priority for the BRI.
- Invest development aid and other financial resources in greening global value chains.

First Steps

- **State-Owned Enterprises (SOEs)**: The State-owned Assets Supervision and Administration Commission (SASAC) should mandate SOEs to assure the sustainability of the commodities they buy that impose major global environmental impacts.
- **Pilots**: The Government of China should launch a pilot program to establish best practices for greening the global value chains for soy, palm oil, and forest products.
- **Development Assistance**: The Ministry of Environmental Protection, the National Development and Reform Commission (NDRC), and the Ministry of Commerce should jointly launch a Green Global Value Chain South-South Cooperation Platform under the newly established South-South Cooperation Fund on Climate Change to support China's major commodity supplier countries in improving the sustainability of commodity production and trade.

Box 9.2. Definition of Key Terms

Soft commodities: Raw materials and their derivatives that are grown or produced by agriculture (crops, livestock) and forestry industries.

Global value chains: Processes by which value is added across different stages from production to consumption and carried out by actors located in different parts of the world [7].

Supply chains: A component of value chains that are composed principally of the logistical linkages at a firm level [7].

Producer countries: Countries that produce a large amount of relevant commodities and often export those commodities.

Consumer countries: Countries that consume a significant amount of commodities and often import those commodities.

Due diligence: A risk management process implemented by a company to identify, prevent, mitigate, and account for how it addresses environmental and social risks and impacts in its operations, supply chains, and investments.

Traceability: The ability to follow a product or its components through stages of the supply chain (e.g., production, processing, manufacturing, and distribution).

Greening: A shorthand term for policies and practices that reduce the negative environmental and social impacts of economic investments, activities, and production processes.

9.2 What Is the Significance of Soft Commodity Value Chains for China?

Global value chains—in which production processes are broken up across countries and among specialized tasks performed by different firms—have become an increasingly important feature of the global economy since 1980. Global value chains now account for about half of all global production [3]. These chains bind the world together, linking the economies and peoples of both developing and developed countries in the trade of soft and hard commodities. The COVID-19 health pandemic and economic crisis, however, have disrupted most global value chains. How quickly these global value chains can be restored after the worldwide lockdowns are lifted remains unclear.

9.2.1 What Are Soft Commodities?

"Soft commodities" are raw materials and their derivatives that are grown or produced by the agriculture and forestry industries. These materials include plant- and animal-derived materials for use as food, fibre, feed, medicines, cosmetics, detergents, and fuels. Soft commodities contrast with "hard commodities," which are raw materials and their derivatives that are extracted or mined, such as metals, oil, and natural gas.

Soft commodities are critical for human development and trade. They provide the world's nutrition, feed for livestock, and raw material for paper, clothing, furniture, and buildings. While some can be domestically produced, many soft commodities are grown in areas that have a comparative advantage for production—such as the right soils, rainfall, and climate. Thus, nations typically rely on global value chains to access the soft commodities they need.

9.2.2 What Are the Challenges of "Business-As-Usual"?

A handful of soft commodities—soybeans, palm oil, beef, forest products (timber, pulp, and paper), coffee, and cocoa—pose significant challenges to sustainable development. A core challenge concerns deforestation, climate change, and biodiversity loss. Many countries that produce these soft commodities have high levels of biodiversity and high rates of deforestation (Fig. 9.1). In fact, soybeans, palm oil, beef, and forest products combined account for anywhere from 40% [4, 5, 73] to more than 50% of the world's tropical deforestation [8, 9]. In major producing countries such as Brazil and Indonesia, the loss of tree cover in the last two decades is closely linked to the production of these soft commodities (oil palm and pulp and paper in Indonesia; beef and soybeans in Brazil) (Fig. 9.2). As such, these soft commodities are the world's leading cause of biodiversity loss and greenhouse gas emissions related to land-use change [10].

A second challenge concerns legality. Revenue from illegally sourced timber, for example, is estimated to be $50 billion–152 billion globally per year [11]. More than 90% of the deforestation in the Brazilian Amazon is illegal and often associated with other crimes, such as drug trafficking and tax evasion [12]. An Indonesian government audit in 2019 found that around 81% of Indonesian oil palm plantations did not meet applicable regulations [13]. A significant proportion of the global supply of soft commodities is linked to illegal logging or land clearing, violation of labour laws, tax avoidance, or corrupt allocation of permits and licenses.

A third challenge concerns the social issues of equality and inclusion. For instance, women tend to experience lower participation rates in these value chains, unequal access to capital and property, and undervaluation of compensation for their work [14–16]. Farmers' livelihoods may be harmed where unsustainable commodity production degrades forests and land or limits their access to high-yielding crop varieties, water, or energy. A lack of recognition of local rights over land and resources—which may be customary or informal—is another key social challenge. Labour-related issues may also arise, such as child labour, slavery, lack of collective bargaining rights, poor wages and benefits, and poor workplace safety and health conditions.

Consequently, "business-as-usual" trade in these soft commodities poses a threat to major international agreements. For example, it contravenes national laws and applicable international law. It threatens to undermine the achievement of numerous SDGs, including Goal 5 (gender equality), Goal 8 (decent work and economic growth), Goal 10 (reduced inequalities), Goal 12 (responsible consumption and production), Goal 13 (climate action), Goal 15 (life on land), and Goal 16 (peace,

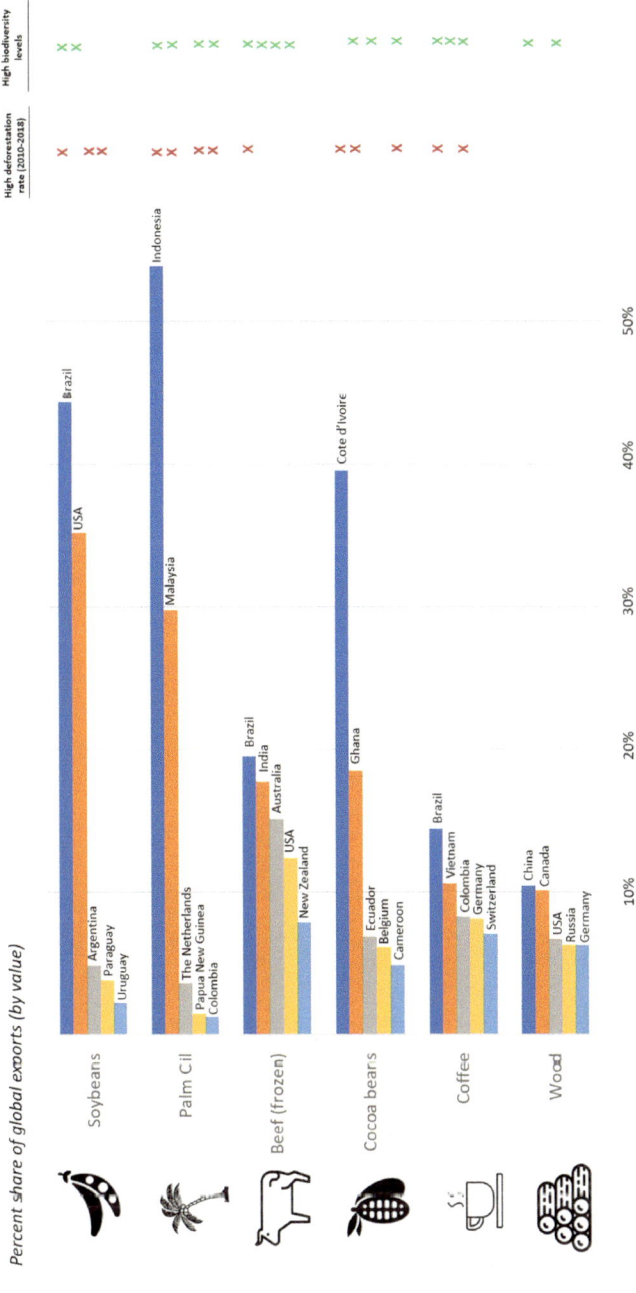

Fig. 9.1 Global exports of soft commodities by top-producing countries (2017)

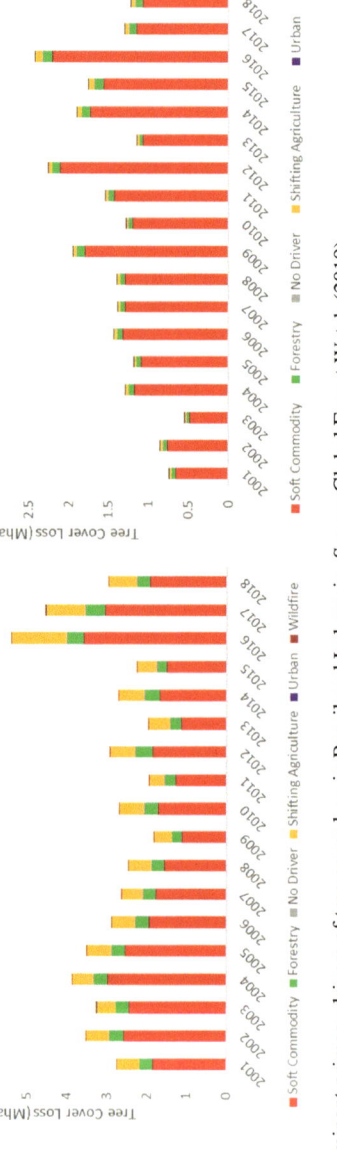

Fig. 9.2 Dominant primary drivers of tree cover loss in Brazil and Indonesia. *Source* Global Forest Watch (2019)

justice and strong institutions). Moreover, continued tropical deforestation by these commodities will make it impossible to achieve the Paris Agreement on Climate Change and the globally agreed goals and targets of the CBD.

Likewise, "business-as-usual" trade in these soft commodities poses economic threats. Recent history showcases a number of high-profile instances where business and economic performance suffered significantly due to engaging in "business-as-usual" practices. Examples include:

- Global wood flooring manufacturer and retailer Lumber Liquidators saw its market capitalization drop by $1.1 billion in the first half of 2015 after being held criminally liable in the United States for importing illegal timber from Russia (through China). It was subsequently exposed in the media for using potentially cancer-causing levels of formaldehyde in its China-sourced laminated flooring [17].
- Cocoa and palm oil firm United Cacao was exposed in 2016 for developing plantations in legally protected forests in the Peruvian Amazon [18]. In early 2017, the London Stock Exchange suspended trading of the firm's stock, its CEO resigned, and its share value fell by 55% [19].
- Five grain trading firms—Cargill, Bunge, ABC Indústria e Comércio SA, JJ Samar Agronegócios Eireli, and Uniggel Proteção de Plantas Ltda—and a number of farmers were fined a total of $29 million by the Brazilian government in 2018 for activities connected to illegal deforestation in Brazil's Cerrado savannah [20].
- One of the world's iconic guitar companies, Gibson Guitar, paid a $300,000 penalty and forfeited the seized wood valued at more than $250,000 in 2012 under a criminal enforcement agreement with the United States government after having been found importing illegally harvested ebony and rosewood from Madagascar and India [21].

Moreover, as the CCICED noted in 2016, goods and ecosystem services that are critical to the global economy may degrade and even disappear if natural resources are unsustainably managed, even in the near term [7]. In addition, business-as-usual presents market, reputational, and compliance risks for the private sector as consumers and governments in both emerging and developed economies increasingly demand products that are more sustainable [4, 5, 73].

9.2.3 Why Is China Important?

China has emerged as the centre of trade in these soft commodity value chains. Driven by demand from the country's rising middle class and limited potential for expanding domestic production commensurate to demand, China is now the world's largest single country importer of soy, beef, and timber, as well as the world's second-largest importer of palm oil (behind India) (Table 9.1). Chinese demand is larger than that from the European Union (EU) and North America for imported soy and pulp and paper. Moreover, Chinese demand is roughly on par with the entire EU for palm

Table 9.1 China's imports of soy, pulp and paper, timber, beef, and palm oil in 2018

Commodities	China's share of global imports (%)	Global rank	Unit
Soy	60	1	USD
Pulp and paper	38	1	USD
Timber[a]	33	1	USD
Beef	17	1	Tonnage
Palm oil	12	2	USD

[a] Includes logs and sawnwood
Source [1, 6] UN Comtrade and USDA

oil (Fig. 9.3) and is projected to grow [4, 5, 73]. Since it is the world's largest or
second-largest importer of these soft commodities, China is a key actor. If China takes
proactive steps in collaboration with the other major markets—the EU, the United
States, and India—the world will be able to transition from the "business-as-usual"
approach toward a more sustainable path for soft commodity value chains.

The BRI is another avenue where China can play a key role in greening global soft
commodity supply chains. In 2019, China's trade with BRI countries exceeded $1.3
trillion and comprised about 30% of China's total trade [22]. By April 2019, China

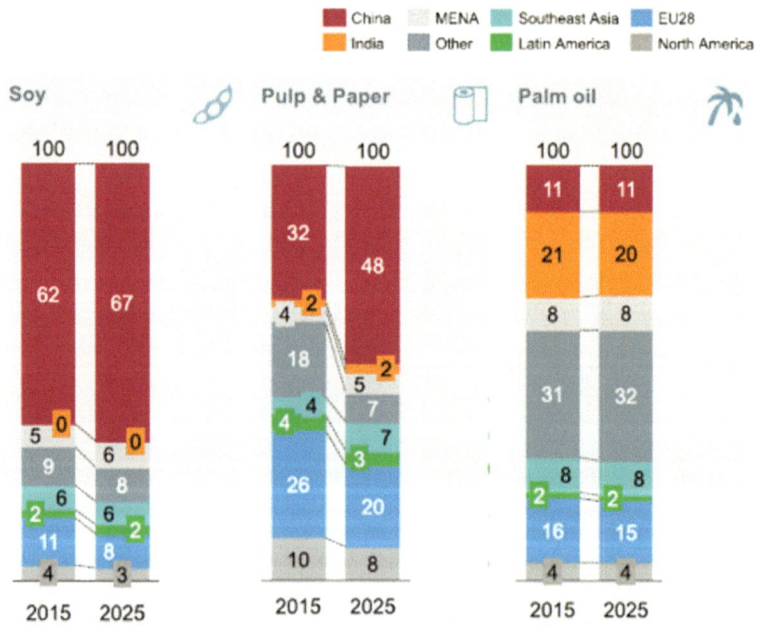

Source: World Economic Forum 2018.

Fig. 9.3 Share of global imports (2015, 2025)

had signed BRI cooperative agreements with 125 countries [23], including many of the world's major soft commodity producer nations in Asia, Africa, and Latin America. Importantly, the Chinese government has signalled an intent to ensure that the BRI advances sustainable, "green" value chains in these countries. In 2017, the Ministry of Ecology and Environment, the Ministry of Foreign Affairs, the Ministry of Commerce, and the NDRC jointly published *Promoting the Green Belt and Road Initiative*. This guidance highlights the need to strengthen value chain management in a manner that promotes green production, green procurement, green consumption, and international cooperation to achieve greener value chains.

9.2.4 What Is a "Green" Soft Commodity?

What are the defining environmental and social characteristics of a "green" soft commodity value chain? *Environmental characteristics* include the efficient use of natural resource inputs, low levels of waste, and low amounts of pollution. More fundamentally, green soft commodity value chain sourcing and production processes do not directly or indirectly cause the degradation, fragmentation, or conversion of natural forests and other important natural ecosystems (e.g., grasslands). This means, for instance, that "green" soy, palm oil, and beef production in producer countries does not involve the clearing and conversion of natural tropical forests and other ecosystems. For wood products, it means that timber is not extracted at an industrial scale from high-conservation value forests (i.e., intact or primary forests). Complementing this, a green soft commodity is one in which productivity per hectare (i.e., yields) of existing agricultural land is high or improving—since boosting yields on existing agricultural land is a key approach to avoiding the need to convert natural ecosystems.

Social characteristics include respect for the internationally recognized rights and interests of Indigenous Peoples, local communities, women, children, and workers. They include protections from discrimination, exploitation, and unsafe or unhealthy working conditions.

A green soft commodity value chain is also a *legal* value chain, in which both national laws and international legal obligations regarding permitting, licensing and harvesting, environmental and social impact assessment, payment of taxes and other fees, participatory decision-making processes, and labour rights and protections are observed according to the national laws and international obligations. And a green soft commodity value chain is a *transparent* value chain, in which all stakeholders have access to relevant information about the legality and sustainability of production and trade processes, from the field to the ultimate market.

Although fully greening soft commodity value chains involves improvements along each stage of the value chain, this study will focus on the production stage—particularly the social and environmental impacts of land acquisition, the conversion of natural ecosystems, and farming and forestry practices. This focus is justified for at least four reasons:

- First, this is the stage of a soft commodity value chain that *has the most impact* on climate change, biodiversity, and land-related rights. That is because it is the growing or extraction of commodities that directly causes the loss or degradation of natural ecosystems and of the rights and livelihoods of Indigenous Peoples and local communities. The loss and degradation of forests, peatlands, and mangroves are major contributors to global greenhouse gas emissions—an important factor in regulating local climate and the leading driver of biodiversity loss [24, 25]. In fact, for many major soft commodities of importance to China, the conversion of land is the commodity's major contribution to greenhouse gas emissions (Fig. 9.4).
- Second, climate change and biodiversity conservation are *high on intergovernmental agendas* for the years 2020 and 2021. The next Conference of Parties to the global agreement on climate change is slated to feature "nature-based solutions," which include forest conservation and more sustainable agriculture. The next Conference of Parties to the UN CBD, to be hosted by China, will set the global agenda for biodiversity conservation for the next decade.
- Third, delinking soft commodities and deforestation is *high on global finance and private sector agendas*. This is evidenced by the incorporation of sustainability standards in global investment firms such as BlackRock and major collaborations on this issue convened by the Consumer Goods Forum, World Economic Forum, and others.
- Fourth, it is most practical to green soft commodity value chains in a *step-by-step approach*. Trying to address every sustainability aspect of a value chain all at the same time could be too overwhelming and thus lead to paralysis. Rather, focusing first on one of the most important and high-profile issues currently could enable governments and companies to take targeted, concrete action now. Investments in improving traceability, for example, can benefit both sustainability goals and value chain cost-effectiveness. And this could have an outsized impact since improving the basics of production will have knock-on benefits in other parts of the value chain. Other sustainability issues can be added to the agenda once sufficient progress is made.

9.3 Why Should China Pursue Green Soft Commodity Value Chains?

As the world's largest importer and consumer of soft commodities, China has the power to catalyze positive change across the global economy. But why would China find it in its self-interest to do so? There are five principal reasons: (1) to ensure consistency with China's vision of an Ecological Civilization, (2) to strengthen supply chain safety and security, (3) to uphold the law, (4) to respond to tomorrow's markets, and (5) to optimize China's international environmental reputation.

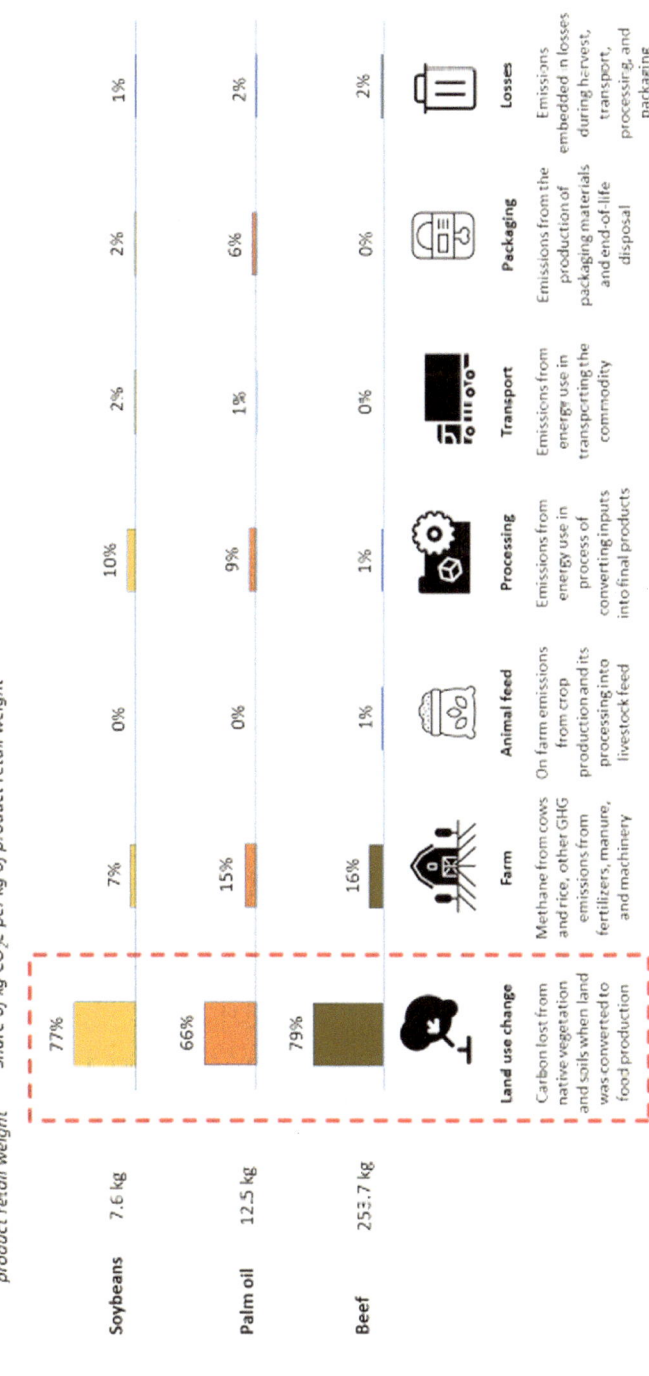

Fig. 9.4 Share of greenhouse gas emissions per stage in the value chain for selected soft commodities

9.3.1   Ensure Consistency with China's Vision of Ecological Civilization

The greening soft commodity value chains is entirely consistent with and supportive of China's vision of an Ecological Civilization, at home and abroad, as laid out by the country's highest leadership. At the 19th National Congress of the Communist Party in 2017, President Xi Jinping stated that: "Taking a driving seat in international cooperation to respond to climate change, China has become an important participant, contributor and torchbearer in the global endeavour for ecological civilization." He continued, noting that:

> The dream of the Chinese people is closely connected with the dreams of the peoples of other countries … We must keep in mind both our internal and international imperatives, stay on the path of peaceful development, and continue to pursue a mutually beneficial strategy of opening up, cultivat(ing) ecosystems based on respect for nature and green development … We should, acting on the principles of prioritizing resource conservation and environmental protection and letting nature restore itself, develop spatial layouts, industrial structures and ways of work and life that help conserve resources and protect the environment.

Embarking on the journey to make Chinese soft commodity value chains "green" would be a concrete manifestation of this vision. It also would help ensure that the 2021 CBD summit in Kunming is a resounding success.

9.3.2   Strengthen Supply Chain Safety and Security

The COVID-19 crisis is placing immense scrutiny on the safety of global trade and the long-term security and stability of global value chains. Greening soft commodity supply chains can be a component of an effective strategy for addressing both challenges.

First, greening soft commodity value chains can help *make global trade safer*. This is because environmental health is linked to human health. A number of recent scientific studies point to a link between the conversion of natural ecosystems, increased human contact with wildlife, the emergence of new (and the spread of old) zoonotic diseases, and epidemics (or even pandemics) harming human health [26]. Examples include Ebola, coronaviruses, Marburg, Zika, and malaria [27, 28] (Box 9.3). In light of COVID-19, the global community, businesses, and citizens will be paying greater attention to ensuring that a country's economic activities—such as what it trades and from where—are not triggering the emergence and spread of zoonotic diseases. Avoiding economic activities that lead to deforestation can reduce this risk of contributing to human diseases.

Second, greening soft commodity value chains can help secure the long-term *stability of supply* (and therefore stability of price) of soft commodities. This is because the long-term availability of soft commodities depends in part on how those resources are managed today. For example, recent studies find that clearing too much of the Amazon for soybeans and cattle will lead to a decline in rainfall in Brazil's "soy belt," thereby reducing the country's soy production in the long term [27, 29]. Soon-to-be-published analysis indicates that the yield shocks could be on the order of 10%, generating losses worth USD 700 million per year [30]. Where unsustainable production leads to social conflict and corruption, the stability of commodity supply and price from that region can fluctuate unpredictably. And for some commodities, such as certain species of timber, overexploitation can lead to the commercial collapse in supply. Thus, sustainable management today ensures availability for tomorrow at stable prices. Conversely, unsustainable management may trigger supply scarcity, unreliability, and volatile prices.

Box 9.3. How Do Green Value Chains Contribute to Human Health?

Green soft commodity value chains can contribute to human health by reducing the risks of zoonotic diseases that spread from animals to humans. More than 60% of emerging infectious diseases are zoonotic in origin, and the majority (70%) of these zoonotic pathogens have emerged because of increased human-wildlife contact—driven by humans and livestock encroaching on natural ecosystems [31, 32].

Deforestation and forest degradation—and exploitation of wild animals—are implicated in the emergence over the past few decades of zoonotic disease outbreaks such as Ebola, SARS, avian flu, and COVID-19 [26]. One study found that Ebola outbreaks in Central and West Africa were significantly associated with forest losses in the previous two years [70]. When forests are cleared for soft commodity production, the buffer zones separating humans from animals or the pathogens that animals harbour are reduced or lost [33].

Establishing green value chains is therefore key to ensuring that economic activities are not causing ecological degradation that increases the likelihood of human exposure to zoonotic viruses.

9.3.3 Uphold the Law

Greening its soft commodity value chains would enable China and Chinese companies to uphold the law. This is important for at least three reasons:

- *It is simply the right thing to do*. It is a longstanding principle of China's foreign policy to respect international law, as well as to respect the sovereignty and laws of other countries (Box 9.4). Laws regarding the production and trade of soft commodities are quickly strengthening around the world. This fast-changing context necessitates the greening of soft commodity value chains if China is to adhere to its longstanding principles and if Chinese enterprises are to remain in legal compliance in the foreign jurisdictions where they do business.

- *Other leading importers are strengthening laws on soft commodities*. Worldwide scrutiny of the legality of soft commodities is rapidly increasing. With respect to timber, for instance, many of the world's major importing nations recently have established laws banning the import of illegally harvested or traded wood products. These nations include the EU, the United States, Australia, Japan, and South Korea, which account for 52% of the world's forest product imports [34]. In addition, since 2017, the UN Convention on International Trade in Endangered Species (CITES) has listed hundreds of timber species, many of which feed the Chinese furniture industry, for protection from illegal trade.

  This trend towards more stringent scrutiny on legality is rapidly spreading to other soft commodities. In 2019, for instance, the European Commission started exploring policies to ensure imports of commodities such as soybeans, palm oil, and beef are not linked to illegal deforestation. An analogous measure is being discussed in the U.S. Congress in 2020. Beyond governments, numerous multinational companies, industry associations, and commercial banks have stepped up efforts to eliminate illegality from their value chains [73]. Likewise, more than 200 endorsers—including governments, companies, and civil society organizations—have supported The New York Declaration on Forests to halt deforestation in agricultural commodity value chains [35].

- *Exporting countries are introducing and enforcing laws on soft commodities*. Several of China's major soft commodity production and trading partners have put in place laws to curtail the illegality (and increase the sustainability) of their soft commodity production and trade (Box 9.4). Moreover, they are taking enforcement action. In 2016, for example, Spanish banking giant Santander incurred a $15 million fine for lending money to farmers illegally destroying Brazilian forests [36, 37] China and Chinese companies could send a signal of supporting the enforcement of these laws and being good trade partners by greening their soft commodity value chains.

Box 9.4. Would Chinese Efforts to Green Its Soft Commodity Supply Chains Interfere with the National Sovereignty of Its Trading Partners?
Chinese efforts to green its soft commodity supply chains would *not* interfere with the national sovereignty of its trading partners. Rather, by greening its soft commodity supply chains, China actually would *support* the national sovereignty of its trading partners. This is because many of China's trading partners already have in place laws that encourage legal and sustainable soft commodity production and trade. Examples include:

- *Indonesia*: Recent Indonesian policies aim to get illegality and deforestation out of its timber and palm oil supply chains. For example, Indonesia's Low Carbon Development Initiative (LCDI)—launched in 2019 as a program of the Ministry of National Development Planning (BAPPENAS)—sets the country's economic development agenda. The LCDI calls for increased supplies of sustainable palm oil and timber via yield increases and using degraded land while avoiding the conversion of natural forests and peatlands [38]. Also, in 2019, the Indonesian president announced a permanent moratorium on new forest clearing for plantations and logging in 66 million hectares of primary forest and peatland 71. In addition, the country has established a National Timber Legality Assurance System to prevent trade in illegally harvested timber. This has enabled a Voluntary Partnership Agreement with the European Union (EU) that ensures only legal timber from Indonesia enters the EU market in return for faster, streamlined processes as timber reaches the EU border [39]. China seeking to "green" its palm oil and timber supply chains would support Indonesia's implementation of these nationally approved economic development plans, government policies, and government trade programs.

- **_Brazil_**: More than 90% of deforestation in the Brazilian Amazon is illegal and often associated with other crimes, such as drug trafficking and tax evasion [12]. Consequently, a number of existing public policies in Brazil focus on preventing illegal deforestation. For example, the Forest Code stipulates the maximum land area per farm that can be cleared for agriculture per biome (e.g., 20% in the Amazon, 65%–80% in the Cerrado) [40]. Any clearing beyond that is illegal, and the products generated on such farms are in violation of the law. In addition, Brazil's Nationally Determined Contribution (NDC) to the Paris Agreement on Climate Change calls for strengthening policies and measures to achieve zero illegal deforestation in the Brazilian Amazon by 2030 [41]. Therefore, China seeking to ensure that the soybeans and beef it imports from Brazil are legal and "green" would support the Brazilian implementation of these laws and commitments.

- **_Africa_**: Many countries in Africa, a growing source of tropical timber for China, have laws in place to eliminate illegal logging and avoid loss of their natural forests. For example, over the past decade, at least eight African countries (Cameroon, Central African Republic, Côte d'Ivoire, Democratic Republic of Congo, Gabon, Ghana, Liberia, Republic of the Congo) have signed or are in the process of negotiating Voluntary Partnership Agreements with the EU to ensure that only legally harvested timber enters European and domestic markets [42]. In 2018, the Republic of the Congo issued Joint Ministerial Decree 9450, which stipulates that new agricultural development greater than 5 ha can only be developed on savannahs and not in forests [43]. The Democratic Republic of the Congo's National REDD + Strategy and Investment Plan steers large-scale agricultural development toward savannahs, as well. Therefore, China seeking to ensure that future timber imports from the Congo Basin are legal and that any future palm oil imports from the region are deforestation-free aligns with government policies and programs of these nations.

9.3.4   Respond to Tomorrow's Markets

China has emerged as a global powerhouse for today's markets. But future economic success rests on China meeting the needs of tomorrow's markets. These markets will increasingly demand greener consumption and greener production [44]. When it comes to soft commodities, these markets are trending toward "green" in three ways:

- *Evolving Chinese consumer preferences*. History shows that, as per capita incomes rise in nations, consumers increasingly care about the social and environmental sustainability of the products they purchase [45]. Thus, rising concern about sustainability typically coincides with a rising middle class. China is no different. For instance, a 2017 survey found that more than 70% of Chinese consumers were willing to pay a 10% premium for sustainably produced goods [46].

- *Globalizing retailer and manufacturer norms*. The business norms of multinational retailers and manufacturers of products containing soft commodities are rapidly shifting towards greater sustainability and are being applied equally across all geographies. Walmart's sustainability policies, for instance, apply to all Walmart stores [47]. These business norms include value chain policies, as well. Walmart is working with its global suppliers to evaluate and share progress on key environmental and social issues in supply chains covering more than 100 product categories, including pulp, paper, and timber products [47]. In 2019, retail giant H&M announced that it would no longer source leather from Brazil due to the role of cattle ranching in Amazon forest fires and deforestation [48]. The company applied this policy to all of its stores worldwide; there was no separate policy for stores in Europe versus those in China. Mars—a major manufacturer of chocolate and other soft commodity-based products—has adopted a comprehensive set of policies to eliminate deforestation from its supply chains [49].

- *Tightening capital market policies*. A growing critical mass of institutional investors is developing investment guidelines to limit access to capital by borrowers whose investments in soft commodity production and trade result in tropical deforestation. In September 2019, for instance, 230 institutional investors representing $16.2 trillion in assets under management called on companies to take urgent action in light of the devastating forest fires in the Amazon: "As investors, who have a fiduciary duty to act in the best long-term interests of our beneficiaries, we recognize the crucial role that tropical forests play in tackling climate change, protecting biodiversity and ensuring ecosystem services" [50].

In light of these reasons, greening soft commodity value chains now can help position China and its companies for the rapidly approaching markets of tomorrow. And acting now would send market signals that ensure an adequate supply of green commodities at competitive prices (Box 9.5).

Box 9.5. Can Green Palm Oil Meet Increased Chinese Demand at a Reasonable Price?

China—and the world—can continue to use palm oil without destroying tropical forests and without paying a large "green premium." In terms of supply, global demand for palm oil certified by the Roundtable on Sustainable Palm Oil (RSPO) equates to ~10% of global palm oil supply, yet ~20% of global supply is already certified [51]. Thus, the palm oil market today could already absorb additional "green" demand.

In terms of price, green palm oil compared to "business-as-usual" palm oil can be quite close. For example, recent pricing from a major palm oil supplier indicates only a 3–4% premium for segregated sustainable crude palm oil and just a 1% premium for non-segregated sustainable crude palm oil. This variation is less than the variation in spot market prices on a weekly basis [52]. As more sustainable supply becomes available, the cost of sustainable production likely will decline and thereby further help ensure cost competitiveness.

As a major palm oil importer, China can play an important role in accelerating growth in the supply of green palm oil. If China were to send a clear preferential sourcing signal that a steadily increasing share of its palm oil will need to be green, the market would respond. Such a "demand signal"—with demand expected by the market to ratchet up over time—would give producers the incentive and time needed to ramp up production in advance, avoiding any potential future shortfalls in supply and keeping prices stable.

Satisfying 100% market demand for a green soft commodity, however, will not happen overnight. For example, the supply of green, deforestation-free palm oil is not sufficient to meet demand if all buyers asked for it today. This is also true for soy. It will take some time for supply to catch up to such complete demand. A signal by China would stimulate suppliers to start working now to meet that future demand. And as both supply and demand for green soft commodities increase, prices will come down.

9.3.5 Optimize China's International Environmental Reputation

China is committed to international development and environment agreements such as the UN Framework Convention on Climate Change (UNFCCC) and its Paris

Agreement on Climate Change, the UN CBD, the CITES, and the UN SDGs. By sending a strong political signal that it will start greening its soft commodity value chains, China can position itself positively as a responsible global player in these landmark agreements.

Business-as-usual soft commodity value chains are a leading cause of damage to ecosystems, biodiversity, and a stable climate. This damage undermines the objectives of each of these international agreements. As the largest importer of soft commodities affecting tropical forests, China has a very important role—complementing the influence of the EU, the United States, and India—to play in minimizing this damage and helping the world fulfill these agreements.

International expectations of China are high as the country prepares to host the COP-15 to the CBD in 2021. The upcoming COP will set the global agenda for biodiversity conservation for the next decade. What vision will China—as the COP President—bring, and what actions of its own can China put on the table to inspire others?

China providing a clear signal that it is embarking on a serious effort to green its soft commodity value chains, along with other major economies, would be an inspiring and well-received response. It would help set the Post-2020 Biodiversity Framework on the path towards a more successful decade and help enshrine "nature-based solutions" as a cornerstone to the Paris Agreement. It would establish China as an important participant, contributor, and torchbearer on the global stage in biodiversity, climate change, and sustainable development. Moreover, it would support soft commodity-producing countries in their own efforts to lift small farmers (including women farmers) out of poverty (Box 9.6) and to meet their own obligations under these international agreements (Box 9.7).

Box 9.6. Does Greening Soft Commodity Supply Chains Hurt Small-Scale Farmers in Producer Countries?

Greening soft commodity supply chains does *not* hurt small-scale farmers in producer countries, as long as proactive policy measures are established to support their economic interests. Done correctly, it can help smallholders boost yields and increase market access.

Small-scale or "smallholder" farmers are important suppliers for some soft commodities. For example, smallholders produce about 40% of the world's palm oil [53]—yet they produce under 12% of the world's soybeans [72]. Smallholders tend to be less productive per hectare and less able to implement new sustainability practices than large farms due to lower access to inputs, finance, and technical know-how. For example, palm oil yields of smallholders in Indonesia are approximately 20–25% lower than those of corporate-managed plantations [69]. Women farmers are particularly disadvantaged in terms of productivity and income, with less access to seeds, fertilizers, finance, and land than men [16].

Shifting to green soft commodity production can benefit smallholders by catalyzing improvements in efficiency (e.g., more judicious use of fertilizers), production per hectare (i.e., yields), access to inputs, and ultimately income. Such improvements in efficiency and yields are a core component of greening soft commodities. Agricultural companies and government programs can support these improvements. Over the past decade, for instance, multinational agribusinesses such as Olam, Sime Darby, Musim Mas, and others have increasingly offered smallholders training, financing, inputs, and administrative support to adopt sustainable cultivation methods and avoid forest clearing. For example, Musim Mas has developed palm oil training for smallholders in Indonesia to boost yields in a sustainable manner while adopting improved health and safety practices [54]. Overseas development assistance from China could complement these private sector interventions, providing technical assistance, inputs, and access to subsidized financing to support the transition to more sustainable agricultural practices.

Box 9.7. Can Greening Supply Chains Help China's Trading Partners Meet International Agreements?
Yes, China greening its soft commodity supply chains can help trading partner countries meet their obligations under several UN agreements. For instance, it would support developing country partners that have committed to reducing emissions from deforestation and forest degradation in their NDCs under the UNFCCC and the Paris Agreement on Climate Change. It would support the implementation of Article 3 of the CBD, which articulates *inter alia* that States have the "responsibility to ensure that activities within their own jurisdiction or control do not cause damage to the environment of other States …". And it would support controlling the trade in rosewood (*hongmu*) species listed by CITES, for which China is overwhelmingly the world's largest importer [55].

9.4 How Can China "Green" Its Soft Commodity Value Chains?

China has an unprecedented opportunity to play a catalytic role in greening the world's soft commodity value chains. Doing so would support the country's own development, business, and diplomatic objectives, as well as make a significant contribution to shared global biodiversity, climate, and SDGs. China can achieve this using three broad strategies. First, establish an ambitious and comprehensive strategy and supporting institutional arrangements at the highest level of government regarding green value chains. Second, adopt policies that require, encourage, or support companies supplying soft commodities to China to progressively green their value chains. Third, incorporate these policies in broader Chinese priorities and initiatives on trade, South–South cooperation, and green finance.

9.4.1 Establish a National Green Value Chain Strategy and Provide Policy and Institutional Support

Three steps to establish an ambitious and comprehensive national policy that is supported institutionally include:

A. Announce a new Chinese policy initiative on greening soft commodity value chains.
B. Establish an Inter-Ministerial National Committee on Value Chain Security and Sustainability.
C. Establish a Global Green Value Chain Institute.

9.4.1.1 Announce a New Chinese Policy Initiative on Greening Soft Commodity Value Chains

At both CBD COP-15 and the Shanghai Expo—two high-profile international events that China will host in 2021 focusing on environment and trade, respectively—China can launch a new initiative signalling a move towards green value chains for key soft commodities for which China is a major importer and which have a significant impact on natural ecosystems. At COP-15, this initiative could be included in the package of Chinese deliverables on biodiversity conservation—at home and abroad—that China announces. The Shanghai Expo could help raise greater awareness of green value chains among importers and exporters worldwide and hence facilitate the implementation of the new policies and initiatives.

China could leverage this national commitment to encourage others to join a multilateral commitment on greening soft commodity value chains. This commitment could be part of a "Kunming Declaration" or another outcome from the High-Level Segment that China is likely to host back-to-back with COP-15. Engaging in this issue with other major economies is important. While China may seek to demonstrate global leadership, the impact will be greater if other major importers are included. As the host of COP-15 and one of the world's largest players in soft commodity markets, China has the power to make this happen.

9.4.1.2   Establish an Inter-ministerial National Committee on Value Chain Security and Sustainability

In order to take this policy commitment forward, China could announce its intent to form a long-term Inter-Ministerial Committee focusing on value chain security and sustainability (Tentative name: National Committee on Value Chain Security and Sustainability). This Committee would address both soft and hard commodities. It could begin by following up on the recommendations of this Special Policy Study on soft commodities and gradually expand to cover other commodity value chains.

An Inter-Ministerial Committee is desirable because commodity value chains cross the jurisdictional and expertise boundaries of ministries. Trade, finance, agriculture, forestry, customs, and environment are all involved to some degree with commodity trade and thus should be represented. The severe disruptions to global value chains brought on by the COVID-19 pandemic further highlight the need for a comprehensive and unified response. Turning political commitment on green value chains into action therefore requires a "whole of government" approach. An Inter-Ministerial Committee could provide this, facilitating cross-sectoral cooperation and releasing policy guidance on the design and implementation of value chain security and sustainability initiatives in China.

The National Committee on Value Chain Security and Sustainability could be led by the Ministry of Ecology and Environment and jointly coordinated with the Ministry of Foreign Affairs, the Ministry of Commerce, China International Development Cooperation Agency, the General Administration of Customs, the Ministry of Agriculture and Rural Affairs, the State Forestry and Grassland Administration, and China Banking and Insurance Regulatory Commission. Depending on the progress of the work, other relevant government agencies could participate.

The main responsibilities of the Committee would include: (i) studying and approving proposed value chain security options and initiatives; (ii) reviewing and formulating proposed policy measures related to value chain security; (iii) coordinating and establishing a cooperation mechanism on value chain security and sustainable development; (iv) assessing and resolving problems in value chain security and sustainable development; and (v) periodically reviewing progress in improving value chain security and linking it to sustainable development objectives.

The Committee would be supported by the work of a new Global Green Value Chain Institute, as discussed below.

9.4.1.3   Establish a Global Green Value Chain Institute

In order to provide the best technical and policy advice to the Inter-Ministerial Committee, China could announce at CBD COP-15 that it will establish a Global Green Value Chain Institute. The institute would engage experts and stakeholders (e.g., governments, companies, financial institutions, research institutions, and civil society organizations) to develop more detailed commodity-specific plans on *what* China and other major economies can do to "green" their global value chains, *how* to do it (including pilot applications), and *who* needs to do what. The institute would initially emphasize soft commodities since they are most relevant to biodiversity and to the CBD. But the institute would address hard commodities as well.

Because the issues of greening value chains involve both environmental and trade issues, it will be crucial to the success of the proposed institute that it be jointly anchored in the Ministry of Environment and Ecology and the Ministry of Commerce. The institute could be either a new organization or a part of the recently established BRI Green Development Institute. Either way, the institute would be responsible to the Inter-Ministerial Committee, befitting its comprehensive cross-sectoral mandate. The institute would be a first of its kind and enable China to develop a centre of excellence on how to achieve legal, secure, and sustainable global value chains, an issue of increasing importance and interest to governments worldwide.

This institute would inform and support policy development and implementation by: (i) conducting scientific research to develop implementation plans of green value chains by type of commodity and sector; (ii) analyzing relevant policy pathways and institutions to determine which parts of government and industry need to be involved in order to achieve particular policy outcomes; (iii) developing guidance to ensure legality standards and requirements for import and export of raw materials and products are met; (iv) supporting sustainable production in producer countries through trade, finance, and development assistance; (v) building a collaborative network of stakeholders and information sharing and communication platform to encourage participation from relevant stakeholders, including industry, enterprises, and civil society, including those working on social and gender-related issues; and (vi) coordinating with international platforms such as the BRI and APEC to create synergies and exchange good practices on green value chains.

To begin, the institute could focus on the three soft commodity-focused policy measures outlined in Sect. 9.4.2 (below), including *strengthening measures to reduce the import of soft commodities from illegal sources; strengthening commodity due diligence and traceability systems; and investing in domestic capacity to rationalize food value chains and improve sustainable diets.* In addition, the institute could work to build soft commodity considerations into broader ongoing Chinese policy arenas, including *trade agreements, South–South cooperation, green finance, and green BRI,* etc. (see Sect. 9.4.3 below).

These six proposed initial areas of work for the institute emerged as high priorities for action in the course of preparing this Special Policy Study, based on their potential *effectiveness* in catalyzing sustainable soft commodity production and their potential *feasibility* for uptake (or relevance) in the Chinese context.

9.4.2 Adopt Mandatory and Voluntary Measures to Green Soft Commodity Value Chains

China should adopt a mix of regulatory and market-based approaches to drive progress towards green soft commodity value chains. This should include measures to achieve three critical outcomes:

A. Strengthen measures to reduce the import of soft commodities from illegal sources.
B. Strengthen due diligence and traceability systems.
C. Invest in domestic capacity to rationalize food value chains and improve sustainable diets.

   In pursuing these policies (especially A and B), China should seek to harmonize its "greening" standards with those of other leading countries.

9.4.2.1 Strengthen Measures to Reduce the Import of Soft Commodities from Illegal Sources

What is it?

China could strengthen its import management of the legality of soft commodities, building on the latest revision of the Forest Law and comparable legality standards in other major markets. Strengthening measures to reduce the import of soft commodities from illegal sources would strongly support the efforts of governments in producing countries aiming to discourage illegal production and the trade of soft commodities. Illustrative examples of illegality include palm oil grown on land where forests were cleared without a permit in Indonesia and soybeans grown on a farm that has cleared more forest than is allowed by the Brazilian Forest Code.

   Mechanisms for implementing such measures would need to be developed in close cooperation with relevant producer countries and might need to be tailored to the specifics of individual commodity value chains. This would normally require the producer country to develop legality standards and verification systems for the production of soft commodities that could be recognized under the Chinese measures.

   Chinese policy would encourage—and eventually require—companies importing soft commodities to China to exercise due diligence to ensure the commodity was produced legally in the source country. A range of incentives could be employed to motivate non-responsive companies to act, ranging from mere warnings at the outset up to civil and criminal penalties when a binding regulatory framework will eventually be developed. This would send a strong signal to foreign exporters that they need to ensure that the soft commodities they ship to China have been produced in accordance with the laws of the country where the commodities originated.

   Given the size and complexity of China's soft commodity imports, measures to ensure the legality of imports would need to be designed and implemented following a clearly articulated and phased approach (e.g., by commodity, by country) to allow

Chinese importers and foreign exporters to review and adjust their sourcing practices in a manner that prevents supply disruptions, and to harmonize Chinese policy with those of producer country governments.

Who needs to act?

Chinese agencies would need to cooperate with counterpart agencies in each relevant producer country to clarify what qualifies as the legal production of specific soft commodities and coordinate with any relevant producer country standards and systems that are in place to verify legality. Within China, multiple ministries would need to collaborate to set due diligence requirements to be applied by companies to verify the legality of imported soft commodities and to define penalties and consequences for importers that fail to exercise due diligence. These ministries would likely include the Ministry of Ecology and Environment, the Ministry of Commerce, the Ministry of Agriculture and Rural Affairs, the National Forestry and Grassland Administration, the State Administration for Market Regulation, the General Administration of Customs, and the Ministry of Foreign Affairs. Working with technical experts, these ministries could also provide tools and training to private sector actors on how to meet the new legality obligations. Given the need for overall policy coordination, the role of the proposed Inter-Ministerial Committee would be critical, supported by the technical work of the proposed institute.

How does this build on existing Chinese efforts?

China is already taking steps toward legal import standards for soft commodities. For example, China has developed a draft national timber legality verification framework and piloted voluntary legality verification standards among a few timber companies in recent years. In December 2019, the Standing Committee of the National People's Congress adopted a revised Forest Law that includes legality requirements for timber product value chains. Article 65 of the revised Forest Law stipulates that "timber trading and processing companies shall establish ledgers to record input and output of raw materials and products. It is forbidden for any organization and individual to purchase, process and transport timber from illegal sources such as knowingly unlawful or wanton." Building on these first steps, China could consider expanding legality due diligence and verification requirements to the import of major soft commodities beyond timber, namely soybeans, palm oil, and beef.

Why is it important?

Ensuring the legality of imported soft commodities is a fundamental feature of green soft commodity value chains for three principal reasons. First, it demonstrates respect for the laws of producer countries and thereby contributes to strong and stable trade and political relationships. Second, it levels the economic playing field for Chinese importers who are obeying the law but who are undercut by cheap, illegal imported products. Third, it provides a clear demonstration of China's support for international norms and agreements on global environmental sustainability and cooperation.

Where is this emerging as the new global norm?

If China were to establish and implement due diligence requirements with respect to the legality of soft commodity imports, it would be joining a growing list of countries ushering in a new era of legal trade. For example, the EU, the United States, Japan, Australia, and South Korea have implemented legality regulations on timber in recent years, and more countries are in the process of developing similar measures. The EU, for instance, requires importers to conduct due diligence to assess and mitigate the risk of illegal timber products entering the EU market. In 2017, South Korea amended its Act on the Sustainable Use of Timbers to regulate the legality of imported and domestically produced timber and timber products. In 2019, the European Commission issued a major communication on *Stepping up EU Action to Protect and Restore the World's Forests*. This policy commits the EU to "promot[ing] trade agreements that include provisions on the conservation and sustainable management of forests and further encourage trade of agricultural and forest-based products not causing deforestation or forest degradation" [56]. As of mid-2020, the U.S. Congress was considering analogous legislative measures.

9.4.2.2 Strengthen Due Diligence and Traceability Systems

What is it?

The Chinese government could encourage companies (both state-owned and non-state-owned enterprises) to strengthen due diligence and traceability systems to achieve greener soft commodity value chains. "Due diligence" is a risk management process implemented by a company to identify, prevent, mitigate, and account for how it addresses environmental and social risks and impacts in its operations, supply chains, and investments. Traceability is the ability to follow a product or its components through the stages of the supply chain (e.g., point of production, processing, manufacturing, and distribution).

A diverse array of tools and approaches (e.g., risk assessment, certification, remote sensing, supplier warranties and reporting, computerized product tracking, blockchain technology) are already available to support due diligence and traceability. The Accountability Framework Initiative guidance on supply chain assessment and traceability [57] can help with the selection of approaches calibrated to the risk associated with a given commodity sourced from a particular region. Very importantly, *a properly designed due diligence and traceability system can reduce costs and facilitate the adequate supply of green soft commodities.* COFCO International's pioneering approach (Box 9.8) provides an example of a potentially promising stepwise risk-based approach. Due diligence and traceability systems can be developed and applied voluntarily or be built into government regulatory controls.

The Chinese government could encourage companies to green their soft commodity value chains through regulations that set standards of due diligence and traceability (including within the proposal discussed in the section above to regulate against the import of illegally-produced commodities). Such regulations could help create a level playing field such that companies that comply with the regulations are not put at a competitive disadvantage against companies trading in cheaper, illegal products.

Box 9.8. A Risk-Based Approach to Due Diligence and Traceability That Reduces Costs
COFCO International and others are exploring the following approach (and variations thereof) when conducting due diligence and traceability of "green" soybean value chains originating in high deforestation-risk areas in Brazil:

- **Take it step-by-step.** Don't try to pursue all aspects of sustainability at once. Rather, start by focusing on a few of the most important and timely issues. Currently, securing "deforestation-free" (and avoiding "deforestation-linked") soft commodities is one of those issues. Over time, pursue additional sustainability issues at a well-managed pace.
- **Use a "cut-off" date for "deforestation free."** Agree not to source a commodity that is linked to deforestation after a "cut-off" date. Agricultural products grown/raised on a tract of land are not considered "green" or "sustainable" if a forest was cleared on that tract of land to make way for the commodity after the cut-off date. These dates can be set via a multi-stakeholder process and can cover a biome or smaller region. For example, the cut-off date for soy in the Amazon biome is 2008.
- **Request supplier boundaries.** Ask suppliers to provide the boundaries of their farms/ranches, or of the jurisdictions (e.g., municipality, district) from which they source.
- **Leverage satellite imagery.** Access historical satellite imagery of the supply location during the year of the cut-off date. At the same time, access recent satellite imagery of the supply location. Compare the imagery. If forests were not there in the cut-off date year, then the commodity grown/raised on that location is deforestation free. If forests were there, then the commodity is not deforestation free. Continue to use recent imagery to monitor the adherence of suppliers to the deforestation-free objective. Today, much of the satellite imagery needed to do this analysis is freely available.
- **Engage suppliers.** Besides informing suppliers of this due diligence and traceability system, work with them to ensure they implement practices that avoid deforestation. One important component is to offer technical and/or financial assistance to boost crop and/or livestock yields on their existing farmland and grazing land. This engagement can be facilitated by involving other supply chain actors, financial institutions, and non-governmental organizations.

This approach is low cost. The necessary data are freely available. The analysis can be done from one's office, and it does not require someone going to a farm/ranch to do an on-site audit or verification. Combining this with a "mass balance" approach further keeps costs low when transporting the commodity to its destination. With a "mass balance" approach, the deforestation-free commodity can be mixed with non-deforestation-free commodities during processing and transport, but the volumes are tracked via ledger or blockchain (as opposed to keeping the tons of deforestation-free commodity physically segregated from the tons of non-deforestation-free commodity).

Who needs to act?

Due diligence and traceability measures require action by companies at all points within a commodity value chain. Pre-competitive, sector-wide collaboration to set standards and harmonize approaches can reduce inefficiencies that would otherwise result if each company developed its own unique approach.

For China to introduce regulations on due diligence and traceability for soft commodities, the Ministry of Ecology and Environment would need to coordinate with the Ministry of Commerce, the State Administration for Market Regulation, the General Administration of Customs, the Ministry of Agriculture and Rural Affairs, the National Forestry and Grassland Administration, and the Ministry of Industry and Information. The Government of China could use or build upon a suite of rapidly improving approaches, technologies, and systems that already exist (Box 9.9). It will be particularly important to engage with business enterprises when further developing and implementing due diligence and traceability systems in order to ensure the approaches used fit business processes and are cost-effective.

How does this build on existing Chinese efforts?

Box 9.9. Examples of Approaches and Tools to Support Due Diligence

A diverse array of existing tools and approaches can support companies to exercise due diligence and comply with related regulations. These include:

- Free online forest monitoring systems that enable companies and regulatory agencies to access publicly available satellite and related data and assess which regions have ongoing deforestation. Companies can overlay this geospatial data with their suppliers' sourcing areas to monitor deforestation and other risks that directly impact their own value chains.

- Voluntary certification systems, based on a sustainability standard governed by a multistakeholder body, that offer third-party verification that commodities were produced in compliance with the standard and that the chain of custody is adequately controlled.

- Mandatory producer-country certification systems that monitor and enforce compliance with regulatory standards for sustainable production and trade of soft commodities.

- "Jurisdictional approaches" wherein an entire geographic or political region (e.g., a state, province, district, municipality) takes action to ensure soft commodities are produced legally and that targets for reduced deforestation or conversion of other ecosystems are met. Some jurisdictions have "produce and protect compacts" whereby farmers agree to avoid expansion into forests in exchange for assistance to improve yields on existing farmland.
- "Risk screening" approaches where retailers, manufacturers, and traders distinguish regions or companies deemed "lower risk" (e.g., a high degree of confidence in legality and no deforestation or human rights breaches) from those deemed "higher risk." Importers can prioritize the use of stricter control measures on higher-risk sources. These include legality verification, certification, greater traceability, and supplier engagement with continued purchasing contingent on suppliers making progress towards full compliance with sustainability standards. Here again, monitoring of land-use change impacts can come from free, publicly available satellite and related data.

China already has elements of due diligence and traceability in its regulations on products such as timber, food, and drugs. China also is at the forefront of digital technologies such as big data and blockchain, which can facilitate the traceability of commodity value chains and can build upon this technological leadership. The Chinese Academy of Forestry has developed a draft national timber legality verification system that would institute measures for verification of legal compliance, from forest management all the way through the value chain ("chain of custody"). This system has been piloted with a few large timber companies in recent years. China could build on this experience, expanding to small- and medium-sized enterprises, imported timber, and other soft commodities. In addition, the Ministry of Commerce has established a National Important Products Traceability System to track the production and distribution of key products such as food, drugs, rare earth minerals, and dangerous products. As a start, the ministry could add soft commodities to this system. Moreover, the proposed BRI Big Data Platform could save and provide data that feeds into traceability efforts.

Why is it important?

Due diligence and traceability are fundamental to green value chains. This is because they enable importers, financiers, the government, and consumers to distinguish those tons or shipments of soft commodities that meet "green" criteria from those that do not. When used in combination, due diligence and traceability can verify a commodity's source location, the chain of custody, and compliance with legality, sustainability, and/or safety standards. They often also make good business sense, enabling companies to better manage logistics and ensure financial discipline throughout the value chain, as well as providing a competitive advantage to companies that can demonstrate they are procuring commodities from known and sustainable sources.

Where is this emerging as a new global norm?

An increasing number of multinational companies are using due diligence and traceability systems to achieve greener soft commodity value chains. For instance, companies such as Cargill, Golden Agri Resources, Louis Dreyfuss, Mondelez, and Walmart use "Global Forest Watch Pro" to monitor their soft commodity supply chains—starting at the farm—to distinguish green from non-green supplies. Food giants like Mars, Unilever, and Wilmar use the Palm Risk Tool to identify sources of palm oil that are at "high risk" of being unsustainably grown. COFCO International now tracks its soybean supply chains from several Brazilian sources. In response to China's Food Safety Law of 2015, Chinese beef processor Kerchin has deployed blockchain and other traceability technologies to track the production and shipping of frozen beef—thereby avoiding the risk of contaminated meat entering its supply chain.

Governments are introducing traceability systems, too. For example, Indonesia uses bar codes to track timber from harvest to port and subsequently grants export permits through an online system. New Zealand and Uruguay have developed national traceability systems for cattle to ensure meat quality, sanitary standards, the transparency of origin, and chain of custody.

9.4.2.3 Invest in Domestic Capacity to Rationalize Food Value Chains and Improve Sustainable Diets

What is it?

China could invest in the technology and manufacturing capacity to produce nutritious, plant-based foods that meet growing domestic (and international) demand for protein. By becoming a plant-based protein manufacturing "powerhouse," China could increase food self-sufficiency, improve citizen health (e.g., lower saturated fat and cholesterol levels in domestic diets), increase food safety (e.g., less risk of contamination), and reduce the risk of zoonotic diseases. The resulting value chain would be less reliant on imports (which is better for stability and for trade balances) and "greener" (e.g., no deforestation and low greenhouse gas emissions). In addition, this investment would create an entirely new 21st-century industry where China could attain global market leadership.

Who needs to act?

Action would be needed by both the public and private sectors. For example, the Ministry of Agricultural and Rural Affairs could coordinate with the State Administration for Market Regulation (as well as the Ministry of Industry and Information) to ensure an adequate domestic supply of raw material and to set any needed national standards for plant-based foods. Public and private investment into food science research and innovation are needed to accelerate plant-based protein product development.

How does this build on existing Chinese efforts?

China's large investment in agricultural technology and land infrastructure makes it well-positioned to meet the demand for plant-based meat ingredients. As one of the world's largest producers of pulses and exporters of plant-based raw materials, China has already developed processing infrastructure that can support this new industry. In 2016, for instance, China had the capacity to process over three quarters of global soy protein isolate and half of textured soy protein. Soybean is currently the most utilized raw material to manufacture plant-based meat products in China. Other candidate raw materials grown in China include konjac, soybean, and fungi.

Chinese investors are already devoting financial resources to startups to advance technological development and begin to scale these products. For example, Bits × Bites—China's first venture capital firm devoted to food tech—has invested in several plant-based startups around the world. Some Chinese plant-based companies, such as Whole Perfect Food and Godly, are starting to become well recognized.

Investing in this growing set of opportunities also fits with China's broader efforts to ensure food security. Amid the COVID-19 pandemic, China has prioritized food security and a stable supply of agricultural products in its efforts to maintain supply chain security and competitiveness. In 2020, China will develop a new national medium-to-long-term food security plan and carry out a response plan to ensure food security under COVID-19 [58]. The NDRC has highlighted diversifying the import of major agricultural products and securing a stable and safe supply of key products, including grain, edible oil, meat, eggs, fruits, and vegetables in its annual draft report to the People's Congress in May 2020 [58].

Why is it important?

Building plant-based protein production capacity would increase Chinese food self-sufficiency, reduce reliance on imported meat and animal feed, improve trade balances, shorten supply chains, increase food safety, reduce risks of zoonotic diseases, reduce environmental impacts of food production, and put China at the forefront of innovation and new markets.

It is particularly relevant in the current food supply chain context. While Chinese protein demand continues to grow, protein supply faces a number of constraints. Notably, African swine fever nearly halved China's hog herd in 2019 alone. While harmless to humans, the fever is deadly to pigs and has resulted in widespread meat shortages and price spikes. Poultry and beef have not been able to make up for the loss in pork production. Rabobank predicts that, in 2020, pork production will continue to decline 10–15% from 2019 levels [59]. Moreover, the trade implications of COVID-19 on the supply of meat (e.g., imported beef) and animal feed are still unclear and might be felt for a long time.

Domestically produced plant-based proteins could make up the supply shortfall and could make supply more secure (since it would be domestic). Likewise, it would avoid bacteria and other contamination issues often associated with conventional food supply chains. And if the plant-based proteins can be made to "look like and taste like" meat (as they increasingly are), then it would meet consumer interest in the flavour of meat.

Where is this emerging as the new global norm?

The plant-based meat industry has rapidly grown over just the past few years. According to the Good Food Institute and the Plant-Based Foods Association, plant-based meat sales in the United States alone have grown 37% from 2017 to 2019. From being negligible just 2–3 years ago, the global plant-based protein market was estimated to be valued at $18.5 billion in 2019 and is likely to reach $40.6 billion by 2025, a compound annual growth rate of 14% [60]. New companies with high market capitalization, such as Impossible Foods and Beyond Meat, have emerged on the market across several continents. Major dine-in restaurants and fast-food retailers are now selling plant-based meat entrées to customers. The moment is ripe for China to have domestic manufacturers supplying its domestic market and then expanding into overseas sales.

9.4.3 Build on Existing Chinese Policy Levers and Initiatives

China can also make rapid progress in greening its soft commodity value chains by building on existing levers and initiatives China has to influence trade, development assistance, and finance. These avenues include:

A. Incorporating green value chain measures into trade agreements.
B. Increasing Chinese South–South development assistance to support green soft commodity value chains.
C. Integrating finance for green soft commodity value chains with green finance and the work of the BRI International Green Development Coalition.

Identifying and accelerating synergies with these existing policy and economic levers would increase the pace and efficiency at which measures get implemented. It also would ensure sufficient engagement with the private sector and across relevant government ministries.

9.4.3.1 Incorporate Green Value Chain Measures into Trade Agreements

What is it?

The Chinese government could incorporate measures to green soft commodity imports in bilateral and multilateral trade agreements.

When it comes to bilateral agreements, for example, China and Indonesia could enter into an agreement wherein Indonesia ensures that all the palm oil exported to China is "deforestation free" and sustainable. In return, China could provide trade incentives such as a "fast lane" at the port of entry or tariff benefits for palm oil that is "deforestation free." A similar agreement could be struck with Brazil on soy and beef.

China could also promote measures to green soft commodity value chains in multilateral trade agreements—in compliance with WTO rules. For example, China could lead the work at WTO on creating tariff benefits for green soft commodity trade. China also could support establishing coordinated sustainability standards for soft commodity production and trade at APEC. Such efforts could be piloted for one or two commodities with existing criteria and mechanisms, such as CITES-listed timber species in the Regional Comprehensive Economic Partnership (RCEP), of which China is one of the key members. Such pilot projects could be communicated to the Committee on Trade and Environment (CTE) at the WTO to ensure that they are developed in compliance with WTO rules. Given the complexity of international trade rules and their importance to China, it will be particularly important—in the work of the proposed Global Green Value Chain Institute— to prioritize research on harmonizing Chinese measures on greening global value chains with the WTO and other international trade rules and regulations.

Such trade agreements could be rolled out in a phased approach, perhaps by piloting in one country/commodity combination first and then moving on to new country/commodity combinations. These trade agreements could also be coordinated with China's South–South cooperation strategy and incorporate capacity-building activities for producer countries (see Sect. 9.4.3.2).

Who needs to act?

Green value chain trade measures would first need to be agreed by the parties to multilateral or bilateral trade agreements. Once agreed, multiple ministries in China would need to collaborate to design implementation mechanisms, coordinate with their counterparts in producer countries, and develop capacity-building activities via South–South cooperation. These ministries include the Ministry of Commerce, the Ministry of Foreign Affairs, the Ministry of Ecology and Environment, the Ministry of Agriculture and Rural Affairs, the National Forestry and Grassland Administration, and the General Administration of Customs. These ministries could establish a coordination mechanism with their counterparts in producing countries to facilitate regular communication, determine verification standards, and identify gaps for technical support on green value chains. The Ministry of Commerce and the Ministry of Ecology and Environment ideally would take the lead in developing proposals to provide trade benefits for "deforestation-free" soft commodities and coordinate standards for legal/sustainable soft commodity production and trade in multilateral trade agreements.

Why is it important?

Incorporating green soft commodity value chain measures into bilateral and multilateral trade agreements is beneficial to China, its trading partners, and overall sustainability. Such trade arrangements would help China realize the benefits articulated in Chap. 3. For trading partners, it would stimulate action and help fulfill their own self-determined national laws and regulations (see Box 9.4). For sustainability, it would align trade agreements with the healthy management of natural resources.

Where is this emerging as a global norm?

Producer countries do not wish to have unilateral "green" measures imposed on them by another country, and China would not seek to do so. The production and trade in soft commodities should only be "greened" by mutual agreements between producer and consumer countries acting as sovereign nations, and in their own respective economic self-interest. This is exactly what China can do, and there are precedents for doing so. For instance, the EU has jointly entered into Voluntary Partnership Agreements with timber exporter countries such as Ghana, Guyana, Indonesia, and Vietnam.

"Stand-alone" or dedicated green commodity agreements are not the only possible model. It is common practice in bilateral and multilateral trade agreements to build in environmental objectives and safeguards. The 2009 Peru-United States Trade Promotion Act (PTPA), for example, is a general trade agreement that liberalizes trade terms across a broad range of economic sectors. However, the PTPA includes a binding Forest Governance Annex that has strong provisions and specific mechanisms to reduce the risk of illegally logged timber from Peru entering the United States.

Multilateral trade agreements also are considering environmental objectives. Forty-six WTO members, including China, have been actively engaged in the WTO Environmental Goods Agreement (EGA) negotiations, which are aimed at eliminating tariffs for environmental products, such as wind turbines and solar water heaters, to help achieve climate goals and the SDGs [61]. The CITES has established formal cooperative arrangements with both the WTO and the International Tropical Timber Organization (ITTO) to further conservation objectives through international trade rules and processes.

China has also been a leader in supporting international cooperation on green value chains through the APEC Cooperation Network on Green Supply Chain (GSCNET), which was formed by the APEC leaders in 2014 [62]. China is host to the first Pilot Center for GSCNET, in Tianjin.

9.4.3.2   Increase Chinese South–South Development Assistance to Support Green Soft Commodity Value Chains

What is it?

China could develop specific lines of bilateral development assistance that support sustainable soft commodity production in countries that supply China. This assistance would include grants, interest-free loans, concessional loans, and technical assistance for practices and technologies that boost commodity yields on existing agricultural land (linked to avoided deforestation), improve traceability, and improve policy design. These assistance programs could also incorporate a gender policy to address issues regarding gender equity and access to resources (Box 9.10). In the spirit of Chinese-led South–South cooperation, this increased development assistance could be combined with the other measures described earlier to make those measures more politically acceptable and to facilitate their implementation.

Who needs to act?

The China International Development Cooperation Agency (CIDCA), in coordination with the Ministry of Ecological Environment, other ministries, and Chinese think tanks, could engage partner countries in South-South cooperation to establish relevant development assistance programs.

How does this build on existing Chinese efforts?

China's development assistance already includes multiple forms of aid to help other countries raise their agricultural productivity. China could increase and/or deliberately target this assistance to countries where improving the productivity of existing farms, ranches, and/or plantations is critical for preventing further deforestation, improving smallholder incomes, improving gender equality, and strengthening sustainability.

China already provides aid to selected countries for developing land-use plans. China could build on this to support major soft commodity producer countries to delineate land suitable for green commodity production and land not eligible for commodity production (e.g., natural forests, peatlands, wetlands). For the latter, China could offer relevant expertise from its own experience implementing the Ecological Redline Policy.

China can also build on the BRI International Green Development Coalition (BRIGC), launched in 2019, to bring together the environmental expertise of all partners to ensure that the BRI brings long-term green and sustainable development to all concerned countries in support of the UN 2030 Agenda for Sustainable Development.

China also already offers trade-related aid and could potentially expand its scope, in compliance with WTO rules, to cover soft commodities produced from countries that seek to meet "green" standards. For example, China provides commodity inspection and other trade-related equipment to developing countries. It could build on this to help strengthen systems that enable developing countries to ensure their commodities are being produced legally and can be reliably traced back to their source. China could reinforce such assistance by offering trade benefits to verified green commodities, such as differentiated tariffs and quotas, removal of non-tariff measures, and mutual recognition of inspection and quarantine systems.

Why is it important?

The transition to green soft commodity production and trade entails doing things differently. It means farmers boosting yields on existing agricultural land while avoiding deforestation. It means implementing more efficient use of inputs and more fair labour practices. It means being able to demonstrate that products at the import border are legal and sustainable. It means providing equal opportunities and fair compensation to empower women in this sector. Doing such things differently requires technical and financial assistance. Via South–South cooperation, China can provide that assistance. In doing so, China helps the sustainable *supply* of the commodity, complementing other measures noted above (Sects. 9.4.2.1 and 9.4.2.2) that help with the *demand* for the commodity.

Box 9.10. Promoting Gender Equality and Women's Empowerment in Global Green Value Chains and International Trade

Gender equality plays an important role in the transition to green soft commodity production and trade. China's development assistance to support sustainable soft commodity production should also take into account gender-related issues.

Most major donors, especially those funding development work, have instituted a gender equality policy or gender action plan and the mechanisms for its implementation. Irish Aid's Gender Equality Policy 2004 lays out two strategic pathways of implementation: (1) mainstreaming and (2) direct support to women's empowerment programs. In its gender strategy, the UK's Department for International Development (DfID) calls for a focus on gender equality to achieve SDG 5 but also to seek it throughout the 17 goals.

These donor policies extend across sectors and can be key to ensuring gender equality in green soft commodity value chains. For instance, the United States Agency for International Development (USAID) developed a handbook [63] for promoting gender-equitable opportunities in agricultural value chains, providing a five-step methodology and illustrative case studies from various agricultural industries. Similarly, to promote gender equality in participation and decision-making within Paraguay's soy and beef sectors, the United Nations Development Programme (UNDP) Green Commodities Programme facilitated a national dialogue between key actors. Its long term aim is to include women in all economic activities and build their capacities to that end [64].

All such policies call for a focus on women's economic empowerment, as well as on women's health, education, and social well-being. The Gender Practitioner Collaborative—a consortium of gender experts representing development and aid organizations—created the Minimum Standards for Mainstreaming Gender Equality. These standards outline benchmarks and foundational steps for development agencies, including developing an organizational culture that promotes gender equality and building staff capacity and budget to support partners in gender mainstreaming.

These policies also institute reporting and tracking mechanisms to ensure their adherence. Global Affairs Canada's and the Swedish International Development Agency's (SIDA) gender equality policy states that conducting a gender analysis is required for all policies, programs, and projects. The EU's gender action plan mandates that all EU actors (European External Action Service, Delegations, Commission services, and Member States) submit reports annually on the progress of mainstreaming gender into their planned activities, including shifting institutional cultures [65].

China's CIDCA could reference these established policies, as well as key lessons learned from its own experiences, to develop and strengthen its gender policy and gender action plans within South-South cooperation assistance programs.

9.4.3.3 Integrate Finance for Green Soft Commodity Value Chains with Green Finance and the Work of the BRI International Green Development Coalition

What is it?

Agencies that regulate the finance sector in China can encourage China's financial institutions to boost forms of finance that support companies to green their value chains. These include:

- *Innovations in trade finance to producers, manufacturers, and traders that meet green performance standards.* These deals offer low interest rates and/or fast-tracked payment of invoices as incentives for borrowers to achieve sustainability and traceability targets. For example, in 2019, COFCO International secured a sustainability-linked loan of US $2.1 billion from a consortium of 20 banks, including Chinese banks, with interest savings based on its performance against environmental, social, and governance targets, including the sustainable sourcing of soy in Brazil [66]. The WTO estimates that between 80 and 90% of global trade is reliant on trade finance [67].

- *Safeguards on investments in infrastructure projects and commodity processing facilities to ensure that these do not directly or indirectly encourage the unsustainable production of soft commodities.* For example, over 100 financial institutions have adopted the Equator Principles to determine, assess, and manage environmental and social risks in projects.
- *Grants and loans to producer countries (see the section above on South–South cooperation) to support the transition to greener production systems and related monitoring, reporting, and verification of progress.*

Financial institutions can set institution-wide sustainability policies for soft commodities and integrate these into their due diligence procedures for credit or asset investments. The finance within the scope of such policies can be conditioned upon, or create incentives for, compliance with environmental and social standards around land acquisition and labour conditions in agriculture and forestry. It can also create incentives for more rigorous monitoring, reporting, verification, and disclosure of risk and control measures relating to legal compliance or sustainability within soft commodity value chains.

Who needs to act?

This will vary with the form of finance, but it includes all financial institutions involved in the following activities and the agencies that regulate them:

- Financial services to entities involved in the production or procurement of soft commodities destined for China.
- Investing or facilitating investments in new or expanded soft commodity production areas or processing facilities.
- Investing or facilitating investments in infrastructure projects that impact ecosystems directly or indirectly facilitates the expansion of forestry or farming by improving access to once remote areas.
- Loans or grants to producer countries as part of South–South cooperation.

The China Banking and Insurance Regulatory Commission (CBIRC) has a broad mandate to regulate the conduct of financial institutions and could play a key role in promoting the adoption of best practices by Chinese banks to help green soft commodity value chains. This could include collaboration with leading banks (e.g., China Development Bank, Export–Import Bank of China, Bank of China, Industrial and Commercial Bank of China, and Agricultural Bank of China) to pilot innovative financial instruments or develop specific guidelines on how to implement best practices in the Chinese context. The Ministry of Ecology and Environment and CBIRC could develop a green finance pilot program with major policy banks to support soft commodity trade.

CBIRC, together with the Ministry of Ecology and Environment and the Ministry of Commerce, could lead a green transformation of financing institutions. Actions could include updating its *Green Credit Guidelines*, accelerating the implementation and performance evaluation of the *Guidelines for Establishing a Green Financial System*, strengthening the environmental risk management and sustainability of financial institutions' investment and financing projects, and incorporating the concept of green soft commodity value chains into the design and development of relevant policies and mechanisms for the green transformation.

Multilateral development banks, including the Asian Infrastructure Investment Bank, can launch supply chain finance programs and partner with other financial institutions to enhance small and medium-sized enterprises' access to funding. For instance, the International Finance Corporation and the Asian Development Bank use their capital to enable and de-risk potential investments.

How does this build on existing Chinese efforts?

Promotion of the uptake of greener financing mechanisms conforms broadly with the current reform and innovation focus of CBIRC. The CBIRC has introduced the *Guidance on Promoting High-quality Development of the Banking and Insurance Industry*. It also plans to strengthen data disclosure requirements and revise both the *Green Credits Guidelines* and *Green Credit Statistical System*, which are recognized by the People's Bank of China for the purposes of its Macro Prudential Assessment. The revision could extend the scope of green credits to include entities that supply or procure soft commodities that meet relevant sustainability and traceability requirements.

Green value chain finance measures can also build on the work of the BRI International Green Development Coalition and the study currently underway on how a "traffic light system" could evaluate the environmental performance of BRI investments. Such a system could support the greening of value chains if it applied to projects involving (a) the production and processing of soft commodities or (b) infrastructure and other projects that could indirectly encourage the expansion of soft commodity production. The system could include safeguards to limit the negative environmental and social impacts of soft commodity production while enabling the uptake of sustainable forestry and farming practices.

Why is it important?

Financial Institutions can play a critical role in creating incentives for sustainable soft commodity production and trade, building on and accelerating the efforts of their corporate clients within soft commodity value chains and of producer countries to achieve the same goals. Through green financing, banks can manage risks associated with unsustainable practices (e.g., compliance, social and market risks to clients, default and reputational risks to banks), generate or seize commercial opportunities, and increase their positive contribution to society. This could also help ensure that companies are receiving the same message from their financiers and investors that they are hearing from the government as to the importance of greening their value chains.

Action from CBRIC and others is needed to coordinate efforts at a sector level and correct system-wide challenges. Without this, individual banks may be slow to implement green reforms due to pressures to pursue short-term profits or out-compete their peers.

Financial institutions are increasingly paying attention to deforestation risk. In 2019, China Asset Management joined 230 institutional investors, representing over USD $16.2 trillion in assets under management, in a landmark statement calling for companies to (a) publicly disclose and implement a commodity specific no-deforestation policy with quantifiable, time-bound commitments covering the entire value chain and all sourcing geographies; (b) establish a transparent monitoring and verification system for supplier compliance with the company's no-deforestation policy, and (c) report annually on deforestation risk exposure and management, including progress towards the company's no-deforestation policy [68].

9.5 Summary of Policy Recommendations

9.5.1 Establish a National Green Value Chains Strategy and Provide Policy and Institutional Support

At both CBD COP-15 and the Shanghai Expo, in 2021, China could launch a new policy signalling a move toward green value chains for key soft commodities for which China is a major importer and which have a significant impact on natural ecosystems. To take this new policy commitment forward, China could announce the intent to form a long-term Inter-Ministerial Committee (Tentative name: National Committee on Value Chain Security and Sustainability) focusing on value chain security, sustainability and green development. This Committee would be responsible for the coordination and implementation of the national green value chain strategy and would address both soft and hard commodities. It could begin by following up on the recommendations of this Special Policy Study on soft commodities.

In order to provide the best technical and policy advice to the Inter-Ministerial Committee, China could establish a technical supporting organization (Tentative name: Global Green Value Chain Institute). The institute would engage experts and stakeholders (e.g., governments, companies, financial institutions, research institutions, civil society organizations) to develop more detailed commodity-specific plans and technical systems, support and promote the implementation of green value chain systems, and provide technical support to key stakeholders.

9.5.2 Adopt Mandatory and Voluntary Measures to Green Soft Commodity Value Chains

The Chinese government could strengthen measures to reduce the import of soft commodities that were illegally harvested or produced in their country of origin. This could build upon the provision regarding the legality of timber in the latest revision of the Forest Law and gradually expand to cover other soft commodities.

The Chinese government could encourage companies to strengthen due diligence and traceability systems to achieve greener soft commodity value chains. A diverse array of tools and approaches are already available to support due diligence and traceability.

China could invest in the technology and manufacturing capacity to produce nutritious, plant-based foods that meet growing domestic (and international) demand for protein, with benefits for human health and food safety and security, while reducing the risk of zoonotic diseases. The plant-based protein industry is a high-growth market in China and globally. The resulting value chain would also be less reliant on imports (which is better for stability and for trade balances) and "greener" (e.g., no deforestation and lower greenhouse gas emissions).

9.5.3   Build on Existing Chinese Policy Levers and Initiatives

The Chinese government could incorporate measures to green soft commodity imports in bilateral and multilateral trade agreements. China could lead the work at the WTO on creating tariff benefits for green soft commodity trade. China also could support establishing coordinated sustainability standards for soft commodity production and trade at APEC, beginning with a few pilot efforts. They could coordinate this work with China's South–South cooperation strategy with key commodity-producing developing countries.

China could develop specific lines of bilateral development assistance that support sustainable soft commodity production in countries that supply China. This assistance might include grants, interest-free loans, concessional loans, and technical assistance for practices and technologies that boost commodity yields on existing agricultural land (linked to avoided deforestation), improve traceability, and improve policy design.

China could integrate the green supply chain strategy with other relevant policies (such as green finance and green BRI) to achieve synergies between these policies and mechanisms. Financial institutions could be encouraged to innovate on investments and financing models for green value chains and incorporate soft commodity green value chain requirements into the due diligence procedures for the extension of credit or asset investment. The Ministry of Ecology and Environment and CBIRC could develop a green finance pilot program with major policy banks to support soft commodity trade. China could also encourage relevant countries to jointly promote global green value chains under the green BRI framework or other international collaborative frameworks.

References

1. USDA (United States Department of Agriculture). 2020. "Foreign Agricultural Service: PSD Online." 2020. https://apps.fas.usda.gov/psdonline/app/index.html#/app/home.
2. National Bureau of Statistics, 2020. "National Bureau of Statistics of China >> Annual Data." Accessed February 22, 2020. http://www.stats.gov.cn/english/Statisticaldata/AnnualData/.
3. World Bank. 2020. "World Development Report 2020: Trading for Development in the Age of Global Value Chains." Washington, DC. https://www.worldbank.org/en/publication/wdr2020.
4. TFA (Tropical Forest Alliance). 2018a. "Emerging Market Consumers and Deforestation: Risks and Opportunities of Growing Demand." https://www.tfa2020.org/en/publication/eme rging-market-consumers-deforestation-risks-opportunities-growing-demand-soft-commod ities-china-beyond/.
5. TFA (Tropical Forest Alliance). 2018b. "TFA 2020 Annual Report 2018." https://www.tfa2020. org/en/annual/report-2018/.
6. UN Comtrade. 2020. "UN Comtrade Database." 2020. https://comtrade.un.org/.
7. CCICED. 2016. "China's Role in Greening Global Value Chains." CCICED Special Policy Study Report. http://www.cciced.net/cciceden/POLICY/rr/prr/2016/201612/P02016 1214521503400553.pdf.
8. Boucher, D. et al. 2011. "The Root of the Problem." Union of Concerned Scientists. https:// ucsusa.org/resources/root-problem.
9. Haupt, F. et al. 2018. "Zero-Deforestation Commodity Supply Chains by 2020: Are We on Track." The Prince of Wales' International Sustainability Unit. https://climatefocus.com/sites/ default/files/20180123%20Supply%20Chain%20Efforts%20-%20Are%20We%20On%20T rack.pdf.pdf.
10. Taylor, R. and C. Streck. 2018. "Ending Tropical Deforestation: The Elusive Impact of the Deforestation-Free Supply Chain Movement." https://www.wri.org/publication/ending-tro pical-deforestation-elusive-impact-deforestation-free-supply-chain-movement.
11. UNEP. 2017. "Environmental Crime – Tackling the Greatest Threats to Our Planet." Our Planet, March. https://wedocs.unep.org/bitstream/handle/20.500.11822/20259/Our%20Planet%20M arch%202017.pdf?sequence=1&isAllowed=y.
12. BCCFA. 2019. "Taxa de desmatamento na Amazônia Legal." 2019. https://www.mma.gov. br/informma/item/15259-governo-federal-divulga-taxa-de-desmatamento-na-amaz%C3% B4nia.html.
13. Mongabay. 2019. "81% of Indonesia's oil palm plantations flouting regulations, audit finds". August 25, 2019. https://news.mongabay.com/2019/08/81-of-indonesias-oil-palm-plantations-flouting-regulations-audit-finds/.
14. Coles, C. and J. Mitchell. 2011. "Gender and agricultural value chains: A review of current knowledge and practice and their policy implications". ESA Working Paper No. 11–05. FAO. http://www.fao.org/3/a-am310e.pdf.
15. Conlon, C. and V. Reca. 2020. "Purchasing Power: The Opportunity for Women's Advancement in Procurement and Global Supply Chains". BSR. https://www.bsr.org/en/our-insights/blog-view/purchasing-power-opportunity-women-procurement-global-supply-chains.
16. Haverhals, M. et al. 2016. "Exploring gender and forest, tree and agroforestry value chains - Evidence and lessons from a systematic review". Infor Brief No. 161. CIFOR. http://www. cifor.org/publications/pdf_files/infobrief/6279-infobrief.pdf.
17. Linnane, C. and T. Kilgore. 2015. "Lumber Liquidators Steps up Campaign to Restore Trust — but Is It Too Late?" MarketWatch. May 7, 2015. https://www.marketwatch.com/story/lum ber-liquidators-steps-up-campaign-to-restore-trustbut-is-it-too-late-2015-05-07.
18. Mongabay. 2016. "Huge Cacao Plantation in Peru Illegally Developed on Forest-Zoned Land." Mongabay Environmental News. July 16, 2016. https://news.mongabay.com/2016/07/huge-cacao-plantation-in-peru-illegally-developed-on-forest-zoned-land/.
19. Chain Reaction Research. 2017. "The Chain: London Stock Exchange Suspends Trading of United Cacao." January 17, 2017. https://chainreactionresearch.com/the-chain-london-stock-exchange-suspends-trading-of-united-cacao/.

20. Spring, J. 2018. "Brazil Fines Five Grain Trading Firms, Farmers Connected to Deforestation." *Reuters*, May 23, 2018. https://www.reuters.com/article/us-brazil-deforestation-bunge-carg-idUSKCN1IO1NV.
21. Ghianni, T. 2012. "Gibson Guitar Settles Probe into Illegal Wood Imports." Reuters, August 6, 2012. https://www.reuters.com/article/us-usa-gibsonguitar-madagascar-idUSBR E8751FQ20120806.
22. China News. 2020. "MOFCOM: The Trade with BRI Countries Grew 6% Last Year." January 21, 2020. http://www.chinanews.com/cj/2020/01-21/9066119.shtml.
23. Xinhua. 2019. "China Has Signed 173 BRI Cooperative Agreements with 125 Countries and 29 International Organizations." April 18, 2019. http://www.xinhuanet.com/2019-04/18/c_1 124385792.htm.
24. Alkama, R. and A. Cescatti. 2016. "Biophysical Climate Impacts of Recent Changes in Global Forest Cover." *Science* 351 (6273): 600–604. https://doi.org/10.1126/science.aac8083.
25. Millennium Ecosystem Assessment. 2004. "Millennium Ecosystem Assessment." 2004. https://www.millenniumassessment.org/en/Condition.html.
26. Evans, T. *et al.* 2020. Links between ecological integrity, emerging infectious diseases originating from wildlife, and other aspects of human health - an overview of the literature. Wildlife Conservation Society.
27. Seymour, F. and J. Busch. 2016. *Why Forests? Why Now?: The Science, Economics, and Politics of Tropical Forests and Climate Change*. Brookings Institution Press.
28. Vidal, J. 2020. "'Tip of the Iceberg': Is Our Destruction of Nature Responsible for Covid-19?" *The Guardian*, March 18, 2020, sec. Environment. https://www.theguardian.com/environment/2020/mar/18/tip-of-the-iceberg-is-our-destruction-of-nature-responsible-for-covid-19-aoe.
29. Lovejoy, T.E., and C. Nobre. 2019. "Amazon Tipping Point: Last Chance for Action." *Science Advances* 5 (12): eaba2949. https://doi.org/10.1126/sciadv.aba2949.
30. Obersteiner, M. Forthcoming. Title TBD.
31. Jones, K.E. *et al.* 2008. "Global Trends in Emerging Infectious Diseases." *Nature* 451 (7181): 990–93. https://doi.org/10.1038/nature06536.
32. Rostal, M.K. 2012. Wildlife: The Need to Better Understand the Linkages. *Current Topics in Microbiology and Immunology* (2012) 365.
33. UNEP. 2016. "UNEP Frontiers 2016 Report: Emerging Issues of Environmental Concern." Nairobi: United Nations Environment Programme. http://www.unenvironment.org/resources/emerging-zoonotic-diseases-and-links-ecosystem-health-unep-frontiers-2016-chapter.
34. FAO. 2020. "FAOSTAT." 2020. http://www.fao.org/faostat/en/#home.
35. New York Declaration on Forests. 2020. "New York Declaration on Forests." 2020. https://for estdeclaration.org/about.
36. Bloomberg. 2016. "Brazil Fines Spanish Bank Santander in Amazon Deforestation." 2016. https://news.bloomberglaw.com/environment-and-energy/brazil-fines-spanish-bank-san tander-in-amazon-deforestation.
37. Chain Reaction Research. 2020. "Financing Deforestation Increasingly Risky Due to Tightening Regulatory Frameworks." https://chainreactionresearch.com/wp-content/uploads/2020/02/Financing-Deforestation-Increasingly-Risky-Due-to-Regulatory-Frameworks.pdf.
38. LCDI. 2019. "Low Carbon Development: A Paradigm Shift Towards a Green Economy in Indonesia." https://www.greengrowthknowledge.org/sites/default/files/downloads/policy-database/indonesia_lowcarbon_development_full%20report.pdf.
39. EU FLEGT Facility. 2012. "Indonesia: Scoping Baseline Information for Forest Law Enforcement, Governance and Trade." 2012. http://www.euflegt.efi.int/documents/10180/23308/Bas eline+Study+7,%20Indonesia+-+Overview+of+Forest+Law+Enforcement,%20Governance+and+Trade/fbbef7de-ead6-4238-b28b-7a3c57fb7979.
40. Soares-Filho, B. *et al.* 2014. "Cracking Brazil's Forest Code." *Science* 344 (6182): 363–364.
41. Government of Brazil. 2016. "Federative Republic of Brazil Intended Nationally Determined Contribution (INDC)." 2016. https://www4.unfccc.int/sites/ndcstaging/PublishedDocuments/Brazil%20First/BRAZIL%20iNDC%20english%20FINAL.pdf.
42. EU FLEGT Facility. 2020. "In Africa | FLEGT." 2020. http://www.euflegt.efi.int/vpa-africa.

43. Arrete N 9450/MAEP/MAFDPRP. 2018. "Arrete N 9450/MAEP/MAFDPRP Portant Orientation Des Plantations Agro-Industrielles En Zones de Savanes." 2018. https://www.docume nts.clientearth.org/library/download-info/arrete-n945-maep-mafdprp-du-12-octobre-2018-portant-orientation-des-plantations-agro-industrielles-en-zones-de-savanes/.
44. CCICED. forthcoming. "Special Policy Study on Green Transition and Sustainable Social Governance." China Council for International Cooperation on Environment and Development.
45. Pampel, F.C. 2014. "The Varied Influence of SES on Environmental Concern." *Social Science Quarterly* 95 (1): 57–75.
46. China Daily. 2017. "Over 70 Percent of Chinese Consumers Aware of Sustainable Consumption." August 23, 2017. https://www.chinadaily.com.cn/business/2017-08/23/content_3100 9090.htm.
47. Walmart Inc. 2020. "Case Study: Encouraging Green Consumption in Retail." Walmart Sustainability.
48. Chambers, A. 2019. "H&M Group Becomes Latest Retailer to Ban Brazilian Leather." ABC News. September 6, 2019. https://abcnews.go.com/International/hm-group-latest-retailer-ban-brazilian-leather/story?id=65429422.
49. Mars. 2019. "Deforestation & Land Use Change Position." 2019. https://www.mars.com/about/policies-and-practices/deforestation-policy.
50. Ceres. 2019. "230 Investors with USD $16.2 Trillion in AUM Call for Corporate Action on Deforestation, Signaling Support for the Amazon." Ceres. 2019. https://www.ceres.org/news-center/press-releases/investors-call-corporate-action-deforestation-signaling-support-amazon.
51. Raghu, A. 2019. "The World Has Loads of Sustainable Palm Oil... But No One Wants It." *Bloomberg.Com*, January 14, 2019. https://www.bloomberg.com/news/articles/2019-01-13/world-has-loads-of-sustainable-palm-oil-just-no-one-wants-it.
52. Markets Insider. 2020. "Palm Oil PRICE Today | Palm Oil Spot Price Chart | Live Price of Palm Oil per Ounce." Markets.Businessinsider.Com. 2020. https://markets.businessinsider.com/com modities/palm-oil-price/usd.
53. Dodson, A. *et al.* 2019. "Smallholders: Key to Building Sustainable Palm Oil Supply Chains." Zoological Society of London: SPOTT. https://www.spott.org/news/smallholders-key-to-bui lding-sustainable-supply-chains/.
54. Musim Mas. 2016. "Training Smallholder Farmers." 2016. https://www.musimmas.com/news/sustainability-journal/2016/training-smallholder-farmers.
55. Treanor, N.B. 2015. "China's Hongmu Consumption Boom." Forest Trends. https://www.for est-trends.org/publications/chinas-hongmu-consumption-boom/.
56. European Commission. 2019. "Stepping up EU Action to Protect and Restore the World's Forests." Communication from the Commission to the European Parliament, The Council, The European Economic and Social Committee and the Committee of the Regions. COM(2019) 352. July 23. Brussels.
57. Accountability Framework. 2019. Core Principles .https://accountability-framework.org/wp-content/uploads/2020/03/Core_Principles-Mar2020.pdf.
58. NDRC (National Development and Reform Commission), 2020. "Report on the Implementation of the 2019 Plan for National Economic and Social Development and on the 2020 Draft Plan for National Economic and Social Development." Xinhua. May 30, 2020. http://www.xin huanet.com/politics/2020lh/2020-05/30/c_1126053830.htm.
59. Alistair Driver. 2019. Asian ASF Crisis to Boost EU Pig Prices in 2020-Rabobank Report. http://www.pig-world.co.uk/news/asian-asf-crisis-to-boost-eu-pig-prices-in-2020-rab obank-report.html.
60. Research and Markets. 2019. "Global Plant-Based Protein Industry Report 2019–2025". December 17. https://www.globenewswire.com/news-release/2019/12/17/1961432/0/en/Glo bal-Plant-based-Protein-Industry-Report-2019-2025-Industry-was-Valued-at-18-5-Billion-in-2019-and-is-Projected-to-Reach-40-6-Billion-by-2025.html#:~:text=The%20global%20p lant%2Dbased%20protein,is%20experiencing%20positive%20growth%20worldwide.

61. WTO (World Trade Organization). 2020a. "Environmental Goods Agreement." WTO. 2020. https://www.wto.org/english/tratop_e/envir_e/ega_e.htm.
62. APEC. 2014. "2014 Leaders' Declaration." APEC. 2014. https://www.apec.org/Meeting-Papers/Leaders-Declarations/2014/2014_aelm.
63. USAID. n.d. "Promoting Gender Equitable Opportunities in Agricultural Value Chains Handbook." Accessed April 13, 2020. https://pdf.usaid.gov/pdf_docs/pnaeb644.pdf.
64. UNDP. 2020. "The Supply Chain in Paraguay, a Space Increasingly Led by Women". January 28, 2020. https://www.greencommodities.org/content/gcp/en/home/media-centre/the-supply-chain--a-space-increasingly-led-by-women.html.
65. Connell, R. 2015. "Meeting at the Edge of Fear: Theory on a World Scale." *Feminist Theory* 16 (1): 49–66.
66. Wragg, E. 2019. "Cofco International Closes US$2.1bn Sustainability-Linked Club Loan." Global Trade Review (GTR). July 22, 2019. https://www.gtreview.com/news/sustainability/cofco-international-closes-us2-1bn-sustainability-linked-club-loan/.
67. WTO (World Trade Organization). 2020b. "The WTO and Word Customs Organization - The Challenges of Trade Financing." 2020. https://www.wto.org/english/thewto_e/coher_e/challenges_e.htm.
68. PRI. 2019. "230 Investors with USD $16.2 Trillion in AUM Call for Corporate Action on Deforestation, Signaling Support for the Amazon." PRI. 2019. https://www.unpri.org/news-and-press/230-investors-with-usd-162-trillion-in-aum-call-for-corporate-action-on-deforestation-signaling-support-for-the-amazon/4867.article.
69. Indonesian Ministry of Agriculture. 2021. "Proven to Survive, the Ministry of Agriculture is Increasingly Boosting Plantation Commodities." 2021. https://ditjenbun.pertanian.go.id/terbukti-bertahan-kementan-makin-genjot-komoditasperkebunan/.
70. Olivero, J., J.E. Fa, R. Real, A.L. Márquez, M.A. Farfán, J.M. Vargas, D. Gaveau, et al. 2017. "Recent Loss of Closed Forests Is Associated With Ebola Virus Disease Outbreaks." Scientific Reports 7 (1): 14291. https://doi.org/10.1038/s41598-017-14727-9.
71. Reuters. 2019. "Indonesia President Makes Moratorium on Forest Clearance Permanent," 2019. https://www.reuters.com/article/us-indonesia-environment-forest-idUSKCN1UY14P.
72. Samberg, L.H., J.S. Gerber, N. Ramankutty, M. Herrero, and P.C. West. 2016. "Subnational Distribution of Average Farm Size and Smallholder Contributions to Global Food Production." Environmental Research Letters 11 (12): 124010. https://doi.org/10.1088/1748-9326/11/12/124010.
73. Tropical Forest Alliance 2020. 2017. "Tropical Forest Alliance Annual Report 2016–2017."

Chapter 10
Green Finance

10.1 Foreword

To reverse the global trends of ecosystem degradation and biodiversity loss, *the Convention on Biological Diversity (CBD)* was signed in 1992 as a milestone. *The Strategic Plan for Biodiversity 2011–2020* issued in the 10th Conference of the parties to the Convention on Biological Diversity (CBD Cop10) in Aichi, Japan, provides 20 targets to halt the loss of biodiversity for the next decade, namely "Aichi targets". The coming CBD CoP15 to be held in Kunming, China will overview the implementation of the Aichi Targets, summarize the progress of global biodiversity conservation over the past decade and establish the structure and goals for the future 10 years of global biodiversity. However, most of Aichi Targets are expected to fail. From the capital investment perspective, the economic development and resource utilization models that are profitability centered are the main lagging force to shift the devastating global trend on biodiversity. Ecological conservation funds that heavily relied on public finance are very limited which makes it difficult to guarantee the full implementation of biodiversity conservation measures in various fields. Under this context, how to maximize the mobilization of financial resources, through the development of ecological conservation finance to support ecological and biodiversity conservation has become a global concern.

Conservation Finance is interrelated to biodiversity finance, climate finance, green finance and sustainable finance (see Box 10.1). Conservation finance aims to leverage and effectively manage economic incentives, policies, and capital in fields of conservation, restoration, efficient use of natural resource and biodiversity protection, to achieve the long-term wellbeing of nature and the services nature provides to society. Conservation finance includes an array of financing mechanisms, such as grants, taxes and fees, debt-for-nature swaps, credit, bonds, trust funds, and payments for environmental services (PES) [1]. Professionals in this field work with stakeholders ranging from local communities to large multilateral finance institutions and philanthropic

© The Author(s) 2022
China Council for International Cooperation on Environment
and Development (CCICED) Secretariat,
Green Consensus and High Quality Development,
https://doi.org/10.1007/978-981-16-4799-4_10

organizations, impact funds, private corporations, and governments. They support conservation efforts that extend across terrestrial, freshwater, coastal, and marine areas to protect ecosystem services and cultural values, and increase direct financial revenues through activities that produce positive biodiversity outcomes [2].

Box 10.1. Conservation Finance and Biodiversity Finance, Climate Finance, Green Finance and Sustainable Finance

Biodiversity finance provides financial incentives, manages capital for sustainable biodiversity management. It includes both private and public financial resources for biodiversity conservation, as well as commercial investments in favor of biodiversity conservation and biodiversity-related capital market transactions. The definition of biodiversity in CBD covers the diversity of ecosystems, species and genetic resources. Thus, ecological protection and biodiversity conservation are closely related. Therefore, conservation finance and biodiversity finance shared high similarity.

Climate finance refers to the financial activities supporting climate change mitigation and adaptation. Although the mitigation of climate change is related to ecological conservation, the emphasis of climate finance and ecological conservation finance is still quite different.

Green finance refers to economic activities that support environmental protection, climate change mitigation, and resource saving. Green finance provides investment, operation, risk management, and other financial services to projects in fields of environmental protection, energy saving, clean energy, green transportation, and green building. According to the catalogue of Green Bond Support Projects published in 2015 by the Green Finance Committee of the China Institute of Finance, comprehensive control of soil erosion, ecological rehabilitation and disaster prevention and control, and construction of nature reserves are all part of green projects. As ecological protection is one of the supporting activities of green finance, conservation finance could be viewed as a part of green finance.

Sustainable finance refers to a financial system that is oriented by long-term and sustainable economic activities, in which environmental and social factors are well taken into account in the investment process.

In general, although climate finance and eco-conservation finance are interrelated, their focus differs from one another. Climate finance sheds light on mitigating climate change, while conservation finance is more on ecological conservation field. They both belong to the category of green finance and are also part of what is known internationally as sustainable development finance, as illustrated in the Figure below.

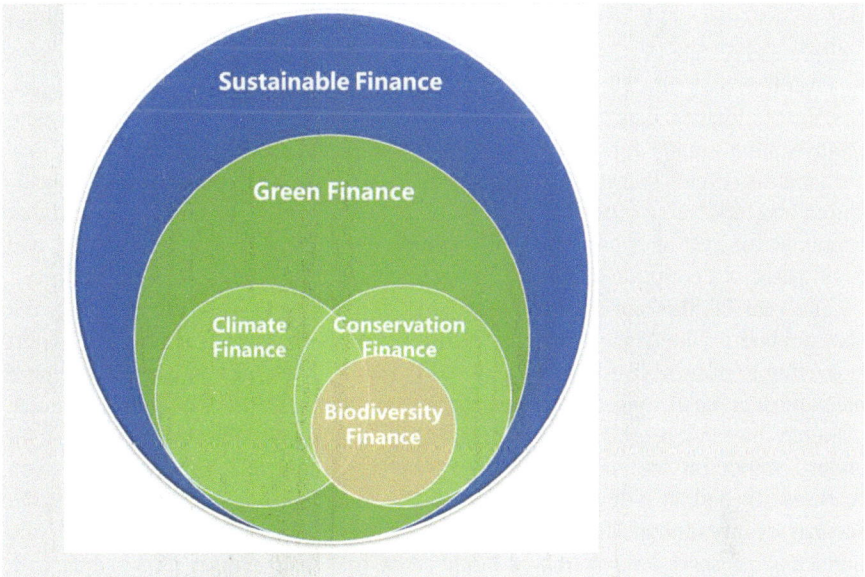

In recent years, conservation finance is growing rapidly while facing challenges at the same time. On the one hand, Funding for conservation efforts globally are broadly acknowledged as insufficient. Between 2014 and 2016 there was a 62% increase in private capital allocated to conservation efforts. With all flow combined, analyses arrive at high estimates of around $120 billion revenues to conservation by 2020. However, this is woefully below the estimated annual requirement of $300–400 billion needed to finance conservation. One the other hand, the economic system that incentivizes and facilitates flows of finance towards conservation has not been established. For instance, for every dollar provided to projects reducing emissions from deforestation, $150 is channeled to activity that drives deforestation. Thus over 99% of our economic engagement with forests is destructive. So even if annual finance flows were increased eight-fold to meet the recommended $400 billion target, it changes the ratio of bad money to good from 150:1 to around 18 or 19:1. We will never succeed by only increasing financing for afforestation and restoration without reducing the overwhelming trend for our broader economic activities to be destructive.

China has a variety of ecosystem types, including forests, wetlands, grasslands and oceans, as well as rich biodiversity. Conservation and biodiversity protection are the important contents of the Eco-civilization in China. In 2015, for example, the value of ecosystem services in China is estimated to be 72.81 trillion yuan, about 1.06 times the GDP of that year [3]. Strengthening conservation and biodiversity protection is not only an important measure to maintain China's ecological security and improve people's well-being, but also an important practice to create social welfare and realize President Xi Jinping's 'Two Mountains' theory.[1]

Over the past years, China and the international community have been working on sustainable finance, especially on climate finance. However, the current ecological security situation in China is not optimistic. The recent large-scale outbreak of new coronavirus reveals that there is still considerable work undone in terms of conservation and biodiversity protection. In particular, establishing a financial system that supports restoring and protecting the natural environment, natural resources, and ecosystems is a common challenge facing China and the international community.

The year 2020 is not only the starting of China's new Five-year Planning, but also dubbed as the 'Super Year for Nature'.[2] Supported by CCICED, this report is striving to address two key challenges: (1) Public financing sources along cannot meet the demand of conservation while a large amount of private capital and financial resources have not been effectively used; (2) The key areas for conservation are often underdeveloped areas or poverty-stricken areas at the same time. The conflict between conservation and local development affects the effectiveness of conservation and the security of investment. The purpose of this study is to unlock more effective private finance to conserve ecosystem and biodiversity for China as soon as possible, with a better use and innovate of financial mechanisms.

10.2 Conservation in China: Policy and Practice

Significant progress has been made in China's ecological and biodiversity conservation in the past decades. To achieve the long-term protection of its important natural ecosystems and resources, China has launched series of laws and regulations, including the *Wildlife Protection Law*, *the Regulations on Protection of Wild Plants*, and *the Regulations on Nature Reserves*, and established 2750 nature reserves (of which 474 are national nature reserves) and 11 pilot national park. In addition, under the support of *the Forest Act*, *the Prairie Act*, and *the Wetlands Act*, China has implemented multiple ecological restoration projects, including 'Natural Forest Resources Protection Project (NFRPP)', 'Returning Farmland to Forests and Grassland Project (RFFGP)', and 'Ecological Protection and Restoration of

---

[1] President Xi Jinping first put forth the idea that "lucid waters and lush mountains are invaluable assets" in his inspection tour to Anji County, Zhejiang Province in August 2005. Colloquially known as the "Two Mountains" theory, it has been the guiding thought of China's ecological civilization.

[2] The coming CBD COP 15 to be held in Kunming, will assess the global biodiversity conservation process over the past decade, and consider adopting the new 'Post-2020 Global Biodiversity Diversity Protection Framework'. As an important step in the planning of the post-2020 biodiversity agenda, the IUCN World Conservation Congress (WCC) to be held in Marseille, France in June 2020, will gather ecological protection experts, social organizations and government representatives from all over the world to discuss the role of nature conservation in achieving the 2030 Sustainable Development Goals.

Mountains-Rivers-Forests-Farmland-Lakes-Grasslands' in key ecological function areas of China (Fig. 10.1).

China has been increasing the amount of investment in ecological and biodiversity conservation (as shown in Table 10.1). Among them, RFFGP and NFRPP are the two largest ecological projects in China, with a total investment of 830 billion yuan. In terms of expenditure, the fund were used in multiple areas, including the construction cost of nature reserves, the implementation costs of ecological restoration projects, and other related operation fees. The fund also covers economic incentive expenditures such as ecological compensation for areas that are prohibited or restricted from development and areas that are ecologically rehabilitated. According to statistics, from 2008 to 2015, China has invested 251.3 billion yuan in transfer payment projects for key ecological zones. In addition, with the development of business for public good, more social capital and the general public has participated in ecological conservation field. Social Institutions that are focusing on ecological protection, such as Ant Financial, SEE Conservation, Paradise Foundation, and China Greening Foundation, began to emerge (as shown in Table 10.1).

Through more than 20 years of efforts, China's vegetation coverage rate has increased significantly, and ecological functions such as water conservation and soil and water conservation have been well improved. The forest coverage rate has increased from 8% in 1970s to 22.96% in 2018, as shown in Fig. 10.2. In the past 20 years, China has added about 25% of the world's new vegetation cover [10]. The increase in vegetation cover has reduced the area of soil erosion from 3.56 million square kilometers in 2000 to 2.524 million square kilometers in 2018 [11]. Desertification and desertified area achieved "double reduction" for 15 years in a row [12].

10.3 Conservation Challenges and Funding Gaps in China

There are still many problems and challenges existing in China's ecological protection work. China is one of the countries with the richest ecosystems and biodiversity in the world; at the same time, China has a large distribution of fragile ecosystems and threatened species. According to a survey in 2017, 10,102 species of vascular plant and 2471 species of Vertebrata (excluding marine fish), which respectively accounted for 29.3% and 56.7% of the total number of species assessed, are in need of nationwide attention and protection. In addition, as the world's second-largest economy, China's impact on global ecosystems has accelerated. The funding gap of Chinese ecological conservation could be reflected in the following three terms.

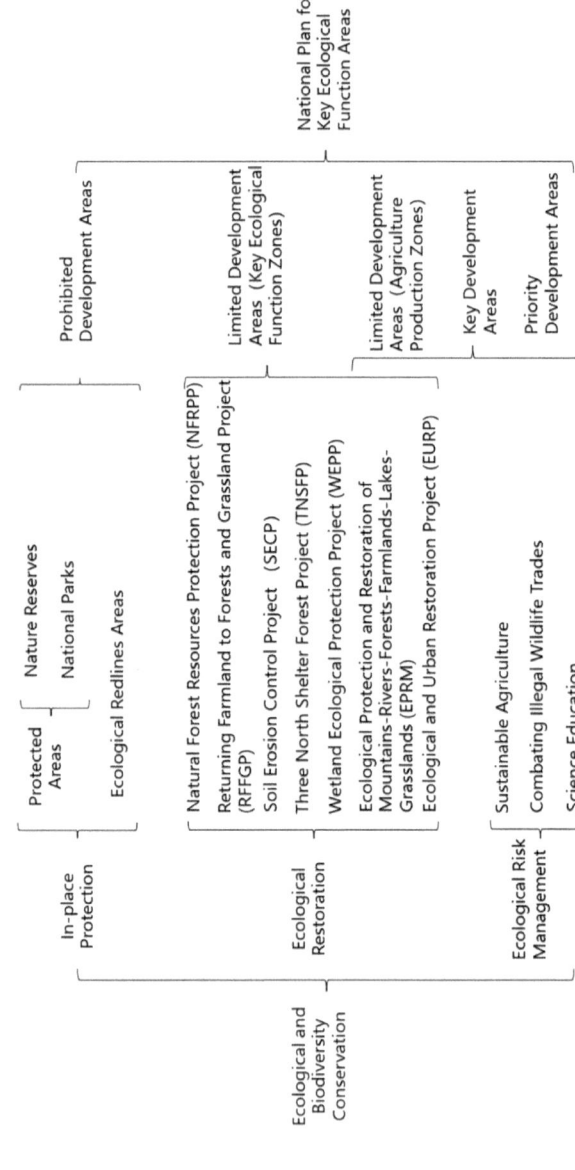

Fig. 10.1   China's practices in ecological and biodiversity conservation

Table 10.1 Investment in major conservation projects in China

Conservation project	Capital investment
RFFGP	By 2019, more than 500 billion yuan had been invested in returning farmland to forests and grassland [4]
NFRPP	By 2017, China had invested 331.355 billion yuan in natural forest protection projects [5]
SECP	Between the 11th five-year plan and the fifteen, the state invested 18.8 billion yuan in SECP [6]. The total investment of the national SECP is estimated to be as high as 22.9 billion yuan from 2017 to 2020
TNSFP	By 2018, the TNSFP had invested 44.3 billion yuan in central and local finance [7]
EPRM	As of 2019, the central government has issued a total of 36 billion yuan in funds for key ecological conservation and rehabilitation [8]
RGLGP	By 2018, China had invested 29.57 billion yuan in the program [9]

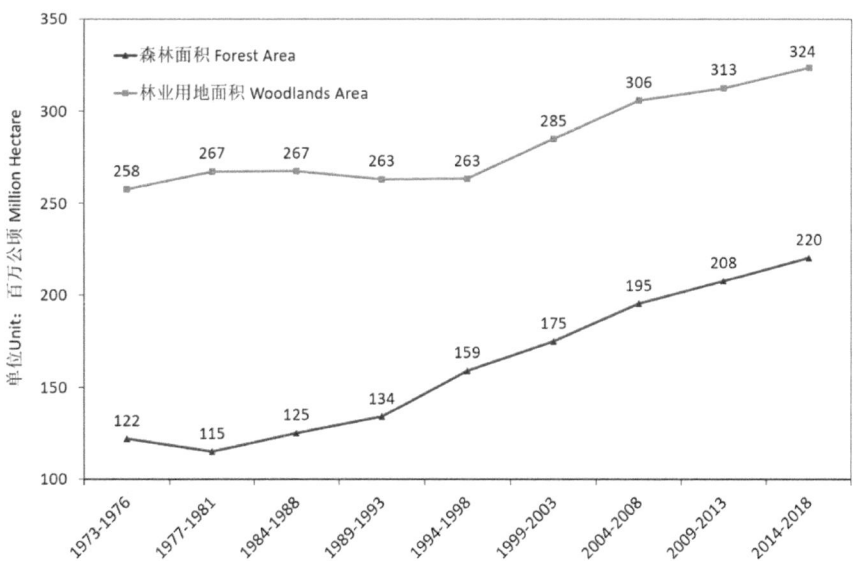

Fig. 10.2 The change of forest and forestry areas in China from 1973 to 2018. *Data Source* Forestry Statistical Yearbook

10.3.1 Conservation Challenges of Terrestrial Ecosystem

10.3.1.1 Low Conservation Efficiency and Insufficient Funding Resource in Nature Reserves

The construction fund for protected lands is insufficient in general. It is estimated that effectively protecting 18% of land areas and 10% of sea areas in China requires an investment around 0.065–0.2% of its GDP each year (at 2011 price levels, about 30.6–95 billion yuan) [13]. But in 2014, the financial outlay for China's nature reserves at all levels was only 8.2 billion yuan, or an average of 6119 yuan per square kilometer, far below the estimated request of 42,000 yuan per square kilometer [14].

The lack of investment in protected areas in China led to a reduction in the conservation effectiveness. On the one hand, most of the protected areas are still patrolled manually, and new technologies are rarely used in activities such as evaluation and monitoring of the protected areas. On the other hand, staff salaries and official expenses of nature reserves at all levels are mainly guaranteed by local governments at the same level. Since most of China's natural reserves located in economically backward areas, financial support for natural reserves here are woefully inadequate. The low wages[3] and the lack of basic social security result in the difficulties of constructing high quality local patrol teams.

10.3.1.2   Severe Ecological Degradation Risks in Wetland System

As indicated in the national wetland survey conducted in 1995–2003 and 2009–2013, the overall trend of wetland system ecological degradation has not been reversed yet. The area of wetlands in China has been reduced by 50.9 million mu in the past decade, which is equivalent to the area of two Beijing cities. The ecological environment of the existing wetland system is not optimistic either. Due to the problems of environmental pollutions, overfishing, reclamation, invasive alien species, and the expanding of infrastructure, more than 50% of the surveyed wetlands are classified as 'poor' ecological condition. The deterioration of wetland ecosystem has directly destroyed the habitat environment of wetland organisms. The two surveys recorded a sharp decline in bird species, with more than half of the bird population declining significantly.

In order to improve the degradation trend of China's wetland system, China began to set up special funds for wetland protection in 2009, but the existing funds are still far from enough compared with the actual demand. The results of the second wetland resources survey show that 69% of the surveyed wetlands in China are threatened, and more than 20% of them need to be restored artificially. Based on the restoration cost of $10,000 to $20,000 per hectare, the future cost of wetland restoration in China could be as high as one hundred billion yuan [15].

10.3.1.3   High Remediation Pressure in Soil Environment

A safe soil environment is an important basis for the healthy and stable development of a regional natural ecosystem. The prevention and control of soil pollution in China is still at a starting stage. Facing the serious pollution in soil environment, China has invested 28 billion yuan from the central government budget since 2016 to prevent

---

[3] About RMB 3,000–4,000 per month.

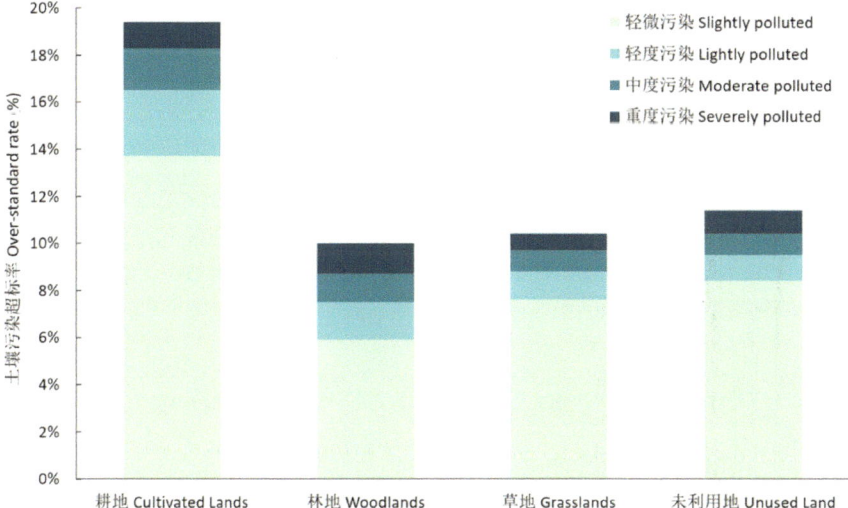

Fig. 10.3 The result of the national soil pollution survey. *Source* First National Soil Pollution Survey (2005–2013)[4]

and control soil pollution nationwide. However, compared the overall investment demand of 7 trillion yuan [16], a big funding gap is still left in China.

The biggest funding pressures are centered in arable land and abandoned mining land, where the soil pollution is extremely serious. According to statistics, the rate of excessive pollution in arable land and abandoned mining land is as high as 19.4% and 34.9% respectively (as shown in Fig. 10.3). In addition, compared with the construction land in urban area, it is especially difficult to identify responsible investors and commercial financing modes for the remediation of these areas.

10.3.1.4 Severe Challenges for Yangtze River and Yellow River

The Yangtze River and the Yellow River not only offer important supports for China's sustainable economic and social development, but also serves as the home of rare and endangered animals and plants. However, with the acceleration of urbanization along the economic belt, the ecological environment of the Yangtze and Yellow River basins is facing severe challenges. The extensive development of the coastal area has led to a series of consequences, including the reduction of forest and grassland coverage, the shrinking of the lakes and wetlands in the middle and lower

---

[4] The national survey of soil pollution divides the soil pollution into five levels. If the content of pollutants doesn't exceed the standard, it is no pollution; if the content of pollutants is between one times to two times (including), it is slightly polluted; if the content is between two times and three times (including) , it is lightly polluted; if the content is between three and five times (including), it is moderate polluted; if the content is above five times, it is severely polluted.

reaches, the continuous decline of the Aquatic Biodiversity Index, and the extinction of many rare species such as the Yangtze River Paddlefish. In order to protect the ecological bottom line of the Yangtze River and the Yellow River, China published the *Ecological and Environment Protection Plan of the Yangtze Economic Belt* and the *Action Plan for the Protection and Restoration of the Yangtze River* in 2017 and 2019. In addition, the ecological protection and high-quality development of Yellow River Basin has been put on the national agenda in 2019. Thus, more funding would be needed for conservation activities in these two rivers, including the protection of natural coastline, the construction of ecological buffer zones of rivers and lakes, the management of eutrophic lakes, and the protection of rare and endemic fish species.

10.3.1.5   Urban Biodiversity Conservation is Still in Its Infancy

With the acceleration of global urbanization, the link between human and nature is declining. The protection and promotion of biodiversity in urban areas will be important to reconstruct the link between human and nature. In recent years, urban biodiversity is gradually becoming the global attention, and its importance in China is increasing as well. For example, China has carried out the 'ecological restoration and urban repair' project in 2017 to improve living environment and control urban ecological problems. On this basis, how to further integrate biodiversity conservation into China's urban planning, construction and management, and fully mobilize the resources and strength of enterprises' and the public's contribution in ecological conservation, is one of the important questions that need to be addressed in China.

10.3.2   Conservation Challenges of Marine Ecosystem

With the rapid development of marine economy, the deterioration trend of marine ecosystem in China is becoming more and more obvious [17]. In 2018, only 23.8% of China's monitored marine ecosystems, including estuaries, bays, tidal wetlands, coral reefs, mangroves and seagrass beds, were in a healthy state, and 71.4% and 4.8% of ecosystems were respectively in Sub-optimal healthy and unhealthy states.[5] The density of phytoplankton in most estuaries and estuaries is high, while the density

---

[5] The health status of marine ecosystem can be divided into three levels: healthy, sub-healthy and unhealthy. Under a healthy state, an ecosystem maintains its natural attributes. Biodiversity and ecosystem structure are basically stable, and the main service functions of ecosystems are functioning normally. The ecological pressure, such as environmental pollution, man-made destruction and irrational exploitation of resources, is within the carrying capacity of the ecosystem. Under sub-health condition, the ecosystem basically maintains its natural attributes. Biodiversity and ecosystem structure have deteriorated to a certain extent, but the main service function of the ecosystem can still play. Ecological pressure, such as environmental pollution, man-made destruction and irrational exploitation of resources, exceeds the carrying capacity of the ecosystem. Under the unhealthy condition, the natural attributes of the ecosystem change obviously. Biodiversity and ecosystem structure have undergone great changes, and the major service functions of the ecosystem have

of fish eggs and larvae is relatively low. In addition, the coverage of coral reef ecosystem is decreasing. The discharge of pollutants from land and near shore and the frequent disturbance of human activities are the main factors affecting China's marine ecological security. In 2018, 12.4% and 14.9% of the 194 river sections were of Class V and Class V respectively. Except for soluble pollutants, solid wastes also have a high distribution density on the sea surface, the beach, and the seabed, and all of them are mainly plastic wastes. In addition, China is one of the countries in the world most severely affected by marine disasters. According to statistics, in 2018, marine ecological disasters such as coastal erosion, red tides, green tides, seawater intrusion and soil salinization occurred frequently in China, causing direct economic losses of 4.777 billion yuan and 73 people dead (including [18]).

In order to solve the outstanding ecological and environmental problems in the marine ecosystem, China has issued the action plan for the comprehensive treatment of the Bohai Sea in November 2018, which ensures that the ecological environment of the Bohai Sea will not deteriorate again by taking measures to control land-based pollution, marine pollution, ecological protection and restoration, and environmental risk prevention. China has invested 7 billion yuan in comprehensive control of the Bohai Sea in 2019. On this basis, China is exploring the construction of marine national parks to enhance the authenticity and integrity of important marine ecosystems and their biodiversity resources. Under the background of the increasingly sound system and mechanism of marine ecological protection, there would be more financial demand in terms of the development and application of marine ecological protection high-tech, as well as the transformation and upgrading of marine industry.

10.3.3 China's Impact on Overseas Ecosystem

The Belt and Road Initiative (BRI) is a global cooperation initiative of China through bilateral and multilateral mechanisms, aiming at building a community of use, destiny and responsibility based on political trust, economic integration and cultural tolerance. Within the countries of the BRI, billions of dollars are invested to build transport infrastructure (about 190 billion USD since 2013) and energy infrastructure and supply chains (about 280 billion USD since 2013), as well as mines and agriculture [19]. Investment in BRI infrastructure can contribute to social and green development. Examples include investments in micro-grids in conjunction with clean energy production through wind and solar, water management, waste-water treatments and sustainable agriculture.

Apart from the economic benefits of investments, investors and society should aim to minimize environmental risks in overseas investments in the BRI. For example, any infrastructure construction can directly lead to breaks in landscape and habitat connectivity, as well as to secondary effects such as spread of invasive animal and

---

been seriously degraded or lost. Ecological pressures such as environmental pollution, man-made destruction and irrational exploitation of resources exceed the carrying capacity of the ecosystem.

plant alien species, wind throws, fires, animal kill (e.g. through road kill), pollution, poaching and microclimates.

As the BRI encompasses many countries and their ecosystems (some studies suggest the BRI infrastructure affects 4138 animal and 7371 plant species and that BRI corridors overlap with 265 threatened species and 46 biodiversity hotspots), [20] three issues stick out for biodiversity and ecosystem protection:

- Investors, project developers and local government should strictly execute stringent environmental impact assessments (EIA) based on Chinese or international best practices (e.g. IFC Performance Standards 6) and stringent project oversight to minimize the negative impacts of projects while protecting ecosystems;
- Investors, project developers and local governments should include transboundary impact assessments (TIA) (e.g. based on UNECE Convention on Environmental Impact Assessment in Transboundary Context) to ensure cross-border impacts of projects are avoided, mitigated and/or compensated fairly across affected countries;
- Investors and project developers should support local communities to manage ecosystem services by providing transparent data on ecosystem impacts, which would allow for a better adjustment of ecosystem protection throughout the project lifecycle;
- Investors and project developers should have environmental liability insurance to be able to compensate for unforeseen events. This should also include end-of life restoration activities.

To support biodiversity protection in the BRI, the 'Green Development Guidance for Belt and Road Initiative Projects' (formerly 'Green Light System') of the BRI International Green Development Coalition (BRIGC), under the supervision of the Chinese Ministry of Ecology and Environment (MEE) aims to provide an evaluation tool and policy tools to ensure that BRI investments are contributing to green development and are minimizing negative environmental impact [21]. Various other initiatives to reduce biodiversity and environmental risks associated with the BRI have been put forward by Chinese government and non-governmental organizations and their international partners as well as financial institutions. The BRI Green Investment Principles (GIP) include 7 principles to encourage financial institutions to invest in projects that meet the Paris Climate Accord and contribute to the UN Sustainable Development Goals [22]; The Chinese Green Credit Guidelines issued by the CBRC (now CBIRC) in 2012 are applicable to international investments of Chinese institutions in the BRI countries [23]. The Guidelines highlight the role of national laws and thus don't encourage financial institutions to go beyond possibly weaker national environmental legal frameworks in BRI countries. Overall, coordinated and applied actions that successfully mitigate adverse environmental impacts of BRI investments, particularly in regard to biodiversity loss, should be accelerated.

To sum up, the lack of available funding is one of bottlenecks to the ecological and biodiversity conservation in China. With the downward pressure on China's economy and the slowdown of fiscal revenue growth, conservation financing mode dominated by public sector has become unsustainable. It has become a matter of priority in China to build channels for private capital and financial resources by forming a diversified, sustainable and high-quality conservation financing system.

10.4 The Financial Model of Conservation and Biodiversity in the World: Experience and Best Practice

The global challenges linked to the steady loss of biodiversity and ecosystem services have, in recent years, begun to take center stage. The 2019 IPBES report leaves no doubt that the risks linked to the degradation of nature and natural resources are so high that the future of human well-being, prosperity and economic stability are under imminent threat. Further, the WEF Global Risk Report 2020 states that the loss of biodiversity ranks as the second most impactful and the third most likely risk for the next decade.

There have long been calls to invest far greater sums of money in conservation activities such as protected areas, species protection and landscape restoration. The sums generated have fallen well below minimum requirements for stemming the downward trend, much less reversing it.[6] The public sector, with rare exceptions, has failed to rise to the challenge, and efforts to attract substantial private capital into conservation activities have hit up against the requirements of private actors for risk management and adequate return over a relatively brief time horizon. But it is not simply a question of directing more money at conservation priorities. Equally, if not more important, is to reduce financial flows to activities that harm biodiversity.

The dominant financial model, followed over the past few decades in most countries of the world, rewards capital owners and shareholders at the expense of the public policy goals reflected in the 2030 Agenda, with its Sustainable Development Goals, the Paris Agreement on climate change, and the targets regularly set by the Convention on Biological Diversity. Too much financial activity still tolerates—and often rewards—the destruction of natural landscapes or the depletion of scarce natural resources. It is time to rethink the financial model with a view to advancing the world that we would like to see come into being.

A recent paper by the Finance for Biodiversity (F4B) initiative proposes a new framing of the conservation finance challenge—a way of thinking about both the challenges and opportunities. It represents an example of the new thinking that is going on at the interface between the pressing needs of nature conservation and the operations of the financial system.

In the sections below, we set out international best practice in conservation finance in three streams:

---

[6] According to the GEF an estimated USD 300 billion–400 billion is needed annually to preserve healthy terrestrial and marine ecosystems, and the clean air, fresh water, and biodiversity on which we all depend. However, only USD 52 billion is currently flowing towards projects supporting conservation, while the private sector manages an estimated USD 300 trillion in assets.

- Efforts to direct more finance—in particular private capital—towards activities compatible with conservation
- The need for rule change to ensure a more favourable environment for conservation; and
- The importance of creating a greater public demand for conservation results—and the attendant intolerance of activities that undermine nature.

Recommendations in these three areas are to be found in Chap. 7.

10.4.1 Expanding Finance Directed at Conservation

A great many efforts are underway internationally to address this shortfall. In response to the growing interest in responsible and impact investment, and the rapidly growing demand for conservation-friendly options, a series of specialized investment houses have sprung up to structure this new market.[7] These include:

- **Fund building efforts**, for instance numerous existing funds such as Althelia and Mirova, Conservation Capital or Encourage Capital. The Meloy Fund, implemented by Conservation International (CI) and RARE, established the first fund for sustainable small-scale fisheries in Southeast Asia to improve the conservation of coral reef ecosystems by providing financial incentives to fishing communities in the Philippines and Indonesia to adopt sustainable fishing practices and rights-based management regimes. Althelia, with the support of CI and with technical and scientific advice from the Environmental Defense Fund, launched the Sustainable Ocean Fund as an impact investment vehicle that can deliver marine conservation, improved livelihoods, and attractive economic returns.
- These efforts are accompanied by efforts from the public sector, or through public–private partnerships, to **clear the obstacles facing conservation finance** and to ensure that those wishing to build this field have access to assistance and advice. This includes:
  - Fund aggregation functions, such as the UNDP Finance for Nature team's efforts to build an Exchange Traded Fund (ETF), or the UK Government's exploration of interventions to build the field (via the International Climate Finance team at the Department for Business, Energy and Industrial Strategy—BEIS). It also includes efforts to create intermediaries between capital seeking responsible nature-based investments and those with projects under development. A prominent example of this is The Nature Conservancy's NatureVest.

---

[7] This section illustrates the different approaches currently underway to enhance conservation finance. More information can be found in the Annex to this chapter.

- These efforts seek to identify **"investor grade" activities** that meet the risk and return profiles required by investors, including the use of public contributions to lower perceived risk and to analyze and disseminate "best practice" in this field.
- They have now moved beyond development of projects for investors to look at a range of issues at the interface between finance and nature conservation, including **refinement of tools** to value the contribution of natural capital to overall economic performance and to bring together professionals in these fields to exchange experience, including:

  - Organizations engaged in building knowledge bases and providing tools to facilitate improved conservation finance related activity, including the Natural Capital Finance Alliance, Forest Trends, Global Canopy and the Biodiversity Finance Initiative. The World Bank's Global Program on Sustainability (GPS) has also generated a good deal of data in this area. Best practice guidance on conservation finance is also available through the Conservation Finance Alliance's 'Conservation Finance: a framework'.
  - **Landscape mapping** initiatives, most notably the Forest Trends-authored State of Private Investment in Conservation 2016, Global Canopy's Little Biodiversity Handbook (presently being updated) and the various publications that are commissioned by and cluster around the CPIC and Crédit Suisse events.
  - **Convening and networks**, especially the CPIC and Crédit Suisse events, that play an important role in community and network building and information exchange. The OECD is also preparing a publication on biodiversity finance to be published in the first half of 2020.

- Related efforts to develop the concept and practice of regarding natural resources and ecosystems as an important form of—and contributor to—**sustainable infrastructure** development. The work of the World Resources Institute, for example on natural water infrastructure, has emphasized the highly cost-effective contribution of natural ecosystems to improved human well-being.
- A major effort is underway to strengthen the base of knowledge and data accessible to finance professionals, both to assess the impact of investments on biodiversity parameters and to strengthen the locally relevant data needed to take sound investment decisions. These include:

  - **Data providers**, such as the UN Biodiversity Lab, Global Forest Watch, ENCORE from the Natural Capital Finance Alliance, the work of CDP and integrations into mainstream data providers such as Bloomberg from organisations such as Sustainalytics. New specialist approaches are also emerging, for example the Sustainable Digital Finance Alliance's recent 'Fintech for Biodiversity Challenge'.

- While there is still an absence of robust and widely accepted **norms and standards**—or even of broadly—accepted definitions of conservation or biodiversity finance—a series of guidelines and standards are emerging in major areas of conservation-related investments and the movement to accelerate the uptake of these is accelerating. The EU has recently added a focus on biodiversity finance in its work on sustainable finance taxonomy. Examples of work in this area include:

- **Guidelines, commitments and standards**, such as the New York Declaration on Forests, and the ZUG Faith Consistent Investing Guidelines. Other initiatives include: Investors Group on Climate Change, Disclosure Insight Action, Principles for Responsible Investment and the Institutional Investors Group on Climate Change.
- Relevant in this area are the Green Investment Principles for the Belt and Road initiative developed by the City of London Corporation's Green Finance Initiative with strong cooperation from China.

Finally, taking a leaf from the book of climate change action, efforts are underway to increase the **transparency** of corporations and investors in terms of the impact of their investments on biodiversity. This effort is both general and specific.

At the general level, a series of international public and private players are considering the establishment of a **Task Force for Nature-related Financial Disclosure (TNFD)**. Modelled on the successful Task Force on Climate-related Financial Disclosure operated by the Financial Stability Board, it would develop a standard and a requirement for corporations and investors to disclose the impact on biodiversity of their actions. This would supply investors with the basis for better assessing the risks to their investments related to biodiversity and ecosystems, and for them and governments to insist on such disclosure as a condition of access to capital. A proposal in this regard is likely to be discussed at CBD COP 15 in early 2021.

In terms of public finance instruments, a renewed interest is being shown in **Debt for Nature Swaps**. First developed in the 1980s, they enabled bilateral debt under the Paris Club to be drawn down or eliminated in exchange for an agreed level of investment in nature conservation by the indebted country in its national currency. This conditional debt draw-down removed a hard currency repayment obligation in exchange for a more manageable deployment of national currency with a conservation benefit. With the prospect of developed and emerging economies once more likely to sink deeply into debt as a result of the COVID-19 pandemic, there is a renewed interest in the potential of Debt for Nature Swaps and related instruments (e.g. Green bonds for nature-related investments, blue bonds for ocean conservation) to offer an incentive for nature conservation in exchange for debt relief.

At the more specific level, efforts are focused on improving the metrics that relate to measuring the impact of corporate activity and investment on biodiversity-related factors. Disclosure must yield reliable and comparable impact data and the metrics for this are still at an early stage. The French Caisse des Dépots Group are developing and refining a 'Global Biodiversity Score' WWF and IUCN have similar tools. However, the set of metrics for measuring the biodiversity impact of economic activity requires rapid further development and alignment among the standards used by different countries.

All of the above describes a wide range of creative efforts underway internationally to channel funding into activities compatible with nature conservation, whether these activities have as their principal purpose the setting aside of natural resources or their exploitation within acceptable limits. It focuses very centrally on matching a growing interest on the part of investors for conservation-compatible activities through increasing the supply and quality of investment opportunities available to investors.

The main barriers for scaling up conservation finance include lack of capacity, small size of projects, the heterogeneous nature of projects, and lack of enabling environment.

Conservation finance needs to include sourcing and structuring investments so they are consistent with the asset-allocation protocols of institutional investors, preferably with competitive, risk-adjusted returns and with the most efficient practicable application of increasingly scarce public and philanthropic credit through blended finance (i.e., DFI loans/guarantees/grants, sovereign loans/guarantees/subsidies, and philanthropic grants). Opportunities to expand use of blended finance will require continued innovation to help countries and private sector partners match the right types of financial instruments to specific projects goals and objectives, including in the natural resource management sectors. Support for project preparation, along with aggregation and bundling of projects that can attract large scale investors will also be needed in many cases.

Increasingly, however, two things are becoming clear. First, despite encouraging growth in finance devoted to conservation-compatible activities, the proportion of investments that seek conservation-related projects remains very small and is likely to plateau at a modest level. Second, many aspects of the policy and regulatory framework within which investment decisions are taken at best do not encourage conservation-friendly investment and, in a great many cases, serve as a strong counterincentive to linking finance and conservation.

10.4.2 Reform and Rule Change

So, a second field of activity in conservation finance is the review and reform of policy and regulatory measures to eliminate obstacles and counterincentives facing the flow of finance to conservation priorities. These obstacles exist at the "hard policy" end of the spectrum—e.g. the laws that impose tight risk-limitation requirements on large-scale institutional investors—to the soft end around perverse incentives (e.g. subsidizing the use of agricultural chemicals or damaging technology on fishing fleets); and even into corporate cultural practices (e.g. rewards to CEOs for short-term growth even when it is at the expense of long-term financial help).

Despite all of the initiatives highlighted above, it is clear that the challenge of aligning finance with conservation cannot be limited to setting aside resources and landscapes from exploitation and restoring those already degraded—the traditional heartland of conservation organizations. Instead, nature is threatened by the form and dimensions of economic activity. Addressing the challenge will require ensuring that regular economic activity does not continue to undermine biodiversity and ecosystem health while seeking to align the economy with the boundaries and requirements of natural systems.

Policy-making is largely the task of public authorities, whether at the international, national or local levels. New policies and rules are crafted, and old ones are reformed in a constant process aimed at advancing public policy goals. As new priorities—like biodiversity conservation—come on-stream, existing policies and regulations are reviewed and updated to ensure that they address these priorities. Many countries have reformed the policy framework in order to address the climate challenge. The process is now beginning for conservation of nature and ecosystems.

With conservation finance a new topic, the process of policy reform is only just taking shape, but all signs are that it will accelerate sharply in the coming years. A high priority is to review existing policies and regulations—particularly those affecting finance—to ensure that they do not offer perverse incentives to undermine biodiversity, as in the examples offered above.

International networks grouping public sector actors are also active. The Network for the Greening of the Financial Sector (NGFS) has grown rapidly since its foundation in December 2017, pulling together central banks and financial supervisors to accelerate the transition towards green finance solutions. Focused first on climate risk, they are now turning their attention to biodiversity. Their guidance to central banks on how responsibly to invest their endowments is already having a strong positive effect. The European Union has adopted an action plan on sustainable finance—including green labelling of financial products and the development of a new EU Biodiversity Strategy.

In addition, the European Commission has put forward a European Green Deal that includes a chapter on biodiversity. And the recently formed Coalition of Finance Ministers for Climate Action, though not yet focused on biodiversity, could provide a model for the future.

However, it is a broadly-shared view that progress, while encouraging, is still inadequate in pace and scale. Policy reforms are welcome but often move at a slow pace due to the complexities of the policy process, the requirement for political compromise and strong resistance from incumbent interests. The CBD process, involving the global community of States, moves at a pace that simply cannot keep up with the rate at which the problems grow more menacing. Given the trends in biodiversity loss and the breakdown in ecosystem services, it is important to move with urgency.

Support is growing for new norms and standards that will, it is hoped, quickly become widespread in public policy and a requirement for corporate value chains. One of the most promising is the proposal to develop a "Net Gain" pledge in which corporations (and, possibly, public works projects) would pledge that biodiversity and ecosystem services would be better off following the activity than it was before—either in the area where the activity takes place or, through offsets, in another location. Such new standards could rapidly become an expectation of investors or governments, including through their public procurement programmes. The idea of a "Net Gain" norm builds on the <u>Mitigation Hierarchy</u> in that it moves from an analytic and measurement approach towards an increasingly directive norm. In cases where net gain is secured through the purchase of biodiversity offsets, the IUCN <u>policy</u> on the subject provides useful guidance.

10.4.3 Positive Disruption

For the challenges to be met at a scale and pace commensurate with the scale of biodiversity loss and the pace of negative trends, action by the public sector and by capital markets must be subject to a far higher level of demand from civil society and the public. This 'disruptive' action rejects the patient and linear approaches to change in favour of moves that have a transformative effect. The attention of these actors is increasingly focusing in on the impact of finance on biodiversity and a wide range of tools is being used. Voluntary action by private actors, often impelled and encouraged by civil society, can be a necessary accelerator of transformation. Earlier in this chapter, the slow and tedious pace of progress in international negotiation and consensus-building was lamented. To stimulate that pace, it has been argued that policy and rule-change are necessary; but the reality is that present policies and rules benefit strong incumbent interests, and these defend their short-term privilege over the wider public benefit of change. The simple fact is that serious change only comes at that point where the advantages of the change are seen by political powers to outweigh the political advantages of the status quo. However, at that stage, change can come quickly.

In the meantime, a range of things are happening which suggest that fundamental shift may be on its way. The decision by eight major insurance companies no longer to insure fishing fleets that indulge in illegal fisheries is one telling example. So, too, are the moves by large-scale institutional investors—for example the decision by the Norwegian State Pension fund to divest from fossil fuel companies. Another example is a certification from the Marine Aquarium Council (MAC) for trade in ornamental fish designed for insurance and air cargo companies. This trade is significant in the Asia Pacific region.

On the civil society end of the spectrum there is a wide range of activities. Some are cooperative—such as the effort to promote and launch a 'No Net Loss' or 'Net Gain' standard for corporations and value chains, mentioned above. Others channel the frustration, fear and impatience of the general population at the slow pace of progress. Movements like the Extinction Rebellion that have recently been attracting attention, particularly in Europe, specifically call for disruptive action; their membership is not just made up of those who block bridges or public squares, but increasingly by doctors and lawyers ready to use the full arsenal of tools at their disposal to push for early change. Many other examples exist or have flared up at times, only to die back as quickly—such as the Occupy Wall Street movement after the global financial crisis in 2008/2009.

Disruptive action can be positive if it levels the playing field and counters the weight of industrial lobbies. It often represents genuine concern asking only to be channeled into positive pathways of change.

Other NGOs are mounting "name and shame" campaigns aimed specifically at financial institutions—for example those who fund land clearance for beef and soy development in Brazil or those that manufacture and promote carcinogenic pesticides in European countries. Fish Tracker, for example, supplies investors with detailed and accurate data on the fishing companies in which they are thinking of investing, thus giving them the chance, through their investment decisions, to reward responsible companies and punish those that undermine ocean biodiversity.

China has seen a growth in civilian actions a large part of which relate to complaints about lack of compliance with environmental regulations or the impact of industrial pollution or to unsustainable or inequitable use of natural resources.

Disruptive action can be positive if it levels the playing field and counters the weight of industrial lobbies. It often represents genuine and widely shared concern asking only to be channeled into positive pathways of change.

10.4.4 China Overseas and International

Developments at the international level must, of course, include the actions of China beyond its frontiers. As a massively important investor, the role of Chinese finance-sector players is a key factor in determining the chances for a rapid transition to sustainable financial practices linked to biodiversity. China's overseas investment takes many forms and operates across a wide spectrum comprising the Chinese government, policy banks, State-Owned Enterprises and private sector financial actors. Especially since its adoption of strong green finance measures, China is looking increasingly both at best practice internationally in the full range of norms and standards linking finance with biodiversity conservation, and extending the standards required in China to the investment of Chinese investors across the globe. Increasingly, the "social license to operate" will depend on value chains that respect and restore biodiversity and invest in the resilience of natural systems. Increasingly, also, those that ignore these requirements risk facing an ever-stiffer backlash.

10.5 Framework, Main Tools and Challenges of China's Conservation Finance

The green financial system in China has developed rapidly since 2015. The concept of green finance has not only permeated the national development plan and policy, but also received a wide response at the local government level. The green financial policy system is becoming more and more abundant, and the construction of relevant market infrastructure is also advancing. For Financial Institutions and investors, more and more attention has been paid to green development and green industry. Green investment action has become a new hotspot, and the active development of green financial instruments and green financial products has become a new option for the industry. As a part of the green financial system, conservation finance and its importance have been gradually recognized by people, while still facing many challenges.

10.5.1 Conservation Funding Sources in China

Despite the gradually increasing in private capital investment, the current funding for China's conservation activities is still heavily dependent on public sectors, including government finance and bank loans based on government credit.

10.5.1.1 Government Finance

Government finance include special funds from the central government, funds, investment, and appropriations (mainly in provincial or local governments). Taking the soil remediation as an example, the specialized fund for preventing soil pollution has been set up in 2016 and its total investment number in 2019 is 5 billion yuan. Any soil remediation projects are capable to share the specialized fund for preventing soil pollution as long as it has been included in the 'program library' of the Ministry of Ecology and Environment (MEE). When the funding has been allocated from central to local, the finance department at provincial level will be jointly with the local environmental protection department in charge of the funding using, and supply some funds depending on the local fiscal situation [8](Table 10.2).

10.5.1.2 Financing with Local Government Credit in Financial Markets

There are two ways of financing with local government credit:

---

[8] Notice from the Ministry of Finance on Printing <Management Method of Special Funds for Preventing Soil Pollution.

Table 10.2   The distribution of the special fund for soil pollution control in 2019

No	Province	Total	No	Province	Total
1	Beijing	184	17	Hubei	26,919
2	Tianjin	1409	18	Hunan	57,462
3	Hebi	29,382	19	Gongdong	31,375
4	Shanxi	4442	20	Guangxi	34,147
5	Inner Mongolia	14,255	21	Hainan	3828
6	Liaoning	6555	22	Chongqing	4907
7	Jilin	2787	23	Sichuan	10,281
8	Heilongjiang	4264	24	Guizhou	42,582
9	Shanghai	1365	25	Yunna	70,804
10	Jiangsu	15,736	26	Tibet	886
11	Zhejiang	29,240	27	Shanxi	15,930
12	Anhui	8692	28	Gansu	12,525
13	Fujian	12,810	29	Qinghai	12,857
14	Jiangxi	12,917	30	Ningxia	1271
15	Shandong	18,434	31	Xinjiang	1127
16	Henan	10,627	Sum		500,000

The first way is bonds issued by local government. They can be divided into general bonds and special bonds. General bonds are included in the public budget to balance the deficit; special bonds are included in the government funding budget, mainly investing in public welfare projects. The special bonds for local governments planned to issue in 2019 increases by nearly 60% compared with 2018 (1.35 trillion yuan). These special bonds not only invest in key national strategies, infrastructure project in extreme poverty areas, and major projects including railway, highway, and water conservancy projects, but also in ecological conservation and environmental protection projects.

The second way is bank loan. These loans are basically invested in companies that are included in the Public–Private Partnership (PPP) programs. The loans are accommodated by policy banks or commercial banks according to the result of risk evaluation.[9]

10.5.1.3   Corporate Financing from Financial Markets

There are two kinds of corporate financing models. First is to borrow from banks or to issue stock or bonds in the financial market based on corporate credit, which is currently the mainstream. Second is project-based financing from the market, mainly

---

[9] According to the National Development and Reform Commission (NDRC), in May 2015, NDRC has established a national-level PPP project database and 1043 projects ....

used for projects with stable cash flow.[10] Industrial Bank, for example, helped a state-owned water company obtain project funding in a relatively low cost by issuing 800 million yuan of green sustainable medium-term bonds in the China Inter-bank Bond Market.

In the capital market, the issuance of Green Bond has expanded rapidly in recent years. In 2016, domestic green bond markets had 51 bonds and 205.2 billion yuan in total. The number has increased to 222.2 billion yuan in 2018 with 139 bonds, and reached 360 billion yuan of green bonds in 2019. With the rapid development of the Green Bond Market, some large enterprises are easier to financing by issuing bonds.

In general, the difficulty of financing for ecological and biodiversity conservation is still widespread.

- Ecological restoration and biodiversity projects often require massive investments. Take soil remediation for example, the general investment amount around one to ten billion yuan for each soil remediation project is totally over the capacity of most private capitals. Due to the lack of business models, well knowledge, and policy guidance, few private equity and venture capital funds have entered the field of conservation investment. In addition, insurance product design and service are facing difficulties due to the lack of specific operability tools and methods, which also hinder the development and innovation of financing model.
- Green funds have limited investment in ecological protection. In recent years, some provincial and municipal governments have established government-led guiding green funds, and some market-oriented green funds have also emerged. But these two types of investment funds are not sufficient in supporting ecological protection investment. There are also some non-profit funds that invest little in this area due to their widespread attention and limited funding sources. Take the China Environmental Protection Foundation as an example. In 2018, the total assets of the foundation were less than 200 million yuan, and the amount of donations accepted in 2018 was only 120 million yuan. At the same time, the foundation's focus is quite extensive, including green recycling, green travel, green innovation, and ecological poverty alleviation. The funds that can support ecological protection is quiet limited.
- Investment from green credit and green bonds are insufficient. Since 2015, green development has been put on the central and local governments' agenda. The central bank, financial supervision department, and local government have introduced measures to encourage and support banks to launch green credit service. Some large and medium-sized banks have increased their green credit. Till June 2019, the green credit balance in 21 banks in China has surpassed 10 trillion yuan. However, regardless of the increasing amount of green credit, the investment is mainly focused in infrastructure such as transportation and energy. Although big banks have already been paying attention to ecological conservation, the green

---

[10] Generally, the funding side will require the funds to be closed operation: the company shall open a bank account for the supervision of the project funds, collecting the project income and the government paid funds. The income would be used with a priority for the loan repayment or repayment deposit.

Table 10.3 Main funding sources for China's conservation projects

Sources	Application frequency
PE/VC	*
Bank loan	****
Issue stock	*
Issue bond	**
Investment fund	*
Charity foundations and NGO	*
Private lending	**
Account receivable	***
Financing lease	*
Trust foundation	*
Non-environmental business subsidies of enterprises and public welfare programs	***

*very few; **few; ***moderate; ****many

finance principle is hard to translate into practice due to the poor analysis ability on conservation projects of loan officers. In terms of the use of green bonds in 2018, only 5.3% of the funds were invested in pollution prevention, ecological protection and climate change, and the amount was only about 10.6 billion yuan.[11]

- Most ecological restoration projects are carried out by small and medium-sized private enterprises, who are difficult in obtaining qualifications on issuing green bonds in the market. Bank loans for ecological protection mainly rely on the credit of enterprises, and most of the loans are working capital loans lasting for one year, which do not match the implementation cycle of investment projects.

- Due to the shortage of supply in the finance market, it is quite common for large upstream enterprises to default on payment to small and medium-sized downstream enterprises in the ecological conservation industry. There are also companies engaged in ecological conservation, which only rely on profits from other businesses to subsidize their ecological conservation investments (Table 10.3).

---

[11] Source: Wind Financial Database, Central University of Finance and Economics.

10.5.2 Challenges in Conservation Financing in China

10.5.2.1 How to Establish a Sustainable Business Model to Attract Private Capital

As a matter of fact, there is still a lack of effective business models in the field of ecological conservation. Conservation finance has the nature of positive externality, while the difficulty of attracting private capital lies in how to establish a stable and sustainable business model. From the perspective of international experience, some conservation projects have certain investment incentives, but the significance of such incentives is not universal. Taking soil pollution governance as an example, recent years, due to intensified legal enforcement[12] under the pressure of legislation and environmental regulation, some enterprises (especially foreign-funded enterprises) take the initiative to control contaminated soil within their capacity.

In first-tier cities and some second-tier cities, the value of the lands is high enough for investors to recover their investment or even gain profits in soil pollution remediation projects through land transfer. Therefore, both local governments and enterprises have the incentive to invest. Even though, the cost of remediation of contaminated soil is still very high so that local governments or enterprises can only manage those issues by installments.[13] The issues are more serious in some third tiers and fourth-tier cities, especially in rural areas. With limit space for land appreciation and widespread local financial constraints, plus the fact that it is hard to identify the liability or the entities are incapable of paying (for example, some mining enterprises have been dissolved or gone bankrupt), the lack of funding has become a huge obstacle to the remediation of contaminated soil.

---

[12] China's soil pollution prevention and control law came into effect on January 1, 2019. "The law stipulates the responsibility of government and land users for the protection of unpolluted land and the treatment of polluted soil." "The competent departments of ecology and environment (CDEE) at or above the municipal level shall, in light of the discharge of toxic and harmful substances and other conditions, formulate a list of key supervised entities for soil pollution in their jurisdictions, which shall be disclosed to the public and updated in due course." "The law requires key regulators of soil pollution prevention to strictly control the discharge of toxic and harmful substances and report the discharge to the CDEE on an annual basis. To formulate and implement the self-monitoring plan and report the monitoring data to the CDEE. Land plots listed in the list of soil pollution risk control and remediation for construction land shall not be used as land for housing, public management and public service." "The holder of the right to the use of the land shall take effective measures to prevent and reduce soil pollution and shall bear the responsibility for the soil pollution caused by it according to law." In particular, the law obliges those liable for soil pollution to remediation. If the liability for soil pollution cannot be ascertained, the land use right holder shall be responsible for the remediation. This means that in addition to the production and operation process, enterprises must monitor soil pollution and have the responsibility to repair; Polluted land can only be transferred if it is restored and turned into "clean land".

[13] According to our research, the cost of general organic pollution repair is 300 ~ 500 yuan/m^3. If the land was polluted by oil tank is leaking, the repair cost is 1000 ~ 2000 yuan/m^3.

10.5.2.2   How to Establish Incentive and Restraint Mechanisms for Financial Institutions and Large Institutional Investors

From the perspective of a few successful cases in China, we found a common feature that financial institutions failed to play a supportive role in deploying comprehensive utilization of resources to raise funds for the operation of conservation projects. For example, in Sishui, Shandong province, a company has invested more than 200 million yuan in the restoration of abandoned mines. This investment is mainly covered through the comprehensive utilization of tailings and waste resources and the development of ecological agriculture and tourism industries. Another example is that a company has invested 6 billion yuan in Kubuqi desert's management over the past 30 years. It has also made up for the cost of desert management and ecological restoration with the profits from agriculture, animal husbandry, health and well ness, ecological industry, photovoltaics, and ecological tourism. In this business model, financial institutions fails to play a role because firstly, in the risk management framework of existing financial institutions, green assets created cannot be valued or used for mortgage guarantee. Secondly, financial institutions lack reorganization and management capability on this integrated business model. Thirdly, the financial institutions lack awareness of the challenges brought by future environmental changes to their own and lack motivation to actively participate in environmental risk management.

Financing is even more complicated for projects involving biodiversity conservation, as biodiversity conservation involves a wider range of sectors and policy tools, and it is harder to measure the externalities. For financial institutions and large institutional investors, on the one hand, identifying and managing environmental risks is a new topic.

10.5.2.3   How to Mobilize the Whole Society to Support Ecological Protection and Biodiversity Conservation Activities

Ecological protection and biodiversity conservation are systematic efforts, which closely relate to public recognition on ecological protection and sciences, investors' social responsibility, scientific research ability, public education, and the role of NGOs. Ecological protection and biodiversity conservation not only require coordination of different policies, but also need further systematic governance and regulatory innovation.

- How to resolve the conflicts among ecological and biodiversity conservation, and the development of local economic and residents' individual interests. As it concerns the vital interests of residents, conservation activities often produce such conflicts as "man and animal fighting for land" and "man and animal fighting for food". Especially for the less developed areas, this kind of conflict is more prominent. For local governments, they lack intentions to address problems of climate change, environmental protection and biodiversity conservation, while institutions and volunteers who are specifically responsible for conserving biodiversity lack the power or capacity to deploy resources.

- How to restrain enterprises' investment on non-green projects or environmentally damaging projects. As the main participants of the market, enterprises play an important role in conservation. At present, many enterprises still take profit and investment payback period as the primary basis for investment decision-making, and rarely consider the impact on the ecosystems and the environment. Only by adjusting current fiscal and finance policies promptly and reducing financial investment activities that are not conducive to ecological protection, including internalization of environmental externalities costs, reduction of output and increase of costs, so as to reduce the return on investment of polluting and damaging ecological projects, can a green capital flow direction be formed in the whole society.

- How to attract philanthropic foundations and other non-profit organizations to participate in ecological conservation activities. As of April 8, 2019, there were 5599 charitable organizations in China, of which 1521 were eligible for public offering. Public attention to charity activities is very high. 20 Internet fundraising information platforms designated by the ministry of civil affairs received more than 8.46 billion hits, followers and participants in 2018. For non-profits organizations, the return on investment and the recoverability of funds are not their focus. The key to attracting them to enter the field of ecological conservation is to let investors to know the use of funds and social value timely. Therefore, transparency and validity of information should be guaranteed. Meanwhile, by relying on their expertise and financial strength in the field of ecological conservation, these institutions can serve as a bridge for the government to open up private capital and financial institutions. These institutions can not only provide professional advice for project planning and program implementation, but also provide innovative models and approaches for the investment and financing participants. In this aspect, some successful experiences from abroad can be used for reference by China.

- How to improve the understanding of ecological conservation in the whole society. The conservation of ecology and biodiversity by governments alone is clearly not enough and requires the participation of society as a whole. At present, China has reached a social consensus on environmental and ecological protection while without being a priority. In China's practice of conservation investment, the funds are mainly invested in ecological restoration activities (after-treatment), with little investment in pre-prevention; the relevant government departments attach great importance to pollution control in special fields such as soil, water and atmosphere, but pay less attention to the conservation of whole ecosystems, including biodiversity. On the other hand, the lack of expertise and capacity of local governments makes it difficult to improve efficiency of the use of ecological protection funds. At the same time, community publicity and education to the public are also relatively weak, so that the volunteers' activities of ecological conservation lack a broad social foundation.

10.5.2.4   How to Incorporate Gender Perspective into the Development of Conservation Finance and Let Women Play a Role

In view of the extensive knowledge and practice of climate finance, a wealth of ideas and suggestions for examining climate finance from a gender perspective have emerged, which are worth learning from, but also reflect significant gaps and problems:

First, the international community has paid long-term attention to climate finance and developed a series of mature financial mechanisms, tools and methods to address the mitigation and adaptation needs of climate change. Based on the specific detailed data and practical cases regarding climate finance sources and channels, as well as the corresponding projects and participants, the integrating of gender perspective is more smooth and systematic in climate finance. However, conservation finance is still in the initial stage of research and practice, and there is still no broad consensus on its basic definitions, standards, concrete mechanisms, tools, and methods. At present, we can only affirm the importance of gender perspective in principle. Further demonstration and implementation are still needed to include gender perspective in the long-term research scope.

Second, the current focus on gender equality in climate finance starts with ensuring women have a voice and priority in accessing climate finance support, and ensuring that available climate fund is used and allocated fully regarding to the needs of female groups (which is also in line with the requirements of most funders). At present, conservation finance is in urgent need of solving the problem of leveraging more funding, especially establishing the close cooperation mechanism with all sectors of society. Therefore, we believe that the gender perspective in conservation finance should focus on how to promote the use of gender equality concepts and principles in investment and financing decision-making in financial sectors, and make full use of women's attention and cognitive advantage on ESG issues, so as to achieve two objectives: (1) to make ecological conservation a priority issue for the entire financial system; (2) based on the existing experience and practice of climate finance, the needs of different vulnerable groups (including women) should be fully considered in the use and allocation of financial resources, and the principle of gender equality should be incorporated into the whole process of planning, implementation, monitoring and assessment of conservation specific projects.

10.6   Overall Plan to Promote Conservation and Biodiversity Finance in China

Ecological Conservation (restoration) has several characteristics including large-scale investment, long payback period and indirect income. The externalities of biodiversity conservation projects are more extensive, involving more individuals, more complex situation, and the environmental benefits are more difficult to measure. At present, financial institutions and large institutional investors generally pay little attention to the protection of ecosystems and biodiversity and lack the proficiency to identify the benefits and risks of projects. In order to make the ecological protection investment more attractive and reduce the damage of traditional financial activities to the ecological environment, a financial framework is needed to establish from the dimensions of financial institutions, administrative departments, relevant policies and financial markets, This study seeks to propose plans on establishing an environment for conservation finance, and stimulate the incentives for financial institutions and large-scale institutional investors from internal motivations and external restraints.

10.6.1   Improve the Efficiency of Using Government Financial Funds

10.6.1.1   Identify Key Areas for Conservation

The premise of financial instruments to involve in conservation investment is that the security and profitability of funds will be guaranteed. This requires: (1) the economic benefits of the conservation activities are clear and measurable; (2) The green assets have liquidity; (3) if the first two cannot be achieved, then the investment requires clear social benefits that are helpful to its branding and business strategies. Examples are the Ant Forest of Ant Financial, and some corporate donations to conservation-related funds.

Not all ecological conservation work is amenable to financial institutions, and some work can only be undertaken by governments. Therefore, the government needs to identify the focus areas that finance can support, and establish an enabling policy framework for the sustained investment of financial institutions and private capital. One potential approach is that the MEE can prioritize the social impact of various conservation measures and the urgency of their capital demand, based on the short-and long-term economic and social values of different conservation work. By this way, financial institutions and private capital will be able to identify the key areas of ecological conservation and biodiversity conservation and to choose the direction that suits them. The government department in charge of investment (NDRC) can cooperate with the MEE to draft investment taxonomy of ecological and biodiversity conservation.

10.6.1.2 Establish a Fiscal and Taxation Policy System that Encourages Financial Institutions and Private Capital to Increase Input in Conservation

China has established a quite complete ecological compensation system and formulated special fund management measures for the conservation and rehabilitation of soil, rivers, lakes and natural forests. The next step requires: (1) to test the effectiveness of these compensation funds and special funds; (2) to link some public funds with financial institutions. This includes clarifying that projects supported by special funds can get preferential policy-oriented financial support and informing financial institutions of these projects supported by special funds; granting appropriate interest rates on loans or issuing green bond to some important ecological rehabilitation and conservation projects; considering granting certain tax concessions to financial institutions and investment institutions that have made outstanding contributions in the ecological field; subsidizing part of insurance premiums to enterprises that purchase green insurance in the field of ecological conservation; and allocating more funds to support voluntary actions and education activities for raising knowledge and awareness among public.

10.6.1.3 Other Supporting Policies

Many local governments have set up government industry investment funds. The funds are mainly focused on scientific and technological innovation and large-scale infrastructure construction, with little allocation to conservation. Requirements about allocating a certain proportion of funds to the ecological field should be defined. The assessment requirements of industrial investment funds should be adjusted based on the characteristics of conservation. Local governments should set up green industry funds to stimulate private investment in environmental enterprises and projects. Policy banks should be required to support the development of ecological and biodiversity conservation.

Considering the tight resources of construction land in China, private investment can be directed to conservation through land policy. Several regions with successful ecological management can be selected to pilot on the award and compensation policy for land restoration, and the restored land can be traded across provinces as cultivated land.

In addition, to address the limited capability of collecting and analyzing the environmental cost of the projects, the government can purchase public services to establish public environmental cost information system, and provide a basis for decision-makers and investors in the whole society.

10.6.2 Strengthen the Coordination Between Macro-financial Policies and Financial Regulatory Policies

10.6.2.1 Monetary and Credit Policies

In the current green finance field, the People's Bank of China has provided some policy support for commercial banks to develop green credit and the construction of green finance reform innovation pilot areas, including (1) re-lending; (2) Green Bill discounting; (3) supporting and open up green channels for enterprises to issue green debt financing instruments; (4) adding green credit indicators in Macro Prudential Assessments; (5) release the 'Green Financial Development Report'; and (6) establish green credit statistical system. In the future, the People's Bank of China should give prominence to ecological and biodiversity conservation in its Green Finance support policies. Ecological and biodiversity conservation can also be listed as a separate indicator for policy support.

10.6.2.2 Financial Regulatory Policy

Guide financial institutions and large institutional investors through the adjustment of financial regulatory policies. These include: (1) expanding the scope of green credit instruments, allowing banks to develop eco-asset-backed credit products on a pilot basis; (2) advocating and promoting ESG investment and establishing ESG evaluation system for listed companies; encourage large institutional investors such as sovereign wealth funds, insurance companies and large fund companies to take the lead in developing ESG investments; (3) establish an evaluation index system for green credit, and launch green rating pilots for commercial banks; (4) allow the issuance of green bonds using ecological assets such as restored land and formed ground attachments.

10.6.3 Exert the Role of the Capital Market to Support Conservation Finance

Chinese capital market is still deficient in supporting conservation investment and financing. On the one hand, some objective factors lead to this issue, including long operation cycle of conservation projects, low or no return, small scale of environmental enterprises and instability of cash flow. On the other hand, there are some problems in the capital market itself, including the issuance process of securities, investment incentive, and the proficiency of institutions. To solve these problems, there are several available measures: (1) Strengthen support and services in the listing of environmental enterprises and set up green channels for IPO, refinancing, M and A and reorganization of such enterprises; (2) Take into account that most of the environmental enterprises are asset-intensive, so equity and bond financing can be moderately loosened up on the use of funds raised; (3) Strengthen support for environmental enterprises to be listed on the National Equities Exchange and Quotations (NEEQ) market; (4) enhance the professional and technical capabilities of financial institutions. Securities Companies, fund companies, and other securities industry institutions may set up Conservation Finance Department and Green Commissioner in reference to the bank, to enhance their capability building in the assessment, screening and investment of ecological conservation projects. Improve the incentive mechanism and link environmental project investment with individual performance; (5) improve the professional capabilities and credibility of third-party assessment agencies by unifying the filing process, clarifying the withdrawal rules, and strengthening the credibility of third-party agencies.

10.6.4 Improve the Basic Conditions of Conservation Finance

The development of conservation finance is a systematic work that requires enabling relevant environment and conditions. Conservation finance serves the protection of ecosystems and biodiversity, thus, if the local government does not have a strong motivation in environment protection, and there are not many conservation projects, conservation finance would have nothing to discuss. In other words, the development of conservation finance depends not only on the idea and behavior of financiers, but also on the dynamics of conservation.

10.6.4.1 Strengthen the Rule of Law and Provide Stable Expectations for the Development of Conservation Finance

China has formed a relatively complete system for ecological and environmental protection policies. However, the existing legal framework is not yet perfect, and there is a certain degree of overlapping and lag in authority between the departmental regulations; In addition, interest conflicts are prominent between central and local government, as well as between the government and the market. Under the pressure of the central government's performance evaluation, local governments have contradictory incentives: On the one hand, there is an urgent need for green development and environmental improvement; on the other hand, in order to maintain a certain economic growth and fiscal revenue, there is also an incentive to tolerate industries that damage the environment but have significant contribution to tax revenues. Thus, local governments are actively developing emerging green industry, while imposes limited environmental regulation on existing enterprises. Under this scenario, we need to further improve the legislation of ecological and environmental protection laws and regulations, clarify and increase the liability of environmental polluters, and strengthen supervision and law enforcement.

10.6.4.2 Addressing the Measurement of the Externality of Ecological Investment

An important reason for the lack of motivation of conservation finance is the distortion of price signals. Therefore, it is necessary to make explicit the hidden benefits of 'lucid waters and lush mountains' and the costs of pollution. The government needs to reconstruct the pricing mechanism by defining the property rights of carbon emission and pollutant discharge, and addressing the charge for the external benefits generated by conservation projects. The reform should provide policy and market signals to reduce the economic value of resource-intensive and carbon-intensive investment, changing the investment preference of financial market participants. The concept of Natural Capital Liability (NCL) that put forward by the British company Trucost could be referred here. NCL has the potential to quantify the environmental costs caused by air pollution emissions, water pollution, and solid waste, and assess the scale of 'negative externalities' that are not reflected in the current market prices. Meanwhile, it is also necessary to establish the concept of 'ecological capital' to quantify the value of ecological assets and reflect its positive externalities into market.[14] On this basis, establish the valuation and trading mechanism of ecological assets to promote their circulation.

10.6.4.3 Establish the Inter-department Coordination and Information Communication Mechanism

First, an inter-departmental coordination mechanism should be established between the ministries of ecology and environment, natural resource, finance, and other departments of macroeconomic and financial regulation. On this basis, promote the concept of conservation finance in the public sectors and ensure the consistency and stability of conservation finance policies. Establish a joint meeting system for ecological and biodiversity conservation, and establish a green channel for private financing in key conservation areas.

Second, comprehensively consider the demand of ecological protection, economic development and social equity in key ecological protection areas, and put forward a package of policies to improve the quality and efficiency of conservation finance. Broaden financing ideas and increase the conservation synergy of conventional investment by carry out the policy coordination.

---

[14] This information system can provide decision-making reference for policy-makers in determining price subsidy, resource tax, and pollution discharge fee, as well as be used by various investors in investment and credit decision-making.

Third, build mutual information communication and sharing platform among departments of industrial management, ecological and environmental protection, and financial regulatory. Use this platform to foster timely communication of technical information, industry standards, and illegal handling situation of ecological and environmental protection. China should make the full use of the power of public supervision and assessment to provide timely feedback on the policy implementation and improve the efficiency of government work.

10.6.5   Fully Utilize the Power from the Whole Society

More attention should be paid to the role of public governance while promoting private investment. The incorporation of public governance could relive public sectors' governance pressure while improving conservation efficiency and expanding financing channels. In foreign countries, there are many successful cases of cooperation among financial institutions, investors and large-scale environmental protection organizations. The latter not only provides available business plan for conservation, but also uses its own grants to develop finance and marketing tools to reduce investment costs. Such 'catalyst' type cooperation allows financial institutions and investors to manage effectively their investment risks, as well as fosters wider participation by sharing experiences. These successful experiences are worth learning from.

In addition, publicity and education should be strengthened to promote the full understanding of the importance of ecological and biodiversity conservation, especially the understanding among financial institutions, large-scale institutional investor executives and practitioners, and the people in the surrounding areas of reserves.

10.7   Policy Recommendations on Promoting Conservation and Biodiversity Finance

The three recommendations below address different aspects of the conservation finance challenge. The first addresses the understanding of conservation finance and the top-level framework. The second addresses the optimal use of policy and market instruments to increase the flow of finance into activities that are favourable to the wise use of nature and natural resources while reducing the flow to harmful activities. The third addresses the governance and institutional framework required for the first two to be successfully accomplished.

10.7.1 Expand the Scope of Green Finance and Provide Strategic Guidance for Private Finance to Enter the Fields of Conservation and Biodiversity

Purpose: To improve the understanding of the importance of conservation finance and development path; and to integrate relevant concepts into major national plans.

10.7.1.1 Short-Term

Existing green finance framework covers conservation, pollution control, efficient use of natural resources (energy saving and emission reduction, etc.), but lack adequate attention to biodiversity protection. Governments, businesses, financial institutions and investors need to raise awareness of conservation biodiversity. Green Finance Framework should be reviewed and incorporate the concept in relation to biodiversity. In addition, relevant contents of conservation finance should be added to the investment guidelines and national plans such as 14th Five-Year Plan.

Analyze the demand and supply potential of conservation finance, and evaluate the effectiveness of existing environmental strategies, conservation finance practices, and public financial support. Develop and apply robust measures to assess the impact of both private and public finance on biodiversity and ecosystems, drawing on the best models currently under development or in use internationally.

Develop and apply the principle that public and corporate development activity must lead to "no net loss" of biodiversity, based on the Mitigation Hierarchy methodology or similar instruments. Where loss is unavoidable, it should be compensated by applying "biodiversity offsets" at least in proportion to the loss incurred in both quantitative and qualitative terms.

Based on a review of best practice in the use of Strategic Environmental Assessment, fully apply SIA methodologies to public policies and programmes, and to both public and private developments in areas of particular natural richness or in fragile ecosystems—and especially in relation to large-scale infrastructure development projects.

Draw on Chinese and international experience to explore the full range of available 'nature-based solutions'.[15] China has many experiences in mangrove protection, desert management, and other areas that fully integrated the protection of biodiversity. It is recommended to study these cases in depth, promote their successful experiences, convert the idea of developing 'large man-made projects' in the past, and leverage the power of natural to conservation.

At the coming CBD, integrate key elements and recommendations of conservation finance and proposals in the relevant declarations and outcomes, and play a leadership role in introducing fresh innovations that will accelerate the political acceptability

---

[15] Nature-based Solutions are defined by IUCN as 'actions to protect, sustainably manage, and restore natural or modified ecosystems that address societal challenges effectively and adaptively, simultaneously providing human well-being and biodiversity benefits'.

and importance of conservation finance. Examples are the development of a No Net Loss standard for corporate value chains, ideally including the following:

- The application of biodiversity mitigation hierarchy for land use development
- Land use planning for biodiversity (to identify high biodiversity value areas to avoid losses and areas to achieve gains)
- Closing markets for the sale and trade in wild animals and all endangered species
- Strategic environmental assessments to identify the actions to deliver no net loss for large-scale infrastructure development
- Natural capital-positive infrastructure planning and ecosystem restoration.

Further, there are now several convergent efforts tending towards the rapid establishment of a Task Force on Nature-Related financial disclosure, with a formal launch possible at the CBD COP. Drawing from the experience of the similar Task Force on Climate-related Financial Disclosure (TCFD) but with more ambitious terms of reference, China is in a unique position to lead and to establish conservation finance as a topic with the same standing and urgency as climate finance.

10.7.1.2   Mid- and Long-Term

Incorporate biodiversity into spatial planning and eco-civilization planning. Identify the key areas for conservation.

10.7.2   Improve the Policy Framework of Conservation Finance and Establish a More Effective Incentive and Restraint Mechanism

Purpose: Internalize both the positive and negative impact of economic activities by improving the design of fiscal and taxation policies, investment policies, and financial policies, and thus provide a stable investment expectations and market environment for financial and investment institutions.

10.7.2.1   Short-Term

Define the criteria of ecological and biodiversity conservation projects to provide a benchmark for the investment of financial institutions and social investors.

Financial regulatory authorities (e.g. PBOC, CBRC, CIRC, and CSRC) should provide guidance to financial institutions, trust funds, insurance asset management companies and other large institutional investors through credit and regulatory policy so as to emphasize on the impact of their activities on the ecosystems and natural resources.

Support commercial banks to consider the effect of loan targets' ecological protection as a factor for pricing. In addition, support the conservation finance innovation activities for the project of 'nature-based solutions', which is different from engineering project in financing with a very low chance of success in practice.

Evaluate the 'green' activities and ESG performance of financial institutions and encourage the capacity building of financial institutions and investors on evaluating and managing the risk of "natural-based solutions".

Clarify the standard of environmental information disclosure and urge listed companies and financial institutions to disclosure the conservation information of their projects. In addition, advocate and promote ESG investment in the capital market.

Encourage investment in conservation finance through fiscal measures. Piloting land and tax incentive policies in areas with significant ecological improvement.[16]

Establish a non-profit environmental cost information system under the support of Government Procurement System and improve the information access and analysis abilities of decision makers and investors in conservation finance.

Guide policy-based financial institutions and government funds, including social security funds and industrial investment funds of governments at all levels, to increase support for conservation investment.

Improve the coordination mechanism between public funds and social capital, and increase the use of blind finance. Reduce the risk of conservation investment with the help of public funds with low return requirements.

Set green thresholds for both foreign procurements and investments, especially in the context of Belts and Roads Initiative, to establish a green supply chain.

Strengthen the application of the precautionary principle, the duty of care principle and the 'no net loss' in the value chain of enterprises.

10.7.2.2 Mid- and Long-Term

Establish the natural capital accounting system to provide standards and methods for the calculation of Green GDP. Applying the Green GDP as an evaluation benchmark in the promotion system for local officials when the accounting system was mature.

Incorporate the measurement and calculation of the stranded assets' risk raised by biodiversity loss in the accounting system.

Regarding the situation that ecological losses are difficult to recover, explore the compensate mechanism based on the principle of 'no net loss', namely compensate ecological losses with 'ecological protection behavior', instead of monetary based compensation. In addition, take the 'no net less' principle as the next step in optimizing the ecological compensation system.

Establish a market-based mechanism for pricing, evaluation and flowing the ecological assets, to promote the appreciation of ecological assets in the flow and

---

[16] Experience from South Africa: Land owners could deduct personal income tax if they could well manage and maintain the biodiversity on the land according to the appointment.

attract more private capital to the field of ecological conservation. Based on the practical experience of carbon emission trading pilot regions such as Beijing, Guangdong, and Shenzhen, introduce the voluntary forestry carbon sink trading projects as offset projects in the national carbon emission trading market. To further internalize the positive externalities of ecological services, consider to integrate the natural service value of wetlands, meadows and other ecosystems in the eco-market when it was well established.

Adjust tax policies to enhance conservation. To ensure the effectiveness of fiscal incentives, the taxation and regulatory system should be coherent and the unreasonable incentives that harms biodiversity should be eliminated. Taxes on activities that undermine biodiversity should aim at behavior change, as conserving biodiversity is more important than increasing tax revenue.

Undertake a comprehensive review of public subsidies. Globally, the scale of unreasonable subsidies to sectors harms climate change and diversity, including fossil fuels, is huge. Some countries are working to adjust subsidy policies. It is recommended to draw on experience from the green ecological reform of agricultural subsidies, by the Ministry of Agriculture and Ministry of Finance. Undertake a comprehensive review of public subsidies in industrial sectors such as mining, manufacturing, power generation, and construction, to identify the subsidies that conflict with China's industry transition and harm the nature and biodiversity to accelerate the green transformation of industries in China [25]. Introduce subsidy reforms that can promote the development of the economy and conserve the biodiversity at the same time. Further optimize the economic leverage of China's fiscal subsidies.

10.7.3   Improve the Infrastructure of Conservation Finance by Strengthening the Systematic Management of Natural Resources and the Eco-environment Protection

Purpose: Optimize the policy and market environment of conservation finance.

10.7.3.1   Short-Term

Strengthen the enforcement of environmental law and justice, especially the enforcement in ecologically sensitive and fragile areas. Up to April 2017, there were 956 environmental resource courts in China, including the trial courts, collegiate bench and circuit courts. Even though, the environmental justice in China still faces many challenges, including the imperfection of legal basis and the difficulty in dealing with inter-regional cases, such as obtaining the evidence cross-regionally. To strengthen the legal enforcement in conservation field, multiple actions should be taken, including clarifying the judicial interpretations of ecological and environmental damage, standardizing the inter-regional environmental jurisdiction system, and improving the evidence collection and enforcement system in environmental justice.

Establish the Corporate Eco-environment Credit System to promote voluntary eco-friendly activities. Incorporate indicators about the ecological compliance into the existing metrics of Corporate Environmental Credit System, which was jointly established by the former Ministry of Environmental Protection and the Development and Reform Commission. Compliance behaviors of land using, resource extraction, deforestation, and the hunting and processing of wild animals and plants are potential indicators here [26]. Linking this new credit system with financial sectors could improve information disclosure of eco-environmental compliance and thus be conducive to the practice of conservation finance (such as green credit). On this basis, consider further establishing an eco-environment credit enhancement mechanism. That is, under the certification of professional third-party organizations, companies that actively reduce the ecological impact of production activities, or carry out eco-environment restoration activity, can be awarded credit points additionally. This mechanism is designed to encourage companies to take eco-friendly measures voluntarily.

Develop the community-based conservation in China to solve challenges of residential livelihoods and community governance simultaneously and comprehensively. In addition, promote and pilot successful conservation financing and business models that integrated with community solutions in poorer areas.

Support the development of consulting industry in conservation finance field to enhance the capacity building of financial and investment institutions in risk identification and decision making. Support and guide existing intermediaries, including credit rating agencies, asset evaluation agencies, accounting firms, law offices, data service companies, and consulting companies to develop third-party services related to conservation finance.

10.7.3.2 Mid- and Long-Term

Promote legislative activities that in line with the development of conservation finance, and explore ways to increase of ecological protection responsibility of companies, banks, investors, and trustees in Commercial Bank Law, Securities Law, Securities Investment Fund Law, and Trust Law.

Improve the environmental impact assessment system and strategic environmental impact assessment mechanism in China. Advance the evaluation methods of ecological impact for construction projects, and scientifically evaluate its systemic impact on species survival, especially the impact on reproduction and genetics. In addition, pay more attention to the evaluation of indirect and long-term impacts.

References

1. "What We Do." Conservation Finance Alliance, www.conservationfinancealliance.org/what-we-do.
2. "WWF Guide to Conservation Finance." WWF, wwf.panda.org/?175961%2Fwwfguidetocon servationfinance+%5B2020-02-20%5D.
3. Ma Guoxia et al., Accounting Research on China's Terrestrial Ecosystem Product Value in 2015, Environmental Science in China, 2017.
4. "Returning farmland to forest and grassland, Ecological and livelihood win-win", 国家林业和草原局政府网 www.forestry.gov.cn/main/435/20190715/102809090429670.html.
5. "Natural forest protection project, what has been gained in 20 years", 人民网 env.people. com.cn/n1/2018/0518/c1010-29999970.html.
6. "The problem of soil erosion in China and its prevention and control measures.", 中国人大网 www.npc.gov.cn/npc/c541/201010/bbcc0908d02a401db646e9509399c058.shtml.
7. "The Information Office held a press conference on the "Comprehensive Evaluation Report on the Construction of the Three North Shelterbelt System in 40 Years", 中国政府网, 24 Dec. 2018, www.gov.cn/xinwen/2018-12/24/content_5351500.htm.
8. "All pilot funds for the 2019 ecological protection and restoration of landscapes, forests, fields, lakes and grasses from the central government have been issued.", 中国政府网 20 Sept. 2019, www.gov.cn/xinwen/2019-09/20/content_5431649.htm.
9. "China has accumulated nearly 30 billion yuan in the project of returning grazing to grassland." , 中国政府网, www.gov.cn/xinwen/2018-07/17/content_5307177.htm.
10. Chen et al. (2019) China and India lead in greening of the world through land-use management. Nature Sustainability, (2) 122–129.
11. Ministry of Water Resources, national soil and Water Conservation Plan (2015–2030); 2018 National Press Conference on soil and water loss dynamic monitoring results.
12. "The area of desertification and desertification in China has been 'double-decreased' for 15 consecutive years.", 中国政府网, www.gov.cn:8080/xinwen/2020-01/06/content_5466784. htm.
13. Protected Area Friendliness, www.baohudi.org/?p=5130.
14. Wang Xiaoxia and Wu Jian, Analysis of the level of financial investment in China's nature reserves, Environmental Protection, 2017.
15. Estimating Wetland Restoration Costs at an Urban and Regional Scale: The San Francisco Bay Estuary Example,2013; Cost Sheet for Reconstructed Wetlands, 2016.
16. Hong Yang, China's soil plan needs strong support, Nature, 2016.
17. Yi Aijun, Discussion on Marine Ecological Security in China, Environmental Conservation, 2018.
18. China maritime disaster bulletin, 2018.
19. Ministry of Commerce People's Republic of China (MOFCOM), "Statistics," Statistics Foreign Trade Cooperation, 2020, http://english.mofcom.gov.cn/article/statistic/foreigntrade cooperation/; Scissors Derek, "China Global Investment Tracker 2019," China Global Investment Tracker (Washington: American Enterprise Institute, January 2020), http://www.aei.org/china-global-investment-tracker/.
20. Alice C. Hughes, "Understanding and Minimizing Environmental Impacts of the Belt and Road Initiative," Conservation Biology 33, no. 4 (August 2019): 883–94, https://doi.org/10. 1111/cobi.13317.
21. Secretariat of BRI International Green Development Coalition, "Joint Research on Green Development Guidance for Belt and Road Initiative (BRI) Projects Was Launched," BRI Green Review (Beijing: BRI International Green Development Coalition, January 2020).
22. Green Finance Leadership Program, "Green Investment Principles (GIP) for the Belt and Road," December 2018, http://www.gflp.org.cn/public/ueditor/php/upload/file/20181201/154359866 0333978.pdf.

23. China Banking and Insurance Regulatory Commission (CBIRC), "绿色信贷指引" [Green Credit Guidelines]" (Beijing: China Banking and Insurance Regulatory Commission (CBIRC), February 24, 2012), http://www.cbrc.gov.cn/chinese/home/docDOC_ReadView/127DE230B C31468B9329EFB01AF78BD4.html.
24. Berger, Joshua. "Global Biodiversity Score: Measuring a Company's Biodiversity Footprint." CDC Biodiversité, a Nature-Based Company. https://ec.europa.eu/environment/biodiv ersity/business/assets/pdf/Assessing_the_footprint_of_economic_activities-Global_Biodivers ity_Score.pdf.
25. "The Ministry of Finance and the Ministry of Agriculture have vigorously promoted the establishment of a green ecological-oriented agricultural subsidy system reform", 中国政府网, www.gov.cn/xinwen/2016-12/19/content_5149900.htm.
26. Ministry of Environmental Protection. "Guiding Opinions of the National Development and Reform Commission on Strengthening the Construction of Enterprises' Environmental Credit System_2016 State Affairs No. 9, 1372 Court Bulletin", 中国政府网 www.gov.cn/gongbao/ content/2016/content_5059107.htm.

Appendix A
Policy Recommendations from the China Council for International Cooperation on Environment and Development (CCICED)

From Recovery to Green Prosperity

Accelerating the transition toward high-quality green development during the 14th Five-Year Plan period

The 14th Five-Year Plan (FYP) period will see the timeframes of the "two centenary goals" converge. Therefore, the choice of the path, policy arrangements, and key targets during the period will set the stage for the medium- and long-term development and goals vital to the realization of the Chinese dream. At present, getting out of the COVID-19 quagmire and getting on with economic recovery have become top priorities for all governments. The international community will closely follow China's economic, social, ecological, and environmental strategies laid down by the 14th FYP, which, in the era of globalization, is not only key to China's sustained and stable growth but also related to global green prosperity and well-being.

CCICED members highly appreciate Chinese President Xi Jinping's reiteration of the new development concept that "green is gold" during important speeches in Zhejiang, Shaanxi, Shanxi, and other provinces. President Xi's great attention and continued emphasis on green development has not only cemented domestic confidence and conviction, but it has also instilled hope in the international community for a successful green transition starting with the green recovery.

CCICED members believe that, during the 14th FYP, China should further advance the comprehensive framework for green development to set up strategic concepts, concrete policy targets, priority areas, and delivery institutions and mechanisms. This will provide a solid foundation for high-quality development and green prosperity and set an example of sustainable development for the rest of the world.

Strategic Concepts: These concepts include ensuring steadfast progress toward Ecological Civilization. This will involve seeking to achieve people-first, high-quality, and green development while promoting socioeconomic green transition

© China Environment Publishing Group Co., Ltd. 2022
China Council for International Cooperation on Environment
and Development (CCICED) Secretariat,
Green Consensus and High Quality Development,
https://doi.org/10.1007/978-981-16-4799-4

by putting into practice the concept that "lucid waters and lush mountains are invaluable assets." In addition, through integrated economic, social, and environmental improvement, this progress will make full use of material capital, human resources, and natural capital harmoniously to reshape the historically conflicting relationship between environmental protection and economic growth into one that is inclusive of both and mutually reinforcing. The ideal includes a shift toward a green development model, including green consumption, to create innovative and sustainable new economic drivers to realize more vigorous, competitive, sustainable, and resilient economic growth. Another key concept is the promotion of effective measures for both COVID-19 and environmental issues, given the close relationship of public health, pollution, and waste management. The ideal will also require the promotion of multi-stakeholder governance, poverty alleviation, and gender mainstreaming and achievement of social equity and justice.

Concrete Policy Targets: These targets include maintaining the strategic focus of building an ecological civilization. Also, the UN 2030 Development Agenda and its Sustainable Development Goals, including actions for mitigation and adaptation to climate change, should be integrated into the 14th FYP. The groundwork for high-quality development in the medium to long terms through green economic recovery plans must be laid. Additional concrete policy targets include setting integrated green indicators around the enhancement of human health and well-being while maintaining or elevating the level of ambition for specific existing indicators. This should provide clear policy signals for a comprehensive and integrated green transition. China should continue contributing to multilateral environmental and developmental processes, fulfill its obligations as a responsible major developing country, and join global green partnerships in fostering a shared future for all life on Earth.

Priority Areas: A critical area is progressing toward a green way of life and production that is people-centric, driven by the innovation of green technologies, supported by sustainable production and consumption, and practiced in urban green development. It is necessary to seize the opportunity of post-pandemic recovery to promote major green technologies that are readily scalable, to strengthen the development of green infrastructure, and to enhance socioeconomic resilience. In addition, it is a priority to implement the strategy of major function-oriented zoning, and promote green urbanization (starting with city clusters and counties) to tap the potential of structural shifts; expand and upgrade domestic consumption, promote green consumption and the greening of soft commodity supply chains, while accelerating green development, transition, and upgrading of manufacturing as engines to green transition. All of this is important, along with promoting the interconnectivity between land and sea and taking holistic nature-based approaches to addressing ecological challenges.

Delivery Institutions and Mechanisms: Green development requires integrated approaches to advance policy coherence. Short-, medium-, and long-term goals should be better aligned. Additionally, focus should be placed on creating synergies among the legislative, judicial, and administrative organs in the practice of ecological civilization, and establishing and improving a modernized environmental governance system. Other mechanisms include exploring more science-based, rational, and practical assessment methods and payment mechanisms for natural capital accounting while formulating policies and plans with a broader vision. It will also be important to integrate environmental considerations into broader economic and social planning and policies, along with establishing and developing green market mechanisms such as the carbon trading market. There also need to be improvements in green standards, green fiscal and taxation systems, and a green finance system. Policy incentives should be aligned with the goals of green development and strengthened by compliance promotion and regulatory enforcement.

The following sections elaborate more detailed recommendations.

Green Economic Recovery: Seize the Opportunity Presented by the Post-pandemic Economic Recovery to Promote Green Development and Pivot Toward Socioeconomic Resilience

After pressing the reset button on economic and social development to survive COVID-19 in 2020, the world once again faces a crossroads. The COVID-19 economic recovery presents a strategic opportunity to advance green development, which will be a testament to the vision and commitment of governments. Recommended concepts and measures of a green recovery include:

1. **Bolster green elements in "New Infrastructure Stimulus".** The economic recovery presents an opportunity to further expand clean energy and avoid high-carbon lock-in. China's "New Infrastructure Stimulus" program should be well designed to strengthen green development by including renewable energy, low-carbon and resilient infrastructure, building efficiency and upgrading, green urban centres, green technologies, and other relevant areas. In order to promote a green and resilient stimulus package, the drive to stimulate the economic recovery should be guided by the principle of "no significant harm" to the environment, ecology, and climate. Economic recovery planning should also apply environmental impact assessments to green recovery programs and projects.

2. **Support green jobs.** These include labour-intensive public works in afforestation and reforestation, wetland and coastal restoration, soil and water decontamination, green building and housing retrofitting, large training programs on sustainable agricultural practices, etc. Regions or provinces that face greater difficulties in managing the green transition should be supported to reskill their workforces.

3. **Integrated measures for community vulnerability reduction.** This would involve strengthening disease prevention by establishing a well-resourced public health infrastructure to provide early warnings and responses to emergencies. Measures should also include tackling the illegal wildlife trade, deterrence of

intensive animal farming, biodiversity/ecosystem loss, and other factors that may increase the threat of zoonotic diseases. These steps will help prevent the next pandemic.

4. **Promote green production and green consumption.** Investing in pollution control, resource and energy efficiency, circular economy upgrades etc. is critical. Promotion of green consumption also involves fostering a revolutionary behavioural change away from over-consumption, toward new ways of working, as well as green, balanced lifestyles.

5. **Support multilateral initiatives and enhance international cooperation.** It is vital to support existing multilateral initiatives such as the World Health Organization/World Food Programme One Health initiative and the UN Decade of Ecological Restoration. This support could include green recovery measures through the G20 and launching of a Bretton Woods-style consultation. This would help build out the green finance system and align the green finance taxonomies through the International Platform for Sustainable Finance (IPSF) and by opening further bilateral and multilateral cooperation channels. In turn, it would set the stage for a green overhaul of the international financial system.

Green Urbanization: Make Progress in Green Urbanization and Rural Revitalization Starting with City Clusters and Counties to Tap the Potential of Structural Shifts

It will be vital to advance urban green transition in line with the principles of green prosperity, low-carbon, intensive and circular development, equity and inclusiveness, as well as security and health. In 2017, China's 20 city clusters accounted for nearly 91% of the country's GDP and 74% of the population. The success of the green transition of the whole country hinges on the green transition of its city clusters. Development of the county economy will be crucial to China's rural revitalization strategy and its urban–rural integration. Urbanization at the county level will contribute to the green development of rural areas and will consolidate the progress in building a well-off society in a balanced way.

1. **Take into account rural factors in planning and work toward integrated urban and rural planning.** The traditional urban–rural dichotomy should be discarded when formulating planning and related policies. They should take a holistic approach to urban and rural planning, and, especially, pay attention to their impacts on the rural economy, ecology, health, society, and culture.

2. **Promote the free movement of factors of production between urban and rural areas.** Planning should encourage the movement of talent from the city to the countryside. This will help maximize the comparative advantages and potential market demand in urban and rural areas and promote rural homestead transfer to urban residents in an orderly manner, facilitating a two-way flow for urban and rural residents.

3. **Align the functional city model with a "nature-loving" city model**. This alignment will integrate biodiversity and ecosystem services into planning and protect biodiversity and natural habitats in urban areas. It will involve formulating policy incentives to materialize the economic value of ecosystem services and increasing the supply of ecosystem services to expand innovative channels for growth in the future.

4. **Accelerate the broad application of green technology**. To accomplish this acceleration, it will be necessary to identify and select forward-looking, comprehensive, innovative, and practical green technologies in all aspects of planning, construction, operation, and maintenance. This will involve a focus on removing institutional obstacles to the promotion of emerging green technologies and establishing a life-cycle assessment framework for green technologies using a whole-life costing methodology that promotes the large-scale application of green technologies that have been proven economically and technically viable. Also important will be the periodic screening and publishing of a list of major innovative green technologies in key areas, such as urban green buildings, green transportation, clean energy, efficient use of water resources, sustainable diet, waste management, land use and planning, and remediation of brown sites.

Green Consumption: Expand and Upgrade Domestic Consumption and Promote Green Consumption as the Transformative Driver of the Green Transition

This will raise awareness of green consumption, starting with a green consumption revolution, and substantially increase the supply of green products and services. In 2019, domestic demand accounted for 89% of China's economic growth, and final consumption expenditures contributed 58% to GDP growth. The growing middle class and the increased population of younger "netizens" have created an opportunity for the upgrading and transition of China's consumption. On the whole, however, the green transition in consumption has been in decline since 2008, despite advances in green production, making the former the weak link in the overall green transformation. The following are recommended;

1. **National Green Consumption Strategic Plan**. This will involve creating a new focus and engine to improve environmental quality and high-quality development by greatly promoting green consumption. Specific measures may include implementing awareness-raising campaigns to promote high-quality green consumption; increasing the supply of green products and green services; ensuring that the values of a frugal, green, healthy and low-carbon lifestyle take hold; and building a green consumption policy system with both incentives and constraints.

2. **Improve and promote systems and mechanisms for green consumption**. Improvements to market-fostering policies and economic incentive policies for

green consumption are vital. Also important will be measures that focus on creating economic incentives and establishing a market-driven system in the aspects of pricing, financing, and taxation, credit, supervision and market credit to encourage the supply of (and demand for) green products and services. A statistical and index system of green consumption should be established, and the development of green product and consumption standards accelerated. The whole-process supervision of standardization and third-party certification of green products and services should be strengthened. It will also be important to implement government preference and mandatory procurement programs to source green and energy-efficient products and set binding rules for public green procurement.

3. **Promote Circular Economy Solutions and Implement Extended Producer Responsibility (EPR)**. This will involve implementing the early 2020 Ministry of Ecology and Environment–National Development Reform Commission joint announcement on plastics, as well as setting guidelines to reduce plastics, and packaging waste in e-commerce, logistics, and related systems; implementing waste separation and sorting to improve the plastic waste recycling system; and reducing and eliminating single-use plastics. Strengthening corporate social responsibility related to green consumption, waste reduction, and improve waste recovery will also be crucial in this respect.

4. **Prioritize green consumption sectors**. To accomplish this, it will be necessary to increase the supply of green products and green services, including clothing, green food, green housing, transportation, and tourism.

- **Green clothing**: Strengthen the environmental labelling and certification of textiles and clothing.
- **Green food**: Promote more sustainable diets, cut food waste in all links from warehousing, food processing and transportation to retail and consumption; fully implement the green takeout scheme; and set up uniform and rigorous certification systems and standards for green organic food.
- **Green building**: Comprehensively promote the design, construction, and operation of green and healthy buildings; strengthen the environmental certification and labelling of green household products—especially low-carbon, energy-efficiency labelling—and increase the supply of energy-efficient green household products.
- **Green mobility**: Encourage low-carbon mobility such as walking, cycling, and public transport, and develop a green policy system for the entire automotive industry chain; roll out a national electric vehicle (EV) recharging infrastructure and formal EV battery recycling infrastructure; enhance the role of taxation in encouraging emission reduction and energy saving in the automobile sector; expand economic incentives for the purchase and use of green vehicles. Supply- and demand-oriented support for the automotive and aviation industries must be linked to ecological requirements. Strengthen rail freight transport and sustainable urban logistics, e.g., digitalization and automation of cargo railway.

- **Green tourism**: Enact and enforce conventions and guidelines on green tourism and consumption; encourage hotels and scenic spots to award green tourism and consumption behaviours; enact and revise the evaluation measures for green services such as green markets, green hospitality, green catering, and green tourism; and incorporate biodiversity conservation into tourism-related standards and certification schemes.

Ecological Integrity: Enhance Ecological Integrity and Connectivity to Tackle Environmental Challenges

The lessons emerging from the COVID-19 pandemic reinforce the importance of comprehensive measures. The concept of "One Health"—which connects and coordinates public health, economic activities, and ecosystems changes (including climate, the sea, and rivers) with the management of other areas—must be strengthened. In tackling the environmental changes, the terrestrial and marine ecosystems must be connected, and climate action should be aligned with biodiversity conservation.

1. **Pursue ambitious climate targets with energy transition at their core to build a low-carbon society**. This will involve building a clean, low-carbon, safe and efficient-energy system while setting more ambitious and binding targets for greenhouse gas emission reductions—e.g., setting an absolute cap on carbon emissions for 2025 and 2030. Emission caps should also include non-carbon dioxide emissions, notably methane and hydrofluorocarbons (HFCs). It will be necessary to update China's Nationally Determined Contributions targets based on the actual circumstances; encourage key regions and sectors to set plans for carbon emission peaking as soon as possible; mainstream climate resilience into national/local government planning and budgets; accelerate a national carbon pricing system; incorporate climate indicators into the Central Environmental Inspection Program; enhance multilateral climate coordination with Europe and with other developing countries through the Ministerial on Climate Action (MoCA) and other initiatives, to forge new global climate leadership; eliminate fossil fuel subsidies and avoid stranded assets by gradually phasing out fossil fuel investments; include environmental and climate protection aspects in financial risk assessments and further include sectors in the Chinese emissions trading system (ETS) to internalize external costs; reinforce economic evaluation of coal power generation and prepare a roadmap for the reduction and eventual phase-out of coal-fired electric power generation. At the same time, it will be important to expand investment in renewable energy-based power infrastructure and pursue green power market reforms. Also critical will be decarbonizing energy-intensive industries (i.e., steel, chemicals, and cement) while expanding large offshore wind power, smart grids, and battery storage capacity as well as rolling out hydrogen economy policies at the national level and promoting the use of fuel cells in transportation and cogeneration. Finally,

ambitious climate targets should include increasing the proportion of sustainable biomass gasification in the energy mix and expanding carbon capture, use, and storage for the power generation, petrochemicals, and metallurgical sectors.

2. **Host a successful UN Convention on Biological Diversity (CBD) 15th Conference of the Parties (COP 15) in 2021 to galvanize ambitious multilateral cooperation and step up national action to protect nature and human well-being.** To do so, work proactively with the international community to raise COP 15 ambition levels by setting clear, quantified targets for land and marine ecosystem restoration and conservation (including ecological corridor and ecological security targets) to build an efficient and stable ecological security network and enhance ecological integrity. This will involve promoting transformative and ecosystem-based approaches to support high-quality green growth; strengthening the conservation of different types of ecosystems with a focus on restoration and regeneration of natural vegetation and ecological processes in priority-degraded areas. Also necessary will be the promotion of nature-based climate adaptation and the prioritizing of adaptation within integrated water and river basin management systems, building codes, infrastructure, and sustainable agricultural systems, within the context of sustainable use to mainstream nature conservation. It will also include the conservation and management of socio-ecological production landscapes as well as tackling alien invasive species as a national priority within the Post-2020 Global Biodiversity Framework. Galvanizing multilateral cooperation will require expanding forests, wetlands, and grasslands as the basis of nature-based climate resilience and prohibiting and actively prosecuting activities that seriously harm biodiversity, such as the illegal wildlife trade, illegal pesticide production and use, illegal fishing, and illegal land conversion. Stopping the habit of eating wild animals and regulating wildlife parts in traditional medicine—as well as promote greater private sector action in biodiversity conservation—will be key in this area.

3. **Enhance comprehensive marine governance to promote the resilience of marine ecosystems and support the sustainable growth of the blue economy.** This will involve the strict controlling of sea reclamation, strengthening the restoration of coastal wetlands and re-establishing key habitats. It will also be necessary to delineate marine ecological red lines and marine protected areas— they have critical long-term benefits for biodiversity and fisheries. Comprehensive marine governance also requires strengthening scientific knowledge, monitoring, and legal enforcement to advance the protection and restoration of ocean ecosystems and high-quality development of ocean economies—as well as fully leveraging the inter-ministerial coordination mechanism and national-level science advisory body for the ocean. Also necessary will be the development of integrated marine management policies based on ecosystems, along with green fishing vessels, green fishing ports, and green marine aquaculture, while establishing a traceability system for seafood and promotion of green shipping.

4. **Improve the assessment methods and payment mechanisms for natural capital and ecosystem services to advance high-quality development of the Yangtze River and Yellow River basins**. To do so, it will be necessary to adhere to the principle of staying within the carrying capacity of resources and environment, boost the application of natural capital accounting in spatial planning, and ensure the integrity, health, and sustainable development of river basin ecosystems. It will also be essential to establish a well-regulated and standardized natural capital accounting system and develop an ecosystem monitoring network. A pricing mechanism for ecological products that consists of market pricing, government pricing, and regulated market pricing will need to be built. Improvements in this area will also require innovating the ecological compensation mechanism and speeding up the process of horizontal compensation for river basins from ecosystems in terms of water resources, water environment, and water ecology.

5. **Incorporate biodiversity conservation into China's Framework for Green Finance and mainstream conservation finance**. To do so, establish a market-based mechanism for the pricing of natural assets and develop financial tools such as payment for ecosystem services, access, and benefit-sharing compensation systems and the introduction of private capital to support the financing of ecological and environmental protection. It will also be important to promote enhanced transparency and disclosure of financial risks related to climate change and ecological risks. This will be helped by the promotion of the "no net loss" principle in major economic plans and an increase in the weight of factors related to ecological conservation in strategic environmental impact assessment, and in the assessment of large-scale infrastructure projects. Incorporating biodiversity conservation also involves compensating ecological losses with "ecological protection behaviour" to complement monetary-based compensation.

Green Belt and Road Initiative (BRI) and the Supply Chain: Achieve Global Green Prosperity Through Green Cooperation

1. **Carry out pilot cooperation projects and promote green development concepts and practices in BRI countries**. To do so, maximize the role of the BRI International Green Development Coalition, the Belt and Road Sustainable Cities Alliance, and other platforms, and carry out joint pilot projects for the construction of the Green Belt and Road and sustainable demonstration countries, cities, and projects. Ecological redlines and nature-based solutions should be promoted among the participating BRI countries. There should be reduced investments in coal-fired power projects accompanied by expansion of clean energy infrastructure projects. It will be vital to give full play to the role of international organizations, professional institutions, multinational companies, and civil organizations, and attract the active participation of social capital in

greening BRI. To do so, strengthen case studies on the green transition of BRI countries and accelerate the broad application of good experience.

2. **Improve green assessment and classification of BRI projects.** This will be accomplished by developing positive and negative lists to provide green development solutions for projects and green credit guidelines for financial institutions, through environmental impact assessment tools, clear standards, and financing safeguards. It will also involve enhancing the application and alignment of green standards and developing more green and high-standard demonstration projects to guide enterprises to effectively bear the responsibility of ecological and environmental risks. Ultimately, it will be necessary to apply more environmentally friendly practices in projects harvesting natural resources, and pursue world-class green mining, forestry, marine transportation and ports, fishing, and aquaculture.

3. **Adopt measures to promote the systematic greening of global soft commodity value chains to avoid deforestation and ecological degradation.** To do so, develop a long-term inclusive mechanism for common efforts to promote the green development of global value chains, including the establishment of a coordinating and supporting institution that involves the participation of multiple players—for example, the government, companies, research institutes, and other organizations. Improving the existing policy and technical support system to avoid incentives for deforestation and promoting a global standard to complement the work of global traders and consumer goods companies to avoid deforestation will be important—as well as implementing relevant demonstration projects and duplicating their experience. It will be valuable to enhance synergies of existing bilateral and multilateral mechanisms to share the theory and practice of green value chains. Other steps to take include research to develop a traceability system for commodity trades and related due-diligence requirements as well as supporting the sustainable production transition through South–South collaboration.

4. **Green overseas development.** It is recommended to explore ways for the China International Development Cooperation Agency to mainstream green practices in all project finance; adopt "do no harm" principles; and increase the proportion of green and environmental assistance in foreign aid for green development in BRI countries.

Appendix B
Progress on Environment and Development Policies in China and Impact of CCICED's Policy Recommendations (2019–2020)

2020 represents a major milestone for China's development and environment. This year, China will win three tough battles against major risks, poverty, and pollution and achieve the goal of building a moderately prosperous society in all respects, On this basis, China will enter its second century of development, opening a new stage of green development in an all-round way, and firmly moving towards the new goal of a beautiful China in 2035.

Over the past year, China has maintained its strategic momentum in building ecological civilization. Taking the high-quality development of national economy as the fundamental goal, it continued its efforts to promote the construction of various ecological civilization systems targeted for improvement of air quality and ecological environment, and stayed determined to win the battle against pollution with successive innovative measures, which has produced effective results.

Over the past year, the Xi Jinping Thought on Ecological Civilization was carried forward, and the new development idea has displayed enormous vitality in practice; green low-carbon and circular development was effectively promoted; efforts have been made to explore, innovate and improve the modern ecological environment governance system; the reform measures on ecological civilization were implemented and have secured good results, injecting strong momentum for environmental protection and pollution control.

Over the past year, the efforts to tackle pollution prevention and control have not diminished. New progress in eco-environmental protection has been made through persistent efforts, China's ecological environment has generally improved, and the ecological environment goal in the 13th FYP period is expected to be fully realized.

Over the past year, the pollution control efforts have been heading toward a new stage. Eco-environmental protection work pays more attention to "targeted, scientific and law-based pollution control", continuously reducing the resistance facing eco-environment governance, and making eco-environment governance a boost to economic transformation. "Targeted, scientific and law-based pollution control"

© China Environment Publishing Group Co., Ltd. 2022   545
China Council for International Cooperation on Environment
and Development (CCICED) Secretariat,
Green Consensus and High Quality Development,
https://doi.org/10.1007/978-981-16-4799-4

becomes an important step in exploring a modern eco-environment governance system.

As a policy direct channel to the Chinese government and the link, bridge and window for China and the international community to carry out environment and development cooperation, CCICED has been adjusting its development and positions according to domestic and international situations, fully mobilizing intellectual resources, including top experts from China and abroad, to explore new ways, new approaches and new means for domestic outstanding issues in the environment and development field in the new era; at the same time, upholding the Going Global principle, it engages in conducting international exchanges and advisory activities, actively participates in international environmental governance, and contributes wisdom and strength to the building of ecological civilization in China and the sustainable development across the world.

I.   Environmental and Development Planning

2020 is the final year of the 13th Five-Year Plan (FYP) period. As indicated by an interim assessment report on the implementation of the *13th Five-Year Plan for Ecological and Environmental Protection* jointly released by the Ministry of Ecology and Environment (MEE) and related ministries and commissions in April 2020, seven of the nine binding targets have been fulfilled, with substantial improvements in eco-environmental quality and great ecological progress, effectively safeguarding the goal of completing the building of a moderately prosperous society in all respects.

The goal of basically building a beautiful China in 2035 leads the new development plan. Based on the good foundation that the 13th FYP targets have been basically accomplished and benchmarking beautiful China in 2035, central and local departments have carried out forward-looking research on the formulation of the 14th FYP in early 2019 and even 2018, which achieved initial results. On March 6, 2020, the National Development and Reform Commission (NDRC) issued the *Evaluation Index System and Implementation Plan for the Construction of a Beautiful China.* The evaluation index system contains five types of indexes, such as air freshness, water cleanness, soil security, ecological health and human settlement tidiness.

The CCICED has brought forward prospective recommendations on the 14th FYP early at its 2017 Annual General Meeting (AGM). Such recommendations include: China should develop a comprehensive long-term strategic plan that covers water, atmospheric, soil and marine pollution control for the next 10–15 years. Under this strategic plan, China should complete the corresponding deployment by 2020 to fit with the timeline of basically realizing socialist modernization in 2035. Moreover, the strategic plan should attach importance to innovation, for example, reducing plastic litter pollution in rivers and oceans by regulating plastic production and source control. In 2019, the CCICED recommended that the 14th FYP should embody and back up the vision of a beautiful China by 2035, efforts to tackle climate change and the global vision 2050 for biodiversity conservation. These CCICED recommendations have produced a positive impact on China's national 14th FYP and the 14th Five-Year Plan for Ecological and Environmental Protection that are being developed.

a. Prospects on the 14th Five-Year Plan for Environment and Development
    At the special meeting on the formulation of the 14th FYP on November 26, 2019, Premier Li Keqiang of the State Council pointed out that the formulation of the 14th FYP should: " be based on China's basic national conditions and development stages, give top priority to development, highlight maintaining economic operations within a reasonable range, promote high-quality development, highlight the people-centered development thinking, focus on solving development problems through reforms and innovations, seek truth from facts, follow the law, take a long-term perspective, and make overall plans."
    Within the framework of the 14th FYP, basic ideas and a preliminary draft of the 14th Five-Year Plan for Ecological and Environmental Protection were basically formed in December 2019. The 14th Five-Year Plan for Ecological and Environmental Protection will promote high-quality economic development and high level ecological and environmental protection in a coordinated manner, which shall not only focus on the next five years, but also shall be geared to the goals by 2035 and the middle of this century. Additionally, it is necessary to make the best use of science and technology, step up efforts to tackle hard-net problems in science and technology, provide strong support for decision-making, management and governance, mobilize enterprises' vitality in innovation and drive the reform of the eco-environmental industry.
    Relevant special plans within the framework of the 14th Five-Year Plan for Ecological and Environmental Protection are being developed at full stretch, such as the *National Marine Eco-environmental Protection Plan*, the *Water Eco-environmental Protection Plan for Key River Basins* and the *14th Five-Year Plan for Air Quality Improvement*. These special plans will provide top-level design for ecological and environmental protection in relevant sectors during the 14th FYP period.

b. Formulation of a plan for eco-environmental protection in the Yellow River basin
    The Yellow River basin is of vital importance to China's socio-economic development and ecological security. Originating from the Qinghai-Tibet Plateau, the Yellow River runs through nine provinces and autonomous regions for a length of 5,464 km, which is China's second longest river only second to the Yangtze River. Ecological protection and high-quality development of the Yellow River basin, like Beijing-Tianjin-Hebei coordinated development, development of the Yangtze River Economic Belt, construction of the Guangdong-Hong Kong-Macao Greater Bay Area and integration of the Yangtze River Delta, is a major national strategy.
    On January 4, 2020, President Xi presided over and addressed the Sixth Meeting of the Central Financial and Economic Affairs Commission. The meeting emphasized that great efforts should be made to protect and manage the Yellow River basin, for which the path of ecological conservation and high-quality development should be taken. Measures should be taken to advance ecological protection and restoration of the Yellow River basin, strengthen pollution control, encourage conservation and intensive utilization of water resources,

drive high-quality development of central cities and city clusters along the Yellow River, build a modern industrial system and carry forward the Yellow River culture.

To promote ecological conservation and high-quality development of the Yellow River basin, MEE has carried out the research on relevant strategy and plan formulation, and drawn up major plans and policies for the river basin to coordinate the settlement of major cross-regional problems, improve cross-regional management coordination mechanisms, strengthen joint prevention and control and joint law enforcement in connection with water eco-environmental restoration. MEE will also promote the establishment of the "three lines and one list" in the nine provinces and autonomous regions along the Yellow River, perfect the eco-environmental zoning system, delineate the ecological red lines, optimize industrial distribution through environmental impact assessment (EIA) of plans and facilitate the adjustment and optimization of the industrial structure in the river basin, thus driving high-quality development there.

c. Eco-friendly and green integration of the Yangtze River Delta

The *Overall Plan for Demonstration Zones for Eco-friendly and Green Integration of the Yangtze River Delta* (the *Plan*) formally approved by the State Council on October 25, 2019 is particularly relevant to promoting high-quality development of the Yangtze River basin by cooperating in protection and preventing excessive development.

The *Plan* states that by 2025, a number of environmental, infrastructure, technology innovation, public service and other major projects will be completed and put into operation, evident improvements seen in eco-environmental protection and construction, innovative development of eco-friendly industries and harmony between man and nature in pilot zones, the main functional framework basically established for integration demonstration zones and ecological quality significantly improved. Innovation in the integration system will bring about replicable experience and system integration of major reforms will unleash the dividend, with the role of these demonstration zones in leading higher-quality integration of the Yangtze River Delta initially exerted. By 2035, a more mature, effective system of green integrated development will be established and become the benchmark for demonstration and leading of high-quality integration of the Yangtze River Delta.

In December 2019, the CPC Central Committee and the State Council issued the *Outline of the Integrated Regional Development of the Yangtze River Delta* (the *Outline*). The *Outline* indicates that by 2025, substantial progress will be made in the integrated development of the Yangtze River Delta, especially in technology innovation industries, infrastructure, ecological environment and public services, just to name a few.

Currently, demonstration zones for eco-friendly, green integration of the Yangtze River Delta have been inaugurated and will adopt a three-level architecture of "Board of Governors + Executive Committee + Development Company".

II. Environmental Protection and Fight against COVID-19

In early 2020, the outbreak of the novel coronavirus (COVID-19) unexpectedly triggered global economic and social systematic risks, for which the world paid a heavy price in treasure and blood. At the same time, this has again aroused global attention to sustainable development, driving post-pandemic green recovery to be a major concern of the international community under the new normal.

a.   COVID-19 prevention and control vs. economic and social development
   COVID-19 has obvious short-term effects on China's economic and social development. The impact on the economy in the first quarter of 2020 was greater than that of the 2008 financial crisis, when negative economic growth was observed. In the second quarter, the economy is expected to register positive growth, but this remains to be seen. According to Zhu Min, Chairman of the National Institute of Financial Research at Tsinghua University and the former Vice President of IMF, even if the government introduces an adequate hedging macro policy, the growth rate will stay at only 5% throughout 2020.[1]
   With the integration of world economy, China's economy "shares weal and woe" with the global economy. The global outbreak of the COVID-19 is about one month later than that in China. The trend of further spread has been formed and the world economy has been greatly impacted, which will have a negative influence on the development of China's economy in the next stage, slow down its economic recovery after resumption of work and production and pose a greater challenge to the attainment of the economic production targets.
   However, there is a positive side of the asynchrony of the COVID-19 outbreak in China and worldwide. China is the first country that has brought the virus under control, and it has become the "home front" for global fight against COVID-19, which can provide other countries with a steady flow of medical and other material support in containing the outbreak and make new contributions to COVID-19 control all over the world. The global outbreak was roughly one month later than the outbreak in China, when the trend of further spread had come into being, giving the world economy a heavy blow. China will further show its irreplaceability in the world when many countries find it difficult to resume work and production, but still have some rigid demand. Moreover, at the time of gradually lifted social isolation and economic recovery as the outbreak is controlled in some countries, the demand for e-bikes—an independent means of transportation in the post-pandemic era—surges, injecting strong impetus into the rapid recovery of relevant industries in China.

b.   Supporting resumption of work and production and promoting green consumption
   MEE released the *Guiding Opinions on Effectively Coordinating COVID-19 Prevention and Control, Economic and Social Development, and Ecological*

---

[1] Impact of the COVID-19 on Global Economy and Finance: Analysis and Prospect. A special issue of State-owned Capital Operation Research Institute on the fight against COVID-19; March 13, 2020.

and Environmental Protection (the *Opinions*) on March 3, 2020. The *Opinions* suggests that China should spare no effort to maintain environmental protection in connection with the prevention and control of the pandemic towards the main goal of 100% environmental monitoring and services for all medical institutions and facilities throughout the country and timely and effective collection, transfer and disposal of 100% medical waste and wastewater.

To help enterprises resume operation and production, the *Opinions* proposes to establish and implement positive lists of EIA and approval and of supervision and law enforcement, increase the working efficiency, give strong support to the resumption of operation and production by enterprises in relevant sectors. China should adopt differentiated eco-environmental regulations and dynamic adjustment on these regulations on the grounds of the status of the prevention and control and relevant requirements of region-specific and risk-based accurate resumption of work and production, thus supporting and guaranteeing pandemic prevention and control and socio-economic development in a coordinated manner.

The key to resumption of work and production lies in stimulation of consumer demand. The 2018 CCICED Policy Recommendations indicated that the government should give full play to the role of existing policy measures in promoting sustainable production and consumption. China should develop and implement categorized targeted strategies for different sectors during the 14th FYP period. How to improve people's well-being and health and constantly reduce ecological footprint should be taken into account; homes, schools and workplaces should be more pleasant; new green jobs should be created; greater importance should be attached to the important role of households in promoting sustainable consumption.

In March 2020, NDRC and other 22 ministries and commissions jointly released the *Implementation Opinions on Promoting Consumption Scale Expansion and Quality Improvement and Accelerating the Establishment of a Strong Domestic Market*, which proposes to "encourage use of green smart products, promoting green consumption by focusing on supply of green products, construction of green public transport facilities, energy-saving and environmentally-friendly buildings and relevant technology innovation, and build green shopping malls". Since March 2020, Zhejiang, Jiangsu, Guangdong, Beijing and Hubei have successively begun to stimulate household consumption by issuing "consumption vouchers" on various digital platforms such as Alipay and WeChat, which has an apparent "multiplier effect" on spurring economic recovery. CCICED SPS on Green Consumption suggests "green"-related elements be integrated into various "e-vouchers", so as to promote green, low-carbon and sustainable development while stimulating economic recovery. Meanwhile, the government can support the introduction of "green consumption vouchers" and relevant supporting measures by formulating general standards for identification of green businesses and consumer behaviors and further expanding the functions of digital living platforms, thus creating new hotspots of online green consumption and facilitating sustainable economic development in all aspects after the pandemic.

c. Strengthening wildlife protection

At the critical moment of global fight against COVID-19, protecting wildlife has drawn much attention. The 2019 CCICED Policy Recommendations proposed alertly to strengthen research on the breeding, cultivation and sustainable use of wild biological resources, promote technological upgrading, reduce consumption of natural and biological resources, etc., and take legal action against illegal wildlife sales and smuggling.

In response to COVID-19, President Xi Jinping emphasized: "We must incorporate biosafety into the national security system from the perspective of protecting people's health, safeguarding national security, and safeguarding the country's enduring peace and stability, systematically plan the construction of a national biosecurity risk prevention and control and governance system and comprehensively improve China's capacity in biosecurity governance."

The National People's Congress (NPC) has launched the revision of the Law on the Protection of Wildlife and plans to incorporate the revision work into the 2020 legislation plan of its Standing Committee and expedite the revision of the Law on Animal Epidemic Prevention and other related laws. The Law on the Protection of Wildlife was systematically revised in 2016, with the principle of protection first, normalized utilization and strict management established, strict provisions on hunting, trading, utilization, transportation of wild animals and consumption of the meat thereof incorporated and, in particular, scientific and reasonable systems established for problems including indiscriminate wild animal meat consumption. However, the COVID-19 outbreak has exposed the loopholes in the administration of breeding and smuggling of wild animals.

Furthermore, the NPC adopted by vote the *Decision on a Complete Ban of Illegal Wildlife Trade and the Elimination of the Unhealthy Habit of Indiscriminate Wild Animal Meat Consumption for the Protection of Human Life and Health* on February 24, 2020, which sets forth specific provisions on an all-out ban of illegal wild animal trade and relevant penalties. In charge of "coordinating the protection of biodiversity" and "supervising wildlife protection", MEE has, upon deliberation, included the enforcement of laws and regulations on wild animal protection in the scope of central supervision of ecological and environmental protection. Qinghai, Beijing, Guangdong, and Shanghai have all strengthened the management of wildlife protection in the form of local legislation.

III. Ecosystem and Biodiversity Conservation

a. The 15th Conference of the Parties to the Convention on Biological Diversity (CBD COP15)

The 2019 CCICED Policy Recommendations proposed to build high-level political momentum through green diplomacy by learning from the successful experience in climate change negotiations in Paris, enlist businesses, the academic community, NGOs and the public to contribute to the post-2020 biodiversity framework and its implementation, publicize the Action Agenda for Nature and People, raise public awareness and catalyze cooperative initiatives.

The CBD COP15 will be held in China, which will deliberate the "post-2020 global biodiversity framework" and define new global biodiversity targets for 2030. The CBD COP15 will develop a new vision, strategic plan and goal for the next decade and point out the direction for global diversity conservation.

China, as one of the world's most biologically diverse countries and one of the earliest members of the CBD, has made numerous efforts in biodiversity conservation and achieved notable results, leading the world in the mainstreaming of biodiversity conservation. It has put forward the concepts of respecting, conforming to and protecting the nature, as well as "lucid waters and lush mountains are invaluable assets". As a result, obvious progress has been made in ecosystem protection and restoration, ecosystem services improved as a whole and remarkable results achieved in wildlife protection. For example, populations of endangered species such as giant panda, Tibetan antelope, crested ibis and snow leopard have increased significantly, with the quality of their habitats improved and endangered categories lowered in the IUCN assessment.

b.   Continuous advancement of the "Green Shield 2019" special action
2020 is the third year of the "Green Shield" special action on natural reserve supervision. Based on 474 national natural reserves, the "Green Shield 2019" mission has been extended to some natural reserves within a 5 km radius of trunk streams, main tributaries and the five lake areas in the 11 provinces on the Yangtze River Economic Belt.

In July 2019, MEE, Ministry of Water Resources (MWR), Ministry of Agriculture and Rural Affairs (MOA), Chinese Academy of Sciences (CAS), National Forestry and Grassland Administration (NFGA) and China Coast Guard of the Chinese People's Armed Police Force jointly launched the "Green Shield 2019" mission to strengthen natural reserve supervision, and carried out on-site inspection of new problems and clue identified by remote sensors of human activities in national and provincial natural reserves in different provinces.

c.   New progress in ecological red line delineation and ecological protection
The 2019 CCICED Policy Recommendations proposed to strengthen the construction of the natural reserve management system dominated by national parks and delineate ecological red lines; formulate and implement a complete set of laws, regulations and market incentive policy measures, and ensure the effectiveness of such implementation; strengthen crosscutting collaboration and remove subsidies that might have adverse impacts on the ecological environment.

Under the *Several Opinions on Delineating and Strictly Observing Ecological Red Lines* promulgated by the State Council in 2017, by the end of 2020, the delineation and demarcation of ecological red lines will be completed nationwide, an ecological red line system basically established, ecological space of China's territory optimized and effectively protected, ecological functions stabilized and the national ecological security pattern perfected.

On August 30, 2019, the General Office of MEE and the General Office of the Ministry of Natural Resources (MNR) issued the *Technical Procedures for Demarcation of Ecological Red Lines*, aiming at guiding the demarcation of ecological red lines nationwide and promoting the implementation and strict management of ecological red lines.

In June 2019, the General Office of the CPC Central Committee and the General Office of the State Council released the *Guiding Opinions on Establishing a System of Natural Reserves with National Parks as the Main Body* (the *Opinions*). The *Opinions* proposes that, by 2020, China should come up with an overall layout and development plan for national parks and various natural reserves, complete the pilot program for a national park system, set up a number of national parks, complete the demarcation of natural reserves and the linkup with ecological red lines, work out a negative list of access for construction projects within natural reserves and establish a unified natural reserve management system by category and level. By 2025, China should establish a sound national park system, complete the consolidation and optimization of natural reserves, improve the laws, regulations and management and supervision systems regarding the natural reserve system, increase the bearing capacity of natural ecosystems and preliminarily build the system of natural reserves with national parks as the main body. By 2035, China should significantly improve the natural reserve management efficiency and the ability of natural reserves to supply ecological products, raise the scale and management of natural reserves to the world advanced level and have in place a natural reserve system with Chinese characteristics.

Since May 2019, NFGA has launched the assessment of the pilot program for the national park system. So far, China has built 11,800 various natural reserves at all levels, accounting for 18% of the land area and 4.6% of the sea area. These natural reserves include 10 pilot national parks, 474 national natural reserves and 244 national scenic spots. China has 14 World Natural Heritage sites, four World Natural and Cultural Heritage sites and 39 world geoparks, all leading the world by quantity.

On November 15, 2019, NDRC released the *Pilot Program for Comprehensive Ecological Compensation*, under which 50 counties (cities and districts) will be selected in national ecological civilization pilot zones, Tibet and other Tibetan areas, and Anhui Province for pilot comprehensive ecological compensation. The pilot program covers four parts, namely innovating the forest ecological benefit compensation system, promoting the establishment of an ecological compensation system for upper and lower reaches in a river basin, developing characteristic industries with ecological advantages and advancing the establishment of an ecological protection compensation system. The goal is, by 2020, to make phased progress in pilot compensation ecological compensation, effectively increase the capital efficiency, enhance the ability of ecological protected areas to promote economic growth, significantly raise the engagement of conservationists and basically establish an ecological protection compensation mechanism matching local economic development.

IV. Energy and Climate

a. Clean, low-carbon energy transition helps improve the environment
The 2019 CCICED Policy Recommendations proposed to step up coal control to win the blue-sky battle with determination; elaborate a national long-term zero-emission strategy toward the eventual phase-out of coal; increase the subsidy and financial support for renewable energy and gradually remove the subsidy for fossil fuels.
As of the end of 2019, China's installed renewable energy power capacity registered 794 million kW, representing a year-on-year increase of 9%. The installed wind and PV power capacity "both" exceeded 200 million kW for the first time. The installed renewable power capacity accounted for roughly 39.5% of the total, a year-on-year increase of 1.1%. Renewable energy generated 2.04 trillion kWh of electric power, increasing by approximately 176.1 billion kWh year on year; the share of renewable energy power in the total was 27.9%, a year-on-year growth of 1.2%.[2]
On April 10, 2020, National Energy Administration (NEA) released the *Energy Law of the People's Republic of China (Exposure Draft)*. The draft explicitly states that energy exploitation and utilization should be adapted to ecological civilization, carry out the new development concept, implement a conservation-first, domestic demand-oriented, green and low-carbon, and innovation-driven energy development strategy, and establish a clean, low-carbon, safe and efficient energy system. For fossil energy, the draft allows market participants that meet access requirements to engage in oil and gas exploration and exploitation while taking protective measures. For non-fossil energy, the draft makes it clear that China has listed renewable energy as a priority area of energy development and developed mid- to long-term total amount targets for renewable energy exploitation and utilization nationwide, the target share of renewable energy in primary energy consumption and binding targets that are included in the plan for national economic and social development and annual plans.

---

[2] www.sinoergy.com. National Energy Administration: In 2019, China's installed renewable energy power capacity registered 794 million KW, representing a year-on-year increase of 9%. http://www.hxny.com/nd-44650-0-46.html.

In terms of promotion of renewable energy, on December 4, 2019, NDRC, NEA and other eight ministries jointly released the *Guiding Opinions on Promoting the Industrialization of Bio-Natural Gas*, which requires the formation of bio-natural gas market of a certain size and a green, low-carbon and clean renewable gas industry by 2025, with an annual bio-natural gas output of more than 10 billion m^3, and steady development of bio-natural gas by 2030. The size of the bio-natural gas will then be world-leading and the annual output will exceed 20 billion m^3, which will account for a certain proportion of the total gas output at home.

China has energetically advanced clean and low-carbon energy transition. On August 19, 2019, NDRC, MOF, MNR, Ministry of Ecology and Environment (MEE), NEA and National Coal Mine Safety Administration (NCMSA) jointly released the *Work Plan for Classification and Disposal of Coal Mines with a Capacity below 300,000 tons/year*, with the purpose of striving to, through three years of efforts, reduce the number of coal mines with a capacity below 300,000 tons/year across the country to less than 800, basically eliminate such coal mines in North and Northwest China (excluding South Xinjiang) and cut the number of such coal mines in other areas by more than 50% in principle compared with the level at the end of 2018. On September 6, 2019, the *Notice of the National Energy Administration on Issuing the Target and Task of the Elimination of Backward Production Capacity in the Coal Power Sector in 2019* was released, which makes clear that 8,664,000 kW of backward production capacity should be eliminated in the coal power sector in 2019.

b. Strengthening energy conservation and emission reduction

China has witnessed a roughly 13.7% decline in the national energy intensity during the first four years of the 13th FYP period. In 2019, in response to the task of atmospheric pollution prevention and control, local governments actively promoted "coal-to-gas" and "coal-to-electricity", and stepped up efforts to solve key problems in energy-saving technologies, with new results achieved on this regard.

On June 13, 2019, NDRC, Ministry of Industry and Information Technology (MIIT) and other five ministries jointly released the *Action Plan on Green and Efficient Refrigeration*, which sets forth a main goal that by 2022, the energy efficiency of refrigeration products such as household air conditioners and variable refrigerant volume (VRV) split air conditioning systems in the market will increase by over 30% and the market share of green and efficient refrigeration products 20%, with 100 billion kWh of electricity saved a year. By 2030, the refrigeration energy efficiency in large public buildings will increase by 30%, the overall refrigeration energy efficiency more than 25% and the market share of green and efficient refrigeration products more than 40%, with 400 billion kWh of electricity saved a year.

On July 26, 2019, Ministry of Transport (MOT) released the *Catalogue of Key Energy-Efficient and Low-Carbon Technologies to be Promoted in the Transportation Sector (2019)*, covering 30 technologies in five fields such as road transportation, highways, ship transportation, channels and ports.

On August 28, 2019, NDRC released the cleaner production evaluation index systems for five sectors: coal mining and processing, zinc sulfate, zinc smelting, sewage treatment and recycling and fertilizer (phosphate fertilizer) manufacturing.

On October 29, 2019, NDRC released the *Overall Plan for the Green Life Initiative*, which requires, through the construction of conservation-oriented government organs, green households, green schools, green communities, green commuting, green shopping malls and green buildings, extensively publicizing and promoting simple and moderate, green and low-carbon, civilization and healthy concepts of life and lifestyles, establishing sound policies and management systems concerning green life, promoting green consumption and pushing for green development.

On March 11, 2020, NDRC and Ministry of Justice (MOJ) released the *Opinions on Accelerating the Establishment of a System of Regulations and Policies for Green Production and Consumption*, which stipulates that by 2025, regulations, standards and policies pertaining to green production and consumption should be further improved, an institutional framework where incentives and restraints are in place basically established, green ways of production and consumption fully practiced in key fields, sectors and links, and the level of green development generally raised in China.

c. Co-control of greenhouse gas (GHG) emissions and atmospheric pollutants

The 2019 CCICED Policy Recommendations proposed that promoting air quality improvement and GHG emission reduction in a coordinated manner was an inevitable approach to high-quality development in China. Both the "Action Plan for Prevention and Control of Atmospheric Pollution" and the "Three-Year Action Plan for Keeping Skies Blue" adopt measures such as controlling new production capacity in energy-intensive and heavily polluting industries, promoting cleaner production, accelerating the adjustment of the energy structure and strengthening constraints for energy conservation and environmental protection. By implementing the Action Plan for Prevention and Control of Atmospheric Pollution, China has seen a significant improvement in its air quality and obvious results in GHG emission reduction.

On June 26, 2019, MEE released the *Plan for Comprehensive Control of Volatile Organic Compounds in Key Sectors*, which states that by 2020, a sound system for prevention and control of pollution caused by volatile organic compounds (VOCs) will be established, and continuous improvements in ambient air quality promoted in coordination with the control of GHG emissions. On July 1, MEE, NDRC, MIIT and MOF jointly issued the *Plan for Comprehensive Control of Atmospheric Pollution Caused by Industrial Furnaces* to guide local authorities to strengthen comprehensive control of atmospheric pollution caused by industrial furnaces, control GHG emissions in a coordinated manner and promote high-quality development of industries.

d. Steady pushing forward the construction of national carbon market

Over the past year, China has steadily pushed the construction of the national carbon emission trading system in line with the tasks and requirements set forth in the *National Carbon Emission Trading Market Construction Plan*, and achieved positive results:

Firstly, in terms of institutional system construction, China has drafted and perfected the *Provisional Regulations on the Administration of Carbon Credit Trading*, an important document laying a legal foundation for carbon trading. Additionally, China has advanced the formulation of an array of institutional documents such as measures for the administration of GHG emission reporting by key emitting units, measures for the administration GHG emission auditing and measures for the supervision and administration of the trading market, so as to safeguard the operation of the national carbon emission trading market.

Secondly, in terms of technical specification system construction, China has improved the technical solution to allocation of allowances in the power generation sector and organized provincial and municipal authorities to submit lists of key emitting units in the power generation sector, promoting the construction of the national carbon market from the technical perspective.

Thirdly, in terms of infrastructure construction, China has optimized the assessment of and further revised the construction plans for the national carbon market registration system and trading system, and carried out the construction of the two systems.

Fourthly, in terms of capacity building, China has carried out carbon trading training for local ecology and environment departments, with a view to provide support for the construction of the national carbon market.

In December 2019, MOF released the *Interim Regulations on Accounting Treatment in Carbon Credit Trading*, which contains explicit provisions on accounting treatment in carbon emission permit trading—accounting treatment for purchase, sales and voluntary cancellation of carbon allowances by key emitting enterprises, asset attributes of carbon emission allowances and Chinese Certified Emission Reductions (CCERs), and the disclosure of information on asset ownership and change.

In January 2020, Hubei Branch of the State Administration of Foreign Exchange (SAFE) released the *Interim Measures for the Administration of Foreign Exchange for Overseas Investors' Participation in Carbon Credit Trading in Hubei*, which allows overseas investors (including institutions and individuals) to open Non-Resident Accounts (NRAs) at domestic banks and participate in carbon emission permit trading through domestic depository banks with accounts dedicated to capital projects. Under the regulation, funds in the NRAs opened by overseas investors for participation in carbon emission permit trading will not use any balance of short-term foreign debts of opening banks, but should be subject to foreign debt registration as required.

V.   Pollution Prevention and Control and Marine Governance

   a.   Atmospheric pollution prevention and control
        The 2019 CCICED Policy Recommendations proposed to step up coal control to win the blue-sky battle with determination; elaborate a national long-term zero-emission strategy toward the eventual phase-out of coal; increase the subsidy and financial support for renewable energy and gradually remove the subsidy for fossil fuels; accelerate the phase-out of bulk coal use by around 2020 in the Beijing-Tianjin-Hebei and Fenhe-Weihe River Plain regions; and give priority to non-fossil fuel energy grid connection.
        In September 2019, ten ministries including MEE and NDRC, together with Beijing and Tianjin Municipal People's Governments, released the *Action Plan for Comprehensive Control of Atmospheric Pollution in the Beijing-Tianjin-Hebei Region and Surrounding Areas in Autumn and Winter during 2019–2020*. The *Action Plan* proposed to stick to the principle of using electricity, gas, coal or thermal power where applicable, promote photo-thermal utilization of solar energy and centralized biomass utilization, increase the support for pricing policies, accelerate the coal-to-electricity program, emphasize the wide application of achievements in digital technologies to improve the capability and efficiency of law enforcement, and make full use of efficient monitoring and surveillance means such as law enforcement APP, automatic monitoring, satellite remote sensing, UAVs and electricity data.
        Through the joint efforts of all regions, the comprehensive treatment of air pollution in autumn and winter has achieved remarkable results, and the degree of heavy pollution has been significantly reduced. In autumn and winter during 2019–2020 (as of February 15, 2020, the same below), $PM_{2.5}$ concentrations in 74 cities averaged 50 ug/m^3, 44% lower than the level in autumn and winter during 2013–2014; each city saw four heavy pollution days on average, a decrease of 15 days compared with the level in autumn and winter in 2013. In the "2 + 26" cities, $PM_{2.5}$ concentrations in autumn and winter during 2019–2020 averaged 77 ug/m^3, a 40% decrease compared with the level in 2013; each city experienced 14 heavy pollution days on average, a decrease of 28 days compared with the 2013 level. In Beijing, $PM_{2.5}$ concentrations in autumn and winter dropped with fluctuations, with the level in autumn and winter during 2019–2020 being 44% lower than that in 2013; the city witnessed eight heavy pollution days,

a decrease of 12 days compared with the 2013 level, indicating a marked improvement in air quality.[3] Currently, the Blue Sky Protection Campaign has entered the decisive phase, when some new problems have appeared. On the one hand, the pollution of particulate matters remains high in key regions and rises rather than declines in some cities. On the other hand, ozone pollution exhibits an obvious growing trend and, in some cities, ozone has even overtaken $PM_{2.5}$ as the primary pollutant.

b. Water pollution prevention and control
    In 2019, critical headway was made in prevention and control of water pollution. In 2019, the proportion of sections with excellent or good surface water quality (Grade I-III) nationwide rose by 3.9% year on year, and the proportion of sections with water quality inferior to Class V dropped by 3.3%. To be specific, the proportion of sections with water quality better than Grade III in the Yangtze River basin rose by 4.2% year on year and that of sections with water quality inferior to Class V declined by 1.2%. The numbers of state-controlled sections with water quality inferior to Class V in the Yangtze River basin and rivers emptying into the Bohai Sea fell from 12 and 10 to 3 and 2 respectively. Water quality in offshore areas was in general steadily improved. The proportion of the area with excellent or good water quality (Class I and II) in the offshore area of the Bohai Sea rose by 12.5% year on year, and that of the area with water quality inferior to Class IV fell by 3.7%. On March 28, 2019, MEE, MNR, Ministry of Housing and Urban–Rural Development (MOHURD), MWR and MOA jointly released the *Implementation Plan for Prevention and Control of Groundwater Pollution*, which explicitly proposes to "control pollution in a systematic and comprehensive way and from the source, with the protection and improvement of groundwater environmental quality at the core, and ensure sustainable utilization of groundwater resources".
    On July 8, 2019, Central Rural Work Leading Group Office, MOA, and MNR jointly released *Guiding Opinions on Promoting Rural Domestic Sewage Treatment, and* MEE, together with MWR and MOA, released the *Guiding Opinions on Promoting the Treatment of Black and Odorous Water Bodies in Rural Areas*, which provide top-level design for deepening rural domestic sewage treatment.
    On January 19, 2020, MEE and MWR jointly released the *Guiding Opinions on Establishing a Joint Prevention and Control Mechanism for Water Pollution Incidents in the Upper and Lower Reaches of Cross-Provincial River Basins* was released, which brings forward eight measures, including establishing a collaboration system, strengthening judgment and early warning, preventing and controlling pollution in a scientific manner,

---

[3] National Center for Joint Air Pollution Prevention and Control. Blue Sky Protection Campaign Experts: Regional air quality has improved significantly, but air pollution prevention and control still has a long way to go. http://sthjt.jl.gov.cn/zwzx/qghb/202003/t20200309_6879783.html.

enhancing information disclosure, implementing joint monitoring, coordinating pollution disposal, carrying out dispute mediation and putting into action basic security.

c.   Soil pollution prevention and control

In June 2019, MOF released the revised *Measures for the Administration of the Funds Earmarked for Soil Pollution Prevention and Control*. 30 provinces (autonomous regions and municipalities directly under the central government) nationwide have preliminarily established an access management mechanism for contaminated sites. For example, Shanghai stipulates that plots that are contaminated or of which remediation fails to meet relevant environmental requirements shall not be assigned or transferred. Henan states that for construction land whose soil environment quality does not meet land use conditions in relevant planning, no building permits involving such construction land shall be issued.

In July 2019, MEE released the *Implementation Opinions on Enforcing the Soil Pollution Prevention and Control Law and Promoting the Settlement of Acute Soil Pollution Problems* in partnership with MOA and MNR, which proposes to implement "one-permit" management by including the responsibilities of key industries and enterprises for soil pollution prevention and control in the pollutant permit system, and step up the formulation of the technical specification for independent monitoring and risk identification by key enterprises subject to regulation.

On January 17, 2020, MOF, MEE and other four ministries jointly released the *Measures for the Administration of the Funds for Soil Pollution Prevention and Control*. These funds should be operated in a market-oriented manner and arranged by finance at the level of provinces, autonomous regions, municipalities and cities specifically designated in the state plan through budgets, established separately or jointly with social capital to exert a guiding, driving and leveraging effect by market-oriented means, e.g., equity investment, guiding social capital to invest in soil pollution prevention and control, and support the development of the soil remediation and treatment industry. These funds are mainly used for soil pollution prevention and control on agricultural land, soil pollution risk control and remediation where soil polluters or owners of land use rights cannot be identified, and other matters stipulated by the government.

d.   Solid waste pollution prevention and control

The 2019 CCICED Policy Recommendations proposed to "reduce the use of plastic products, completely eliminate disposable plastic products, reduce the use of plastics in the upstream packaging industry, implement garbage sorting and realize the recycling of plastic waste". It also proposed to "revise the government procurement law. Government procurement should give priority to encouraging green transportation, green buildings, reduction of waste and deforestation and other nature-based products and services".

On May 8, 2019, MEE released the *Guidelines for the Formulation of the Implementation Plan for Pilot Construction of "Zero-Waste Cities"* and the *Index System for the Construction of "Zero-Waste Cities" (for Trial Implementation)*.

On September 9, 2019, the Comprehensively Deepening Reform Commission of the CPC Central Committee deliberated and adopted the *Opinions on Further Enhancing Plastic Pollution Control*. On January 19, 2020, NDRC and MEE published the *Opinions on Further Enhancing Plastic Pollution Control* (the *Opinions*). The *Opinions* explicitly states that by the end of 2020, China will ban or limit the production, sales and use of some plastic products first in some regions and in some sectors, with consumption of disposable plastic products significantly reduced and substitute products promoted by the end of 2022.In 2019, MEE, together with relevant departments, fully implemented the *Plan for Prohibiting the Entry of Foreign Garbage and Advancing the Reform of the Solid Waste Import Administration System*. On the basis of two consecutive years of remarkable achievements in 2017 and 2018, significant progress of the goal of reform in 2019 has been successfully achieved. In 2019, the total import volume of solid waste in China was 13.478 million tons, a decrease of 40.4% year-on-year.[4]

e.   Marine environmental governance

The 2018 CCICED Policy Recommendations proposed to strengthen legal protection for marine and coastal ecosystems; formulate a national action plan for the prevention and control of marine litter pollution; strengthen research on global emerging marine environmental issues such as ocean acidification, marine plastics and microplastics. The 2019 CCICED Policy Recommendations put forward: China should strengthen comprehensive ocean governance, actively participate in global ocean governance, and enhance marine ecological protection and governance capacity.

In terms of land-based pollution control, in February 2019, MEE released the *Work Plan for Investigation and Rectification of Sewage Outlets into the Bohai Sea*, laying a foundation for effectively managing and controlling land-based pollution sources and improving the ecological and environmental quality of the Bohai Sea. In June 2019, MEE released the *Work Plan for the Campaign on Rectification of State-Controlled Sections with Water Quality Inferior to Class V in Rivers Emptying into the Bohai Sea*, which lists eliminating state-controlled sections with water quality inferior to Class V in rivers emptying into the sea as a marked achievement of the Bohai Sea battle.

In terms of marine pollution control, MOA, MOT and MOHURD have worked together in the control of mariculture pollution, control of ship

---

[4] Ministry of Ecology and Environment. MEE regular press conference in January 2020. http://www.mee.gov.cn/xxgk2018/xxgk/xxgk15/202001/t20200117_760049.html.

and port pollution and prevention and control of marine litter pollution respectively. In terms of ecological protection and restoration, MNR, NFGA, MOA and MEE have taken "protective" and "restoration" measures at the same time. In terms of environmental risk prevention, MEE has strengthened oil spill risk control in offshore oil exploration and development activities, studied and established a linkage mechanism for collaboration support and emergency response to oil spills in partnership with oil-related enterprises like China National Offshore Oil Corporation (CNOOC); MOT has strengthened capacity building for emergency response to ship pollution incidents; Ministry of Emergency Management (MEM) has perfected the Beijing-Tianjin-Hebei collaborative accident and disaster response mechanism, with ecological and environmental security of the Bohai Sea further improved.

VI.   Environmental Governance and Rule of Law

a.   Law formulation/revision
The 2018 CCICED Policy Recommendations put forward that it is necessary to accelerate the legislation for the protection of the Yangtze River, and the legislation should reflect the systematic comprehensiveness, heterogeneity and special pertinence of the protection of the Yangtze River Basin.

On December 23, 2019, the *Yangtze River Protection Law of the People's Republic of China (Draft)* was submitted for the first time to the 15th Meeting of the 13th NPC Standing Committee for deliberation. The Environmental Protection and Resources Conservation Committee of the NPC explained the draft law. The Draft defines the geographical scope to which it applies as relevant county-level administrative areas in the entire Yangtze River basin on the basis of the scope of the relevant 19 administrative areas through which the river flows and in the light of the physical geographical conditions in the basin. With respect to specific areas and problems, the draft law sets out detailed systems and measures on management of national land space, ecological and environmental remediation, protection and utilization of water resources, promotion of green development, and law enforcement and supervision.

On December 28, 2019, the 15th Meeting of the 13th NPC Standing Committee adopted by vote the revised *Forest Law*, which will take effect from July 1, 2020. The revised Forest Law proposes to make clear forest ownership, strengthen the protection of forest ownership and mobilizes the public enthusiasm for afforestation; pursue forest classification and management, highlight the functions of non-commercial and commercial forests, and foster stable, healthy, quality and efficient forest ecosystems; strengthen the guiding role of planning, combine development planning with special planning, and determine scientific structure and layout of forest conservation and utilization; strengthen forest conservation, reasonably define the responsibilities of the government, the sector and forest managers, and protect forests, woods and forest-lands with the strictest legal system; reform the forest felling management system, combine the reforms to streamline administration and delegate power, improve regulation and upgrade services, and enhance the development vitality of forestry; increase support, perfect the forest ecological benefit compensation system, and guarantee investment in forest ecosystem protection and restoration; define the target-oriented responsibilities, intensify supervision and inspection, and implement the responsibility and assessment systems for fulfilling forest conservation and development targets.

On April 29, 2020, the 17th Meeting of the 13th NPC Standing Committee deliberated and adopted the revised *Law on Prevention and Control of Environmental Pollution by Solid Waste*, which took effect from September 1, 2020. The revised *Law on the Prevention and Control of Environmental Pollution by Solid Waste* proposes to establish a sound household waste sorting system, strengthen the responsibility of producers of industrial solid waste and further tighten the punishments on environmental polluters, and reiterates the ban on imported waste. For waste electric and electronic products, it encourages producers to conduct ecological design and establish recycling systems, thus promoting recycling of resources. It raises fines on a number of offences to RMB 1 million and adds corresponding penalties to some acts on which no specific penalties were imposed.

b. Advancement of reforms to streamline administration and delegate power, improve regulation and upgrade services

In order to effectively complete the aforesaid reforms in the ecological and environmental sector, MEE has successively released three documents. In 2018, it released the *Guiding Opinions on Further Deepening Reforms to Streamline Administration and Delegate Power, Improve Regulation and Upgrade Services in the Ecological and Environmental Sector and Promoting High-Quality Economic Development*, which puts forward 15 measures, including accelerating the reform of examination and approval system, strengthening environmental supervision and law enforcement, enhancing the ability of high-quality service development, promoting the development of environmental protection industry, and improving the ecological environment economic policies.. In January 2019, MEE released the *Opinions on Supporting and Serving Green Development of Private Enterprises* in partnership with the All-China Federation of Industry and Commerce (ACFIC), which comes up with 18 measures to support the green development of private enterprises. The *Opinions on Further Deepening Ecological and Environmental Regulatory Services and Promoting High-Quality Economic Development* enacted in September 2019 attaches more importance to serving "stability on six fronts" (employment, the financial sector, foreign trade, foreign investment, domestic investment and expectations). These measures, oriented towards greater vitality, fairness and convenience, are focusing on optimizing the business environment, actively serving the green development of enterprises, and working together to promote high-quality economic development and high-level protection of the ecological environment, to achieve positive results in environmental, economic and social dimensions.

c. Further development of ecological and environmental protection supervision

Marked by the *Regulations on Central Supervision of Ecological and Environmental Protection* released by the General Office of the CPC Central Committee and the General Office of the State Council, supervision of ecological and environmental protection has been formally established as an important part of the institutional construction and normal mechanism arrangement of the modern ecological governance system. Supervision of ecological and environmental protection includes routine, special and "look-back" supervision, which will become effective means whereby the central government directly supervises local governments to drive ecological progress, rectify ecological and environmental deficiencies, address problems left over by history and realize high-quality economic development.

In June 2019, the General Office of the CPC Central Committee and the General Office of the State Council released the *Regulations on Central Supervision of Ecological and Environmental Protection* (the *Regulations*), which explicitly provides for the supervision work, ranging from the top-level design to specific operations. The *Regulations* makes clear the promotion of the leading group for supervision. Supervision of environmental protection was in the charge of the Leading Group for the Supervision of Ecological and Environmental Protection under the State Council, with specific organization and coordination led by the former Ministry of Environmental Protection (MEP). The office of the Leading Group was at the MEP, named the National Supervision Office of Environmental Protection. The *Regulations* makes it clear that the supervision work will be coordinated and pushed forward by the Central Leading Group for the Supervision of Ecological and Environmental Protection, of which the head and deputy head will be determined by the CPC Central Committee and the State Council upon deliberation. The Central Leading Group is composed of the General Office of the CPC Central Committee, the Organization Department of the CPC Central Committee, the Propaganda Department of the CPC Central Committee, the General Office of the State Council, MOJ, MEE, National Audit Office (CNAO) and the Supreme People's Procuratorate.

The central environmental supervision system, essentially a specific measure of including ecological progress in the five-sphere integrated plan, reflects that ecological and environmental issues have been elevated to have political significance. The establishment and further development of the system has displayed the unbreakable will and steadfast determination of the Chinese government to promote ecological progress and strengthen ecological and environmental protection, which plays an important role in promoting the in-depth development of supervision of ecological and environmental protection in accordance with the law.

d. Promoting the development of a modern environmental governance system

In March 2020, the General Office of the CPC Central Committee and the General Office of the State Council officially released the *Guiding Opinions on Building a Modern Environmental Governance System* (the *Guiding Opinions*). As an important part of China's efforts to build a modern governance system, this document is not only a perfect summary of decades of exploration experience in ecology and environment, but also a milestone guiding future efforts to build a beautiful China.

The *Guiding Opinions* explicitly proposes to maintain benign interactions among government governance, social regulation and enterprise autonomy, improve relevant systems and mechanisms, strengthen governance from the source and make concerted efforts by upholding the centralized and unified leadership of the Party, enhancing the leading role of the government, deepening the principal role of enterprises and better mobilizing social organizations and the public to participate, thus providing a strong system guarantee for promoting fundamental ecological and environmental improvements, driving ecological progress and building a beautiful China.

The *Guiding Opinions* propose to establish sound leadership responsibility system, enterprise responsibility system, universal action system, regulatory system, market system, credit system and system of laws, regulations and policies, ensuring that the responsibilities of all types of entities will be fulfilled, increasing the enthusiasm of market players and the public for participation and establishing an environmental governance system featuring clear orientation, scientific decision-making, effective execution, efficient incentives, multi-stakeholder participation and benign interactions by 2025.

VII. Regional and International Engagement

a. New advances in greening the "Belt and Road Initiative"

The 2018 CCICED Policy Recommendations stated that with its strong emphasis on infrastructure, the BRI requires careful consideration of climate impacts and long-term ecological changes. The 2019 CCICED Policy Recommendations proposed to effectively align biodiversity protection with the BRI, strengthen the development of Green BRI to promote biodiversity protection, build relevant platforms to share best practices in environmental protection, biodiversity conservation and sustainability impact assessment, attach importance to nature-based solutions, carry out natural capital assessment and set related indicators. These policy recommendations have been adopted by relevant government departments and enterprises in practice and achieved good results.

Over the six years since the BRI was presented, China has entered into cooperation on ecological and environmental protection with BRI-related countries and regions, and achieved positive results in a number of areas. First, cooperation mechanisms continue to be improved. Currently, MEE has concluded bilateral documents on ecological and environmental cooperation with 33 BRI countries. Second, participants in the cooperation platform have increased. The Coalition has labeled more than 130 organizations as partners. Third, government communication continues to be deepened. The Thematic Forum on Green Silk Road of the 2nd Belt and Road Forum for International Cooperation, China-ASEAN Environmental Cooperation Forum and other thematic exchange activities have boosted cooperation and exchanges among relevant countries and regions. Fourth, the cooperation has constantly yielded fruitful results. Exchanges and communication with relevant countries and regions in environmental policies, standards, regulations and technologies have promoted information sharing, knowledge sharing and meeting sharing in connection with building a green "Belt and Road", thus informing decision-making concerning green development of the "Belt and Road".

b. International cooperation on climate change

The Chinese government continues to constructively participate in global climate governance in a highly responsible manner. It has played a positive, constructive role in global negotiations on climate change, firmly upheld multilateralism and enhanced dialogues and exchanges on climate change with other countries; promoted global climate governance in partnership with other parties in accordance with the principles of equity, "common but differentiated responsibilities" and respective capabilities as defined by the *United Nations Framework Convention on Climate Change* (UNFCCC); facilitated positive results in the negotiations on the implementation rules of the *Paris Agreement* and contributed Chinese initiatives and proposals to the United Nations Climate Action Summit. On November 6, 2019, China and France released the *Beijing Call for Biodiversity Conservation and Climate Change.* On November 23, environment ministers of China and Japan had exchanges on issues such as climate change and COP15.

In the video speech of the 11th Petersburg Climate Dialogue on April 27–28, 2020, Mr. Huang Runqiu, the new Minister of the MEE, stated that China will unswervingly uphold multilateralism and, together with all other countries, promote the fulfillment of the Paris Agreement in a comprehensive, balanced and effective way. China will also implement the proactive national strategy on addressing climate change and make its due contributions to the global fight against climate change.

China has been promoting South-South cooperation on climate change and doing its best to provide assistance for other developing countries. As of September 2019, China had signed over 30 MOUs on South-South cooperation on climate change with other developing countries, under which they would work together to build low-carbon demonstration zones, launch climate change mitigation and adaptation projects, and organize training sessions on South-South cooperation on climate change.

c. Active participation in global ocean governance

The 2018 CCICED Policy Recommendations proposed to mobilize partnerships for action on plastic pollution. The 2019 CCICED Policy Recommendations stated that China should enhance comprehensive ocean governance, actively participate in global ocean governance and improve governance capabilities for marine ecological protection. The Chinese government has taken a series of actions to enhance comprehensive ocean governance.

The Fourth Session of the UN Environmental Assembly (UNEA-4) in 2019 brought forward the initiative of launching innovative waste management, e.g., "zero waste", in some countries and regions. More importantly, it adopted resolutions concerning global plastic pollution, such as *Addressing Single-Use Plastic Products Pollution* and *Marine Plastic Litter and Microplastics*, and those promoting sustainable production and consumption, including *Innovative Solutions for Environmental Challenges and Sustainable Consumption and Production* and *Promoting Sustainable Practices and Innovative Solutions for Curbing Food Loss and Waste*, stressed dealing with severe plastic pollution and food waste problems by transforming consumption and production habits, and establishing a sound waste management system.

China attaches great importance to marine litter and plastic pollution control, and has vigorously participated in international processes in response to marine litter and plastic pollution, joined the UNEP Regional Seas Programme, carefully observed the *Basel Convention on the Control of Transboundary Movements of Hazardous Wastes and Their Disposal*, facilitated the introduction of documents such as the *East Asia Leaders' Statement on Combatting Marine Plastic Debris* and the *Implementation Framework of the G20 Action Plan on Marine Litter*, with the aim of working together to promote global marine litter and plastic pollution prevention and control. Meanwhile, China have engaged in bilateral cooperation and established a cooperation mechanism for marine litter prevention and control with Japan, Canada and the United States.

VIII.   Conclusions

At present, the international economic and social development environment is undergoing profound and complex changes. China is facing more complex and severe external environment, and unstable and uncertain factors are increasing. In particular, the COVID-19, the global public health security incident, has caused severe damage to the world's economic development, and China's economic and social development has also been deeply affected.

As a major country with outstanding ecological and environmental problems in the world, China is faced with such challenges as thoroughly solving historical debt problems, and effectively dealing with emerging ecological and environmental problems. At the same time, the economic downturn, trade sluggishness, logistics obstruction, and difficulty in resuming work caused by the COVID-19, have made the eco-environmental governance increasingly complex. How to properly handle the relationship between the intensity, depth and effectiveness of governance and move towards a high-quality development path is a great challenge facing China in the future.

The international community calls for clean, green and just transition, so as to create new job opportunities, accelerate the post-epidemic "green recovery" of the economy, and push the world to embark on a more sustainable and inclusive road. "Green recovery" will provide new opportunities for China's eco-environment governance and tackling climate change in the post-epidemic era. Under the epidemic, the proposition of the idea of constructing "new infrastructure" offers strong technological supports for the high-quality development, which, together with initiatives like the Green Development of Belt and Road Initiative and the China International Import Expo, forms a new pattern of high-quality development in a new era.

Over the past year, the forward-looking recommendations of the CCICED in the areas of wildlife resource protection, ecological redline delineation, coal use, plastic pollution control, green consumption, etc. have been highly valued by the Chinese government, providing important enlightenments for the future construction of ecological civilization. China cannot achieve high-quality development and ecological civilization without international cooperation. In addition to the cooperation with established powers, China has been redoubling efforts to strengthen cooperation with countries on ecological and environment protection and with African countries to intensify environmental protection and promote South-South Cooperation. China has also seen new progresses in marine ecosystem conservation, climate change tackling, biodiversity protection, and other issues of global concern.

2020 will be a year of milestone significance for China, as China is set to finish building a moderately prosperous society in all respects, but severe environmental problems may undermine its efforts to accomplish this great mission. 2020 is also the last year that will mark the successful conclusion of the 13th Five-Year Plan. Furthermore, it is the year to launch the 14th Five-Year Plan. China has not yet reached the turning point to see fundamental improvements in environmental quality, and will still face very arduous tasks in the coming period. Since the founding of CCICED, 28 years have passed. All stakeholders have great expectations for CCICED to deliver "innovative" and "inspiring" policy research findings. CCICED is required to attach great importance to the following three aspects: first, better grasp its own new positioning in the new era; second, have finger on the pulse of major issues and the development tendencies of environmental protection and economic development at home and abroad in a more accurate manner, in order to offer appropriate solutions; and third, give better play to the "two-way communication" platform.

Appendix: Overview on the Relevance of China's Environmental and Development Policies and CCICED Policy Recommendations During 2019–2020

Field	Time of Release of Policy	Policy Progress (2019–2020)	Content
Planning for Environment and Development	November 2019	At the special meeting on the formulation of the 14th FYP on November 26, 2019, Premier Li Keqiang of the State Council listened to the NDRC report on the formulation of the 14th FYP and proposed the general requirements for plan formulation in next steps. Premier Li pointed out that we should take a long-term perspective and a holistic approach……A number of critical projects should be launched to improve weak links, push ahead industry upgrading, enhance sustainability and improve people's lives, with emphasis placed on exerting the role of social forces to upgrade infrastructure, enhance industry innovation and competitiveness, promote the improvement of ecological environment and ensure a better life for people	The 2017 CCICED Policy Recommendations proposed that China should formulate a long-term comprehensive strategic plan with coverage of water, air and ocean pollution for the next 10 or 15 years. It was pointed out that the plan should be fully implemented before 2020, tailored to the agenda for basic realizing socialist modernization of Chinese society. In addition, the plan is also suggested to place focus on innovation, and reduce plastic litter pollution in rivers and oceans by regulating plastic production and source control The 2019 CCICED Policy Recommendations proposed that the 14th FYP should echo and support the vision of Beautiful China 2035, and the global vision of climate action and biodiversity protection. The national 14th FYP and the plan for environmental protection during the 14th FYP period that are currently under drafting and review process have made full reference to these suggestions
	April 2019	A workshop on the national 14th FYP for ecological and environmental protection was held by MEE to probe into how to maintain and implement the concept of "ecology first and green development" based on the strategic positioning of ecological and environmental protection in national economic and social development during the 14th FYP period; by 2035, ecological and environmental quality should be fundamentally improved and the "Beautiful China" strategic goal should be basically attained; efforts should be made to actively plan strategic tasks reflecting environmental quality improvement	

(continued)

(continued)

Field	Time of Release of Policy	Policy Progress (2019–2020)	Content
	December 2019	Under the framework of the master plan of the 14th FYP, the outlines of the 14th FYP for ecological and environmental protection began to be drafted in December 2019. The Plan will emphasize coordinated actions to advance high-quality economic development and high-level environment protection; give full play to the role of science and technology and tackle key problems in science and technology to provide strong support for decision making, management and governance, and motivate business innovation to bring about revolution of the ecological environment. The Plan is committed to improving environmental quality and addressing prominent environmental problems, with focus on setting scientific, targeted, feasible and effective objectives and planning	
	October 2019	The State Council approved the Overall Plan for Demonstration Zones for Eco-friendly and Green Integration of the Yangtze River Delta (the Plan). The Plan sets the goals that by 2025, a number of environmental, infrastructure, technology innovation, public service and other major projects will be completed and put into operation, evident improvements seen in eco-environmental protection and construction, innovative development of eco-friendly industries and harmony between man and nature in pilot zones, the main functional framework basically established for integration demonstration zones and ecological quality significantly improved. Innovation in the integration system will bring about replicable experience and system integration of major reforms will unleash the dividend, with the role of these demonstration zones in leading higher-quality integration of the Yangtze River Delta initially exerted. By 2035, a more mature, effective system of green integrated development will be established and become the benchmark for demonstration and leading of high-quality integration of the Yangtze River Delta	
	December 2019	The CPC Central Committee and the State Council issued the Outline of the Integrated Regional Development of the Yangtze River Delta (the Outline). The Outline indicates that by 2025, substantial progress will be made in the integrated development of the Yangtze River Delta, especially in technology innovation industries, infrastructure, ecological environment and public services, just to name a few	

(continued)

(continued)

Field	Time of Release of Policy	Policy Progress (2019–2020)	Content
Environmental protection and the COVID-19 epidemic	Feburary 24, 2020	The NPC adopted by vote the Decision on a Complete Ban of Illegal Wildlife Trade and the Elimination of the Unhealthy Habit of Indiscriminate Wild Animal Meat Consumption for the Protection of Human Life and Health on February 24, 2020, which sets forth specific provisions on an all-out ban of illegal wild animal trade and relevant penalties. In charge of "coordinating the protection of biodiversity" and "supervising wildlife protection", MEE has, upon deliberation, included the enforcement of laws and regulations on wild animal protection in the scope of central supervision of ecological and environmental protection	The 2019 CCICED Policy Recommendations proposed to strengthen research on the breeding, cultivation and sustainable use of wild biological resources, promote technological upgrading, reduce consumption of natural and biological resources…, and take legal action against illegal wildlife sales and smuggling
	Feburary 2020	The National People's Congress launched the revision of the Law of The People's Republic of China on Protection of Wildlife, proposing to include it into the legislative work plan of the NPC standing committee and accelerating the progress of revising Law on Animal Epidemic Prevention	
Ecosystem and Biodiversity Conservation	October 2020	The CBD COP15, under the theme of "Ecological Civilization: Building a Shared Future for All Life on Earth", is scheduled to be held in Kunming, China in 2021, which will deliberate the "post-2020 global biodiversity framework" and define global biodiversity conservation targets for 2030	The 2019 CCICED Policy Recommendations proposed to draw on the success of Paris climate negotiation to gather high-level political wills through green diplomacy, calling for joint participation of business communities, academic circles, social organizations and the general public in making and implementing biodiversity protection framework for the post-2020 period, in an effort to advertise the Human and Nature Action Agenda, raise public awareness and advance concerted action The 2018 CCICED Policy Recommendations proposed to actively promote the implementation of CBD, and play a leading role in the development of post-2020 global biodiversity conservation targets
	July 2019	MEE, MWR, MOA, CAS, NFGA and China Coast Guard of the Chinese People's Armed Police Force jointly launched the "Green Shield 2019" mission to strengthen natural reserve supervision. The mission mainly involves on-site inspection of new problems and clue identified by remote sensors of human activities in national and provincial natural reserves in different provinces, especially problems such as quarrying and sand mining, industrial and mining land, tourist facilities and utilities in core areas in those reserves	

(continued)

(continued)

Field	Time of Release of Policy	Policy Progress (2019–2020)	Content
	June 2019	The General Office of the CPC Central Committee and the General Office of the State Council released the Guiding Opinions on Establishing a System of Natural Reserves with National Parks as the Main Body (the Opinions). The Opinions proposes to establish a system of natural reserves with national parks as the main body and with Chinese characteristics, promote scientific setup of various natural reserves, create new systems, mechanisms and models for natural ecosystem protection and construct healthy, stable and efficient natural ecosystems, thus laying a solid foundation for maintaining national ecological security and realizing sustainable economic and social development, and an ecological foundation for building a great modern socialist country that is wealthy, democratic, civilized, harmonious and beautiful	The 2019 CCICED Policy Recommendations proposed to strengthen the construction of Natural Reserve Management System targeted on national parks, and set up red lines for ecological protection. Besides, it also suggested to develop and implement an all-round legal system and market incentive measures, while ensuring the effectiveness of implementation, and to enhance cross-department concerted action and cancel subsidies that may have harmful influences on the ecological environment
	November 2019	Governmentally, MEE, together with departments concerned and local authorities, has pushed forward the assessment and demarcation for the delineation of ecological red lines. By scientifically assessing the delineation of local ecological red lines and making reasonable adjustments in an orderly manner, MEE can ensure that these red lines are authoritative, scientific and executable, thus laying a foundation for the establishment of an ecological red line system by the end of 2020. Locally, Shenyang began the construction of its ecological red line supervision platform in August 2018, which was accepted in November 2019 and has been put into trial operation recently. It is China's first city-level ecological red line supervision platform. At the same time, Shenyang has become the first pilot city for the construction and interconnection of ecological red line supervision platforms	The 2014 CCICED Policy Recommendations suggested that the State Council accelerate the formulation of the Administrative Measures of Ecological Protection Red Lines and clarify the definition, significance, standard-making methods and regulation system of the red lines
Energy and Climate	December 2019	The presidents of China and Russia had a video meeting, witnessing the formal operation of the east-route natural gas pipeline. The China-Russia east-route natural gas pipeline, once completed, enables stable supply of 38 million m3 of clean and high-quality gas resources to Northeast China, the Bohai Rim and the Yangtze River Delta every year, equivalent to roughly 13.6% of China's gas consumption in 2018. This will benefit more than 400 million people in the nine provinces and municipalities along the pipeline and effectively alleviate and improve atmospheric pollution there	The 2019 CCICED Policy Recommendations proposed to step up coal control to win the blue-sky battle with determination; elaborate a national long-term zero-emission strategy toward the eventual phase-out of coal; increase the subsidy and financial support for renewable energy and gradually remove the subsidy for fossil fuels

(continued)

(continued)

Field	Time of Release of Policy	Policy Progress (2019–2020)	Content
	April 10, 2020	NEA released the Energy Law of the People's Republic of China (Exposure Draft), arousing wide attention and a heated discussion from home and abroad. The draft explicitly states that energy exploitation and utilization should be adapted to ecological civilization, carry out the new development concept, implement a conservation-first, domestic demand-oriented, green and low-carbon, and innovation-driven energy development strategy, and establish a clean, low-carbon, safe and efficient energy system. It repeatedly emphasizes energy structure optimization and clearly proposes to give priority to renewable energy development, maintain safe and efficient development of nuclear power, increase the share of non fossil energy and promote clean, efficient utilization and low-carbon development of fossil energy; encourages efficient and clean exploitation and utilization of energy resources, supports prior exploitation of renewable energy, rational exploitation of fossil energy resources, development of distributed energy according to local conditions, advances the substitution of fossil energy with non-fossil one and of high-carbon energy with low-carbon one and supports the development and application of new fuels and industrial raw materials that can replace petroleum and gas	The 2019 CCICED Policy Recommendations proposed to step up coal control to win the blue-sky battle with determination; elaborate a national long-term zero-emission strategy toward the eventual phase-out of coal; increase the subsidy and financial support for renewable energy and gradually remove the subsidy for fossil fuels. The 2018 CCICED Policy Recommendations highlighted the necessity of tightening control on coal use and expanding the growth of energy efficiency
	2019	In 2019, China has deeply promoted the reduction of structural coal overcapacity, organized the classification and disposal of coal mines with an annual capacity below 300,000 tons, and shut down more than 450 outdated coal mines. Coal-fired power generation units with a combined capacity of 20 million kW were eliminated and shut down, outperforming the specified target	The 2018 CCICED Policy Recommendations highlighted the necessity of tightening control on coal use and expanding the growth of energy efficiency
	2017–2021	As a backbone power station of the "West-East Electricity Transmission Project" during the 13th FYP period, the Baihetan Project that is still under construction is the second largest hydro-power station in China only next to the Three Gorges Hydro Project, and is the world's largest hydro-power station; its installed capacity is 16 million kWh; after completion, it is expected to form a Yangtze River Clean Energy Passage with the Yangtze River Three Gorges Station and the Gezhouba Station. The Baihetan hydro-station can reduce consumption of standard coal by 20 million tons per year and cut carbon emissions by around 40 million tons	

(continued)

(continued)

Field	Time of Release of Policy	Policy Progress (2019–2020)	Content
	2019–2020	In the second half of 2019, problems in the gas source guarantee for the "coal-to-gas" project in northern China were tackled. Clean heating in winter in the north has been expedited, with an increased clean heating area of roughly 1.5 billion m2 and a clean heating rate of 55%. Approximately 100 million tons of bulk coal was replaced and the clean heating rate in the "2+26" key cities reached 75%, outperforming the mid-term target. China has also strengthened electric energy substitution in different sectors, which is expected to increase by roughly 200 billion kWh; gasoline and diesel for vehicles that meet China VI emission standards have been fully supplied. The demand for natural gases in the north is growing as a result of the "Coal-to-Gas" project. To tackle that, the PetroChina has been making every attempt to ensure sufficient supplies, and has tried its best to support the "coal-to-gas" project in the "2+26" Cities and Fenhe-Weihe River Plain, with full delivery of the volume as agreed in the contract to ensure a warm winter for the people	
	January 2020	The MIIT released the Notice on the Issuance of Key Work Plan of Industrial Energy-saving Monitoring (the Notice). The Notice was pointed out that special energy-saving monitoring shall be conducted for key industries of intensive energy consumption. To be specific, special monitoring will be carried out for the implementation of compulsory energy consumption limit per product for enterprises with heavy energy consumption in sub-industries, such as those in the petrochemical industry specializing oil refinement, p-xylene, soda ash, polyvinyl chloride, sulfuric acid, tire, methanol, those in the non-ferrous metal industry specializing in gold smelting, rare earth smelting & processing, aluminum alloy, copper and copper alloy processing, and those in the construction material industry specializing in calcined gypsum, sintered wall materials, asphalt-based waterproof roll materials, rock wool, slag wool and their products, and those in light industries specializing in sugar and beers	The 2018 CCICED Policy Recommendations highlighted the necessity of tightening control on coal use and expanding the growth of energy efficiency

(continued)

(continued)

Field	Time of Release of Policy	Policy Progress (2019–2020)	Content
	March 2020	NDRC and other 22 ministries and commissions jointly released the Implementation Opinions on Promoting Consumption Scale Expansion and Quality Improvement and Accelerating the Establishment of a Strong Domestic Market, which proposes to "encourage use of green smart products, promoting green consumption by focusing on supply of green products, construction of green public transport facilities, energy-saving and environmentally-friendly buildings and relevant technology innovation, and build green shopping malls"	The 2018 CCICED Policy Recommendations pointed out that the government shall give full play to existing policies and measures in promoting sustainable production and consumption, and develop and implement targeted industry-specific strategies for the 14th FYP period. In addition, it was also suggested that governments fully consider how to improve the well-being and healthy lives of the public while reducing carbon footprints, make homes, schools and workplaces more comfortable and easy, create more green job opportunities, and place more attention to the families' role in promoting sustainable consumption
	March 2020	On March 17, 2020, NDRC and MOJ released the Opinions on Accelerating the Establishment of a System of Regulations and Policies for Green Production and Consumption, which stipulates multiple tasks such as promoting green design, enhancing clean industrial production, developing recycling industrial economy, strengthening control of industrial pollution, advancing the development of clean energies, facilitating the green development the agricultural and service industry, boosting consumption of green products and advocating green lifestyles	The 2019 CCICED Policy Recommendations stated that the green consumption is one of the key measures to construct ecological civilization, and should be included into the state's 14th FYP as a key task of ecological civilization construction
	October 2019	The Manual on Green Lifestyle, which is the first of its kind in China, was released. The Manual centers around green lifestyle and provides guidance for promoting green lifestyle across the country from various aspects, like green production and green consumption	The 2019 CCICED Policy Recommendations proposed to launch the Green Lives Campaign, spurring demands for green products and fully unleashing the exemplary role of prominent public figures in leading green consumption into a social trend, with a focus on promoting green consumption as a life style to bring benefits to public health and the environment
	2019	In terms of local practice, Shenzhen City has made positive headway in promoting coordinated control of GHG emissions and atmospheric pollutants. "Homology" analysis indicates that of road traffic emissions, carbon accounts for 49% and PM2.5 41%; of non-road traffic emissions, carbon accounts for 12% and PM2.5 11%; of emissions from electric and thermal power sectors, carbon accounts for 23% and PM2.5 8%; of emissions from non-energy industries, carbon accounts for 3% and PM2.5 15%	The 2019 CCICED Policy Recommendations proposed that promoting air quality improvement and GHG emission reduction in a coordinated manner was an inevitable approach to high-quality development in China The 2014 CCICED Policy Recommendations proposed to intensify control on multiple source pollution and multiple pollutants in a synergistic manner. To be specific, efforts should be redoubled to control such pollutants as SO2, NOX, PM2.5, VOC and NH3 from industrial sources, civil and rural non-point sources, motor vehicle and non-road machinery

(continued)

(continued)

Field	Time of Release of Policy	Policy Progress (2019–2020)	Content
Pollution Prevention and Control	December 2018	Air pollution prevention and control. In view of the current situation and tasks of protecting the blue sky, efforts will be made to further adjust, optimize, and enhance the monitoring of supports for designated areas; focus will remain on the 39 cities of the key areas, which are Beijing-Tianjin-Hebei Region and its surrounding areas, and the Fenhe-Weihe Plain, with more focus on non-key areas like Jiangsu, Anhui, Shandong and Henan provinces; the key factor PM2.5 will be closely followed, with attention to other factors like Ozone etc.; efforts will be continued to ensure the emergency response to and joint prevention and control of severe pollution in autumns and winters, and to carry out the "Treat the Winter Issues in Summers" and ozone control; close watch will be kept not only on key industries like steel and thermal power generation, but also on special industries that cause ozone pollution like the petrochemical industry	The 2019 CCICED Policy Recommendations proposed to step up coal control to win the blue-sky battle with determination; elaborate a national long-term zero-emission strategy toward the eventual phase-out of coal; increase the subsidy and financial support for renewable energy and gradually remove the subsidy for fossil fuels; accelerate the phase-out of bulk coal use by around 2020 in the Beijing-Tianjin-Hebei region and Fenhe-Weihe River Plain; and give priority to non-fossil fuel energy grid connection The 2014 CCICED Policy Recommendations proposed to intensify control on multiple source pollution and multiple pollutants in a synergistic manner. To be specific, efforts should be redoubled to control such pollutants as SO2, NOX, PM2.5, VOC and NH3 from industrial sources, civil and rural non-point sources, motor vehicle and non-road machinery
	February 2020	Tianjin's headquarter on the battle against pollution has recently issued the 2020 Work Plan of Tianjin Municipality for Winning the Battle against Pollution. The Work Plan makes it clear that in 2020, the core target of the battle to protect blue skies is to control the annual average PM2.5 concentration at around 48ug/m3 and the proportion of days with excellent or good air quality at 71% or above. Tianjin will further emphasize "three time nodes", namely before summer, in autumn and winter and before heating, continue to adjust "four structures", including industrial, layout, energy and transportation structures, and unremittingly advance "control of five types of projects", such as coal-fired, industrial, motor vehicle, fugitive dust and new projects	
	March 2019	Water pollution prevention and control. The Ministry of Ecology and Environment and the Ministry of Natural Resource jointly issued the Notice on Issuance of the Implementation Plan for the Prevention and Control of Groundwater Pollution. It was pointed out that as of 2020, the typical source of underground water pollution has been preliminarily monitored, and the aggravation trend of underground water pollution has been preliminarily brought under control. Besides, it is expected that by 2025, the typical source of underground water pollution is effectively monitored, and the aggravation trend of underground water pollution is effectively controlled; it is expected that as of 2035, with hard efforts, the overall underground water quality across the country can be improved, and the ecosystem can basically restore its function	

(continued)

(continued)

Field	Time of Release of Policy	Policy Progress (2019–2020)	Content
	2019	Water pollution prevention and control. In 2019, the proportion of sections with excellent or good surface water quality (Class I-III) nationwide rose by 3.9% year on year, and the proportion of sections with water quality inferior to Class V dropped by 3.3%. To be specific, the proportion of sections with water quality better than Class III in the Yangtze River basin rose by 4.2% year on year and that of sections with water quality inferior to Class V declined by 1.2%. The numbers of state-controlled sections with water quality inferior to Class V in the Yangtze River basin and rivers emptying into the Bohai Sea fell from 12 and 10 to 3 and 2 respectively. Water quality in offshore areas was in general steadily improved. The proportion of the area with excellent or good water quality (Class I and II) in the offshore area of the Bohai Sea rose by 12.5% year on year, and that of the area with water quality inferior to Class IV fell by 3.7%	
	2019	2019 marked the year that China's Soil Pollution Prevention and Control Law entered into effect. In July 2019, MEE released the Implementation Opinions on Enforcing the Soil Pollution Prevention and Control Law and Promoting the Settlement of Acute Soil Pollution Problems in partnership with MOA and MNR, which contains detailed tasks, measures and division of labor by focusing on acute problems affecting "food and housing" safety of the public, and promotes the prevention and control of soil pollution according to law. 31 provinces (autonomous regions and municipalities directly under the central government) and Xinjiang Production and Construction Corps have published the lists of totally 5,927 key enterprises subject to soil environmental monitoring	The 2010 CCICED Policy Recommendations proposed to push forward the protection of soil environment to safeguard public health and ecological security, and to develop a special law for preventing soil pollution and protecting the soil environment
	2019	Under the revised Measures for the Administration of the Funds Earmarked for Soil Pollution Prevention and Control released by MOF in June 2019, central earmarked funds will give appropriate support for the establishment of provincial soil pollution prevention and control funds	

(continued)

(continued)

Field	Time of Release of Policy	Policy Progress (2019–2020)	Content
	September 2019	Marine environmental protection. On September 9, 2019, the Comprehensively Deepening Reform Commission of the CPC Central Committee deliberated and adopted the Opinions on Further Enhancing Plastic Pollution Control	The 2018 CCICED Policy Recommendations proposed to strengthen legal protection for marine and coastal ecosystems; formulate a national action plan for the prevention and control of marine litter pollution; strengthen research on global emerging marine environmental issues such as ocean acidification, marine plastics and microplastics
	January 2020	On January 19, 2020, NDRC and MEE published the Opinions on Further Enhancing Plastic Pollution Control. The Opinions explicitly states that by the end of 2020, China will ban or limit the production, sales and use of some plastic products first in some regions and in some sectors, with consumption of disposable plastic products significantly reduced and substitute products promoted by the end of 2022	The 2019 CCICED Policy Recommendations proposed to "reduce the use of plastic products, completely eliminate disposable plastic products, reduce the use of plastics in the upstream packaging industry, implement garbage sorting and realize the recycling of plastic waste". It also proposed to "revise the government procurement law. Government procurement should give priority to encouraging green transportation, green buildings, reduction of waste and deforestation and other nature-based products and services"
Environmental Governance and Rule of Law	April 2020	The 17th Meeting of the 13th NPC Standing Committee deliberated and adopted the revised Law on Prevention and Control of Environmental Pollution by Solid Waste, which took effect from September 1, 2020	The 2014 CCICED Policy Recommendations highlighted the necessity of accelerating the building of the ecological civilization system and the progress in ecological protection system reform, so as to improve the country's environmental governance capacities
	September 2019	The General Office of the CPC Central Committee and the General Office of the State Council released the Regulations on Central Supervision of Ecological and Environmental Protection (the Regulations), which, in the form of a statute for the first time, explicitly provides for the supervision work, ranging from the top-level design to specific operations	
	July 2019	The second round of the central supervision of ecological and environmental protection was fully launched, in which eight central supervision groups for ecological and environmental protection were set up and dispatched to six provinces (municipalities) such as Shanghai, Fujian, Hainan, Chongqing, Gansu and Qinghai, and two central enterprises, namely China Minmetals Corporation (Minmetals) and China National Chemical Corporation Ltd. (ChemChina), respectively for supervision	

(continued)

(continued)

Field	Time of Release of Policy	Policy Progress (2019–2020)	Content
	March 2020	The General Office of the CPC Central Committee and the General Office of the State Council officially released the Guiding Opinions on Building a Modern Environmental Governance System (the Guiding Opinions). The Guiding Opinions propose to establish sound leadership responsibility system, enterprise responsibility system, universal action system, regulatory system, market system, credit system and system of laws, regulations and policies, ensuring that the responsibilities of all types of entities will be fulfilled, increasing the enthusiasm of market players and the public for participation and establishing an environmental governance system featuring clear orientation, scientific decision-making, effective execution, efficient incentives, multi-stakeholder participation and benign interactions by 2025	
Regional and International Engagement	March 2020	As indicated by the practice of BRI projects, green and environmental protection has become symbols of these projects. For example, Hasyan Clean Coal Fired Power Station, built by Harbin Electric Corporation (HE) and funded by the Silk Road Fund, is the first clean coal-fired power station in the Middle East. HE employed advanced combustion, denitration, dedusting and desulfurization technologies to ensure that the dust, sulfide and nitride indicators emitted during the operation of the power station were better than similar generator units in the world, so as to reduce atmospheric pollutant emissions. Apart from environmental protection, biodiversity conservation is also an integral part of the project. To guarantee normal reproduction of hawksbill turtles, an endangered species, during the construction, Chinese builders controlled construction lighting in the nighttime to avoid disturbing the spawning, transferred their eggs to safe hatcheries	The 2019 CCICED Policy Recommendations proposed to effectively align biodiversity protection with the BRI, strengthen the development of Green BRI to promote biodiversity protection, build relevant platforms to share best practices in environmental protection, biodiversity conservation and sustainability impact assessment, attach importance to nature-based solutions, carry out natural capital assessment and set related indicators. These policy recommendations have been adopted by relevant government departments and enterprises in practice and produced positive results

(continued)

(continued)

Field	Time of Release of Policy	Policy Progress (2019–2020)	Content
	September 2019	China has been promoting South-South cooperation on climate change and doing its best to provide assistance for other developing countries. As of September 2019, China had signed over 30 MOUs on South-South cooperation on climate change with other developing countries, under which they would work together to build low-carbon demonstration zones, launch climate change mitigation and adaptation projects, and organize training sessions on South-South cooperation on climate change. Since 2019, China has actively promoted the consultation and implementation of the cooperation on low-carbon demonstration zones with Cambodia, Laos, Kenya, Ghana and Seychelles, the implementation of the donation projects for climate change mitigation and adaptation with over 10 countries including Ethiopia, Egypt and Guinea, and the consultation on new projects with Botswana, Uruguay and the Philippines, and held nine training sessions on South-South cooperation on climate change	
	2019	China vigorously participated in international processes in response to marine litter and plastic pollution, joined the UNEP Regional Seas Programme, carefully observed the Basel Convention on the Control of Transboundary Movements of Hazardous Wastes and Their Disposal, facilitated the introduction of documents such as the East Asia Leaders' Statement on Combating Marine Plastic Debris and the Implementation Framework of the G20 Action Plan on Marine Litter, with the aim of working together to promote global marine litter and plastic pollution prevention and control. Meanwhile, China has propelled bilateral cooperation. For example, China has established a cooperation mechanism for marine litter prevention and control with Japan, Canada and the United States	The 2018 CCICED Policy Recommendations proposed to mobilize partnerships for action on plastic pollution The 2019 CCICED Policy Recommendations stated that China should enhance comprehensive ocean governance, actively participate in global ocean governance and improve governance capabilities for marine ecological protection. The Chinese government has taken a series of actions to enhance comprehensive ocean governance

Appendix C
CCICED Phase VI Composition (as of December 2020)

Executive Members

Mr. HAN Zheng	**Chairperson** Vice Premier of the State Council, P.R.China
Mr. HUANG Runqiu	**Executive Vice Chairperson** Minister, Ministry of Ecology and Environment, P.R.China
Mr. Jonathan Wilkinson	**Executive Vice Chairperson** Minister, Environment and Climate Change Canada
Mr. XIE Zhenhua	**Vice Chairperson** Special Advisor on Climate Change Affairs, Ministy of Ecology and Environment, P.R.China
Mr. ZHOU Shengxian	**Vice Chairperson** Former Minister of the Ministry of Environmental Protection, P.R.China
Mr. Achim Steiner	**Vice Chairperson** Administrator, The United Nations Development Programme
Ms. Inger Andersen	**Vice Chairperson** Executive Director, The United Nations Environment Programme
Mr. Vidar Helgesen (2017–2020)	**Vice Chairperson** Special Envoy to the High-level Panel on Building a Sustainable Ocean Economy; Former Minister, Ministry of Climate and Environment of the Kingdom of Norway
Mr. Erik Solheim	**Vice Chairperson** Advisor, World Resouce Institute
Mr. ZHAO Yingmin	**Secretary General of CCICED** Vice Minister, Ministry of Ecology and Environment, P.R.China

(continued)

© China Environment Publishing Group Co., Ltd. 2022
China Council for International Cooperation on Environment
and Development (CCICED) Secretariat,
Green Consensus and High Quality Development,
https://doi.org/10.1007/978-981-16-4799-4

(continued)

Chinese Members	
Mr. LIU Shijin	**Chinese Chief Advisor** Deputy Director, Committee for Economic Affairs of the National Committee of the Chinese People's Political Consultative Conference; Vice Chairman, China Development Research Foundation
Mr. HAN Wenxiu	Vice Minister, the Office of the Central Financial and Economic Affairs Committee
Mr. YANG Weimin	Deputy Director, Committee for Economic Affairs of the 13th National Committee of the Chinese People's Political Consultative Conference; Former Vice Minister, the Office of Central Leading Group on Financial and Economic Affairs
Mr. MA Zhaoxu	Vice Minister, Ministry of Foreign Affairs
Mr. XIN Guobin	Vice Minister, Ministry of Industry and Information Technology
Mr. YU Weiping	Vice Minister, Ministry of Finance
Mr. WANG Hong	Vice Minister of the Ministry of Natural Resources, Director of State Oceanic Administration
Mr. WANG Shouwen	Vice Minister, Ministry of Commerce
Mr. ZHOU Wei	Chief Engineer, Ministry of Transport
Mr. CHEN Yulu	Vice Governor, the People's Bank of China
Mr. CHEN Li	Member, the Overseas Chinese Affairs Committee of the 13th National People's Congress; Former Vice President, Chinese Academy of Governance
Mr. WANG Feng	Member, Supervisory and Judicial Affairs Committee of the National People's Congress; Former Deputy Director, State Commission Office of Public Sectors Reform
Mr. XU Xianping	Counsellor, the State Council; and Specially-appointed Professor, Guanghua School of Management, Peking University
Mr. QIU Baoxing	Counsellor, the State Council
Ms. LI Xiaolin	Former President, the Chinese People's Association for Friendship with Foreign Countries
Mr. TANG Huajun	Member, the Leading Party Group of the Ministry of Agriculture; President, the Chinese Academy of Agricultural Sciences; and Academician, the Chinese Academy of Engineering
Mr. ZHANG Yaping	Vice President, Member of the Leading Party Group; and Academician, the Chinese Academy of Sciences
Mr. CAI Fang	Vice President, Chinese Academy of Social Sciences
Mr. HAO Jiming	Professor, Department of Environmental Engineering, Tsinghua University; and Academician, the Chinese Academy of Engineering
Mr. XUE Lan	Dean of Schwarzman College in Tsinghua University; Co-Chair of the Leadership Council of the UN Sustainable Development Solution Network (UNSDSN); Professor at School of Public Policy and Management at Tsinghua University

(continued)

(continued)

Mr. SHU Yinbiao	Chairman and Secretary of the Leading Party Group, China Huaneng Group.,Ltd.;Former Chairman and Secretary of the Leading Party Group, State Grid Corporation of China; Chairman, International Electrotechnical Commission
Mr. FU Yuning	Chairman, China Resources (Holdings) Co., Ltd.
Mr. QIAN Zhimin	Chairman of the Board, State Power Investment Corporation Limited
Mr. WANG Xiaokang	President, China Industrial Energy Conservation and Clean Production Association; and Former Chairman, China Energy Conservation and Environmental Protection Group
Mr. WANG Tianyi	Executive Director and the Chief Executive Officer, China Everbright International Limited
Ms. Marjorie YANG	Chairman, Esquel Group

International Members

Mr. Scott Vaughan	**International Chief Advisor** Former President and CEO, International Institute for Sustainable Development
Mr. Arthur Hanson	Senior Advisor and former president of International Institute for Sustainable Development
Mr. Joachim von Amsberg	Vice President, Asian Infrastructure Investment Bank
Mr. Peter Bakker	President, World Business Council for Sustainable Development
Mr. Francesco La Camera	Director-General, the International Renewable Energy Agency
Mr. Srun Darith	Secretary of State, Ministry of Environment, Cambodia
Mr. John J. DeGioia	President, Georgetown University
Mr. Jan Hendrik Dronkers	Secretary General, the Ministry of Infrastructure and Water Management, the Netherlands
Mr. Richard Florizone	President and CEO, the International Institute for Sustainable Development
Mr. Hans Friederich	Fommer Director General, International Network for Bamboo and Rattan (relieved from office from March 31)
Mr. Stephen P. Groff	CEO, National Development Fund of Saudi Arabia
Ms. Kate Hampton	CEO, Children's Investment Fund Foundation
Mr. Eric Heitz	Former CEO and Co-founder, the Energy Foundation
Mr. Stephen Heintz	President, the Rockefeller Brothers Fund
Mr. Nuritdin Inamov	Director, Department for International Cooperation, Ministry of Natural Resources and Environment of the Russian Federation
Ms. Naoko Ishii	Professor and Director for Center for Global Commons, University of Tokyo; Former CEO and Chair, Global Environment Facility

(continued)

(continued)

Mr. Rodolfo Lacy	Director for the Environment Directorate, Organization for Economic Co-operation and Development
Mr. Marco Lambertini	Director General, World Wide Fund for Nature
Mr. Yong Li	Director General, The United Nations Industrial Development Organization
Mr. Ajay Mathur	Director General, The Energy and Resources Institute of the Republic of India
Mr. Michael McElroy	Gilbert Butler Professor, Environmental Studies; Harvard University
Ms. Kathleen McLaughlin	Chief Sustainability Officer and President, Walmart Foundation
Mr. Dirk Messner	President, German Federal Environment Agency
Mr. Andrew Metcalfe	Secretary, Department of the Agriculture, Water and Environment of the Commonwealth of Australia
Mr. Hideki Minamikawa	President, Japan Environmental Sanitation Center
Mr. Oliviero Montanaro	Directorate General, Nature and Sea Protection, Ministry of the Environment, Land and Sea, Italy
Ms. Jennifer Morris	Chief Executive Officer, the Nature Conservancy
Ms. Nosipho Ngcaba	Director-General, Department of Environmental Affairs of the Republic of South Africa
Mr. Bruno Oberle	Director General, the International Union for Conservation of Nature
Mr. Félix Poza Peña	Chief Sustainability Officer, Inditex Group
Mr. Jonathan Pershing	Director of Environment, Hewlett Foundation
Ms. Diane Regas	President and Chief Executive Officer, the Trust for Public Land
Mr. Frank Rijsberman	Director General, Global Green Growth Institute
Ms. Åsa Romson	Expert in environmental law & policy, IVL Swedish Environmental Research Institute; Former Deputy Prime Minister and Minister for Climate and the Environment, the Kingdom of Sweden
Ms. Gwen Ruta	Executive Vice President, Environmental Defense Fund
Mr. Ahmed M. Saeed	Vice President (Operations 2), Asian Development Bank
Mr. Andrew Steer	President and CEO, World Resources Institute
Mr. Mark Tercek	Former CEO, the Nature Conservancy
Mr. Frans Timmermans	First Vice President, European Commission
Mr. Juergen Voegele	Vice President for Sustainable Development, the World Bank
Mr. Jan-Gunnar Winther	Director, Centre for the Ocean and the Arctic, Nofima; Specialist Director, Norwegian Polar Institute
Mr. Seung-Joon Yoon	Professor, Seoul National University; Former President, Korea Environmental Industry and Technology Institute

Chinese Special Advisors

Mr. ZHANG Yong	Director-General, Bureau of General Affairs, the Office of the Central Financial and Economic Affairs Committee
Mr. FAN Bi	Invited Research Fellow, China Center for International Economic Exchanges
Mr. LI Junfeng	Director, Academic Committee of the Energy Research Institute of the National Development and Reform Commission; Former Director General, National Center for Climate Change Strategy and International Cooperation
Mr. LI Pengde	Deputy Director General, China Geological Survey
Mr. JI Yongjun	Deputy Director General, Department of American and Oceanian Affairs of the Chinese People's Association for Friendship with Foreign Countries
Mr. HU Baolin	Honorary Dean of Research Institute of China Green Development of Tianjin University; Former Deputy Director, Executive Office of The Three Gorges Project Construction Committee of the State Council
Ms. DONG Xiaojun	Deputy Director, Division of Economics of the Chinese Academy of Governance
Mr. ZHANG Yongsheng	Deputy Director, Institute for Urban and Environmental Studies, Chinese Academy of Social Sciences
Mr. ZHANG Yuanhang	Dean, College of Environmental Sciences and Engineering of Peking University; and Academician of the China Academy of Engineering
Mr. HE Kebin	Dean and Professor, School of Environment, Tsinghua University; and Academician of the China Academy of Engineering
Mr. ZHAO Zhongxiu	President, Shandong University of Finance and Economics
Mr. YE Yanfei	Senior Inspectorate Advisor, Policy Research Bureau of the China Banking and Insurance Regulatory Commission
Mr. CHEN Xinjian	Vice President, Industrial Bank
Mr. MA Jun	Chairman, Green Finance Committee of China Society for Finance and Banking; and Former Chief Economist, Research Bureau of the People's Bank of China
Ms. LIU Kun	General Manager, Medical and Health Division, China General Technology (Group)
Mr. LIU Tianwen	Founder, Chairman and CEO, iSoftStone Holdings Ltd.
Mr. ZHAI Qi	Executive Secretary General, China Business Council for Sustainable Development

International Special Advisors

Mr. Iskandar Abdullaev	Deputy Director, Central Asia Regional Economic Cooperation Institute; Former Executive Director, The Regional Environmental Center for Central Asia
Mr. Knut Halvor Alfsen	Former Head Research Director, Center for International Climate and Environmental Research
Mr. Howard Bamsey	Chair, Global Water Partnership; Former Executive Director, Green Climate Fund Secretariat
Mr. Manish Bapna	Executive Vice President and Managing Director, World Resources Institute
Mr. Dimitri de Boer	China Country Director, ClientEarth
Mr. Guillermo Castilleja	Senior fellow, the Gordon and Betty Moore Foundation
Ms. Galit Cohen	Deputy Director General for Policy and Planning, Ministry of Environmental Protection of the State of Israel
Mr. Stephan Contius	Commissioner for the 2030 Agenda for Sustainable Development, the Federal Ministry of the Environment, Nature Conservation and Nuclear Safety, Germany
Ms. Lucie Desforges	Director General of Bilateral Affairs and Trade, Environment and Climate Change Canada
Mr. Yannick Glemarec	Executive Director, Green Climate Fund
Ms. Isabel Hilton	CEO and Editor, China Dialogue
Ms. Lisbeth Jespersen	Former Head of International Partnerships and Fundraising, IDH, the Sustainable Trade Initiative; Former Head of Secretariat, Global Green Growth Forum
Mr. Johan C.I. Kuylenstierna	Vice Chair, Swedish Climate Policy Council; Adjunct Professor, Stockholm University
Ms. Bernice Lee	Research Director - Global Economy and Finance, Chatham House
Mr Zafar Makhmudov	Executive Director, the Regional Environmental Centre for Central Asia
Ms. Jane McDonald	Executive Vice President and Chair of the IISD Experimental Lakes Area Board, the International Institute for Sustainable Development
Ms. Désirée McGraw	Former President and Head, College of Pearson College UWC
Mr. Hans Mommaas	Director-General, PBL Netherlands Environmental Assessment Agency
Mr. Ismo Tiainen	Director-general, Administration and International Affair, Ministry of the Environment, the Republic of Finland
Mr. Hau Sing Tse	Executive Director, African Development Bank for Canada, China, South Korea and Kuwai
Mr. Dominic Kailash Nath Waughray	Managing Director and Head of Centre for Global Public Goods, World Economic Forum

(continued)

(continued)

Mr. ZHANG Hongjun	Partner, Holland & Knight
Mr. ZHANG Jianyu	Vice President and China Project Director, Environmental Defense Fund
Mr. ZOU Ji	CEO & President of Energy Foundation China

Deputy Secretary Generals and Assistant Secretary General

Mr. GUO Jing	**Deputy Secretary General** Director General of International Cooperation Department, Ministry of Ecology and Environment
Ms. Zhou Guomei	**Deputy Secretary General** Director General, Foreign Environmental Cooperation Center, Ministry of Ecology and Environment
Mr. Li Yonghong	**Assistant Secretary General** Deputy Director General, Foreign Environmental Cooperation Center, Ministry of Ecology and Environment

Secretariat and SISO

ZHANG Huiying, LIU Kan, GAO Lingyun, WANG Ran, ZHU Jianlei, Yao Ying, FEI Chengbo, ZHAO Haishan, CHEN Xinying, LI Gongtao, LIU Qi
   Joe Zhang, Tristan Easton, Samantha Zhang, Brice Li, Cesar Henrique Arrais